CROSSING THE BOUNDARIES: GAUGE DYNAMICS AT STRONG COUPLING

PROCEEDINGS OF THE WORKSHOP IN HONOR OF THE 60TH BIRTHDAY OF MISHA SHIFMAN

EDITORS

MARCO PELOSO
University of Minnesota

ARKADY VAINSHTEIN
Fine Theoretical Physics Institute

CROSSING THE BOUNDARIES: GAUGE DYNAMICS AT STRONG COUPLING

PROCEEDINGS OF THE WORKSHOP IN HONOR OF THE 60TH BIRTHDAY OF MISHA SHIFMAN

FTPI, University of Minnesota 14 – 17 May 2009

World Scientific

NEW JERSEY • LONDON • SINGAPORE • BEIJING • SHANGHAI • HONG KONG • TAIPEI • CHENNAI

Published by

World Scientific Publishing Co. Pte. Ltd.
5 Toh Tuck Link, Singapore 596224
USA office: 27 Warren Street, Suite 401-402, Hackensack, NJ 07601
UK office: 57 Shelton Street, Covent Garden, London WC2H 9HE

British Library Cataloguing-in-Publication Data
A catalogue record for this book is available from the British Library.

CROSSING THE BOUNDARIES
Gauge Dynamics at Strong Coupling
Proceedings of the Workshop in Honor of the 60th Birthday of Misha Shifman

ISBN-13 978-981-4304-02-3
ISBN-10 981-4304-02-6

Printed in Singapore by B & Jo Enterprise Pte Ltd

FOREWORD

This volume contains the proceedings of the workshop "Crossing the Boundaries: Gauge Dynamics at Strong Coupling," hosted by the William I. Fine Theoretical Physics Institute at the University of Minnesota, May 14-17, 2009. The Workshop honored the 60th birthday of Professor Misha Shifman and his outstanding achievements in the field of gauge dynamics. The meeting attracted a fascinating group of researchers working on the cutting edge of dynamics of gauge theories, including supersymmetric and string theories. Their talks covered a wide area of recent developments in the field.

The volume begins with Misha Shifman's lecture on "Large Extra Dimensions," which presents a pedagogical introduction to the paradigm. Also included is his introduction to the book "Vacuum Structure and QCD Sum Rules," which Misha edited. Although the book was published in 1992, the text of the introduction had never been published before.

The next section is devoted to non-Abelian vortices – stringy topological objects in supersymmetric gauge theories which also go by the name of heterotic strings. This is an area of research in which Misha has been very active in recent years. Section 3 continues the discussion of topological excitations in quantum field theories.

The hot topic of duality between gauge theories and their string description, which reduces to the gravity dual in a certain limit, is the subject of section 4. In the following section interesting aspects of supersymmetric field theories are discussed.

Section 6 returns to the discussion of gravity duals in application to QCD. Its title "Holographic QCD," refers to the extra, holographic coordinate which appears in the dual description. The next section, also devoted to QCD, discusses various aspects of hadronic phenomena including: duality to supersymmetric theories, form factors, high energy scattering and the nature of confinement-deconfinement transitions.

In section 8 the issue of dark matter axions is discussed. Additionally some unusual extensions of particle theory such as "Unparticle Physics," and "The Lee-Wick Standard Model," are presented. Contributions not fitting the above classifications are collected in section 9. The last section contains photographs taken during the event and a full list of participants.

Nearly all speakers contributed their presentations to this volume. We are grateful to Adi Armoni and Gia Dvali who were unable to attend but sent their contributions regardless. Many of talk slides are available online. These files, along with additional information about the workshop can be found ar URL *http://www.ftpi.umn.edu/shifmania/* .

Finally, we would like to thank two other organizers, Keith Olive and Mikhail Voloshin. Our special thanks go to the FTPI administrator Marie Larson, who was a major force in organizing the meeting, and to Poline Tylevich for her graphic design of the book.

Marco Peloso and Arkady Vainshtein
October 12, 2009

CONTENTS

* The presenting author.

SECTION 1.

MISHA SHIFMAN'S LECTURES

Large Extra Dimensions
Becoming acquainted with an alternative paradigm

M. Shifman*

William I. Fine Theoretical Physics Institute
University of Minnesota, 116 Church Street S.E.
Minneapolis, MN 55455, USA
**E-mail: shifman@umn.edu*

This is a colloquium style pedagogical introduction to the paradigm of large extra dimensions. This talk was delivered at various universities including UCLA (1999), Technion (2000), DESY (2001), University of Pisa (2001), UBC (2002), NYU (2002), University of Nantes (2003), Bern University (2003), Max-Planck Institute/Munich University (2003), University of Padova (2003), University of Minnesota (2005).

1. Road Map (Instead of Introduction)

They say God does not exactly know how parts of his Creation work. When he sees a nice theory which he likes he says: "OK, let it be so..."

Today theoretical high energy physics deals basically with two options: (i) Grand Desert stretching from $\sim 10^2$ GeV to $\sim 10^{16}$ GeV, with no new physics inside; and (ii) Large Extra Dimensions paradigm various versions of which predict new physics at a much lower scale of energies. If the first option is realized, this would mean that high-energy physics in the future will face a serious menace of becoming a non-empirical science: experiments at energies in the ballpark of $\sim 10^{16}$ GeV are impossible in terrestrial conditions.

The LED paradigm was born from the desire "to have new physics around the corner," in an attempt to keep high-energy physics as an experiment-based discipline. One may hope that God will like it.

The topic of large extra dimensions (LED) experienced an explosive development since the mid-1990's. Since then thousands of works dedicated to this subject were published. The reason why the LED paradigm attracted so much attention is due to the fact that it brings the scale of new fundamental physics from inaccessible 10^{16} or 10^{19} GeV down to 10 or 100 TeV or so.

A comparison with a huge country the exploration of which is not yet completed is in order here. This lecture presents a bird's eye view of the territory, giving a brief and nontechnical introduction to basic ideas lying behind the large extra dimension paradigm and a particular braneworld model.

The task of describing the large extra dimensions paradigm is "multidimensional" in itself. First, there exist three main scenarios which sometimes compete and sometimes complement each other. Second, each scenario starts from a general design of a basic model, while phenomenological consequences come later. Moreover, some scenarios predict new macrophenomena, such as modifications of gravity at distances comparable with the observed Universe size.

Below we will focus on the simplest LED scenario – that of Arkani-Hamed–Dimopoulos–Dvali (ADD) – limiting our forays into alternative scenarios to a minimum. We start from a brief review of fundamental regularities of our world in the context of the paradigm that had existed before the advent of LED. The latter was based on the standard model and its supersymmetric version, supersymmetric grand unification and great desert. We will refer to this paradigm as to the great desert paradigm, or, sometimes, good old paradigm. (We hasten to add, though, that it was not particularly old or particularly good.) Then, after familiarizing ourselves with the history of the topic, we will discuss the very same regularities as they are interpreted from the standpoint of the LED paradigm.

2. Genesis/Glimpses of history

2.1. *Kaluza–Klein Theory*

The story starts in the 1920's. At that time time Theodor Kaluza and Oscar Klein, who were working on the unification of Einstein's gravity and elec-

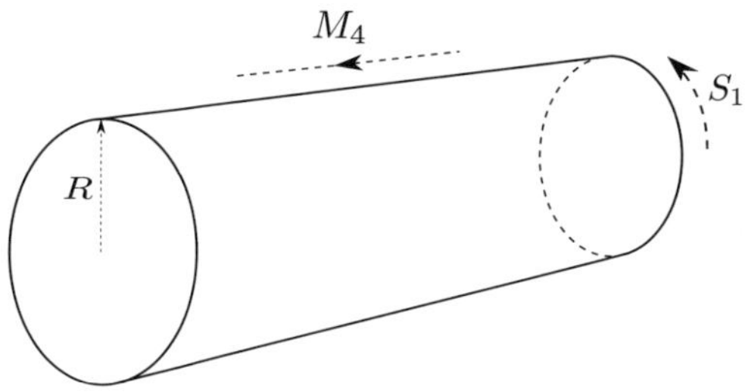

Fig. 1. The Kaluza–Klein set-up.

tromagnetism, invented the Kaluza–Klein (KK) mechanism.[1,2] Its essence is as follows. Assume that our world, rather than being four-dimensional, is in fact $(4+n)$-dimensional, $n \geq 1$, but the extra dimensions are compact. An illustration is presented in Fig. 1, where $n = 1$, so that our world is a direct product of the four-dimensional Minkowski space M_4 and a circle S_1 with the radius R. All fields are defined on this "cylinder." For a scalar field the single-valuedness on the cylinder can be written as

$$\Phi(x_\mu,\ Z) = \Phi(x_\mu,\ Z + 2\pi R)\,, \qquad \mu = 0,1,2,3\,. \tag{1}$$

Here and below we will use small Latin letters for "our" four coordinates reserving capital Latin letters for extra dimensions. The $2\pi R$ periodicity in Z means that one can present the field $\Phi(x_\mu,\ Z)$ as a Fourier series,

$$\Phi(x_\mu,\ Z) = \sum_{k=0,\pm 1,\ldots} \phi_k(x_\mu) e^{ikZ/R}\,. \tag{2}$$

The expansion coefficients depend only on "our" coordinates x_μ, and are often referred to as *modes*. As we will see shortly, the zero mode corresponding to $k = 0$ will play a special role. Modes with $k \neq 0$ are shown in Fig. 2. Each zero mode is accompanied by non-zero modes $k \neq 0$ which are referred to as the KK excitations or the KK tower.

From the four-dimensional point of view, the modes $\phi_k(x_\mu)$ represent a tower of regular four-dimensional fields, the so-called Kaluza–Klein tower. Let us start from the five-dimensional wave equation, assuming that the five-dimensional field $\Phi(x_\mu,\ Z)$ is massless,

$$\Box_5\, \Phi(x_\mu,\ Z) \equiv \left(\partial_\mu^2 - \frac{\partial^2}{\partial Z^2}\right) \Phi(x_\mu,\ Z) = 0\,. \tag{3}$$

Substituting the Fourier decomposition (2) we see that each mode ϕ_k sat-

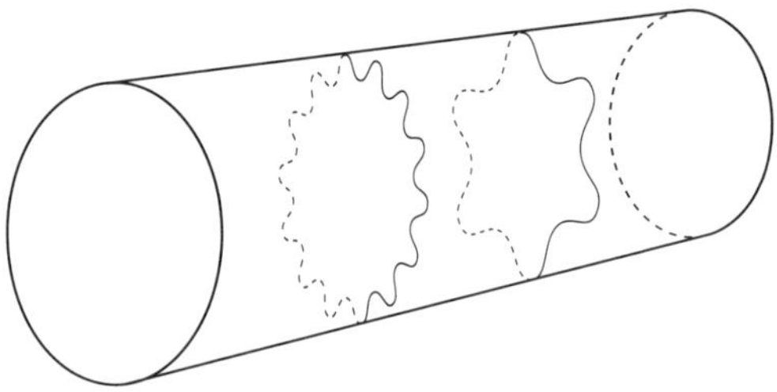

Fig. 2. The non-zero modes in the KK decomposition.

isfies the four-dimensional wave equation

$$\left(\Box_4 + \frac{k^2}{R^2}\right)\phi_k(x_\mu) \equiv \left(\partial_\mu^2 + \frac{k^2}{R^2}\right)\phi_k(x_\mu) = 0\,. \tag{4}$$

The zero mode ϕ_0 remains massless, while all other modes become *massive* four-dimensional fields, with $|k|/R$ playing the role of the mass term,

$$m_k = \frac{|k|}{R}\,. \tag{5}$$

Compactification of $(4+n)$-dimensional fields with spin exhibits another interesting phenomenon and leads to a richer KK tower, which includes a spectrum of four-dimensional spins. Consider for example the metric tensor G_{MN} in five dimensions. The vectorial indices corresponding to higher-dimensional world here and below are denoted by capital Latin letters. From the four-dimensional standpoint $G_{\mu\nu}$ is the four-dimensional metric tensor, $G_{5\mu} = G_{\mu 5}$ is a four-vector, while G_{55} is a scalar. So, the KK tower includes the zero and non-zero modes of spin-2, spin-1 and spin-0 fields.

In the KK picture one assumes that "R is small and $1/R$ is large" compared to some currently available energy scale. Moreover, all our four-dimensional world, in its entirety, including all experimental devices and all potential observers, is made from the zero modes. Then, given the energy limitation, the non-zero mode quanta cannot be produced, and we perceive our world as four-dimensional. Only when accessible energy becomes higher than $1/R$ can we directly discover KK excitations, a signature of the extra dimension(s).

The infancy of the Kaluza–Klein scenario was eventful. Suffice it to mention that Schrödinger, Gordon and Fock worked on its development in the 1920's, while important contributions in the 1930's were due to Pauli,[3] and Einstein and Bergmann.[4] In particular, in considering compactification of two extra dimensions into sphere S_2 (see Fig. 3), Pauli discovered the Yang–Mills theory long before Yang and Mills. Since he did not know what to do with massless vector fields he never published anything on this discovery. However, gradually the interest to the Kaluza–Klein scenario languished, probably because of the absence of realistic applications in model-building of that time.

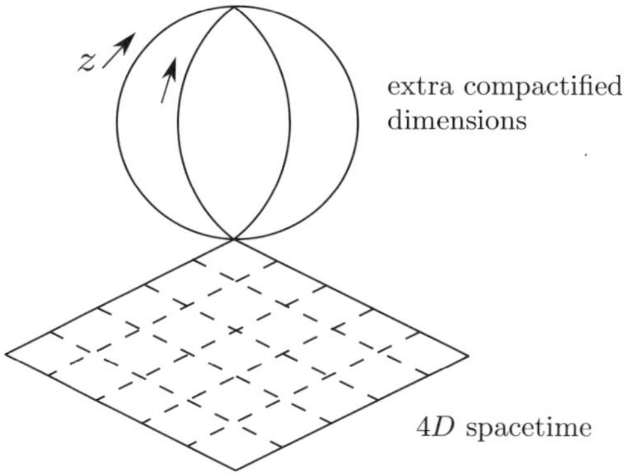

Fig. 3. Pauli considered compactification of the six-dimension space onto $M_4 \times S_2$.

A long period of a relative hibernation of the Kaluza–Klein theory gave place to a revival in the 1980's. The dawn of a "new era" was marked by Witten's no-go theorem.[5] Witten noted that fermions cannot be chiral if one starts from any eleven-dimensional manifold of the type $T = M_4 \times K$ where K is a compact manifold admitting the symmetry of the standard model (SM), namely, SU(3)×SU(2)×U(1). This was a sad conclusion since it meant that no realistic model could be based on the Kaluza–Klein theory since the fermions in our world are definitely chiral. Fortunately, it was negated, just a few years later, in the first superstring "revolution."

2.2. *Strings*

Consistent superstring theory exists in ten dimensions. Nonsupersymmetric string is consistent in 26 dimensions. Our world is four-dimensional. In 1975 Joel Scherk and John Schwarz suggested[6] to consider superstrings in a product space of our conventional four-dimensional space-time and a six-dimensional compact manifold whose size is of the order of $M_{\rm Pl}^{-1}$.

The celebrated paper of Candelas *et al.*,[7] which opened the superstring revolution of 1985, demonstrates that if the six compact dimensions form the so-called Calabi–Yao manifold, then in the low-energy limit one recovers a $E(8) \times E(8)$ gauge theory which includes SM, with three generations of chiral fermions that are observed in nature. The Calabi–Yao compactification, conceptually, continues Pauli's line of reasoning. As we have already mentioned, in the 1930's Pauli observed that the KK model on $M_4 \times S_2$ produces three gauge bosons of SU(2) in the low-energy limit. The occurrence of these bosons is due to isometries of the sphere S_2. Of course, the six-dimensional Calabi–Yao manifold is much more contrived. Geometry of the Calabi–Yao manifold is so complicated that the explicit form of the metric is not known even now.

The typical sizes of the compact dimensions in the Calabi–Yao manifold are of the order of 10^{-33} cm. Needless to say, there is no way to observe such extra dimensions in a direct human-designed experiment.

Systematic searches for string-inspired realistic models of the SM type began in the 1990's.[8] Since from the "human" standpoint, extra dimensions in the Calabi–Yao scenario can be viewed as an auxiliary mathematical construction, it was suggested to replace the compactified coordinates by a more formal construction — internal free fermions propagating on the string world sheet. One can then completely abandon the geometrical description of the compactification and formulate it entirely in terms of free fermions on the string world sheet and their boundary conditions. One can then extract the physical spectrum, as well as the assignment of the quantum numbers under the four-dimensional gauge group.

Following this procedure realistic three-generation models were constructed.[8] They differed in their detailed phenomenological properties, but some elements were in common. In particular, an SO(10) grand unification, with an $SO(6)^3$ flavor symmetries and a hidden E_8 gauge group were typical.

In the current millennium this topic — searches for string-inspired realistic models of the SM type — experienced a dramatic development based on D-brane engineering.[9] The advent of D branes[10] allowed one to find string/D-brane models yielding just the SM massless fermion spectrum, with relative ease. One of the features predicted by the D-brane-based models is that the SM global symmetries — such as baryon and lepton numbers — are gauged symmetries whose anomalies are canceled (by a Green-Schwarz mechanism) only in the case of three quark-lepton generations. Proton stability and the Dirac nature of neutrino masses follow naturally.

This direction *per se* — string-inspired phenomenology — seems promising. What is important for our narrative is that typical sizes of the compact dimensions in the string/D-brane scenario are of the order of 10^{-33} cm;[a] hence, the masses of the excited states in the KK tower are of order of 10^{19} GeV. Such energies are (and will be) inaccessible to any terrestrial experiment. In other words, if physics were to be described by such scenarios, the KK tower would be unobservable, and the KK theory would have no practical implications.

2.3. *Parallel development: localizing on topological defects*

Independent ideas which later formed one of the pillars of the LED paradigm emerged and were developed in the 1980's-90's. The idea of localizing matter on topological defects was formulated, in the most clear-cut form,[b] by Rubakov and Shaposhnikov[14] in the paper entitled "Do We Live Inside A Domain Wall?".

It is convenient to explain the essence of this suggestion in a simplified setting where "our" world is assumed to be (1+2)-dimensional, while the coordinate z is treated as an "extra dimension." Assume that the underlying microscopic theory has several discrete degenerate vacua which are labeled by distinct values of an order parameter. Call two such vacua — they can be chosen arbitrarily — vacuum I and II. There exists a static field configuration, a domain wall, which divides the three-dimensional space in two parts, say, on the left hand-side our system is in the vacuum I while on the right-hand side in the vacuum II (Fig. 4).

The domain wall represents a transitional domain and is topologically stable. Once created, it cannot be destroyed. The thickness of the domain wall δ depends on details of the microscopic theory. At distances $\gg \delta$ the domain wall can be viewed as a two-dimensional surface.

One can excite the domain wall field configuration by pushing the wall at a certain point or by pumping in energy in any other way. All possible excitations naturally fall into two categories. Some of them are localized on the wall (their spatial extension in the perpendicular direction is of the order of δ). These are usually associated with *zero modes*. Being considered from the (1+2)-dimensional point of view, the zero modes represent massless particles which can propagate only along the wall surface.

[a] It would be fair to add that some work on introducing larger extra dimensions in the string context had been done in the 1990's. For instance, Hořava and Witten pointed out[11] that a single extra dimension of the size $\sim 10^{-28}$ cm could eliminate the gap between the scale of grand unification and the Planck scale, see Sect. 3.4.

[b] Similar ideas were discussed around the same time in Refs. [12,13].

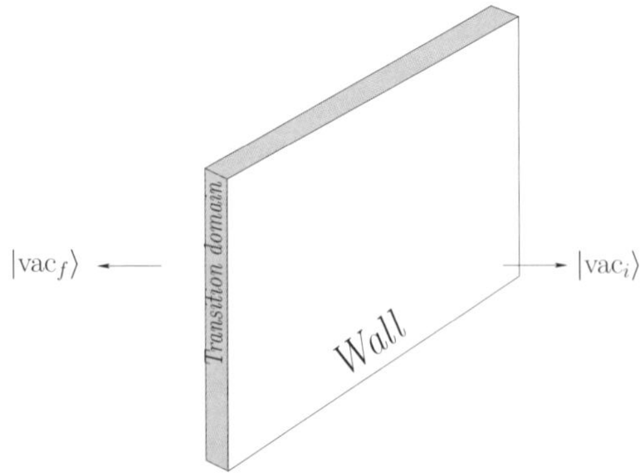

Fig. 4. A domain wall separating two distinct degenerate vacua.

Other excitations are delocalized and can escape in the bulk (i.e. in the perpendicular direction). They are represented by *nonzero* modes with typical energy eigenvalues of the order of $1/\delta$. From the (1+2)-dimensional standpoint each nonzero mode is a particle with mass $M_n \sim 1/\delta$.

Assume that all matter that we see around is made of the zero modes trapped on the domain wall surface. Then "our world" will be confined to the wall surface and will be effectively (1+2)-dimensional. To discover the third (perpendicular) spatial dimension an observer made of the zero modes will have to have access to energies larger than $1/\delta$.

An obvious distinction between the KK scenario and localization on the domain walls (or other topological defects) is the mass scale of the excited modes. In the KK model it is related to the inverse size of the extra dimension, while in the case of the domain walls the extra dimension is infinite, and the mass scale is set by the inverse thickness of the wall. This distinction turns out to be crucial in physical applications.

The existence of at least one zero mode is easy to demonstrate. Indeed, the underlying microscopic theory has four-dimensional translational invariance. The domain wall breaks, spontaneously, the invariance with respect to translations in the z direction. Physics becomes dependent on the distance to the wall in the perpendicular direction. Correspondingly, in accordance with the Goldstone theorem, there emerges a Goldstone boson which is confined to the wall surface. If the profile of the order parameter describing the wall (we will call it $\phi(z)$) is known, then the profile of the translational zero mode is given by the derivative $d\phi/dz$, see Fig. 5.

The (1+2)-dimensional Goldstone boson appearing in this way has spin

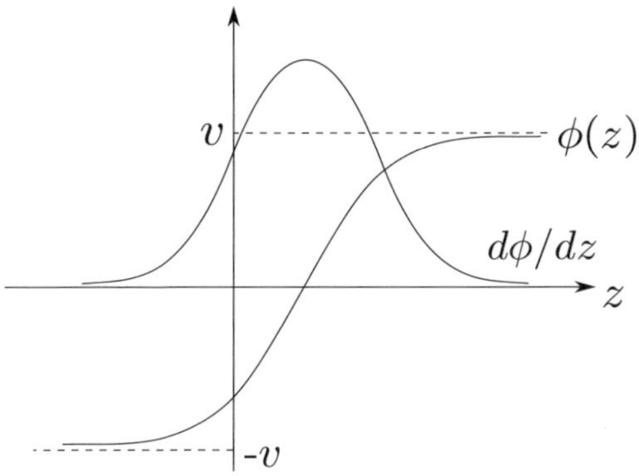

Fig. 5. The profile function of the domain wall determines localization of the translational zero mode.

zero. In fact, the original work of Rubakov and Shaposhnikov was motivated by the desire to have a Higgs boson whose mass is protected by the Goldstone theorem from being dragged in the ultraviolet by quadratic divergences, typical of the scalar particle masses in field theory. The novelty of the idea and its potential were not recognized till mid-1990's since shortly after the Rubakov–Shaposhnikov publication supersymmetry gained the role of a universal saviour.

In general, localization of spin-zero bosons presents no problem, at least, at a conceptual level. Assume that the microscopic theory has a global symmetry group G which remains unbroken both in the vacuum I and II. Assume that on the given wall solution the symmetry G is (spontaneously) broken down to H. Then the Goldstone bosons corresponding to the broken generators will be confined to the wall; their interactions will be described by a coset G/H sigma model.

Localization of the spin-1/2 particles is also a long-known phenomenon. Fermion fields coupled to the wall can have zero modes too. The number of such zero modes is regulated by the Jackiw–Rebbi index theorem.[15] For fermion fields one must consider the mass matrix as a function of z. If in piercing the wall (i.e. in passing from $z = -\infty$ to $z = \infty$) k eigenvalues of the fermion mass matrix change sign, then one will have k fermion zero modes. The thickness of the profile of the fermion zero modes is of the order of the inverse fermion mass in the bulk. With all bulk fermions massive, localization of the fermion zero modes on the wall clearly presents no problems.

Localizing non-Abelian gauge fields on the wall (in the framework of field theory) is a more complicated task. A working mechanism was found in 1996.[16] The basic idea is as follows. Assume that in the bulk, outside the wall, we have a gauge theory (with the gauge group G) in the confining phase. The bulk dynamical scale parameter is $\Lambda_{\rm bulk}$. Assume that inside the wall the gauge group is G' where $G' \in G$, the dynamical scale is Λ', and $\Lambda' \ll \Lambda_{\rm bulk}$. Then the gauge fields of the G' theory will be localized on the wall. Indeed, at energies $\Lambda' \ll E \ll \Lambda_{\rm bulk}$ the chromoelectric fields of the G' theory cannot escape in the bulk since, due to confinement, the lightest states in the bulk (glueballs) have masses $\sim \Lambda_{\rm bulk}$. One can dualize this picture. Assume that the bulk G theory is Higgsed. Then the probe magnetic charges are confined in the bulk through formation of chromomagnetic flux tubes. If appropriate condensates vanish inside the wall, then the chromomagnetic flux can spread freely inside the wall.

In string theory localization of the gauge fields is achieved with no special effort, provided one identifies domain walls with D branes. The gauge bosons are then represented by open strings with the end points attached to the branes (Fig. 6). Thus, they are naturally confined to the brane surface. (Let us parenthetically note that the closed strings representing gravitons can freely propagate in the bulk in this picture. We will return later to the discussion of this feature.)

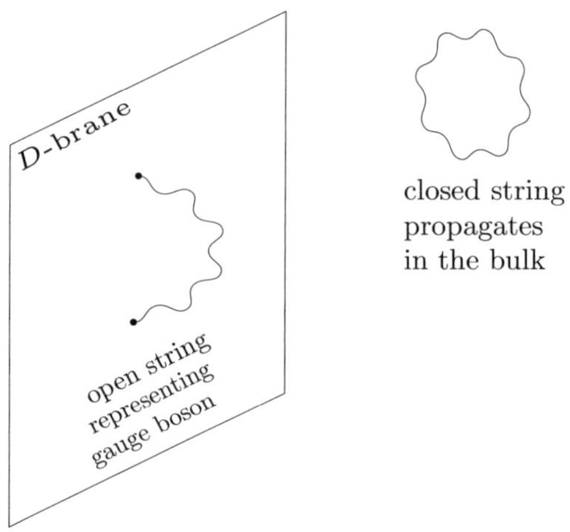

Fig. 6. The gauge bosons are represented by open strings and are attached to the D-brane, while the gravitons represented by the closed strings "live" in the bulk.

As was mentioned, the idea of localization on topological defects was in a rather dormant state till mid-1990's. In 1996 it was revived in the supersymmetric context.[17]

Why supersymmetry?

The reason for invoking supersymmetry is two-fold. First, in nonsupersymmetric theories the existence of degenerate discrete vacua requires spontaneous breaking of some discrete symmetry. On the other hand, in supersymmetric theories the vacuum degeneracy is hard to avoid, it is typical. Indeed, all supersymmetric vacua must have the vanishing energy density and are thus degenerate. Therefore, domain walls are more abundant.

The second motivation is the ease with which one can localize simultaneously spin-0 and spin-1/2 fields on critical (or BPS saturated) domain walls.

2.4. *What to do with gravity?*

One cannot hope to successfully describe our world without including gravity. Unlike spin-0,1/2 and 1 fields, no field-theoretic mechanisms ensuring *bona fide* localization of gravity on domain walls are known. Moreover, if one approaches the problem from the string theory rather than field theory side, a drastic distinction between, say, gauge fields and gravity is obvious too. The gauge fields are represented by open strings with the endpoints attached to D branes. Thus, they are naturally localized on D branes. At the same time, gravity is represented by closed strings which can freely propagate in the bulk, see Fig. 6. In order for the LED paradigm to take off, a fresh idea regarding what to do with gravity was needed. It was not before long that it was put forward.

3. Flat compact extra dimensions

Historically the first was the compact extra dimension model which goes under the name ADD, where ADD stands for Arkani-Hamed, Dimopoulos and Dvali, its inventors.[18] In the March 1998 paper, and a follow-up publication of the same authors with Antoniadis,[19] a marriage between the Kaluza–Klein scenario and localization on domain walls was suggested. Compactifying extra dimensions a lá Kaluza–Klein solves the problem of gravity, while localizing all other fields on the wall solves the hierarchy problem.

But let us begin from the very beginning.

3.1. *The Arkani-Hamed–Dimopoulos–Dvali (ADD) scenario*

The model starts from the assumption that the space-time is $(4+n)$-dimensional, $n \geq 1$, while its geometry is factorized,

$$M_{\text{world}} = M_4 \times K_n \,. \tag{6}$$

All SM particles are localized on a $(1+3)$-dimensional domain wall (3-brane) representing M_4 in the above expression (Fig. 7). At the same time, gravity spreads to all $4+n$ dimensions (Fig. 8).

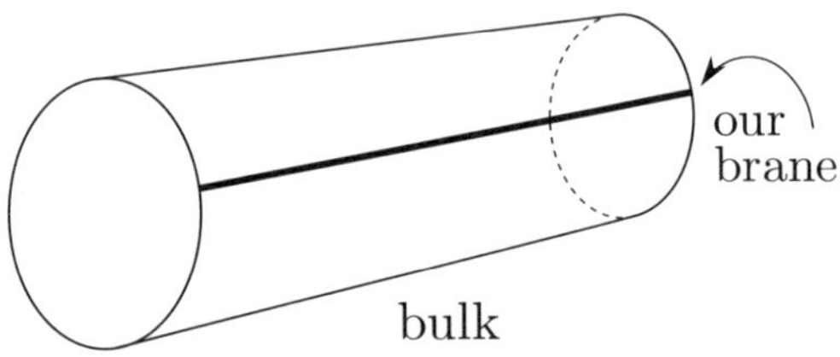

Fig. 7. "Our world" is the 1+3 dimensional domain wall which is shown by a solid line in this figure. Perpendicular directions are compact.

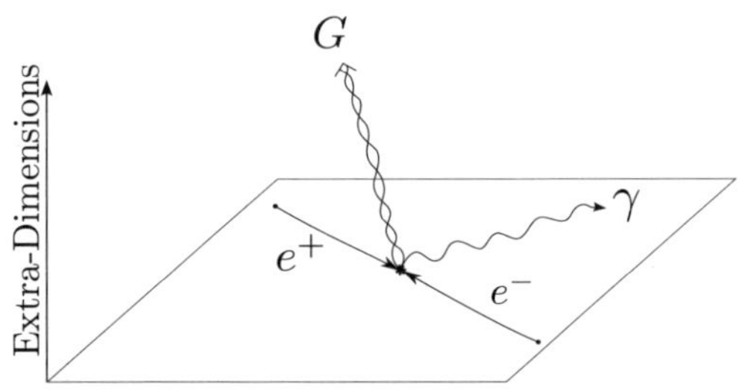

Fig. 8. We, and everything around us, are made from the zero modes localized on the domain wall. Gravity escapes in the bulk.

If the thickness of the domain wall is chosen to be $\delta \lesssim (10 \text{ TeV})^{-1}$ then at energies $\lesssim 10$ TeV physics is effectively four-dimensional in all experiments except those with gravity.[c] Since gravity escapes in the bulk it be-

[c] One may ask where does this mass scale, 10 TeV, come from? We will answer this question shortly.

comes four-dimensional only at distances $r \gg R$. At distances $r \lesssim R$ gravity is $4 + n$ dimensional.

We want $R \gg \delta$. The reason behind this requirement will become clear momentarily. Let us ask ourselves what values of the size of the compact dimensions R are compatible with what we know about our world today. Constraints on R are surprisingly lax. They follow from the fact that in the ADD scenario gravity becomes effectively four-dimensional at distances $\gtrsim R$. Experimentally, gravity is well-studied at large distances (where it is certainly four-dimensional) and known to a much lesser extent at short distances. In fact, below 0.1 mm or so the gravitational force has not been measured, and one cannot rule out that at such distances it is $(4 + n)$-dimensional.[d]

Let us assume that the size of extra dimensions $R \sim 0.1$ mm and see to what consequences this assumption leads. If in the future gravity will be proved to be four-dimensional at such distances one can always downsize extra dimensions making $R \sim 0.01$ mm or less.

3.2. *Fundamental scale and the size of extra dimensions*

Somewhat symbolically, the action can be written as

$$S = \frac{M_{\mathrm{f}}^{2+n}}{2} \int d^4x \int_0^{2\pi R} d^n Z \sqrt{G}\, R_{4+n} + \int d^4x \sqrt{g}\,(T + \mathcal{L}_{\mathrm{SM}}(\Phi_{\mathrm{SM}})) \quad (7)$$

where M_{f} is the *bona fide* fundamental constant of gravity, G and R_{4+n} are the $(4 + n)$-dimensional metric and scalar curvature, respectively, g is four-dimensional metric, and, finally, $\mathcal{L}_{\mathrm{SM}}$ is a Lagrangian describing all SM fields (which, remember, are trapped on the brane). A typical mass scale associated with $\mathcal{L}_{\mathrm{SM}}$ will be denoted as M_{SM},

$$M_{\mathrm{SM}} \sim 100 \text{ GeV}.$$

The constant T has to be adjusted in such a way that the overall cosmological term, which includes T plus all quantum loops, vanishes. This is the usual fine-tuning condition on the cosmological constant. The ADD scenario adds nothing new in this respect.

Now let us apply the Kaluza–Klein mode expansion to the graviton field and keep, for the time being, only the zero mode, neglecting all states in

[d] In 1998 measurements of the gravity extended down to 1 mm. Dedicated experiments[20] performed after the ADD suggestion gave rise to a considerable theoretical activity improved the above bound by an order of magnitude.

the KK tower with masses of the order of $1/R$. Since the zero mode is Z independent, we can perform the Z integration in the first term on the right-hand side thus obtaining

$$\frac{M_{\rm f}^{2+n}}{2}\int d^4x \int_0^{2\pi R} d^n Z\,\sqrt{G}\,R_{4+n} \longrightarrow \frac{1}{2}\,M_{\rm f}^{2+n}\,V_n \int d^4x\,\sqrt{g}\,R\,, \tag{8}$$

where g and R are four-dimensional metric and scalar curvature evaluated on the zero mode, and V_n is the volume of extra dimensions,

$$V_n = (2\pi R)^n\,. \tag{9}$$

The expression on the right-hand side of Eq. (8) is applicable at distances $r \gg R$.

At such distances the gravitational potential takes the standard Newton form

$$V(r) = -\frac{G_N m_1 m_2}{r}\,, \tag{10}$$

with the Newton constant G_N determined by Eq. (8).

$$G_N = (M_{\rm f}^{2+n}\,V_n)^{-1}\,. \tag{11}$$

At the same time, if $r \ll R$ we must return to Eq. (7) which implies the following static potential

$$V(r) = -\frac{m_1 m_2}{M_{\rm f}^{2+n}\,r^{1+n}}\,. \tag{12}$$

It is quite obvious that, upon inspecting Eq. (11), a four-dimensional observer will interpret $M_{\rm f}^{2+n}\,V_n$ as the Planck scale,

$$M_{\rm f}^{2+n}\,V_n = M_{\rm Pl}^2 \sim \left(10^{19}\ {\rm GeV}\right)^2\,. \tag{13}$$

He or she will think that this is the fundamental scale at which gravity becomes of order unity and needs quantization. The "visible" fundamental scale, as established by such an observer, is separated from accessible energies $M_{\rm SM}$ by a huge interval, thus creating an enormous hierarchy of scales.

In fact, in the ADD scenario the genuine fundamental scale is $M_{\rm f}$; this is the energy at which $4+n$ dimensional gravity becomes strong. $M_{\rm f}$ is related to the "visible" fundamental scale $M_{\rm Pl}$ as follows:

$$M_{\rm f} = \frac{1}{2\pi R}\,\left(2\pi R\,M_{\rm Pl}\right)^{\frac{2}{n+2}} \tag{14}$$

If $n=1$ and $M_{\rm f} \sim 10$ TeV, then $R \lesssim 10^{12}$ cm which is definitely inconsistent with the Newton law well-established at such distances. Thus, a single extra dimension is ruled out in the ADD scenario. Then it is natural to assume that $n = 2$. If so, and $M_{\rm f} \sim 10$ TeV, we can use the formula

$$R = \frac{1}{2\pi\, M_{\rm f}} \left(\frac{M_{\rm Pl}}{M_{\rm f}}\right)^{2/n} \tag{15}$$

to deduce that $R \sim 0.1$ mm.

Thus, two or more extra dimensions in the ADD framework are not inconsistent with the existing data. What is important is that even at larger n the size of the extra dimensions R is large compared with δ^{-1} and the more so with $M_{\rm Pl}^{-1}$. There are no theoretical motives behind the choice of $M_{\rm f} \sim 10$ TeV. The only reason is the desire to have new physics in the accessible range of energies. If a typical scale of new physics is higher than 10 TeV, one can adjust $M_{\rm f}$ appropriately.

Several simplifying assumptions were made in the course of the above consideration, namely,

(i) The wall thickness δ (which is assumed to be $\delta \sim M_{\rm f}^{-1}$) is neglected;

(ii) The wall shape fluctuations are neglected (these are the Nambu–Goldstone bosons which couple to matter derivatively);

(iii) All extra dimensions are assumed to have equal size R;

(iv) It is postulated that only gravity propagates in the bulk.

One or more of the above assumptions can be easily lifted. For example, one can consider several extra dimensions with individual sizes, or let escape in the bulk other fields in addition to gravity (but only those which carry no charges with respect to SM). This will change technical details, leaving conceptual foundations of the approach intact.

3.3. *Phenomenological implications*

Some basic phenomenological regularities that we observe in Nature are:

(i) Proton stability;
(ii) Mass hierarchies;
(iii) The existence of three and only three generations;
(iv) The lightness of neutrinos;
(v) The lightness of (yet to be discovered) Higgs particles;
(vi) A peculiar pattern of quark mixing angles (the CKM matrix elements); a peculiar pattern of the neutrino mixing angles.

In the years that elapsed after the creation of the standard model and before the advent of the LED paradigm these regularities received more or less satisfactory explanations. Some of them were understood at a conceptual level, while for others detailed technical explanations were worked out. At the very least, we believed that the above regularities presented no mysteries that could shake foundations of physics. For instance, the proton stability was explained by a very high mass scale of unification/strong gravity, of the order of $M_{\rm Pl}$. The lightness of the left-handed neutrino was thought to be due to a *seesaw* mechanism (see below), and so on.

With the advent of the LED paradigm a drastic rethinking of particle physics, and in particular, flavordynamics, was inevitable. Everything had to be questioned anew, and novel explanations had to be invented. Needless to say, they were suggested before long.

Below we will outline some mechanisms discussed in this context. The aspect which we will emphasize here is a natural conversion of dynamical regularities into geometric ones within the LED framework.

3.4. *Hierarchy of scales*

At terrestrial energies gravity is exceedingly weak. The gravitational interaction is characterized by the Newton constant $G_N \equiv M_{\rm Pl}^{-2}$. While a typical electroweak scale is $M_{\rm SM} \sim 100$ GeV, the Planck scale $M_{\rm Pl} \sim 10^{19}$ GeV, so that there is a huge hierarchy of scales, $M_{\rm Pl}/M_{\rm SM} \sim 10^{17}$.

In the standard model *per se*, the electroweak scale is not stable, since quantum loops drag the Higgs boson mass (which is supposed to be of the order of $M_{\rm SM}$) to the Planck scale. In the desert paradigm supersymmetry plays a protective stabilizing role — superpartners cancel quadratic divergences in the Higgs boson mass above the supersymmetry breaking scale $M_{\rm SUSY}$. Although supersymmetry is not yet discovered experimentally, the general belief is that $M_{\rm SUSY} \sim$ few hundred GeV, i.e. quite close to $M_{\rm SM}$.

A minimally supersymmetrized version of SM (minimal supersymmetric standard model, or MSSM for short) has three gauge groups, U(1), SU(2), and SU(3), and, correspondingly, three gauge coupling constants, α_1, α_2, and α_3. These coupling constants run; the running formula being logarithmic in energy. This slow logarithmic running introduces another huge scale — the scale of grand unification. In MSSM the scale of grand unification $M_{\rm GUT} \sim 10^{16}$ GeV. It turns out that all three gauge coupling constants unify to within a few percent at $M_{\rm GUT}$, see Fig. 9. The grand unification scale is, in turn, quite close to the Planck scale $M_{\rm Pl}$, where gravity becomes strong,

$$M_{\rm GUT} \sim 10^{-3} M_{\rm Pl}\,. \tag{16}$$

There is a vast desert between $M_{\rm SM}$ and $M_{\rm SUSY}$ on the one hand, and $M_{\rm GUT}$ and $M_{\rm Pl}$ on the other hand, stretching in energy over 16 to 17 orders of

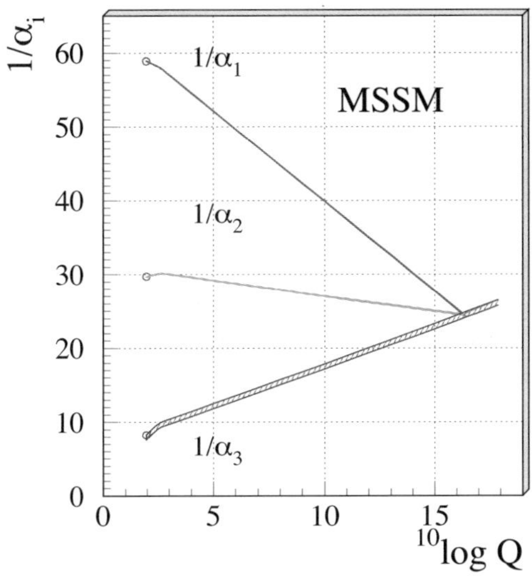

Fig. 9. Evolution of the inverse of the three gauge coupling constants in MSSM.

magnitudes.

In the LED scenario there is only one fundamental scale, $M_{\rm f} \sim 10$ TeV. That's the scale where gravity must become strong, and all interactions unify.[e] Where has the hierarchy gone?

The enormous hierarchy did not disappear. The energy hierarchy is converted into a geometric hierarchy of the "transverse" sizes: the radius of the extra dimension versus the brane thickness. Why the compact dimensions are so large in the LED paradigm? It is impossible to answer this question without complete understanding of the compactification dynamics, which is well beyond the scope of this lecture.

3.5. *Proton stability*

Why this issue is potentially dangerous for the very existence of the ADD scenario?

To properly set a reference point, let us start from the desert paradigm, where two mechanisms were considered in connection with the problem of the proton decay. First, in the grand unification theories (GUT's), there

[e] Beyond $M_{\rm f}$ physics is $4+n$ dimensional, and the law of running of the gauge couplings is non-logarithmic.[21]

exist gauge bosons with mass $\sim M_{\rm GUT}$ and leptoquark quantum numbers (in the Russian literature they are sometimes called "elephants"). These elephant bosons mediate quark annihilation into leptons, see Fig. 10, leading to the proton decay ($B-L$ is still conserved). Experimentally the proton lifetime is known to exceed 10^{32} years. The appropriate suppression of the "elephant" contribution is due to the fact that their masses are very large, $\gtrsim M_{\rm GUT}$.

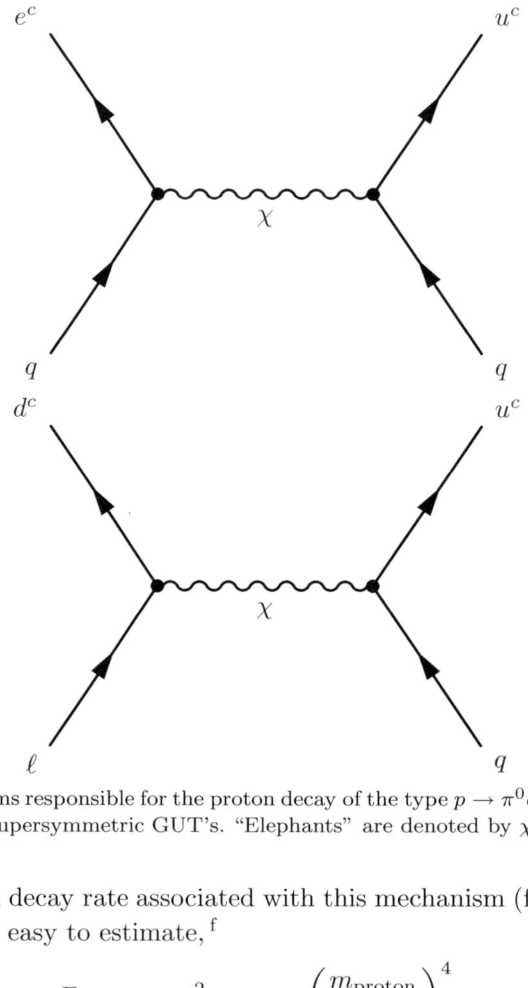

Fig. 10. Diagrams responsible for the proton decay of the type $p \to \pi^0 e^+$, or $p \to K^0 e^+$, or $p \to K^+ \nu$ in supersymmetric GUT's. "Elephants" are denoted by χ.

The proton decay rate associated with this mechanism (for a review see e.g. [22,23].) is easy to estimate,[f]

$$\Gamma_{\rm proton} \sim \alpha^2\, m_{\rm proton} \left(\frac{m_{\rm proton}}{M_{\rm GUT}}\right)^4 , \tag{17}$$

[f] The estimate given below refers to the so-called dimension-six operators, see Fig. 10; dimension-five operators result in a different (more contrived) expression, which is close numerically, however.

where α is the common value of the three gauge couplings at the unification scale. Given Eq. (16), the proton lifetime is predicted to be longer than 10^{33} years, which is compatible with experiment.

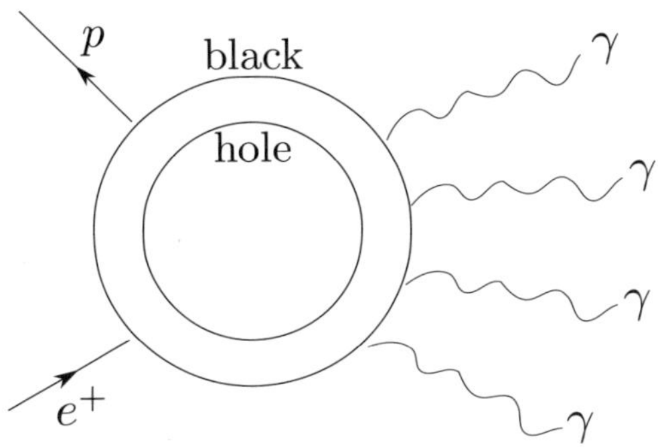

Fig. 11. Proton decay through a virtual black hole.

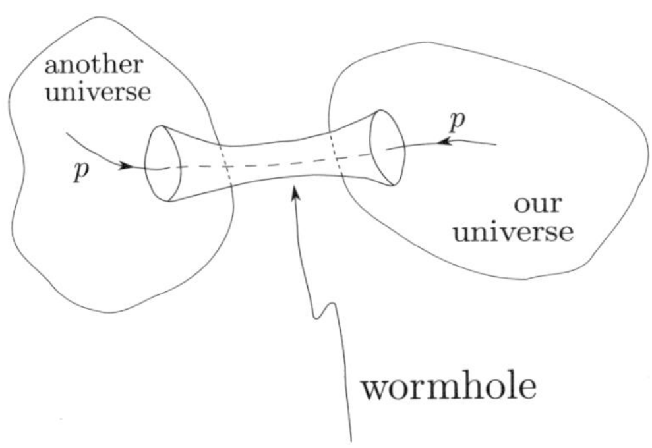

Fig. 12. A proton escaping from our universe via a wormhole.

The second mechanism discussed in connection with the proton decay is possible non-conservation of global quantum numbers, such as the baryon number B or lepton number L, in the presence of quantum gravity. The argumentation goes as follows [24]: global charges can be swallowed by black holes — say, virtual black holes which are formed non-perturbatively at the distance scales where gravity becomes strong — which then eventually evaporate (Fig. 11). Or a wormhole may suck in a global charge from our universe (Fig. 12) and spit it out into another one.[25] In both cases such event will be interpreted by "our" observer as a proton decay. In the "old" four-dimensional picture the contribution of this mechanism is of the order of

$$\Gamma_{\text{proton}} \sim m_{\text{proton}} \left(\frac{m_{\text{proton}}}{M_{\text{Pl}}} \right)^4 , \tag{18}$$

a rate which is even smaller than that in Eq. (17). What makes the proton instability due to gravity effects phenomenologically acceptable is the huge value of the mass scale where gravity becomes strong.

With the advent of the LED paradigm this situation dramatically changes. First, the extended desert (14 orders of magnitude in energy!) preceding the unification point of the gauge couplings disappears. Indeed, in $4 + n$ dimensions, at energies $E \gtrsim 1/\delta$, the logarithmic law of running gives place to a power law. (Here δ is the wall thickness which is usually assumed to be related to M_{f}.) If unification of the gauge couplings takes place,[21] it occurs at energies close to M_{f}. With the fundamental scale M_{f} as low as 10 TeV in the ADD scenario, a disaster seems inevitable. With new interactions and particles (mediating the proton decay) as light as 10 TeV protons will decay immediately unless special protective mechanisms are found.

The proton instability due to quantum gravity is, at the very least, as severe a problem as the one discussed above. If strong gravity occurs at the scale $M_{\text{f}} \sim 10$ TeV, virtual black holes will be abundant and they will destroy all globally conserved quantum numbers (such as the baryon number) much faster than in 10^{32} years. This is seen from Eq. (18) where M_{Pl} is to be replaced by M_{f}. Needless to say, this would be a mortal blow to the ADD scenario.

One of possibilities to suppress the proton decay is to associate a gauge symmetry with the baryon number, more exactly, a discrete gauge symmetry .[g] This mechanism of protection was invented long ago.[26] As was noted by Krauss and Wilczek,[26] "neither black-hole evaporation, wormholes, nor anything else can violate discrete gauge symmetries." With the advent of the LED paradigm people used it first to guarantee the proton stability. This method was shown to be viable, i.e. its phenomenological implications are compatible with experimental data.[23]

Here we will focus on an alternative, geometric protection, one of many

---

[g] What is a discrete gauge symmetry? One first gauges the baryon charge in a regular way coupling it, say, to a U(1) gauge boson. Then this continuous gauge symmetry must be spontaneously broken by a Higgs mechanism down to a discrete subgroup.

manifestations of a universal geometrical idea which, being combined with the ADD idea, brings lavish fruits. It goes under the name *fat branes*, or branes with a substructure, and was put forward by Arkani-Hamed and Schmaltz [27] in the context of the problem of proton stability. [h]

Let us examine the transverse structure of the domain wall. A hypothetical slice through the wall is schematically shown in Fig. 13. The overall wall thickness is δ. A crucial observation is that quarks and leptons need not be localized at one and the same point on the Z axis, and, moreover, the localization width need not coincide with δ. Assume that quarks are localized on the left edge of the wall, with the localization width $\ell \ll \delta$, while leptons are localized on the right edge. If W and other gauge bosons are smeared everywhere inside the wall — i.e. their degree of localization is δ — this will ensure that "normal" SM quark/lepton decays proceed as they should. At the same time, the baryon number changing transition of quarks into an appropriate lepton will be suppressed by an exponentially small overlap of the wave functions in the Z direction. The suppression factor in the amplitude is proportional to

$$\int d^n Z\, \Psi_q^3(Z)\, \Psi_l^*(Z) \sim e^{-\delta/\ell}\,, \tag{19}$$

where the subscripts q and l stand for quarks and leptons, respectively. A relatively modest geometric hierarchy of $\delta/\ell \sim 30$, after the exponentiation, will suppress the proton decay to the acceptable level, with no protection in the form of additional (gauged) conservation laws.

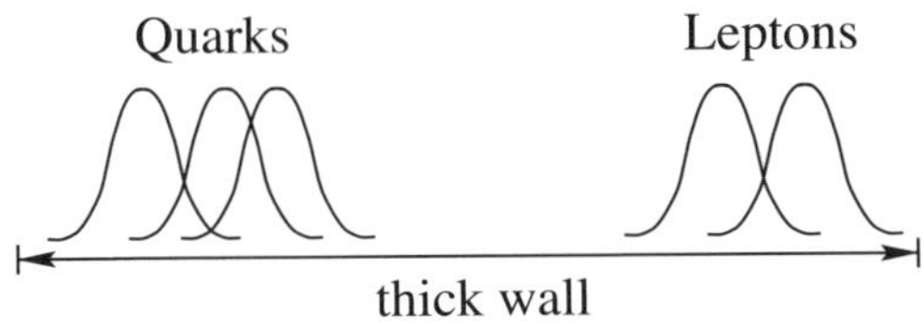

Fig. 13. A layered domain wall with a substructure.

A possible substructure of the domain wall defining our brane world may provide geometric solutions to other problems from flavordynamics, in particular those listed in Sect. 3.3. For instance, the four-layer structure shown

[h] These authors also considered the hierarchy problem following the same line of reasoning.

in Fig. 14 may explain [28] not only the fact of three distinct generations, but also the pattern of quark masses and mixing angles.[28,29] Four domains correspond to localizations of the Higgs field and quark-lepton fields belonging to generations three, two, and one. It is clear that the suppression of mass terms due to overlap of the wave functions similar to Eq. (19) will depend on the number of the generation at hand — the closer it is to the Higgs brane, the larger is the overlap with the Higgs field Z-profile, implying a larger mass mass term. In this way a typical pattern of the quark masses in three generations naturally emerges.

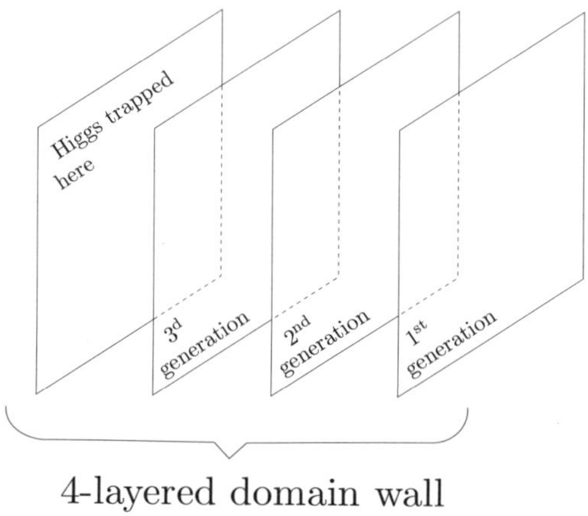

Fig. 14. Four-layer structure of the domain wall generates a reasonable pattern of the quark masses and mixing angles.

3.6. *The lightness of the left-handed neutrino*

The masses of "our" left-handed neutrinos are believed to lie in the ballpark of 10^{-2} or 10^{-3} eV. How does the good old paradigm cope with such small masses?

The standard explanation relies on the seesaw mechanism invented in the late 1970's.[30] The essence of the seesaw mechanism is as follows. Let us assume that the Dirac neutrino mass is described by the term $\mu\,\bar{\nu}_R\,\nu_L$ where the mass term μ is of the order of $M_{\rm SM}$, its natural order of magnitude. The right-handed neutrino which is (nearly) sterile has a Majorana mass term $M\,\bar{\nu}_R\,\nu_R^c$ where the superscript c stands for the charge conjugation. It is natural to assume that $M \sim M_{\rm GUT}$. Then, upon diagonalization of the mass matrix, one finds that the true left-handed neutrino is a mixture of

ν_L and ν_R^c (the admixture of ν_R^c is very small, $\sim \mu/M$) and its mass is

$$m_{\nu_L} \sim \frac{\mu^2}{M}\,. \tag{20}$$

With the above assumptions regarding μ and M, we get "our" neutrino mass in the right ballpark. The lightness of "our" neutrino is due to the enormity of $M_{\rm GUT}$.

As we remember, in the ADD scenario the *bona fide* fundamental scale $M_{\rm f}$ is much lower. Gone with $M_{\rm GUT}$ is the seesaw mechanism. A question that immediately comes to one's mind is: "Can one engineer a LED-based mechanism (preferably, geometrical) that would naturally explain the lightness of the left-handed neutrino?"

The answer to this question is positive, and, in fact, there exists more than one solution. One of the nicest ideas belongs to the authors of the ADD scenario themselves.[31] The right-handed neutrino carries no charges with respect to the SM gauge bosons. Therefore, nothing precludes us from letting the right-handed neutrino escape to the bulk, unlike the left-handed neutrino, which has to be localized on the wall. The very existence of the wall may be responsible for the disparity between ν_R and ν_L. Indeed, the topology of the wall solution (i.e. wall vs. anti-wall) is typically correlated with the chirality of the fermion zero modes. Thus, it is quite natural to expect that a wall traps ν_L, while ν_R is free to go in the bulk. An anti-wall would trap ν_R.

If the right-handed neutrino wave function is smeared all over the bulk then the neutrino mass term, which is proportional to the overlap of Ψ_{ν_L} and $\Psi^*_{\nu_R}$, takes the form

$$\int d^4x\, H(x)\, \Psi_{\nu_L}(x)\, \Psi^*_{\nu_R}(x, Z=0)\,, \tag{21}$$

where $H(x)$ is the Higgs field wave function. Due to the fact that Ψ_{ν_R} is totally delocalized in the extra dimensions,

$$\Psi_{\nu_R}(x, Z) \sim \frac{1}{\sqrt{V_n}}\,, \tag{22}$$

so that our neutrino mass gets a natural suppression factor $V_n^{-1/2}$,

$$m_\nu \sim \frac{v}{\sqrt{V_n M_{\rm f}^n}} \sim \frac{v\, M_{\rm f}}{M_{\rm Pl}}\,, \tag{23}$$

where v is the expectation value of the Higgs field and we used Eq. (13). Given the uncertainty in numerical factors, this estimate places the neutrino mass in the right ballpark.

Thus, the lightness of the left-handed neutrino is due to the same reason why gravity is weak — large volume of the extra dimensions.

3.7. *Downside of the ADD scenario*

Precision electroweak measurements firmly established the validity of the standard model. What will come beyond the standard model? Although theoretical speculations are abundant, experimental support is scarce. In fact, the only semiquantitative achievement in this direction is the success of the gauge coupling unification.

The gauge interactions unified by the standard model are SU(3)×SU(2)×U(1). The first is the color group, the last two represent electroweak interactions. At low energies the corresponding three gauge couplings are very different in their values, see Fig. 9.[i] The logarithmic running brings them closer, and eventually all three intersect — nearly exactly at one point — $M_{\rm GUT}$, which, in turn, turns out to be rather close to $M_{\rm Pl}$. Figure 9 illustrates this statement. It is important to note that the intersection occurs only provided that SM is supersymmetrized. Thus, the above success may be viewed, simultaneously, as a semidirect indication that supersymmetry is relevant to nature. It would be a pity to loose this encouraging indication.

The unification and the desert paradigm are closely connected. Being honest, we should admit that the ADD scenario erases the above success. The logarithmic running stops at $M_{\rm f}$, where the three gauge couplings are still very far from each other. Will a power-like running, which replaces the logarithmic one above $M_{\rm f}$, unify the couplings, and at what scale? With the loss of the great desert the answer to this question becomes model-dependent and almost completely devalued.[21]

4. A few words on other scenarios

Next, we will outline, very briefly other scenarios, in which extra dimensions are instrumental in understanding the world we observe around us. Chronologically they appear later than ADD and are "ideologically" related.

---

[i] The SU(2) and U(1) couplings are measured with high precision; the accuracy of the SU(3) coupling is not that high, mainly due to theoretical uncertainties in its determination.

4.1. *Warped Extra Dimensions*

In the ADD scenario gravity of the domain wall as such plays little role. Given that the brane tensions are small, this is a good approximation. However, this need not be the case. In the scenario suggested by Randal and Sundrum (RS)[32,33] the brane-induced gravity strongly warps extra dimensions which is instrumental in ensuring an appropriate localization.

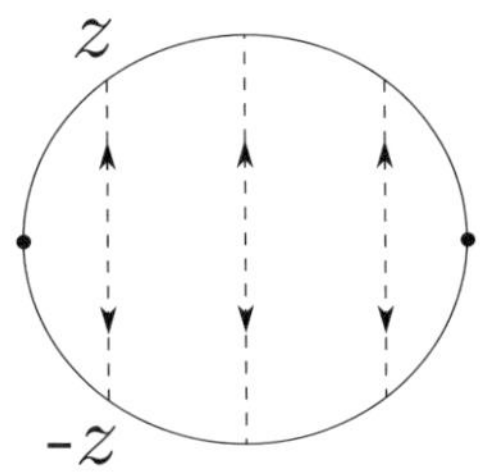

Fig. 15. Graphical representation of S_1/Z_2. The dashed lines indicate the points of the circle which are to be identified.

A general observation on which the RS construction is based is as follows: if one has 3-branes in five dimensions, one can, in principle, balance the gravitational effects of the above branes by a five-dimensional bulk cosmological constant to get a theory in which an effective cosmological constant of our four-dimensional world will vanish. Our universe will seem static and flat for an observer on our brane.[14] The price to pay for this fine-tuning is a highly curved five-dimensional background. This phenomenon is called "off-loading." We off-load curvature from the brane on which we live onto the bulk. Needless to say, there is no theoretical rationale for the above fine-tuning. We just take the fact that the world we observe is (nearly) flat as given.

Let us discuss some basic elements of the RS construction. One starts from a finite length extra dimension[32] assumed to be an S_1/Z_2 orbifold. What is this? Take a sphere S_1 and identify opposite points, as shown in Fig. 15. The points 0 and π are called fixed points since they are identified with themselves. Two branes are placed at these points in the extra dimension. The solution of five-dimensional gravity – I will not derive it here – is self-consistent (and consistent with the fine tuning discussed above) provided that one of the branes has a positive tension while the tension of the other brane is negative (and the same in the absolute value).

Since we want the Lorentz invariance to be preserved in our four-dimensional world we have to assume that the induced metric (its $\mu\nu$ part) at every point along the extra dimension is proportional to the four-dimensional Minkowski metric $\eta_{\mu\nu}$, while the components of the five-dimensional metric depend only on the fifth coordinate Z in the following way:

$$ds^2 = e^{-A(Z)} dx^\mu dx^\nu \eta_{\mu\nu} - dZ^2 \,. \tag{24}$$

The degree of warping along the extra dimension depends on the factor $e^{-A(Z)}$, which is therefore called the warp factor. In the RS solution the warp factor depends on Z exponentially,

$$A(Z) = 2\, k\, |Z| \tag{25}$$

where k is a constant of dimension of mass which sets the scale of warping. It turns out that it is related to the five-dimensional cosmological constant Λ_{5D} and the five-dimensional Planck constant $M_{5D}^{\rm Pl}$ by the formula [32]

$$k^2 = -\frac{\Lambda_{5D}}{\left(M_{5D}^{\rm Pl}\right)^3} \,. \tag{26}$$

This relation implies, in particular, that Λ_{5D} is negative. Thus, the five-dimensional space-time one deals with in the RS scenario is anti-de Sitter.

Now finally we are in a position to understand the impact of two branes – a positive tension brane at $Z = 0$ and a negative tension brane at $Z = b$ – as well as the impact of warping. Equations (24) and (25) show that the induced metric on the negative tension brane is exponentially smaller than that on the positive tension brane provided that $kb \gg 1$,

$$g_{\mu\nu}^{\rm ind}\Big|_{Z=b} = e^{-2kb}\,\eta_{\mu\nu} \,. \tag{27}$$

This suppression sets the scale for all other mass parameters. Indeed, consider the Higgs field part of the action. Proceeding to the canonically normalized kinetic term of the Higgs field we see that the physical value of its vacuum condensate is "warped down" to

$$v_{\rm phys.Higgs} = e^{-kb} v_0 \,, \tag{28}$$

which shows, in turn, that all masses following from (28) are exponentially suppressed on the negative tension brane (but not on the positive tension brane). Then it is natural to refer to the positive tension brane at $Z = 0$ as the *Planck brane*, since the fundamental mass scale there is of the order of the Planck scale. The negative tension brane is said to be the *TeV brane* which follows from Eq. (28) where we use $kb \sim 17$ and $M_{5D}^{\rm Pl} \sim M_{4D}^{\rm Pl}$.

In fact, there are two popular versions of the RS scenario. The one discussed above (RS1), has a finite size extra dimension with two branes, one at each end. The second version (RS2) is similar to the first, but one brane has been placed infinitely far away, so that there is only one brane left in the model. The particles of the standard model are placed on the Planck brane. This model was originally of interest because it represented an infinite five-dimensional model resembling, in many respects, a four-dimensional model.

For a more detailed consideration the reader is referred to numerous reviews, for instance [34].

4.2. *Braneworlds with Infinite Volume Extra Dimensions*

Here we will briefly outline a model proposed by Dvali, Gabadadze, and Porrati (DGP) [35] in 2000 in which, strictly speaking, gravity never becomes four-dimensional, no matter how far in the infrared we go. Therefore, this is not a genuine compactification. Unlike the ADD scenario in this construction gravity only approximately imitates four-dimensional behavior at large distances. A residual "tail" associated with extra dimensions produces an effect which could, in principle, reproduce the cosmic acceleration of dark energy.[36] Then the four-dimensional cosmological constant need not be non-vanishing. The DGP mechanism allows the volume of the extra space to be infinite,

$$V_N \equiv \int d^N\, Z\sqrt{G} \to \infty\,, \tag{29}$$

where N is the number of extra dimensions and G is the metric tensor in the $4+N$ dimensional space. For orientation below we will assume $N = 1$, although other versions of the DGP scenario, with $N > 1$, were also considered.

The basic idea behind this scenario is as follows. We start from $4+N$-dimensional space, with $4+N$-dimensional gravity governed by the standard Einstein–Hilbert action. A four-dimensional domain wall is embedded in N dimensions. As in the ADD scenario, all matter fields are assumed to be localized on the wall. The central point of the DGP construction is the emergence of the induced four-dimensional Einstein–Hilbert term on the wall due to a loop of virtual matter localized on the wall, see Fig. 16. The bare action has no such term. However, as soon as the localized matter fields are coupled to $4+N$ dimensional gravity, we obtain 4-dimensional gravity action on the wall from loops. This mechanism described by Sakharov [37] long ago is usually referred to as Sakharov's induced gravity.

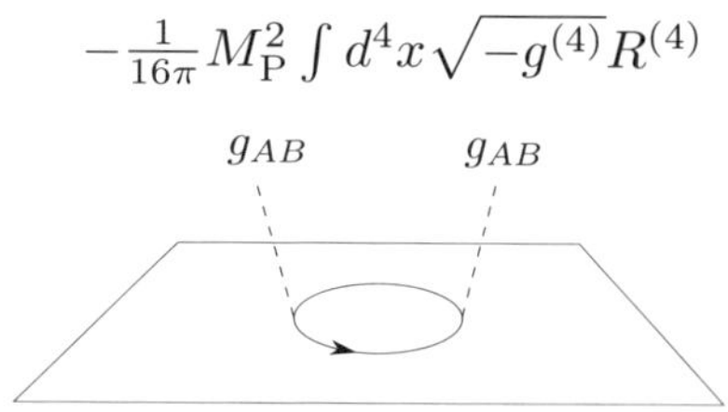

Fig. 16. One-loop contribution of the matter fields to the effective action of gravitons. The matter fields are localized on the brane yielding the four-dimensional Einstein–Hilbert term which is also localized on the brane.

Then the DGP action takes the form

$$S = \frac{\left(M_{5D}^{\rm Pl}\right)^3}{2}\int d^4x \int_{-\infty}^{+\infty} dy\sqrt{G}\,R_5$$

$$+\int d^4x\sqrt{g}\left[\frac{\left(M_{4D}^{\rm Pl}\right)^2}{2}R_4 + \Lambda + \mathcal{L}_{\rm matter}(\Phi_{\rm SM},\Psi_{\rm SM})\right], \qquad (30)$$

where G and g stand for five- and four-dimensional metrics, respectively,

$$g = G(Z=0)\,,$$

R_5 and R_4 are the corresponding Ricci tensors, while $\Phi_{\rm SM}$, $\Psi_{\rm SM}$ is a generic notation for all boson and fermion matter fields. The parameter $\left(M_{4D}^{\rm Pl}\right)^2$ is obtained in Eq. (30) as an ultraviolet cut-off in loops of the brane-confined matter, representing the scale of energies at which the standard-model physics drastically changes. In the DGP scenario it is assumed that

$$M_{5D}^{\rm Pl}/M_{4D}^{\rm Pl} \gg 1\,.$$

Let us discuss the value of the domain wall tension. One assumes that the $4+N$-dimensional theory is supersymmetric, and that supersymmetry is spontaneously broken only on the brane. Non-breaking of supersymmetry in the bulk is only possible due to the infinite volume of the extra space; supersymmetry breaking is not transmitted from the brane into the bulk since the breaking effects are suppressed by an infinite volume factor. Then, the bulk cosmological term can be set to zero, without any fine-tuning. As for the four-dimensional cosmological constant Λ in Eq. (30), the natural value of Λ can be as low as TeV^4, since the brane tension can be protected above this value by $\mathcal{N}=1$ supersymmetry. Please, note that Λ must be fine-tuned in such a way that $\Lambda + \langle\mathcal{L}_{\rm matter}\rangle = 0$. This is a usual fine-tuning of the cosmological constant. The DGP construction adds nothing new in this respect.

With the action (30) gravity from the sources localized on the domain wall will propagate both, in the bulk and in the brane. Interplay of these two modes of propagation leads to quite a peculiar gravitational dynamics on the brane. Unlike the ADD scenario, in which gravity deviates from its four-dimensional form only at short distances, in the DGP case deviations from the four-dimensional Newton law occur both at short and large distances. Despite the fact that the volume of extra space is infinite, an observer on the brane measures four-dimensional gravitational interaction up to distances

$$r_c \sim \frac{\left(M_{4D}^{\rm Pl}\right)^2}{\left(M_{5D}^{\rm Pl}\right)^3}\,. \tag{31}$$

At distances larger than r_c the Newton potential losses its four-dimensional form. In order for the late-time cosmology to be standard we must require r_c to be of the order of the universe size, i.e. $r_c \sim H_0^{-1} \sim 10^{28}$ cm. This, in turn, requires $M_{5D}^{\rm Pl}$ to be very small. Small $M_{5D}^{\rm Pl}$ means that gravity in the bulk is strong. However, the large Einstein–Hilbert term on the brane "shields" matter localized on the brane from strong bulk gravity.

4.3. *Universal extra dimensions*

For completeness I mention the universal extra dimension (UED) scenario.[38] It assumes compactification of the ADD type. However, all standard matter fields are free to propagate through all of the extra-dimensional space (which is essentially flat), rather than being confined to a brane. The size of the extra dimensions is assumed to be in the ballpark of TeV^{-1}. To my mind this scenario lacks elegant theoretical features that can be found in ADD, RS and DGP. Phenomenology that ensues was discussed in the literature, see e.g. the review papers [39].

References

1. Th. Kaluza, Sitzungsber. Preuss. Akad. Wiss. Phys. Math. Klasse 966 (1921) [Reprinted, with the English translation, in *Modern Kaluza–Klein Theories*, Eds. T. Appelquist, A. Chodos, P.G.O. Freund (Addison-Wesley, Menlo Park, 1987), p. 61]. Theodor Kaluza, of Königsberg University, submitted his paper to Einstein in 1919. Einstein apparently had serious doubts as to its merits — he forwarded it for publication in the above journal only in 1921.
2. In 1926 Oscar Klein who at that time was in Copenhagen, and the Russian physicist H. Mandel independently rediscovered Kaluza's theory (O. Klein,

Z.F. Physik, **37**, 895 (1926) and Nature, **118**, 516 (1926); H. Mandel, Z.F. Physik, **39**, 136 (1926)). Klein's papers are reprinted in *Modern Kaluza–Klein Theories*, Eds. T. Appelquist, A. Chodos, P. G. O. Freund (Addison-Wesley, Menlo Park, 1987), p. 76 and 88.

3. W. Pauli, Ann. d. Physik, **18**, 305, 337 (1933).
4. A. Einstein and P. G. Bergmann, Ann. Math. **39**, 810 (1931) [Reprinted, with the English translation, in *Modern Kaluza–Klein Theories*, Eds. T. Appelquist, A. Chodos, P. G. O. Freund (Addison-Wesley, Menlo Park, 1987), p. 89].
5. E. Witten, Nucl. Phys. B **186**, 412 (1981).
6. J. Scherk and J. H. Schwarz, Phys. Lett. B **57**, 463 (1975).
7. P. Candelas, G. T. Horowitz, A. Strominger and E. Witten, Nucl. Phys. B **258**, 46 (1985).
8. A. E. Faraggi, *Superstring phenomenology: A personal perspective,* in *Proc. 2nd International Conference on Physics Beyond The Standard Model,* (June 1999, Tegernsee, Germany), "*Beyond the Desert 1999*", Eds. H. V. Klapdor-Kleingrothaus and I. V. Krivosheina (Bristol, IOP, 2000), p. 335-357 [hep-th/9910042].
9. D. Cremades, L. E. Ibañez and F. Marchesano, Nucl. Phys. B **643**, 93 (2002) [hep-th/0205074]; *More about the standard model at intersecting branes,* in *Proc. Int. Conf. on Supersymmetry and unification of fundamental interactions, DESY, Hamburg 2002,* Eds. P. Nath and P. Zerwas, Vol. 1, p. 492 [hep-ph/0212048]; *Towards a theory of quark masses, mixings and CP-violation,* hep-ph/0212064.
10. J. Polchinski, Phys. Rev. Lett. **75**, 4724 (1995) [hep-th/9510017].
11. P. Hořava and E. Witten, Nucl. Phys. B **475**, 94 (1996) [hep-th/9603142].
12. K. Akama, Lect. Notes Phys. **176**, 267 (1982) [hep-th/0001113].
13. M. Visser, Phys. Lett. B **159**, 22 (1985) [hep-th/9910093].
14. V. A. Rubakov and M. E. Shaposhnikov, Phys. Lett. B **125**, 136 (1983).
15. R. Jackiw and C. Rebbi, Phys. Rev. D **13**, 3398 (1976) [Reprinted in *Solitons and Particles*, Eds. C. Rebbi and G. Soliani (World Scientific, Singapore, 1984), page 331.]
16. G. R. Dvali and M. A. Shifman, Phys. Lett. B **396**, 64 (1997), [E] B **407**, 452 (1997) [hep-th/9612128].
17. G. R. Dvali and M. A. Shifman, Nucl. Phys. B **504**, 127 (1997) [hep-th/9611213].
18. N. Arkani-Hamed, S. Dimopoulos and G. R. Dvali, Phys. Lett. B **429**, 263 (1998) [hep-ph/9803315].
19. I. Antoniadis, N. Arkani-Hamed, S. Dimopoulos and G. R. Dvali, Phys. Lett. B **436**, 257 (1998) [hep-ph/9804398].
20. C. D. Hoyle *et al.*, Phys. Rev. Lett. **86**, 1418 (2001) [hep-ph/0011014]; J. Chiaverini, S. J. Smullin, A. A. Geraci, D. M. Weld and A. Kapitulnik, Phys. Rev. Lett. **90**, 151101 (2003) [hep-ph/0209325]; E. G. Adelberger [EOT-WASH Group Collaboration], hep-ex/0202008; For reviews see G. Landsberg, hep-ex/0105039; J. C. Long and J. C. Price, hep-ph/0303057.

21. K. R. Dienes, E. Dudas and T. Gherghetta, Phys. Lett. B **436**, 55 (1998) [hep-ph/9803466]; Nucl. Phys. B **537**, 47 (1999) [hep-ph/9806292].
22. S. Wiesenfeldt, *Proton decay in supersymmetric grand unified theories,* DESY-THESIS-2004-009.
23. P. Nath and P. Fileviez Pérez, Phys. Rept. **441**, 191 (2007) [arXiv:hep-ph/0601023].
24. Ya. B. Zeldovich, Pis'ma Zh. Eksp. Teor. Fiz. **24**, 29 (1976) [JETP Lett. **24**, 25 (1976)]; Phys. Lett. A **59**, 254 (1976).
25. L. F. Abbott and M. B. Wise, Nucl. Phys. B **325**, 687 (1989); S. B. Giddings and A. Strominger, Phys. Lett. B **230**, 46 (1989); Nucl. Phys. B **321**, 481 (1989); S. R. Coleman and K. M. Lee, Nucl. Phys. B **329**, 387 (1990).
26. L. M. Krauss and F. Wilczek, Phys. Rev. Lett. **62**, 1221 (1989).
27. N. Arkani-Hamed and M. Schmaltz, Phys. Rev. D **61**, 033005 (2000) [hep-ph/9903417].
28. G. R. Dvali and M. A. Shifman, Phys. Lett. B **475**, 295 (2000) [hep-ph/0001072].
29. C. Matti, Eur. Phys. J. C **48**, 251 (2006) [arXiv:hep-ph/0606158].
30. M. Gell-Mann, P. Ramond and R. Slansky, *Complex Spinors And Unified Theories,* in Proc. Stony Brook Supergravity Workshop (September 1979) *Supergravity*, Eds. P. van Nieuwenhuizen and D. Z. Freedman (North Holland, Amsterdam, 1979), p. 315-321; T. Yanagida, Prog. Theor. Phys. **64**, 1103 (1980).
31. N. Arkani-Hamed, S. Dimopoulos, G. R. Dvali and J. March-Russell, Phys. Rev. D **65**, 024032 (2002) [hep-ph/9811448].
32. L. Randall and R. Sundrum, Phys. Rev. Lett. **83**, 3370 (1999) [hep-ph/9905221].
33. L. Randall and R. Sundrum, Phys. Rev. Lett. **83**, 4690 (1999) [hep-th/9906064].
34. C. Csaki, *TASI lectures on extra dimensions and branes,* arXiv:hep-ph/0404096, in *Particle Physics and Cosmology*, Eds. H. Haber and A. Nelson (World Scientific, Singapore, 2004), p. 605.
35. G. R. Dvali, G. Gabadadze and M. Porrati, Phys. Lett. B **485**, 208 (2000) [arXiv:hep-th/0005016].
36. C. Deffayet, G. R. Dvali and G. Gabadadze, Phys. Rev. D **65**, 044023 (2002) [arXiv:astro-ph/0105068]; G. Dvali, G. Gabadadze and M. Shifman, Phys. Rev. D **67**, 044020 (2003) [arXiv:hep-th/0202174].
37. A. D. Sakharov, Sov. Phys. Dokl. **12**, 1040 (1968) [reprinted in Sov. Phys. Usp. **34** 394 (1991)].
38. T. Appelquist, H. C. Cheng and B. A. Dobrescu, Phys. Rev. D **64**, 035002 (2001) [arXiv:hep-ph/0012100].
39. G. D. Kribs, *Phenomenology of extra dimensions,* arXiv:hep-ph/0605325, in *Physics in* $D \geq 4$, Ed. J. Terning, C. Wagner and D. Zeppenfeld, (World Scientific, Singapore, 2006), p. 633; D. Hooper and S. Profumo, Phys. Rept. **453**, 29 (2007) [arXiv:hep-ph/0701197].

Vacuum Structure and QCD Sum Rules: Introduction

M. Shifman*

William I. Fine Theoretical Physics Institute
University of Minnesota, 116 Church Street S.E.
Minneapolis, MN 55455, USA
**E-mail: shifman@umn.edu*

This note was written as an introduction to the volume *Vacuum Structure and QCD Sum Rules* (North-Holland, 1992). It has never been published.

The story of this article is as follows. Shortly after my arrival to the United States I got a letter from Professor Hector Rubinstein who at that time was the Coordinating Editor of *Current Physics — Sources and Comments*, a serial North-Holland publication of reprint volumes on high-energy physics. Each volume was supposed to present a selection of essential source material on a given important topic compiled by an independent specialist editor, supplemented by extensive editorial commentaries and mini-reviews. Professor Rubinstein invited me to make a volume devoted to the QCD sum rules.[a] After some hesitations I agreed. The work turned out to be much more time and labor-consuming than I had expected. In several cases (Chapter 1, 8 and in some other cases) I failed to find among the published papers the one which I could consider satisfactory; then I had to write the corresponding sections myself.

Hector's idea was to put a historical introduction/foreword, which I was supposed to write, in the beginning of the book. By deadline I had only an unpolished and unfinished draft. Other urgent commitments prevented me from completing the article in time. Therefore, I had to settle for a short technical foreword which appeared in print in 1992. The original rather long

[a] This project was aborted in 1992. In fact, the Volume *Vacuum Structure and QCD Sum Rules* (North-Holland, Amsterdam, 1992) of which I was the editor, was the tenth and the last in the series.

Foreword has never been published. I used edited fragments of this article in various later publications. Presented below is the original draft.[b]

⋆ ⋆ ⋆ ⋆ ⋆

The purpose of this Introduction is to give a general perspective of the development of the method which goes under an awkward name[c] *the QCD sum rules* – one of the most productive tools for calculating hadronic parameters in Quantum Chromodynamics. The second task is to convey the historic flavor of the exciting time when the theory of hadrons was making its first steps.

It is, perhaps, worth starting from the second point. It seems fair to say that QCD was born after the talk of Gell-Mann and Fritzsch[1] (see also Ref. [2]) in which the color-octet gluons were introduced. The next step is certainly the discovery of asymptotic freedom in 1973.[3] In the first few years, roughly speaking till 1975, the theorists' attention was almost totally focused on hard processes in which the short-distance physics plays the dominant role. This topic became hot and fashionable, piles of new papers appearing daily. The main achievement of this period is that people learned how to reliably isolate the short distance contributions governed by the small coupling constant and how to generalize electrodynamical perturbative calculations to non-Abelian theories.

A recent paper of Polyakov[4] presenting his understanding of the development of our field in the seventies is entitled "A View from the Island" which gives a good idea of our place in the scientific process in Moscow in those days. The isolation was almost complete, and we could not compete with the western theorists in most fashionable directions where the results seemed to be on the surface, for obvious reasons.

Because of the total censorship preprints and journals from the West used to come with enormous delays, and our own papers could be submitted to the Western journals only after a complicated procedure of getting clearance from half a dozen of different instances, typically a waste of a few months. Publication of preprints was an adventure due to bureaucratic limitations. For instance, one and the same paper could not be issued as a preprint and, then, in a journal, and the preprint version could not be longer than 25 typewritten pages (or 35, I do not remember exactly). So we

[b] The list of references is slightly updated.
[c] Also known as the Shifman–Vainshtein–Zakharov (SVZ for short) sum rules.

had a lot of "fun" trying to muddle and deceive our censors by making different titles in the preprint and journal versions of one and the same work, changing the order of the authors or serializing preprint publications like a detective story in a popular newspaper (unlike the latter case, though, we had to make an impression that each successive part is not connected to the previous). Quite often these tricks created a terrible mess, to say nothing about wasted efforts. Occasionally, in the most important cases, one would risk to bypass the standard procedure by using different "illegal" channels, mostly our Western friends. By the way, any contacts with the latter were also severely constrained. There is a famous story about one of the scientific bosses in Dubna who was instructing the Soviet participants of a conference before its opening. He said: "Well..., we had to organize this conference, and even invite some foreigners. Unfortunately, to my deep regret, this time it will be impossible to completely avoid contacts..." Conferences in the West were open for a handful of specially selected, through a humiliating procedure; and even those few which took place in the Soviet Union were not always accessible. I remember, for instance, that a colleague of mine from ITEP and I were not granted permission to participate in "Neutrino –77" in Baksan. The decision to throw us out from the list of participants was made in the very last moment, and the only reason behind this decision I could think of was the fact that we both were Jewish.

Some words of the Soviet newspeak made me sick. One of such words was *Glavlit.* As I just said, to publish a scientific paper was much more than just typing the manuscript and mailing it to the publisher. There was a long latent period, associated with getting all sorts of clearances. First, the so-called Expert Commission (a group of authorized fellow physicists in the given institution) was supposed to study the paper and recommend its publication. According to the official rules they had to certify that no new discoveries were reported, because if they were, the Expert Commission had to recommend to classify the paper right away. Of course, people tended to stretch the official rules, otherwise not a single breakthrough paper would have ever appeared in the Soviet Union.

At the next stage the paper would go to the so called Regime Department whose task was to check that no references to classified work or undesirable persons were made, no subversive ideas put forward, and so on. With all this paperwork done, the decision to allow (disallow) publication was to be made by the Director of the Institute. This is not the end of the story, however. All materials intended for publication had to be cleared through the so-called GLAVLIT, the almighty agency whose sole obligation was to ensure total Censorship in the country. If, at the previous stages the author would have at least some minimal control over what was going on with his (her) paper, GLAVLIT was a total black box.

The process of getting all clearances could extend anywhere from weeks to many months, and the paper was officially nonexistent until the very end. The author could not even refer to it in his/her further work.

This is a long saga, and I could easily write many pages on this topic, but now it is clearly time to stop. To make the long story short I will only say that making our results known was a difficult, nervous and time-consuming part of our job. This largely predetermined the choice of topics we could work on and formed in the ITEP theory group and elsewhere a very special atmosphere, now gone forever.

ITEP of 1970s had one of the best groups in the world, an excellent collection of enthusiasts whose attitude to physics was totally "non-commercial." People were always eager to discuss with each other every interesting scientific question emerging during the seminar talks or elsewhere, and these discussions quite often would last till midnight. You could easily find experts in any conceivable field or direction who would gladly share their knowledge with you. Our common enemies and common isolation created, as a counter-reaction, very strong friendly and scientific ties, as the only way of survival, and helped develop protective values; the tacit understanding that good physics was above all was among these values.

The only drawback I can think of now, in retrospect, is the general negative attitude to field theories in the very end of 1960s and the beginning of 1970s when I just appeared in ITEP. The reason is obvious, of course: the influence of Landau and his discovery of the zero-charge property in QED [5] – the influence which was alive and very strong in the ITEP theory group in those days. The attitude to the field theory as to something absolutely not serious was so deeply rooted that the fact of the *anti*screening of the gauge coupling constant in non-Abelian theories which was reported in ITEP at least twice [6,7] in the *sixties* has not been appreciated and recognized.[8] A restructuring in minds started only after the very same fact, the surprising fall-off of the gauge charge at short distances, became known from Ref. [3].

So, our start was relatively slow. By 1974, however, we were fully submerged in this subject, and shortly after it became clear that the cavalry attacks do not help to solve the problem of confinement and that the widespread expectations of a Messiah who would come soon and teach us the mystery of the solution had to be tempered. The fact that instantons,[9] a beautiful theoretical construction which was a breakthrough in the qualitative understanding of the QCD vacuum,[10] turned out to be useless in the quantitative sense because of the infrared divergences was a serious blow. So we adopted a less ambitious approach (by "we" I mean Valya Zakharov, Arkady Vainshtein and myself). The idea was to start from short distances where the quark-gluon dynamics was essentially perturbative and we felt ourselves on a firm ground, and then to extrapolate to larger distances (where the hadronic states are presumably formed) including non-perturbative effects "step by step" and using some kind of an approximate procedure to extract information on hadronic properties. Of course, this idea was rather vague at first, the program as we know it now has been crystallizing gradually.

It is rather difficult to identify the work which, for us, was the first crucial step. With some reservations I might say that the first step has been done in Ref. [11]. Perhaps now the sum rule for the charmed particle photoproduction obtained in [11] does not seem impressive but this analysis carried important elements which where later laid in the foundation of the sum rule method. A spectacular success came after we united our efforts with V. Novikov, L. Okun and M. Voloshin. It turned out that a whole variety of the charmonium parameters are predictable essentially from pure duality, and for about a year we played the game of constructing the charmonium widths and mass ratios from simple numbers. In 1977 a review article[12] was submitted. At about that time it became clear that the progress was limited; the method presented in [12] could not be reliably generalized to such typical representatives of the hadronic family as, say, ρ mesons or nucleons. And the desire to get access to these hadrons was strong.

The remainder of this story, including its culmination – introduction of the gluon condensate[13] in fall 1977 – is described elsewhere.[14] It was a hot summer of 1977, just before the vacations. Our big six-head strong collaboration ceased to exist. We — Valya, Arkady and myself — were leisurely discussing something when the first hints appeared. The conjecture was that the vacuum was actually something like a gluon "medium," and the basic particle properties are due to the quark interaction with this medium which can be conveniently parametrized by certain quark and gluon condensates. We worked out the first implications of the gluon condensate in the fall of 1977. At first, we were discouraged by a wrong sign of the gluon condensate term in our "show-case" sum rule, the one for the ρ meson. Then we suddenly understood that this sign could be compensated by the four-quark condensate — a real breakthrough. The accuracy of our results turned out to be much higher that any of us could have expected *a priori*.

The basic elements of the approach were already visible in the first JETP Letters publication,[13] with new important findings (e.g. Borelization) coming later, one by one. We worked at a feverish pace for the whole academic year, accumulating a large number of results for the hadronic parameters.

I can't help mentioning an episode that occurred in the spring of 1978 when we were done with this work. The episode could have been funny were it not so nerve-wrecking for me. When we decided that this stage of the work was over, I collected all my drafts (hundreds sheets of paper with calculations and formulas), organized them in a proper order, selected all formulas and expressions we might need for the paper and the future work, carefully rewrote them in a thick notebook (remember, we had no access to photocopying machines), destroyed the drafts, put the the notebook in my briefcase and went home. It was about midnight, and I was so exhausted that I fall asleep while on the metro train. A loud voice announcing my stop awoke me, and I jumped out of the train, leaving the briefcase were it was, on the seat. By the time I realized what have happened the train was gone, and gone with it forever my calculations ... I have never recovered my briefcase with the precious notebook... After a few agonizing days it became clear that the necessary formulas and expressions had to be recovered anew. Fortunately, Arkady had kept lots of his own calculations. He never throws out anything. Therefore, the problem with his drafts was to dig out "informative" sheets of paper from the "noise". (This was hindered by the fact that he was in Novosibirsk while we were in Moscow.) Some of my drafts survived in the drawers of a huge desk which I had inherited from Sudakov. Moreover, many crucial calculations were discussed so many times by us, over and over again, that I remembered them by heart. Nevertheless, I think it took a couple of uneasy weeks to recover the contents of the lost notebook.

In late spring 1978 the question of how all this wealth could be published became of prime concern to us. As usual, we had a lot of funny adventures in preparing the manuscript, typing it and issuing preprints. Needless to say, it was a serial publication, see above. As for the journal article, Nuclear Physics was a natural candidate, but previously we had bad luck with this journal: our paper on penguins[15] was buried in the editorial office for more than two years. We were too tired, however, to invent anything new and decided to try our luck again. The report occupying the whole issue of Nuclear Physics [16] appeared a year later.

Beginning in 1980, the approach we suggested was tested, with a remarkable success, in analyzing practically every static property of all established low-lying hadronic states. "Classic" mesons were supplemented by baryons.[17] The approach was then developed in various directions — three-point functions, inclusion of external electromagnetic and other auxiliary fields, light-cone modifications and so on. This allowed one to expand the range of applicability, shifting the emphasis from the calculation of masses and coupling constants of "classic" resonances to such problems as magnetic moments,[18] form factors [19] at intermediate momentum transfers, weak decays,[20] structure functions [21] of deep inelastic scattering at intermediate x, and many others. A wealth of low-energy parameters – dozens, if not hundreds – were obtained with a typical theoretical uncertainty of 10 to 30% essentially from the gluon and quark condensates and a couple of other vacuum mean values introduced later. At the initial stage the method was essentially unchallenged since the lattice calculations were lagging far behind.

At present, after the spectacular advances of lattice QCD, the numerical aspect of the sum rule approach is somewhat overshadowed, in many instances, although even now the QCD sum rules are quite competitive in problems with complicated kinematics. What is probably even more important, theoretical insights and tools developed in connection with this method paved the way to numerous advances in related areas.

Let me start from the operator product expansion (OPE). The idea of factorization of short and large distances, the central point of OPE, dates back to classical Wilson's work where it was put forward in connection with theories of strong interaction with conformal invariance at short distances. Shortly after, Wilson formulated a very general procedure of the renormalization-group flow (see e.g. [22]) which became known as the Wilsonean renormalization group. Wilson's formulation makes no reference to perturbation theory, it applies both to strongly and weakly coupled theories. The focus of Wilson's work was on statistical physics, where the program is also known as the block-spin approach. Starting from the microscopic degrees of freedom at the shortest distances a, one "roughens" them, step by step, by constructing a sequence of effective (composite) degrees of freedom at distances $2a$, $4a$, $8a$, and so on. At each given step i one constructs an effective Hamiltonian, which fully accounts for dynamics at distances shorter than a_i in the coefficient functions.

Surprisingly, in high-energy physics of 1970s the framework of OPE was narrowed down to a very limited setting. On the theoretical side, it was discussed almost exclusively in perturbation theory, as is seen, for instance, from Refs. [23]. On the practical side, in applications, people did not go beyond the leading power effects. The OPE-based analysis of Christ *et al.* [24] of deep inelastic scattering (DIS) was implemented in QCD at the level of leading twists. Asymptotic freedom implies that the power asymptotic behavior is determined by canonic dimensions of the operators involved. The asymptotic behavior is modified by perturbative logarithms. This explains both the Bjorken scaling in DIS as well as logarithmic deviations from it. Similar logarithmic corrections were found in weak decays (e.g. penguins [25]). In essence, that was all one could find on the theoretical market of the day.

We were the first to adapt [26] the general Wilson construction to QCD, extending OPE to systematically include power suppressed effects. A consistent Wilsonean approach requires introduction of a normalization point μ which plays the role of a running parameter separating "hard" contributions included in the coefficient functions and "soft" contributions residing in local operators occurring in the expansion. The degree of locality is regulated by μ itself.

Prevalent at those days was a misconception that the OPE coefficients are determined exclusively by perturbation theory while the matrix elements of the operators involved are purely nonperturbative. Attempts to separate perturbation theory from "purely nonperturbative" condensates gave rise to multiple paradoxes and inconsistencies (see e.g. [27]) which questioned the very possibility of using the OPE-based methods in QCD. We had to spend much effort untangling these spurious paradoxes each time the claim of the failure of OPE has been made.[26] Surprisingly, such claims are being made till present. This became a perennial problem reminding of *perpetuum mobile* seekers.

Certainly, Wilsonean OPE, being just a book-keeping procedure, cannot fail. The question of its practical implementation and extracting consequences, especially in the Minkowski domain, is a different story. The coefficient functions summarize the hard contributions, which, to a good approximation, are given by perturbation theory. However, strictly speaking, perturbation theory does not exhaust QCD dynamics even at short distances. If there are small-size nonperturbative fluctuations, the coefficient functions acquire nonperturbative parts. An instructive example is provided by instantons. We found [26] that in most cases their contribution is suppressed and can be neglected. That is why our sum rules were so successful in the family of the "classic" mesons. But not in all cases! We managed to identify [28] some exceptional channels and "exceptional hadrons" (which will *not* be discussed in this book).

So much was said and done in the construction of consistent OPE in QCD because, after we started the process in connection with the QCD sum rules, OPE in QCD gained a life of its own! The very same OPE constitutes the basis of the heavy quark mass expansions.[29] The heavy quark theory accounts for most significant developments in QCD in the recent years. This is a branch of QCD where there still exists a direct live feedback from experiment, which gives a special weight to any advancement in theoretical understanding. One of the most elegant results established [30] in the heavy quark physics, on the basis of OPE, is the absence of the $1/m_Q$ correction to the total inclusive decay widths of the heavy-flavor hadrons. This theorem will undoubtedly make its way into text books, let alone its practical importance for the precision determination of V_{cb} from data.

On the technical side, the progress of the OPE-based methods we are witnessing now would hardly be possible without the computational technique based on the background field method. Back in 1978, when we entered this field, people used a very awkward approach, a direct evaluation of the Feynman graphs. It worked okay as far as the leading power terms in OPE were concerned. We needed to go beyond the leading terms. It became clear almost immediately, that one cannot go further without perfecting the computational technique. The background field formalism was known for a long time in QED, were it had been developed mainly by J. Schwinger. It is extremely efficient and ideally suited for OPE-based calculations. Somehow the QCD practitioners paid no attention to it. So, we adapted it to make it usable in the QCD environment. In fact, two versions were engineered: one is based on the Fock–Schwinger gauge for the background field, another is more general. They nicely supplement each other in various problems. The review paper [31] (see Chapter 3) summarizes our work on the background field formalism; exploiting it became a routine practice from then on.

⋆ ⋆ ⋆ ⋆ ⋆

References

1. H. Fritzsch and M. Gell-Mann, in Proc. XVI Int. Conf. on High Energy Physics, Batavia 1972 (J. D. Jackson and A. Roberts, eds.), Vol. 2, p. 135.
2. H. Fritzsch, M. Gell-Mann and H. Leutwyler, Phys. Lett. B **47**, 365 (1973).
3. D. Gross and F. Wilczek, Phys. Rev. Lett. **30**, 1343 (1973); H. D. Politzer, Phys. Rev. Lett. **30**, 1346 (1973).
4. A. M. Polyakov, *A view from the island,* arXiv:hep-th/9211140, published in *The Rise of the Standard Model: Particle Physics in the 1960s and 1970s,*

Eds. L. Hoddeson, L. Brown, M. l. Riordan and M. Dresden (Cambridge University Press, 1997).
5. L. D. Landau, in *Niels Bohr and the Development of Physics*, ed. W. Pauli (Mc-Graw Hill, New York, 1955).
6. V. S. Vanyashin and M. V. Terentyev, ZhETF **48**, 565 (1965) [Sov. Phys. - JETP **21**, 375 (1965)].
7. I. B. Khriplovich, Yad. Fiz. **10**, 409 (1969) [Sov. J. Nucl. Phys. **10**, 235 (1970)].
8. The paper of Vanyashin and Terentyev is entitled *The vacuum polarization of a charged vector field*. Let me quote a passage from the concluding part of the paper: "Unlike the ordinary electrodynamics where the renormalized charge is smaller than the bare one our case gives $e^2 > e_0^2$... As is well-known, in electrodynamics the restriction $0 \leq Z \leq 1$ on the Z factor (charge renormalization) follows from the most general principles... In our case it is impossible, generally speaking, to get the restriction $0 \leq Z \leq 1$ starting from these principles... Nevertheless, the inequality $Z > 1$ seems extremely undesirable." The first coefficient of the Gell-Mann-Low function obtained in [6] (for SU(2)) is 20/3, instead of the correct value 22/3 which we would get now. The difference is due to the contribution of the longitudinal polarizations, which in the modern language would correspond to the Higgs fields. Vanyashin and Terentyev worked before the Higgs mechanism; they introduced the W-boson mass by hand. This is okay at one loop.
In Khriplovich's work a non-covariant (Coulomb) gauge was used which is ghost-free. In this gauge the charge renormalization is given by that of the D_{00} component of the vector Green function. The logarithmically divergent part of D_{00} is presented in Eq. (61) of ref. [7]. After taking into account an unconventional normalization of the coupling constant there we read off the famous 22/3! Moreover, we additionally learn from this expression that 22/3 is actually $8 - \frac{2}{3}$. The logarithms are of two different types. One is non-dispersive, corresponds to the instantaneous Coulomb interaction and is responsible for antiscreening (*eight*). The second logarithm is normal dispersive; in full accordance with the general principles it gives the negative contribution (*minus two thirds*) characteristic of screening.
9. A. Belavin, A. Polyakov, A. Schwarz and Yu. Typkin, Phys. Lett. B **59**, 85 (1975) [reprinted in *Instantons in Gauge Theories*, Ed. M. Shifman (World Scientific, Singapore, 1994), page 22].
10. C. G. Callan, R. F. Dashen and D. J. Gross, Phys. Lett. B **63**, 334 (1976) [reprinted in *Solitons and particles*, Eds. C. Rebbi and G. Soliani (World Scientific, Singapore, 1984), p. 770]; R. Jackiw and C. Rebbi, Phys. Rev. Lett. **37**, 172 (1976) [reprinted in *Solitons and particles*, Eds. C. Rebbi and G. Soliani (World Scientific, Singapore, 1984), p. 766].
11. M. Shifman, A. Vainshtein and V. Zakharov, Phys. Lett. B **65**, 255 (1976).
12. V. Novikov, L. Okun, M. Shifman, A. Vainshtein, M. Voloshin and V. Zakharov, Phys. Rep. **41**, 1 (1978).
13. A. Vainshtein, V. Zakharov and M. Shifman, Pis'ma v ZhETF **27**, 60 (1978) [JETP Lett. **27**, 55 (1978)].

14. M. Shifman, Current Contents (Eng. Techn. Appl. Sci.) **23**, No 34, 9 (1992).
15. M. Shifman, A. Vainshtein and V. Zakharov, Nucl. Phys. B **120**, 316 (1977).
16. M. Shifman, A. Vainshtein and V. Zakharov, Nucl. Phys. B **147**, 385; 448; 519 (1979).
17. B. L. Ioffe, Nucl. Phys. B **188**, 317 (1981); (E) **191**, 591 (1981);
Y. Chung, H. G. Dosch, M. Kremer and D. Schall, Phys. Lett. B **102**, 175 (1981); Nucl. Phys. B **197**, 55 (1982).
18. B. L. Ioffe and A. V. Smilga, JETP Lett. **37**, 298 (1983); Nucl. Phys. B **232**, 109 (1984);
I. Balitsky and A. Yung, Phys. Lett. B **129**, 328 (1983).
19. V. Nesterenko and A. Radyushkin, Phys. Lett. B **115**, 410 (1982);
B. L. Ioffe and A. V. Smilga, Nucl. Phys. B **216**, 373 (1983).
20. B. Blok and M. Shifman, Sov. J. Nucl. Phys. **45**, 135; 301; 522 (1987); Sov. J. Nucl. Phys. **46**, 1310 (1987); for a review and a representative list of references see A. Khodjamirian and R. Rückl, hep-ph/9801443, published in *Heavy Flavors*, 2nd Edition, Eds., A. J. Buras and M. Linder (World Scientific, Singapore, 1998).
21. V. Belyaev and B. L. Ioffe, Int. J. Mod. Phys. A **6**, 1533 (1991).
22. K. Wilson and J. Kogut, Phys. Rep. **12**, 75 (1974).
23. K. Symanzik, Commun. Math. Phys. **23**, 49 (1971); C. Callan, Phys. Rev. D **5**, 3302 (1972); W. Zimmermann, in *Lectures on Elementary Particles and Quantum Field Theory*, Eds. S. Deser, M. Grisaru, and H. Pendelton (MIT Press, Cambridge, MA, 1971), Vol. 1.
24. N. Christ, B. Hasslacher, and A.H. Mueller, Phys. Rev. D **6**, 3543 (1972).
25. M. A. Shifman, A. I. Vainshtein and V. I. Zakharov, Nucl. Phys. B **120**, 316 (1977); Sov. Phys. JETP **45**, 670 (1977)
26. V. A. Novikov, M. A. Shifman, A. I. Vainshtein and V. I. Zakharov, Nucl. Phys. B **249** 445 (1985).
27. F. David, Nucl. Phys. B **209**, 433 (1982); Nucl. Phys. B **234**, 237 (1984).
28. V. A. Novikov, M. A. Shifman, A. I. Vainshtein and V. I. Zakharov, Nucl. Phys. B **191**, 301 (1981).
29. M. A. Shifman and M. B. Voloshin, Sov. J. Nucl. Phys. **41**, 120 (1985); Sov. Phys. JETP **64**, 698 (1986); Sov. J. Nucl. Phys. **45**, 292 (1987); H. Georgi, Phys. Lett. B **240**, 447 (1990).
30. I. I. Bigi, N. G. Uraltsev and A. I. Vainshtein, Phys. Lett. B **293** 430 (1992).
31. V. A. Novikov, M. A. Shifman, A. I. Vainshtein and V. I. Zakharov, Fortsch. Phys. **32**, 585 (1985).

SECTION 2.

NON–ABELIAN VORTICES

Non-Abelian Vortices: Results in New Directions

K. Konishi*

Department of Physics, "E. Fermi", University of Pisa
Largo Pontecorvo, 3, Pisa, 56127 Italy
**E-mail: konishi@df.unipi.it*
www.df.unipi.it/ konishi/

Two new results on non-Abelian vortices are illustrated. In the first part we discuss the construction of non-Abelian vortices in theories with general gauge groups. The second part is dedicated to the *fractional vortices and lumps*, which occur in a wide variety of Abelian and non-Abelian generalizations of the Higgs model.

Keywords: Non-Abelian vortex, sigma-model lumps, duality, confinement.

1. Introduction

The last several years have witnessed a remarkable progress in our understanding of non-Abelian vortices, vortices carrying a continuous, non-Abelian moduli (zero modes), in various non-Abelian gauge systems in Higgs phase. Such vortices may have many applications in physics, and in particular, are believed to be key to the understanding of non-Abelian monopoles, Goddard-Nuyts-Olive-Weinberg (GNO) duality and possibly of quark confinement in Quantum Chromodynamics. The aim of this talk is to discuss the most salient features of the two new results concerning these non-Abelian vortices by our Pisa-Cambridge-Tokyo collaboration.[1,2]

2. Non-Abelian vortices with general gauge groups

The first result [1] concerns the construction of non-Abelian vortex solutions in theories with a general gauge group $G' \times U(1)$, where $G' = SU(N), SO(N), USp(2N)$, etc. The case of $SU(N)$ theories has been discussed extensively in the last several years. To be concrete, let us work with simple models which have the structure of the bosonic sector of $N = 2$ supersymmetric models. The model contains a FI (Fayet-Iliopoulos) like term in the $U(1)$ part, ensuring that the system develops stable vortices. The crucial aspect of the model however is that we work in a completely Higgsed vacuum, with an unbroken color-flavor diagonal symmetry. The secret of non-Abelian vortex lies in the so-called color-flavor locked phase.

The Lagrangian density reads

$$\mathcal{L} = \mathrm{Tr}_c\Big[-\frac{1}{2e^2} F_{\mu\nu} F^{\mu\nu} - \frac{1}{2g^2} \hat{F}_{\mu\nu} \hat{F}^{\mu\nu} + \mathcal{D}_\mu H \left(\mathcal{D}^\mu H\right)^\dagger$$
$$- \frac{e^2}{4} \left|X^0 t^0 - 2\xi t^0\right|^2 - \frac{g^2}{4} \left|X^a t^a\right|^2 \Big] ,$$

with the field strength, gauge fields and covariant derivative denoted as

$$F_{\mu\nu} = F^0_{\mu\nu} t^0 , \quad F^0_{\mu\nu} = \partial_\mu A^0_\nu - \partial_\nu A^0_\mu , \quad \hat{F}_{\mu\nu} = \partial_\mu A_\nu - \partial_\nu A_\mu + i \left[A_\mu, A_\nu\right],$$
$$A_\mu = A^a_\mu t^a , \quad \mathcal{D}_\mu = \partial_\mu + i A^0_\mu t^0 + i A^a_\mu t^a .$$

A^0_μ is the gauge field associated with $U(1)$ and A^a_μ are the gauge fields of G'. The matter scalar fields are written as an $N \times N_{\mathrm{F}}$ complex color (vertical) – flavor (horizontal) mixed matrix H. It can be expanded as $X = HH^\dagger = X^0 t^0 + X^a t^a + X^\alpha t^\alpha$, $X^0 = 2\,\mathrm{Tr}_c\left(HH^\dagger t^0\right)$, $X^a = 2\,\mathrm{Tr}_c\left(HH^\dagger t^a\right)$, where the traces with subscript c are over the color indices. e and g are the $U(1)$ and G' coupling constants, respectively, while ξ is a real constant.

We choose to work in a maximally "color-flavor-locked vacuum" of the system,

$$\langle H \rangle = \frac{v}{\sqrt{N}} \mathbf{1}_N , \qquad \xi = \frac{v^2}{\sqrt{2N}} , \tag{1}$$

We have taken $N_{\mathrm{F}} = N$ which is the minimal number of flavors allowing for such a vacuum. Note that, unlike the $U(N)$ model studied extensively in the last several years, the vacuum is not unique in cases with a general gauge group, even with such a minimum choice of the flavor multiplicity. This difference reflects the fact that groups such as $SO(N) \times U(1)$ and $USp(N) \times U(1)$ form strictly smaller subgroups of $U(N)$.

The existence of a continuous vacuum degeneracy implies the emergence of vortices of semi-local type; this aspect will be crucial in the discussion of the fractional vortices in the second part of this talk. However, for now we stick to the particular vacuum Eq. (1), and consider vortices and their moduli in this theory. The standard Bogomol'nyi completion reads

$$T = \int d^2x\, \mathrm{Tr}_c \Big[\frac{1}{e^2} \left| F_{12} - \frac{e^2}{2} \left(X^0 t^0 - 2\xi t^0\right) \right|^2 + \frac{1}{g^2} \left| \hat{F}_{12} - \frac{g^2}{2} X^a t^a \right|^2$$
$$+ 4 \left|\bar{\mathcal{D}} H\right|^2 - 2\xi F_{12} t^0 \Big] \geq -\xi \int d^2x\, F^0_{12} , \tag{2}$$

where $\bar{\mathcal{D}} \equiv \frac{\mathcal{D}_1 + i\mathcal{D}_2}{2}$, $z = x^1 + ix^2$. In the BPS limit one has

$$T = 2\sqrt{2N}\pi\xi\nu = 2\pi v^2\nu \, , \qquad \nu = -\frac{1}{2\pi\sqrt{2N}} \int d^2x \, F_{12}^0 \, , \tag{3}$$

where ν is the $U(1)$ winding number of the vortex. This leads immediately to the vortex BPS equations

$$\bar{\mathcal{D}}H = \bar{\partial}H + i\bar{A}H = 0 \, , \tag{4}$$

$$F_{12}^0 = e^2 \left[\mathrm{Tr}_c \left(HH^\dagger t^0 \right) - \xi \right] \, , \qquad F_{12}^a = g^2 \, \mathrm{Tr}_c \left(HH^\dagger t^a \right) \, . \tag{5}$$

The matter BPS equation (4) can be solved by the Ansatz

$$H = S^{-1}(z, \bar{z}) H_0(z) \, , \qquad \bar{A} = -iS^{-1}(z, \bar{z}) \bar{\partial} S(z, \bar{z}) \, , \tag{6}$$

where S belongs to the complexification of the gauge group, $S \in \mathbb{C}^* \times G'^{\mathbb{C}}$. $H_0(z)$, holomorphic in z, is called the *moduli matrix*, which contains all moduli parameters of the vortices.

A gauge invariant object can be constructed as $\Omega = SS^\dagger$. This can be conveniently split into the $U(1)$ part and the G' part, such that $S = s\, S'$ and analogously $\Omega = \omega \, \Omega'$, $\omega = |s|^2$, $\Omega' = S'S'^\dagger$. The tension (3) can be rewritten as

$$T = 2\pi v^2 \nu = 2v^2 \int d^2x \, \partial\bar{\partial} \log \omega \, , \qquad \nu = \frac{1}{\pi} \int d^2x \, \partial\bar{\partial} \log \omega \, , \tag{7}$$

and ν determines the asymptotic behavior of the Abelian field as

$$\omega = ss^\dagger \sim |z|^{2\nu} \, , \qquad \text{for } |z| \to \infty \, .$$

The minimal vortex solutions can be written down by making use of the holomorphic invariants for the gauge group G' made of H, which we denote as $I^i_{G'}(H)$. If the $U(1)$ charge of the i-th invariant is given by n_i, $I^i_{G'}(H)$ satisfies

$$I^i_{G'}(H) = I^i_{G'} \left(s^{-1} S'^{-1} H_0 \right) = s^{-n_i} I^i_{G'}(H_0(z)) \, ,$$

while the boundary condition is $I^i_{G'}(H)\Big|_{|z|\to\infty} = I^i_{\rm vev} \, e^{i\nu n_i \theta}$, where $\nu \, n_i$ is the number of the *zeros* of $I^i_{G'}$. This leads then to the following asymptotic behavior

$$I^i_{G'}(H_0) = s^{n_i} I^i_{G'}(H) \overset{|z|\to\infty}{\longrightarrow} I^i_{\rm vev} z^{\nu n_i} \, .$$

It implies that $I^i_{G'}(H_0(z))$, being holomorphic in z, are actually polynomials. Therefore $\nu \, n_i$ must be positive integers for all i:

$$\nu \, n_i \in \mathbb{Z}_+ \qquad \to \qquad \nu = \frac{k}{n_0} \, , \qquad k \in \mathbb{Z}_+ \, ,$$

with $n_0 \equiv \gcd\left\{n_i \,|I^i_{\rm vev} \neq 0\right\}$. The $U(1)$ gauge transformation $e^{2\pi i/n_0}$ leaves $I^i_{G'}(H)$ invariant and thus the true gauge group is

$$G = \left[U(1) \times G'\right]/\mathbb{Z}_{n_0} \ ,$$

where $\mathbb{Z}_{n_0}$ is the center of the group G'. The minimal winding in $U(1)$ found here, $\frac{1}{n_0}$, corresponds to the minimal element of $\pi_1(G) = \mathbb{Z}$: it represents a minimal loop in the group manifold G. As a result we find the following non-trivial constraints for H_0

$$I^i_{G'}(H_0) = I^i_{\rm vev}\, z^{\frac{kn_i}{n_0}} + \mathcal{O}\left(z^{\frac{kn_i}{n_0}-1}\right) \ .$$

2.1. *GNO quantization for non-Abelian vortices*

In general it is not easy to construct all the solutions for a given theory. But certain special solutions can be found readily. It turns out that each such solution is characterized by a *weight vector of the dual group*, and are labelled by a set of integers ν_a $(a = 1, \cdots, {\rm rank}(G'))$

$$H_0(z) = z^{\nu \mathbf{1}_N + \nu_a \mathcal{H}_a} \in U(1)^{\mathbb{C}} \times G'^{\mathbb{C}} \ , \tag{8}$$

where $\nu = k/n_0$ is the $U(1)$ winding number and $\mathcal{H}_a$ are the generators of the Cartan subalgebra of $\mathfrak{g}'$. H_0 must be holomorphic in z and *single-valued*, which gives the constraints for a set of integers ν_a

$$(\nu \mathbf{1}_N + \nu_a \mathcal{H}_a)_{ll} \in \mathbb{Z}_{\geq 0} \quad \forall l \ .$$

Suppose that we now consider scalar fields in an r-representation of G'. The constraint is equivalent to

$$\nu + \nu_a \mu_a^{(i)} \in \mathbb{Z}_{\geq 0} \quad \forall i \ , \tag{9}$$

where $\vec{\mu}^{(i)} = \mu_a^{(i)}$ $(i = 1, 2, \cdots, \dim(r))$ are the weight vectors for the r-representation of G'. Subtracting pairs of adjacent weight vectors, one arrives at the quantization condition

$$\vec{\nu} \cdot \vec{\alpha} \in \mathbb{Z} \ , \tag{10}$$

for every root vector α of G'.

Equation (10) is formally identical to the well-known GNO *monopole* quantization condition,[3] as well as to the naïve vortex flux quantization rule.[4] There is however a crucial difference here, from these earlier results. Because of an exact flavor (color-flavor diagonal $G_{\rm C+F}$) symmetry, broken by individual vortex solutions, our vortices possess continuous orientational moduli. These zero modes are normalizable, unlike those encountered in the earlier attempts to define quantum "non-Abelian monopoles".

G'	$\tilde{G}'$
$SU(N)$	$SU(N)/\mathbb{Z}_N$
$U(N)$	$U(N)$
$SO(2M)$	$SO(2M)$
$USp(2M)$	$SO(2M+1)$
$SO(2M+1)$	$USp(2M)$

These non-Abelian modes of our vortices—they are a kind of Nambu-Goldstone modes—can fluctuate and propagate inside the vortex only. In systems with a hierarchical symmetry breaking,

$$G_0 \to G = G' \times U(1) \to \mathbf{1},$$

where our $G = G' \times U(1)$ model emerges as the low-energy approximation, these orientational zero modes get absorbed by finite mass monopoles at the vortex ends. This process endows the monopoles with fully quantum-mechanical non-Abelian (GNO-dual) charges, as has been suggested by the author and others in several occasions,[5] but we shall not dwell on this question further here.

The solution of the quantization condition (10) is that

$$\tilde{\vec{\mu}} \equiv \vec{\nu}/2 \ ,$$

is any of the *weight vectors* of the dual group of G'. The dual group, denoted as $\tilde{G}'$, is defined by the dual root vectors[3] $\vec{\alpha}^* = \vec{\alpha}/(\vec{\alpha}\cdot\vec{\alpha})$. Examples of dual pairs of groups G', $\tilde{G}'$, are shown in Table 1. Note that (9) is stronger than (10), the l.h.s. must be a nonnegative integer. This positive quantization condition allows for a few weight vectors only. For concreteness, let us consider scalar fields in the fundamental representation, and choose a basis where the Cartan generators of $G' = SO(2M), SO(2M+1), USp(2M)$ are given by

$$\mathcal{H}_a = \mathrm{diag}\Big(\underbrace{0,\cdots,0}_{a-1},\frac{1}{2},\underbrace{0,\cdots,0}_{M-1},-\frac{1}{2},0,\cdots,0\Big)\ , \tag{11}$$

with $a = 1,\cdots,M$. In this basis, special solutions H_0 have the form for $G' = SO(2M)$ and $USp(2M)$

$$H_0^{(\tilde{\mu}_1,\cdots,\tilde{\mu}_M)} = \mathrm{diag}\left(z^{k_1^+},\cdots,z^{k_M^+},z^{k_1^-},\cdots,z^{k_M^-}\right)\ , \tag{12}$$

while for $SO(2M+1)$

$$H_0^{(\tilde{\mu}_1,\cdots,\tilde{\mu}_M)} = \mathrm{diag}\left(z^{k_1^+},\cdots,z^{k_M^+},z^{k_1^-},\cdots,z^{k_M^-},z^{k}\right)\ , \tag{13}$$

where $k_a^{\pm} = \nu \pm \tilde{\mu}_a$.

For example, in the cases of $G' = SO(4), USp(4)$ with a $\nu = 1/2$ vortex, there are four special solutions with $\vec{\tilde{\mu}} = (\frac{1}{2}, \frac{1}{2}), (\frac{1}{2}, -\frac{1}{2}), (-\frac{1}{2}, \frac{1}{2}), (-\frac{1}{2}, -\frac{1}{2})$

$$H_0^{(\frac{1}{2},\frac{1}{2})} = \mathrm{diag}(z, z, 1, 1) = z^{\frac{1}{2}\mathbf{1}_4 + 1\cdot\mathcal{H}_1 + 1\cdot\mathcal{H}_2}, \tag{14}$$

$$H_0^{(\frac{1}{2},-\frac{1}{2})} = \mathrm{diag}(z, 1, 1, z) = z^{\frac{1}{2}\mathbf{1}_4 + 1\cdot\mathcal{H}_1 - 1\cdot\mathcal{H}_2}, \tag{15}$$

$$H_0^{(-\frac{1}{2},\frac{1}{2})} = \mathrm{diag}(1, z, z, 1) = z^{\frac{1}{2}\mathbf{1}_4 - 1\cdot\mathcal{H}_1 + 1\cdot\mathcal{H}_2}, \tag{16}$$

$$H_0^{(-\frac{1}{2},-\frac{1}{2})} = \mathrm{diag}(1, 1, z, z) = z^{\frac{1}{2}\mathbf{1}_4 - 1\cdot\mathcal{H}_1 - 1\cdot\mathcal{H}_2}. \tag{17}$$

These four vectors are the same as the weight vectors of two Weyl spinor representations $\mathbf{2} \oplus \mathbf{2}'$ of $\tilde{G}' = SO(4)$ for $G' = SO(4)$, and the same as those of the Dirac spinor representation $\mathbf{4}$ of $\tilde{G}' = Spin(5)$ for $G' = USp(4)$.

The weight vectors corresponding to the $k = 1$ vortex in various gauge groups are shown in Fig. 1. In all cases the results found are consistent with the GNO duality.

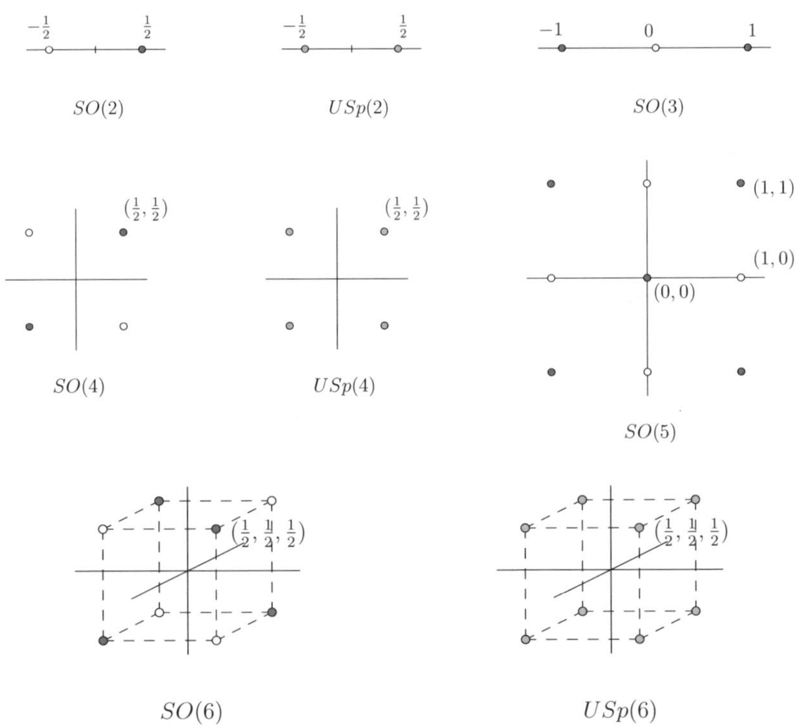

Fig. 1. The special points for the $k = 1$ vortex.

3. Fractional vortices and lumps

Another interesting new result concerns the fractional vortex and lumps.[2] We have pointed out above that in a general class of models the vacuum is not unique, even if the Fayet-Iliopoulos term is present and even if the number of the flavors is the minimum possible for a "color-flavor" locked vacuum to exist in the theory. In other words, there is a nontrivial vacuum degeneracy, or the vacuum moduli $\mathcal{M}$. In the first part of the talk, we were mainly interested in the *vortex moduli* $\mathcal{V}$, in a single maximally color-flavor locked vacuum. Here we are going to consider all possible vortices—the vortex moduli $\mathcal{V}$—on all possible points of the vacuum moduli $\mathcal{M}$ at the same time.

There are in fact two crucial ingredients which lead to the fractional vortex: the *vacuum degeneracy* and the *BPS saturated* nature of the vortices. The first point was emphasized just above: the situation is schematically illustrated in Fig. 2. Even if we restrict ourselves to the minimally winding vortex solutions only, the vortices represent non-trivial fiber bundles over the vacuum moduli $\mathcal{M}$.

The BPS saturated nature of the vortices, on the other hand, implies that the vortex equations linearize. The matter equations of motion are solved by the moduli-matrix Ansatz. The other equations–the gauge field equations–reduce, in the strong coupling limit or anyway sufficiently far from the vortex center, to the vacuum equations for the scalar fields. In other words, the vortex solutions approximate the sigma model lumps.

3.1. *Structures of the vacuum moduli*

A vortex solution is defined on each point of $\mathcal{M}$, in the sense that the scalar configuration along a sufficiently large circle (S^1) surrounding it traces a non-trivial closed orbit in the fiber F (hence a point in $\mathcal{M}$). The existence of a vortex solution requires that

$$\pi_1\,(F, f) \neq \mathbf{1}\ , \tag{18}$$

where f is a point in M. The field configuration on a disk D^2 encircled by S^1 traces $\mathcal{M}$, apart from points at finite radius where it goes off M (hence from $\mathcal{M}$). In other words it represents an element of $\pi_2(\mathcal{M}, p)$, where p is the gauge orbit containing f, or $p = \pi(f)$: π is the projection of the fiber onto a point of the base space $\mathcal{M}$. The exact sequence of homotopy groups for the fiber bundle reads

$$\cdots \to \pi_2\,(M, f) \to \pi_2\,(\mathcal{M}, p) \to \pi_1\,(F, f) \to \pi_1\,(M, f) \to \pi_1\,(\mathcal{M}, p) \to \cdots$$

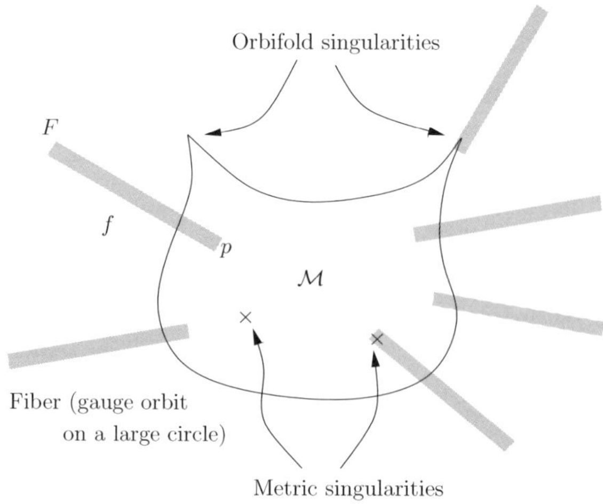

Fig. 2. Vacuum moduli $\mathcal{M}$, fiber F over it, and possible singularities

where $\pi_2(M/F, f) \sim \pi_2(\mathcal{M}, p)$. See Fig. 3.

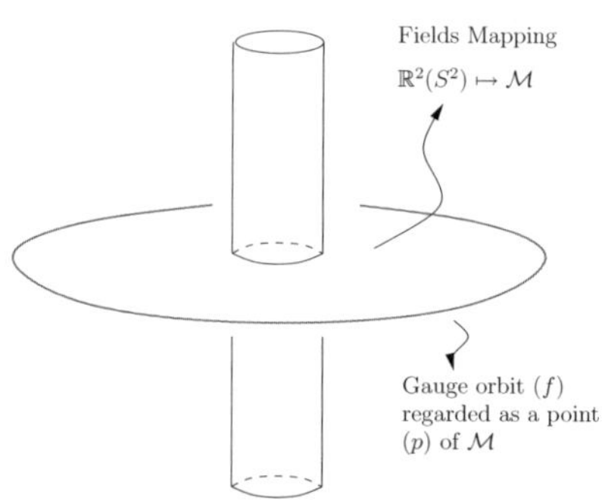

Fig. 3.

Given the points f, p and the space $\mathcal{M}$, the vortex solution is still not unique. Any exact symmetry of the system broken by an individual vortex solution gives rise to vortex zero modes (moduli), $\mathcal{V}$. Our main interest here however is the vortex moduli which arises from the non-trivial vacuum moduli $\mathcal{M}$ itself. Due to the BPS nature of our vortices, the gauge field equation

$$F_{12}^I = g_I^2 \left(q^\dagger T^I q - \xi^I \right) ,$$

reduces, in the strong-coupling limit (or in any case, sufficiently far from the vortex center), to the vacuum equation defining M. This means that a vortex configuration can be approximately seen as a non-linear σ-model (NLσM) lump with target space $\mathcal{M}$, as was already anticipated. Various distinct maps $S^2 \mapsto \mathcal{M}$ of the same homotopy class correspond to physically inequivalent solutions; each of these corresponds to a vortex with the equal tension

$$T_{\min} = -\xi^I \int d^2x \, F_{12}^I > 0 \ ,$$

by their BPS nature. They represent non-trivial *vortex moduli.*

The semi-local vortices of the so-called extended-Abelian Higgs (EAH) model arise precisely this way. In an Abelian Higgs model with N flavors of (scalar) electrons, $M = S^{2N-1}$, $F = S^1$, $\mathcal{M} = S^{2N-1}/S^1 = \mathbb{C}P^{N-1}$, and the exact homotopy sequence tells us that $\pi_2(\mathbb{C}P^{N-1})$ and $\pi_1(S^1)$ are isomorphic: each (i.e. minimum) vortex solution corresponds to a minimal σ-model lump solution.

In most cases discussed in our paper [2], however, the base space $\mathcal{M}$ will be various kinds of *singular manifolds*: a manifold with singularities, unlike in the EAH model. The nature of the singularities depends on the system and on the particular point(s) of $\mathcal{M}$. Our BPS degenerate vortices represent (generalized) fiber bundles with the singular manifolds $\mathcal{M}$ as the base space.

3.2. *Two mechanisms for fractional vortex–lump*

There are two distinct mechanisms leading to the appearance of a fractional vortex. The first is related to the presence of orbifold singularities in $\mathcal{M}$. For example, let us consider a $\mathbb{Z}_2$ point p_0 such as the one appearing in a simple $U(1)$ model with two scalars, one of which has charge 2. At this singularity, both $\pi_2(\mathcal{M}, p)$ and $\pi_1(F, f)$ make a discontinuous change. The minimum element of $\pi_1(F_0, f_0)$ is half of that of $\pi_1(F, f)$ defined off the singularity, and similarly for $\pi_2(\mathcal{M}, p_0)$ with respect to $\pi_2(\mathcal{M}, p)$, $p \neq p_0$. Even though the exact homotopy sequence continues to hold on and off the orbifold point, the vortex defined near such a point will look like a doubly-wound vortex, with two centers (if the vortex moduli parameters are chosen appropriately). Analogous multi-peak vortex solution occurs near a $\mathbb{Z}_N$ orbifold point of $\mathcal{M}$.

Another cause for the appearance of fractional peaks is simple and very general: a deformed sigma model geometry. This phenomenon can be best seen by considering our system in the strong coupling limit. Even if the base point p is a perfectly generic, regular point of $\mathcal{M}$, not close to any

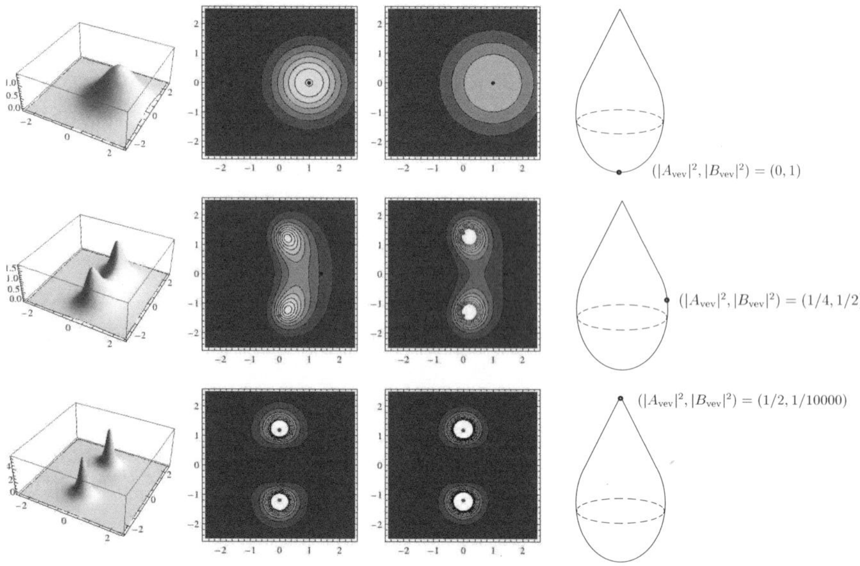

Fig. 4. The energy (the left-most and the 2nd left panels) and the magnetic flux (the 2nd right panels) density, together with the boundary values (A, B) (the right-most panel) for the minimal vortex.

singularity, the field configurations in the transverse plane (S^2) trace the whole vacuum moduli space $\mathcal{M}$. The energy distribution reflects the non-trivial structure of $\mathcal{M}$ as the volume of the target space is mapped into the transverse plane, $\mathbb{C}$

$$E = 2 \int_{\mathbb{C}} \frac{\partial^2 K}{\partial \phi^I \partial \phi^{\dagger \bar{J}}} \partial \phi^I \bar{\partial} \phi^{\dagger \bar{J}} = 2 \int_{\mathbb{C}} \bar{\partial} \partial K \ .$$

The field configuration may hit for instance one of the singularities (conic or not), or simply the regions of large scalar curvature. Such phenomena thus occur very generally if the underlying sigma model has a deformed geometry. At such points the energy density will show a peak, not necessarily at the vortex center. Even at finite coupling, the vortex tension density will exhibit a similar substructure.

3.3. *Some models*

A simple model showing the fractional vortex is an extended Abelian Higgs model, with two scalar fields A and B with charges 2 and 1, respectively. Depending on the point of $\mathcal{M}$ (which is $\mathbb{C}P^1$) the minimum vortex shows doubly-peaked substructure clearly, see Fig. 4. The fractional vortex structure in this model nicely illustrates the first mechanism discussed above: the point $B = 0$ is a Z_2 orbifold point, since there the only nonvanishing field, A, having charge 2, must wind only half of the $U(1)$ to be single-valued.

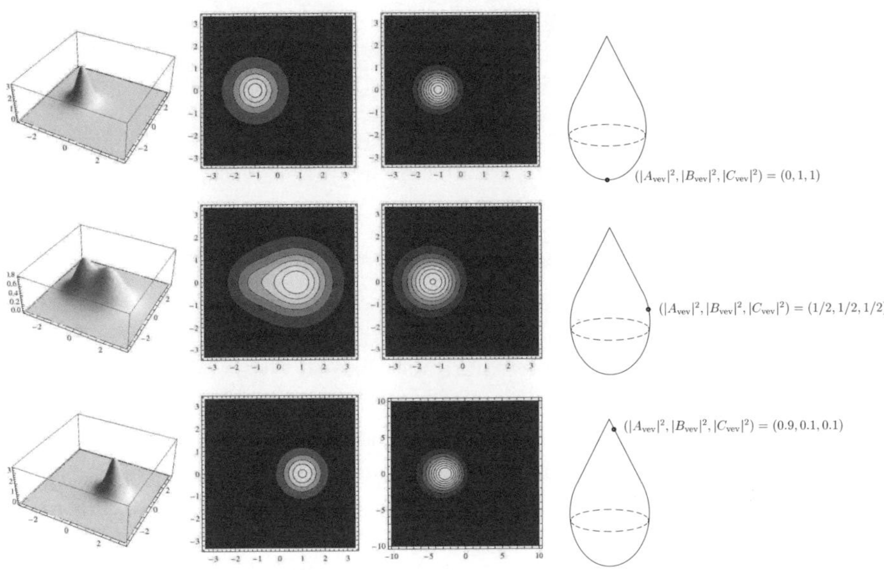

Fig. 5. The energy density (left-most) and the magnetic flux density $F_{12}^{(1)}$ (2nd from the left), $F_{12}^{(1)}$ (2nd from the right) and the boundary condition (right-most).

Another interesting model is a $U_1(1) \times U_2(1)$ gauge theory with three flavors of scalar electrons $H = (A, B, C)$ with charges $Q_1 = (2, 1, 1)$ for $U(1)_1$ and $Q_2 = (0, 1, -1)$ for $U(1)_2$. Even though the model has the same $\mathbb{C}P^1$ as the vacuum moduli $\mathcal{M}$ as the first model, the vortex properties are quite different. This model turns out to provide a good example of fractional vortex of the second type (deformed sigma model geometry).

Fractional vortex occurs also in non-Abelian gauge theories, such as the one with gauge group $G = SO(N) \times U(1)$. An illustrative example of fractional vortex in an $SO(6) \times U(1)$ model is shown in Fig. 6.

Note and Acknowledgments

The last, but the most important message of this talk is "Happy 60th Birthday, Misha!" In wishing him many more fruitful years of scientific

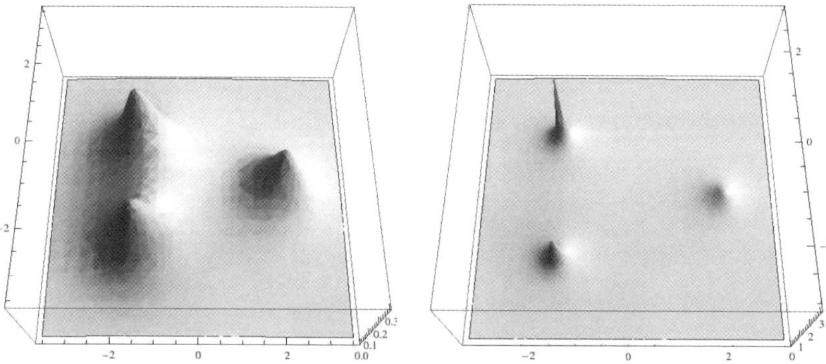

Fig. 6. The energy density of three fractional vortices (lumps) in the $U(1) \times SO(6)$ model in the strong coupling approximation. The positions are $z_1 = -\sqrt{2} + i\sqrt{2}, z_2 = -\sqrt{2} - i\sqrt{2}, z_3 = 2$. The two figures correspond to two different choices of certain size moduli parameters.

activity and of happiness, I would also like to thank Misha Shifman for so much he has done for our community, and on a more personal level, for our long-standing friendship, (indirect) collaboration, occasional fights (naturally, only on scientific questions!), and mutual stimulus.

Many thanks also to all of the organizers of this splendid "Shifmania" Workshop.

The work reported here was a fruit of a collaboration with my friends, M. Eto, T. Fujimori, S. B. Gudnason, T. Nagashima, M. Nitta, K. Ohashi, and W. Vinci. I thank them all.

References

1. M. Eto, T. Fujimori, S. B. Gudnason, K. Konishi, T. Nagashima, M. Nitta, K. Ohashi and W. Vinci, JHEP **0906**, 004 (2009) [arXiv:0903.4471 [hep-th]]; M. Eto, T. Fujimori, S. B. Gudnason, K. Konishi, M. Nitta, K. Ohashi, W. Vinci, Phys. Lett. B **669**, 98 (2008) [arXiv:0802.1020 [hep-th]].
2. M. Eto, T. Fujimori, S. B. Gudnason, K. Konishi, T. Nagashima, M. Nitta, K. Ohashi and W. Vinci, Phys. Rev. D **80**, 045018 (2009) [arXiv:0905.3540 [hep-th]].
3. P. Goddard, J. Nuyts and D. Olive, Nucl. Phys. B **125**, 1 (1977); F. A. Bais, Phys. Rev. D **18**, 1206 (1978); B. J. Schroers and F. A. Bais, Nucl. Phys. B **512**, 250 (1998) [arXiv:hep-th/9708004]; Nucl. Phys. B **535**, 197 (1998) [arXiv:hep-th/9805163]; E. J. Weinberg, Nucl. Phys. B **167**, 500 (1980) 500; Nucl. Phys. B **203**, 445 (1982).
4. K. Konishi and L. Spanu, Int. J. Mod. Phys. A **18**, 249 (2003) [arXiv: hep-th/0106175].
5. M. Eto, L. Ferretti, K. Konishi, G. Marmorini, M. Nitta, K. Ohashi, W. Vinci and N. Yokoi, Nucl. Phys. B **780**, 161 (2007) [arXiv: hep-th/0611313].

K-Vortex Dynamics in $\mathcal{N} = 1^*$ Theory and in its Gravity Dual

R. Auzzi*

Department of Physics, Swansea University,
Singleton Park, Swansea, SA2 8PP
**E-mail: r.auzzi@swansea.ac.uk*

We study magnetic flux tubes in the Higgs vacuum of the $\mathcal{N} = 1^*$ mass deformation of $SU(N_c)$, $\mathcal{N} = 4$ SYM and its large N_c string dual, the Polchinski-Strassler geometry. Choosing equal masses for the three adjoint chiral multiplets, for all N_c we identify a "colour-flavour locked" symmetry, $SO(3)_{C+F}$ which leaves the Higgs vacuum invariant. At weak coupling, we find explicit non-Abelian k-vortex solutions carrying a $\mathbb{Z}_{N_c}$-valued magnetic flux, with topological winding $0 < k < N_c$. These k-strings spontaneously break $SO(3)_{C+F}$ to $U(1)_{C+F}$ resulting in an S^2 moduli space of solutions. The world-sheet sigma model is a nonsupersymmetric $\mathbb{CP}^1$ model with a theta angle $\theta_{1+1} = k(N_c - k)\theta_{3+1}$ where θ_{3+1} is the Yang-Mills vacuum angle. We find numerically that k-vortex tensions follow the Casimir scaling law $T_k \propto k(N_c - k)$ for large N_c. In the large N_c IIB string dual, the $SO(3)_{C+F}$ symmetry is manifest in the geometry interpolating between $AdS_5 \times S^5$ and the interior metric due to a single D5-brane carrying D3-brane charge. We identify candidate k-vortices as expanded probe D3-branes formed from a collection of k D-strings. The resulting k-vortex tension exhibits precise Casimir scaling, and the effective world-sheet theta angle matches the semiclassical result. S-duality maps the Higgs to the confining phase so that confining string tensions at strong 't Hooft coupling also exhibit Casimir scaling in $\mathcal{N} = 1^*$ theory in the large N_c limit.

Keywords: Solitons; Gauge/Gravity Correspondence; Confinement.

1. Introduction

Supersymmetric gauge theories, following the works of [1], have provided a large class of examples where condensation of monopoles is the mechanism for confinement of electric charges. In these theories the confined, heavy, coloured sources are held together by Abelian strings (Abrikosov-Nielsen-Olesen solitons [2]). In contrast, in pure Yang-Mills theory with $SU(N_c)$ gauge group for instance, heavy external charges are expected to be confined by chromoelectric flux tubes which annihilate in groups of N_c, which we refer to as $\mathbb{Z}_{N_c}$-strings. One example where the dynamics of $\mathbb{Z}_{N_c}$ strings may be accessed at weak coupling, is presented by the so-called $\mathcal{N} = 1^*$ theory [3–8] which is a mass deformation of $\mathcal{N} = 4$ theory preserving $\mathcal{N} = 1$ supersymmetry (SUSY). This kind of $\mathbb{Z}_k$ string has been defined "non-Abelian", as opposed to the Abelian Strings, whose topology is described by a $\mathbb{Z}$ or $\mathbb{Z}^k$ topological number.

In recent years, certain special flux tubes at weak coupling have been encountered in gauge theories (with and without SUSY) with $U(N_c)$ gauge group and N_f flavours in the fundamental representation [9,10] and extensively studied therein [11–13]. The crucial feature of all these strings at weak coupling is the presence of orientational moduli associated with rotations within a colour-flavour locked symmetry. We will refer to these as "non-Abelian" flux tubes. The interested reader can find excellent reviews in [14,15]. In the context of $\mathcal{N} = 1^*$ theory with $SU(2)$ gauge group, non-Abelian vortices in the Higgs vacuum were constructed and studied first in [16]. Here we will generalize this picture to higher N_c and we will explore the flux tube dynamics in the large N_c limit, using the string dual proposed by Polchinski and Strassler in [7].

The strings that we will describe are "non-Abelian" in both of the two senses, they are $\mathbb{Z}_{N_c}$ objects and have internal orientational moduli associated with rotations within a colour-flavour locked symmetry. This is a short version of the paper [17], in collaboration with S. Prem Kumar.

2. Theoretical Setting

In the language of $\mathcal{N} = 1$ supersymmetry, the $\mathcal{N} = 1^*$ theory contains an $\mathcal{N} = 1$ vector multiplet W_α and three chiral multiplets (Φ_1, Φ_2, Φ_3), transforming in the adjoint representation of the gauge group which we take to be $SU(N_c)$. The theory is obtained by a relevant, mass deformation of $\mathcal{N} = 4$ supersymmetric Yang-Mills theory. The superpotential of $\mathcal{N} = 4$ SYM reads,

$$\mathcal{W} = \frac{1}{g_{\rm YM}^2} \,\mathrm{Tr}([\Phi_1, \Phi_2]\Phi_3)\,. \tag{1}$$

The superpotential can be deformed by adding $\mathcal{N} = 1$ SUSY preserving mass terms for the adjoint matter fields,

$$\Delta\mathcal{W} = \frac{1}{g_{\rm YM}^2} \sum_{i=1}^{3} \frac{1}{2}\, m_i \mathrm{Tr}(\Phi_i^2)\,. \tag{2}$$

This is a relevant deformation and the resulting theory exhibits nontrivial dynamics in the infrared, resulting in a rich phase structure. In the UV however, the theory flows to $\mathcal{N}=4$ SYM, with the gauge coupling remaining a freely adjustable parameter. Thus the $\mathcal{N}=1^*$ theory has, in addition to three complex mass parameters, a dimensionless, tunable complexified gauge coupling

$$\tau = \frac{4\pi i}{g_{\rm YM}^2} + \frac{\theta_{3+1}}{2\pi}\,. \tag{3}$$

In this paper we will be mainly interested in the case where the masses of the three adjoint chiral multiplets are equal:

$$m_1 = m_2 = m_3 = m\,. \tag{4}$$

With this choice, the superpotential term $\Delta\mathcal{W}$ breaks the $SO(6)_R$ global symmetry of the $\mathcal{N}=4$ theory to an $SO(3)$ subgroup under which the complex chiral multiplets (Φ_1,Φ_2,Φ_3), transform as a triplet. In the $\mathcal{N}=1^*$ theory, this $SO(3)$ acts as an ordinary global symmetry, and not as an R-symmetry.

The mass deformation above results in a large set of vacuum configurations determined by the F-flatness conditions (modulo complex gauge transformations),

$$\Phi_i = -\frac{1}{m}\,\epsilon_{ijk}\,\Phi_j\Phi_k\,. \tag{5}$$

As is well-known,[4,5] the solutions to these equations may be classified in terms of all N_c-dimensional representations the $SU(2)$ algebra. Each such classical ground state then splits into a certain number quantum vacua depending on the non-Abelian gauge symmetry subgroup left unbroken by the classical solution.

In the following we will focus on the Higgs vacuum of the theory. The VEVs of the adjoint scalars in the Higgs vacuum are proportional to the generators of the irreducible $SU(2)$ representation with dimension N_c,

$$\Phi_l = i\,m\,J_l\,, \qquad (l=1,2,3)\,. \tag{6}$$

The VEVs of the adjoint scalars in the Higgs vacuum break the $SO(3)$ global symmetry and the $SU(N_c)$ gauge symmetry. However, it is always possible to find a combined global colour-flavour rotation which is unbroken.[16] This combined $SO(3)_{C+F}$ global symmetry subgroup can be understood as follows. Any global $SO(3)$ rotation of the triplet (Φ_1,Φ_2,Φ_3) can be undone by a global colour transformation whose generators are chosen to be proportional to the VEVs of the adjoint scalars i.e., the N_c dimensional $SU(2)$ generators.

A combination of the above flavour and colour rotations are unbroken by the scalar VEVs. The existence of this $SO(3)_{C+F}$ symmetry allows the determination of the worldsheet sigma model of vortices (magnetic flux tubes) in the Higgs vacuum, as we see below.

3. The Vortex as a Soliton

Since the $\mathcal{N} = 1^*$ theory has only fields transforming in the adjoint representation of the gauge group, the Lagrangian is invariant under transformations in the center $\mathbb{Z}_{N_c}$ of $SU(N_c)$. In the Higgs vacuum, magnetic charges valued in $\mathbb{Z}_{N_c} = \pi_1 \left[SU(N_c)/\mathbb{Z}_{N_c}\right]$ are confined by magnetic flux tubes, also carrying a $\mathbb{Z}_{N_c}$ charge. Since the fluxes are defined modulo N_c, they annihilate in groups of N_c.

The classical equations of motion for the bosonic fields read

$$\partial_\mu F^{\mu\nu} - i[A_\mu, F^{\mu\nu}] = \frac{i}{2}\sum_{l=1}^{3}\left([D^\nu \Phi_l, \Phi_l^\dagger] + [D^\nu \Phi_l^\dagger, \Phi_l]\right), \tag{7}$$

$$D^\mu D_\mu \Phi_i = (m\, w_i^\dagger - \epsilon_{ijl}[w_j^\dagger, \Phi_l^\dagger]) + \frac{\partial V_D}{\partial \Phi_i}. \tag{8}$$

Below we list explicit vortex solutions which satisfy

$$\Phi_i = -\Phi_i^\dagger, \qquad i = 1, 2, 3. \tag{9}$$

so that the D-term contribution to the potential is identically zero when evaluated on the solution, and the resulting equations of motion are somewhat simpler. For general N_c, there are $N_c - 1$ distinct topological sectors labeled by an integer k with $0 < k < N_c$.

For generic winding $1 < k < N_c$ the flux carried by the corresponding k-vortex is proportional to

$$Y_k = \mathrm{Diag}\left(\underbrace{\frac{k}{N_c}, \ldots, \frac{k}{N_c}}_{N_c - k\ \text{elements}}, -\frac{N_c - k}{N_c}, \ldots, -\frac{N_c - k}{N_c}\right), \tag{10}$$

and away from the vortex core, the adjoint scalars behave as

$$\Phi_{1,2}(r,\varphi) = e^{iY_k\varphi}\, \Phi_{1,2}(r, \varphi = 0)\, e^{-iY_k\varphi} \tag{11}$$

while Φ_3 has only a radial dependence. Under the effect of this rotation, the components $(\Phi_{1,2})_{N_c-k,N_c-k+1}$ and $(\Phi_{1,2})_{N_c-k+1,N_c-k}$ wind around the vortex with phase $e^{i\varphi}$ and $e^{-i\varphi}$ respectively. We believe that these are the solutions of lowest tension in each topological sector k, as each field winds at infinity exactly once. The solutions also display an obvious vortex/anti-vortex symmetry, which is evident under the replacement $k \to N_c - k$.

3.1. *Ansatz for $N_c = 4$*

In the following we will discuss explicitly the solution for $N_c = 4$ and $k = 1, 2$. The solution for the $k = 3$ string can be obtained by the $k = 1$ one by a spatial rotation.

<u>*$k = 1$ solution:*</u> The ansatz follows the general pattern described earlier,

$$\Phi_1 = \frac{mi}{2} \begin{pmatrix} 0 & \sqrt{3}\psi_1 & 0 & 0 \\ \sqrt{3}\psi_1 & 0 & 2\psi_2 & 0 \\ 0 & 2\psi_2 & 0 & \sqrt{3}\psi_3 e^{i\varphi} \\ 0 & 0 & \sqrt{3}\psi_3 e^{-i\varphi} & 0 \end{pmatrix}, \tag{12}$$

$$\Phi_2 = \frac{mi}{2} \begin{pmatrix} 0 & -i\sqrt{3}\psi_1 & 0 & 0 \\ i\sqrt{3}\psi_1 & 0 & -i2\psi_2 & 0 \\ 0 & i2\psi_2 & 0 & -i\sqrt{3}\psi_3 e^{i\varphi} \\ 0 & 0 & i\sqrt{3}\psi_3 e^{-i\varphi} & 0 \end{pmatrix},$$

$$\Phi_3 = \frac{mi}{2} \begin{pmatrix} 3\kappa_1 - 2\kappa_3 & 0 & 0 & 0 \\ 0 & \kappa_2 + 2\kappa_3 & 0 & 0 \\ 0 & 0 & -\kappa_2 + 2\kappa_3 & 0 \\ 0 & 0 & 0 & -3\kappa_1 - 2\kappa_3 \end{pmatrix}.$$

The following expression is used for the gauge field:

$$A_x = \frac{-y}{r^2}\left((1-f)Y_1 + \sum_{\ell=1}^{2} g_\ell(r)\lambda_\ell\right), \quad A_y = \frac{x}{r^2}\left((1-f)Y_1 + \sum_{\ell=1}^{2} g_\ell(r)\lambda_\ell\right), \tag{13}$$

where

$$Y_1 = \frac{1}{4}\,\mathrm{Diag}(1,1,1,-3)\,. \tag{14}$$

which yields $\exp(2\pi i Y_1) = \mathrm{Diag}(e^{\pi i/2}, e^{\pi i/2}, e^{\pi i/2}, e^{\pi i/2})$. The non-Abelian flux carried by the vortex is proportional to Y_1. The g_ℓ's are functions of r vanishing both at $r = 0$ and at $r \to \infty$, and λ_ℓ are a basis of diagonal matrices with satisfying,

$$\mathrm{Tr}\, Y_1 \lambda_\ell = 0\,, \qquad \mathrm{Tr}\, \lambda_i \lambda_\ell = \frac{1}{2}\,\delta_{i\ell}\,. \tag{15}$$

We choose

$$\lambda_1 = \frac{1}{\sqrt{12}}\,\mathrm{Diag}(1,1,-2,0)\,, \qquad \lambda_2 = \frac{1}{2}\,\mathrm{Diag}(1,-1,0,0)\,. \tag{16}$$

The string profile can then can be found by the minimization of the energy functional. We do not write the explicit form as it is quite lengthy.

<u>$k = 2$ *solution:*</u> We now turn to the $k = 2$ vortex solution. The relevant configuration for the scalars is now obtained by applying an $SU(4)$ rotation to the Higgs vacuum VEVs, generated by

$$Y_2 = \frac{1}{2}\,\mathrm{Diag}(1,1,-1,-1) \tag{17}$$

with $\exp(2\pi i Y_2) = \mathrm{Diag}(-1,-1,-1,-1)$. The chromomagnetic flux is also proportional to Y_2. The explicit ansatz is then,

$$\Phi_1 = \frac{mi}{2}\begin{pmatrix} 0 & \sqrt{3}\psi_1 & 0 & 0 \\ \sqrt{3}\psi_1 & 0 & 2\psi_2 e^{i\varphi} & 0 \\ 0 & 2\psi_2 e^{-i\varphi} & 0 & \sqrt{3}\psi_3 \\ 0 & 0 & \sqrt{3}\psi_3 & 0 \end{pmatrix}, \tag{18}$$

$$\Phi_2 = \frac{mi}{2}\begin{pmatrix} 0 & -i\sqrt{3}\psi_1 & 0 & 0 \\ i\sqrt{3}\psi_1 & 0 & -i2\psi_2 e^{i\varphi} & 0 \\ 0 & i2\psi_2 e^{-i\varphi} & 0 & -i\sqrt{3}\psi_3 \\ 0 & 0 & i\sqrt{3}\psi_3 & 0 \end{pmatrix},$$

$$\Phi_3 = \frac{mi}{2}\begin{pmatrix} 3\kappa_1 - 2\kappa_3 & 0 & 0 & 0 \\ 0 & \kappa_2 + 2\kappa_3 & 0 & 0 \\ 0 & 0 & -\kappa_2 + 2\kappa_3 & 0 \\ 0 & 0 & 0 & -3\kappa_1 - 2\kappa_3 \end{pmatrix}.$$

The gauge fields are still given by Eq. (13), with Y_1 replaced by Y_2 and

$$\lambda_1 = \frac{1}{2}\,\mathrm{Diag}(1,-1,0,0), \qquad \lambda_2 = \frac{1}{2}\,\mathrm{Diag}(0,0,1,-1). \tag{19}$$

As before the vortex profiles can be found numerically. The numerical results for the profiles function can be found in [17].

3.2. *K-String Tensions*

With the explicit ansätze at hand for general N_c and k, we can compute their tensions, albeit numerically. We will then compare these results with the known tensions in a different parametric regime for $\mathcal{N} = 1^*$ theory wherein the vortex strings are almost BPS. It should be pointed out that when $m_1 = m_2 = m_3 = m$, the strings are *far from* BPS. Nevertheless we will see that the numerical values of the tensions approach the BPS values as N_c is increased. In the limit

$$m_1 = m_2 = m\,, \quad \text{and} \quad m_3 << m\,. \tag{20}$$

the vortex becomes an almost BPS object, due to the fact that in the limit $m_3/m \to 0$, $\mathcal{N} = 2$ supersymmetry is restored. The theory may then be viewed as softly broken $\mathcal{N} = 2^*$ theory. The $\mathcal{N} = 2^*$ theory, with $m_1 = m_2 = m$ and $m_3 = 0$, is realized in the Coulomb phase due to Φ_3 obtaining a VEV. Adding a mass m_3 for Φ_3 at the appropriate point on the Coulomb branch moduli space results in complete Higgsing of the theory due to electric degrees of freedom becoming light and condensing. The profile functions of the vortices in this limit are simpler, because it is consistent to take the profiles $\kappa_j(r)$ (equivalently, Φ_3) as constant. For a BPS vortex the tension is exactly proportional to the field condensates:

$$T^{BPS}_{N_c,k} = 2\pi\,\frac{m m_3}{g^2_{\rm YM}}\,k(N_c - k)\,. \tag{21}$$

This behavior is the so-called 'Casimir scaling' of k-string tensions.

The case we have focussed attention on this paper is far from the BPS limit with,

$$m_1 = m_2 = m_3 = m\,. \tag{22}$$

Using our ansatz above we have numerically evaluated the vortex tension functional $T_{N_c,k}$ for $2 \le N_c \le 6$. The main conclusion that we can draw from this numerical data is that for large N_c the k-string tension $T_{N_c,k}$ in the theory with $m_3 = m$ approaches the BPS tension formula given by Eq. (21). The relative deviation from the BPS formula is a quantity of order $1/N_c$. There does not appear to be an obvious explanation for this result. We also note also that for fixed N_c the ratios in the table are, to a very good approximation, independent of k.

The numerical results for string tension ratios $T_{N_c,k+1}/T_{N_c,k}$ are even more striking. For $N_c = 4$ we find the following numerical result,

$$\frac{T_{N_c=4,\,k=2}}{T_{N_c=4,\,k=1}} = 1.334 \tag{23}$$

while the prediction from Casimir scaling is 4/3. For $N_c = 5$ we find

$$\frac{T_{N_c=5,\,k=2}}{T_{N_c=5,\,k=1}} = 1.501 \tag{24}$$

while the Casimir scaling prediction is 3/2. Finally, for $N_c = 6$:

$$\frac{T_{N_c,\,k=2}}{T_{N_c=6,\,k=1}} = 1.6008\,, \qquad \frac{T_{N_c=6,\,k=3}}{T_{N_c=6,\,k=1}} = 1.801, \tag{25}$$

while the Casimir scaling values are 8/5 and 9/5.

3.3. *Effective World-Sheet Theory*

The general class of vortex solutions presented above have the property that they are invariant only under a $U(1)$ subgroup of the colour-flavour locked $SO(3)_{C+F}$ transformations. This unbroken $U(1)$ is a rotation acting on the pair (Φ_1, Φ_2) which can be undone by a gauge transformation. The moduli space of inequivalent solutions is thus $\mathbb{CP}^1 \simeq SU(2)_{C+F}/U(1)$. The associated moduli correspond to the orientational modes of the magnetic flux in the string solution.

The following effective theory [17] can be written for the vortex orientational zero modes:

$$S_{1+1} = \int dz\, dt \left(B_{N_c,k} (\partial_s \vec{n})^2 - k(N_c - k) \frac{\theta_{3+1}}{8\pi} \epsilon^{sr} \epsilon^{abc} n^a \partial_s n^b \partial_r n^c \right) , \tag{26}$$

This corresponds to a bosonic $\mathbb{CP}^1$ sigma model with an effective $\theta_{1+1} = k(N_c - k)\theta_{3+1}$. The vortices are non-BPS and the effective world-sheet theory is non-supersymmetric. Thus there are no fermionic super-orientational zero modes. The resulting world-sheet dynamics is different from that of BPS vortex strings in $\mathcal{N} = 2$ SQCD [9,10] for example.

It is well-known that the value of the theta angle has a strong effect on the IR dynamics of the $\mathbb{CP}^1$ model.[18,19] First of all the $\mathbb{CP}^1$ model is asymptotically free and so is a strongly coupled theory. This is interesting: the four dimensional field theory is weakly coupled, but the dynamics on the vortex is highly quantum. When $\theta_{1+1} = 0$ and $\theta_{1+1} = \pi$, the model is exactly solvable. Specifically, the spectrum at $\theta_{1+1} = 0$ is known to consist of a single massive $SO(3)$ triplet with an exact S-matrix [18] and the theory has a mass gap. This picture continues to be valid for generic non-zero values of θ_{1+1}. When θ_{1+1} hits π, however, something drastic happens. The theory has massless excitations and flows to a $c = 1$ conformal fixed point [19] described by the $SU(2)$ Wess-Zumino-Witten model at level $k = 1$. The spectrum now consists of massless $SU(2)$ doublets. The picture, therefore, is that at generic θ_{1+1}, the doublets are confined and bound into meson-like excitations, transforming as a triplet of $SO(3)$.

4. The Vortex as a D-brane in the String Dual

4.1. *The String Dual*

The string theory dual of $\mathcal{N} = 1^*$ theory was constructed by Polchinski and Strassler [7] by considering an appropriate deformation of Type IIB string theory on $AdS_5 \times S^5$ background. The undeformed $AdS_5 \times S^5$ background with N_c units of Ramond-Ramond five form flux is dual to the large N_c limit of $SU(N_c), \mathcal{N} = 4$ SYM. The relation between gauge theory and string theory parameters is as follows. The string coupling g_s and the radius of curvature of AdS space are related to the gauge coupling and the 't Hooft coupling respectively as

$$4\pi g_s = g_{\mathrm{YM}}^2\,, \qquad \frac{R_{\mathrm{AdS}}}{\sqrt{\alpha'}} = (4\pi g_s N_c)^{1/4} \gg 1\,, \qquad C_0 = \frac{\theta_{3+1}}{2\pi} \tag{27}$$

where C_0 is the Type IIB RR scalar.

The $\mathcal{N} = 1^*$ mass deformation of the $\mathcal{N} = 4$ theory is achieved by switching on a non-normalizable mode for the three-form flux $G_3 = F_3 - \tau H_3$, with $\tau = i/g_s + C_0/2\pi$, the unperturbed Type IIB coupling. The Polchinski-Strassler dual geometry was obtained by treating the G_3 flux as a perturbation and solving the Type IIB supergravity equations of motion to linear order in this perturbation. The rich infrared physics of $\mathcal{N} = 1^*$ theory was captured in the string dual using two central ingredients: the Myers dielectric effect [20] and the action of $SL(2,\mathbb{Z})$ duality on the vacua of the theory.

In the Higgs vacuum, with $m_1 = m_2 = m_3 = m$, the metric in the interior matches onto the geometry generated by a D5-brane wrapped on an S^2 carrying N_c units of D3-charge. The D3-brane worldvolume coordinates are x^μ, $(\mu = 0, 1, 2, 3)$ wherein the field theory lives. The six transverse directions are denoted as

$$w^i = x^{7,8,9}\,, \qquad \text{and} \qquad y^i = x^{4,5,6}. \tag{28}$$

The D3 branes spread out along the w^i directions with $y^i = 0$ and the resulting D5-brane wraps a round sphere of radius

$$r_0 = \pi\alpha' m N_c\,. \tag{29}$$

The string frame metric of the Polchinski-Strassler solution [7] corresponding to the Higgs vacuum (D5-brane) with equal masses for the adjoint chiral multiplets and with $\theta_{3+1} = 0$, is

$$ds^2_{\mathrm{string}} = Z_x^{-1/2}\eta_{\mu\nu}dx^\mu dx^\nu + Z_y^{1/2}(dy^2 + y^2 d\Omega_y^2 + dw^2) + Z_\Omega^{1/2} w^2 d\Omega_w^2\,, \tag{30}$$

where

$$Z_x = Z_y = \frac{R_{\rm AdS}^4}{\rho_+^2 \rho_-^2}\,, \qquad Z_\Omega = \frac{R_{\rm AdS}^4\, \rho_-^2}{\rho_+^2(\rho_-^2+\rho_c^2)^2}\,, \qquad \rho_\pm = \sqrt{y^2+(w\pm r_0)^2}\,. \tag{31}$$

The parameters in the metric are

$$R_{\rm AdS}^4 = 4\pi g_s N_c \alpha'^2\,, \qquad \rho_c = (\alpha' m)\sqrt{g_s N_c \pi}\,. \tag{32}$$

and r_0 is as defined in (29). The dilaton is non-constant, approaching its asymptotic value g_s as $w^i, y^i \to \infty$, but vanishing close to the D5/D3 brane,

$$e^{2\Phi} = g_s^2 \frac{\rho_-^2}{\rho_-^2+\rho_c^2}\,, \qquad C_0 = \theta_{3+1} = 0\,. \tag{33}$$

The Polchinski-Strassler (including the R-R and NS-NS potentials) background has a manifest $SO(3)$ isometry acting on the S^2 in the w-plane. This naturally gets identified with the colour-flavour locked $SO(3)_{C+F}$ symmetry of the Higgs phase of $\mathcal{N}=1^*$ theory.

4.2. *The Vortex as a D-Brane*

Magnetic flux tubes in the Higgs vacuum have a natural brane interpretation as bound states of D1-branes with the D5-brane. Such D1-D5/D3 bound states involving the single D5-brane in the Higgs vacuum are possible due to the non-vanishing B_2 potential on the 5-brane, responsible for the D3-charge. Following [7], we model a magnetic vortex as a D1-brane probe in this geometry. The DBI action for the probe D1 brane in the geometry reads,

$$S_{\rm DBI} = \frac{1}{2\pi\alpha'}\int d^2\xi \left\{ e^{-\Phi}\sqrt{(-\det(G_{ab}+B_{ab}+2\pi\alpha' F_{ab}))}\right\}. \tag{34}$$

Let us consider a D1-brane oriented in the x_0, x_1 directions and the embedding $(\xi_0,\xi_1)=(x_0,x_1)$. This is minimized when,

$$w = \frac{r_0+\sqrt{r_0^2-2\rho_c^2}}{2} \approx r_0 - \frac{\rho_c^2}{2r_0}\,, \qquad y=0\,. \tag{35}$$

The tension of the D1-brane at this location is

$$T_{\rm D1} \approx \frac{2 r_0 \rho_c}{g_s R_{\rm AdS}^2}\frac{1}{2\pi\alpha'} = \frac{N_c m^2}{2 g_s} = \frac{2\pi N_c m^2}{g_{\rm YM}^2}\,. \tag{36}$$

It is straightforward to read from the Born-Infeld action also the vortex effective theory:

$$\mathcal{L}_{1+1} = \frac{\pi N_c}{g_{\rm YM}^2}(\partial_s \vec{n}_w)^2 + \frac{N_c \theta_{3+1}}{8\pi}\,\epsilon^{sr}\,\vec{n}_w\cdot(\partial_s\vec{n}_w\times\partial_r\vec{n}_w)\,. \tag{37}$$

At the leading order in N_c this is perfectly consistent with the field theory result.

It is also possible to study k-string configurations, in the limit of large k and N_c, with k/N_c fixed. The corresponding configuration, in analogy with [22] and [23], is given by a $D3$ brane with fluxes. The details of the calculation, which for $\mathcal{N} = 1^*$ involve a D3 brane configuration formed by a "spherical icecap" and by a disc, are discussed in [17]. The result for the tension confirm in an analytic way the Casimir scaling formula that we have found numerically for the field theory solitons:

$$T_{\rm D3} = \frac{m^2}{2g_s} k\,(N_c - k)\,. \tag{38}$$

Also the vortex effective theory can be extracted from the Born-Infeld action. The result is exactly the same as the one found in field theory:

$$S_{1+1} = \int dz\, dt \left(B_{N_c,k}(\partial_s \vec{n})^2 - k(N_c - k)\, \frac{\theta_{3+1}}{8\pi}\, \epsilon^{sr}\, \epsilon^{abc}\, n^a \partial_s n^b \partial_r n^c \right). \tag{39}$$

5. Conclusions

We have studied solitonic k-vortices in the Higgs vacuum of the $SU(N_c)$ $\mathcal{N} = 1^*$ theory with equal adjoint masses, transforming under an $SO(3)_{C+F}$ symmetry group. We have found that for every k and N_c the vortex world-sheet theory is a non-supersymmetric S^2 sigma model. Perhaps the most interesting feature of the two dimensional world-sheet theory is the relation between its theta angle and the four dimensional one, $\theta_{1+1} = k(N_c - k)\theta_{3+1}$. This has very specific implications for the IR dynamics of the sigma model which is asymptotically free and for general values of θ_{1+1}, has a mass gap with the spectrum consisting of a triplet of $SO(3)$. When $\theta_{1+1} = \pi$ however, the theory is integrable and the spectrum consists of massless doublets of $SO(3)$ and the theory flows to a $c = 1$ conformal fixed point.

Yet another feature of our non-Abelian string solutions is that their tensions, evaluated numerically, are extremely well approximated by the Casimir scaling law,

$$T_{N_c,k} = \frac{2\pi m^2 k(N_c - k)}{g^2_{\rm YM}}\,,$$

for large N_c. This numerical calculation (done for $N_c = 2, 3, 4, 5, 6$) is confirmed for large N_c by an analytical calculation on the string theory side.

References

1. N. Seiberg and E. Witten, Nucl. Phys. B **426**, 19 (1994) [Erratum-ibid. B **430**, 485 (1994)] [arXiv:hep-th/9407087].
2. A. Abrikosov, Sov. Phys. JETP **5**, 1174 (1957); H. B. Nielsen and P. Olesen, Nucl. Phys. B **61**, 45 (1973).
3. C. Vafa and E. Witten, Nucl. Phys. B **431**, 3 (1994) [arXiv:hep-th/9408074].
4. R. Donagi and E. Witten, Nucl. Phys. B **460**, 299 (1996) [arXiv:hep-th/9510101].
5. N. Dorey, JHEP **9907**, 021 (1999) [arXiv:hep-th/9906011].
6. N. Dorey and S. P. Kumar, JHEP **0002**, 006 (2000) [arXiv:hep-th/0001103].
7. J. Polchinski and M. J. Strassler, "The string dual of a confining four-dimensional gauge theory," arXiv:hep-th/0003136.
8. M. J. Strassler, Prog. Theor. Phys. Suppl. **131**, 439 (1998) [arXiv:hep-lat/9803009]; "Millennial messages for QCD from the superworld and from the string," arXiv:hep-th/0309140.
9. A. Hanany and D. Tong, JHEP **0307**, 037 (2003) [arXiv:hep-th/0306150].
10. R. Auzzi, S. Bolognesi, J. Evslin, K. Konishi and A. Yung, Nucl. Phys. B **673**, 187 (2003) [arXiv:hep-th/0307287].
11. D. Tong, Phys. Rev. D **69**, 065003 (2004) [arXiv:hep-th/0307302].
12. M. Shifman and A. Yung, Phys. Rev. D **70**, 045004 (2004) [arXiv:hep-th/0403149].
13. A. Hanany and D. Tong, JHEP **0404**, 066 (2004) [arXiv:hep-th/0403158].
14. M. Shifman and A. Yung, Rev. Mod. Phys. **79**, 1139 (2007) [arXiv:hep-th/0703267].
15. D. Tong, Annals Phys. **324**, 30 (2009) [arXiv:0809.5060 [hep-th]].
16. V. Markov, A. Marshakov and A. Yung, Nucl. Phys. B **709**, 267 (2005) [arXiv:hep-th/0408235].
17. R. Auzzi and S. Prem Kumar, JHEP **0812**, 077 (2008) [arXiv:0810.3201 [hep-th]].
18. A. B. Zamolodchikov and A. B. Zamolodchikov, Annals Phys. **120**, 253 (1979).
19. A. B. Zamolodchikov and A. B. Zamolodchikov, Nucl. Phys. B **379**, 602 (1992).
20. R. C. Myers, JHEP **9912**, 022 (1999) [arXiv:hep-th/9910053].
21. J. Madore, Class. Quant. Grav. **9**, 69 (1992).
22. C. G. Callan, A. Guijosa, K. G. Savvidy and O. Tafjord, Nucl. Phys. B **555**, 183 (1999) [arXiv:hep-th/9902197].
23. C. P. Herzog and I. R. Klebanov, Phys. Lett. B **526**, 388 (2002) [arXiv:hep-th/0111078].

The Many Faces of the Heterotic Vortex

S. Bolognesi

William I. Fine Theoretical Physics Institute, University of Minnesota, 116 Church St. S.E., Minneapolis, MN 55455, USA

We address the problem of non-Abelian super-QCD, with a Fayet-Iliopoulos term, as seen from the vortex worldsheet perspective. Together with the FI term ξ, also a mass μ for the adjoint superfield Φ enters into the game. This mass allows the interpolation between $\mathcal{N}=2$ and $\mathcal{N}=1$ super-QCD. We distinguish, inside the parameter space spanned by ξ and μ, four different corners where some quantitative statements can be made. We focus on two questions: 1) Is the quantum vortex BPS or non-BPS? 2) What is the phase of the internal non-Abelian moduli? We find that the answer to these questions depends upon the choice of the linear term in the superpotential.

We then address the problem of multiple non-Abelian vortices, in the presence of the $\mathcal{N}=1$ deformation. We show that the tension of the "Abelian" vortex is not affected by the heterotic deformation, and in particular it remains BPS-saturated when μ is also switched on.

1. Introduction

$\mathcal{N}=2$ theories in 3+1 dimensions can be "squeezed" inside 1+1 vortices by the Fayet-Iliopoulos term.[3] Although the $3+1$ theory becomes essentially trivial (as $\sqrt{\xi} \gg \Lambda$ is in the weak coupling Higgs phase), non-trivial strong dynamics survive inside the core of the non-Abelian vortices. It would certainly be interesting to see how far we can extend this idea to the $\mathcal{N}=1$ case.

The possibility of lowering the supersymmetries, but still keeping the BPS property of the string, was suggested in [5]. That can happen, classically, if we introduce a superpotential $W(\Phi)$ so that one root of its derivative W' coincides with the mass of the quarks. The theory on the vortex worldsheet preserves half of the supercharges, and that means that it is $\mathcal{N}=(0,2)$ or $\mathcal{N}=(2,0)$ depending on the vortex orientation[7]; hence the name heterotic.

We shall focus here on U(n) $\mathcal{N}=2$ SQCD with n flavors. We consider two deformations. One is a superpotential for the adjoint fields

$$\int d^2\theta\ \sqrt{2}\mu\ \mathrm{Tr}\left(\frac{\Phi^2}{2} - a\Phi\right) + \mathrm{h.c.}\ , \qquad (1)$$

with mass μ and linear term a, and the other is the Fayet-Iliopoulos term $-2\xi \int d^2\theta d^2\bar{\theta}\ \mathrm{Tr}\, V$. In the space of the deformation parameters, there are four corners where the vortex theory can be analyzed:[1] **(A)**: $\sqrt{\xi}\ , \mu \ll \Lambda$. This is the region where the theory is essentially $\mathcal{N} = 2$ undeformed until energies well below the dynamical scale. Quantum effects can be treated using the Seiberg-Witten (SW) solution. **(B)**: $\sqrt{\xi} \ll \Lambda \ll \mu$. When the mass μ becomes much greater than the dynamical scale, we can use the results of pure $\mathcal{N} = 1$ SQCD. The light degrees of freedom are the gauge singlet mesons and baryons. **(C)**: $\mu, \Lambda \ll \sqrt{\xi}$. As the FI term becomes larger, the $3+1$ dynamics is Higgsed at high energy, much above the dynamical scale. Dynamics survive only inside the $1+1$ worldsheet of the non-Abelian vortex. **(D)**: $\Lambda \ll \sqrt{\xi} \ll \mu$. The gauge $3+1$ dynamics is Higgsed at the weak coupling, but not all particles in the bulk acquire mass. Some of them have mass well below $\Lambda_{\mathcal{N}=1}$. The worldsheet dynamics is generically entangled with those.

Since the various corners are separated by regions where their respective approximations fail, we cannot expect quantitative agreement between the various results. But we shall find a coherent explanation with respect to these two basic questions: 1) Is the vortex BPS or is the supersymmetry broken? 2) What is the phase of vortex ground state?

For the generic value of the coefficient a, the vortex is non-supersymmetric and the phase on its worldsheet is confining. For the particular case $a = 0$, we have n degenerate ground states, with non-zero energy. For a particular choice $a \sim \Lambda e^{\frac{i\,2\pi k}{n}}$, with $k = 1, \ldots n$, one of the n strings ground states reaches zero energy.

The theory at finite μ and small FI term $\sqrt{\xi} \ll \Lambda$ (regions (A) and (B)) is known to be confining. Quarks are bounded into gauge invariant mesons or baryons. The hope is also to detect signals of confinement inside the flux tubes (regions (C) and (D)). When μ and ξ are both non-vanishing, but $a = 0$, strings are no-more BPS, but still there are n degenerate ground states. This degeneracy is a consequence of the breaking of the residual axial symmetry, $Z_{2n} \to Z_2$. Kinks between these n degenerate strings are thus *not* confined on the non-Abelian string worldsheet, and genuine one-particle states in this reduced $1+1$ theory. A conclusion could thus be that confinement on the worldsheet is not visible; the phase transition between small and large FI obscures the physics of confinement from the vortex theory. Luckily the previous conclusion is not the right one, or at least is not the end of the story.

Not much has been explored so far about the effect of the heterotic deformation on the multiple vortices moduli space. The Abelian string is a particular element in the moduli space of n fundamental non-Abelian strings, with coincident positions and orthogonal internal orientations. It has no internal structure and thus is presumably *not* lifted by the heterotic deformation.[2] In Section 6, we shall present an argument for that. This also implies that n fundamental non-Abelian strings are bounded together to form the Abelian one. Kink in the strings are confined into gauge-invariant states. Transition from well-separated non-Abelian vortices, to their Abelian bound-state, can happen through a quantum brane fusion.[4]

2. Perturbed $\mathcal{N} = 2$

Here we consider the perturbed $\mathcal{N} = 2$ region. We can use the known results about the $\mathcal{N} = 2$ strong dynamics, obtained with the Seiberg-Witten technique, and then add the perturbations to the low-energy effective Lagrangian. The SW curve for $n_c = n_f = n$ is the hyperelliptic surface defined by

$$y^2 = \frac{1}{4}\det(z-\phi)^2 - \Lambda^n z^n = \frac{1}{4}\left(z^n - \Lambda^n\right)^2 \tag{2}$$

where the values of ϕ label the point in the moduli space of vacua, and the last passage is the factorization at the root of the baryonic branch, located at the Z_n invariant vacuum $\phi = e^{i\pi/n}\mathrm{diag}(\Lambda, \Lambda\omega_n, \Lambda\omega_n^2, \ldots, \Lambda\omega_n^{n-1})$, where ω_n is the nth root of unity $e^{i2\pi/n}$. Note that all the roots are doubled. The low-energy theory consists of n $\mathrm{U}(1)_j$ gauge theories, with $j = 1, \ldots, n$, each one with its own charged hypermultiplet E_j, $\widetilde{E}_j$.

Now we examine SUSY breaking (from $\mathcal{N} = 2$ down to $\mathcal{N} = 1$) in the effective low-energy theory after introduction of a superpotential (1). The low-energy superpotential is

$$\mathcal{W} = \sqrt{2}\left(\sum_{j=1}^{n} \widetilde{E}_j A_j E_j + \mu(u_2 - a u_1)\right) . \tag{3}$$

Each $\mathrm{U}(1)_j$ is Higgsed by the condensation of the respective hypermultiplet E_j, $\widetilde{E}_j$, and each admits a formation of an ANO vortex. The tension of these vortices is proportional to the length of the $\mathrm{SU}(2)_R$ triplet $(-\mathrm{Re}\,\mathcal{W}', \mathrm{Im}\,\mathcal{W}', \xi)$:

$$T_j = 4\pi\sqrt{|\mathcal{W}'(\mathrm{root}_j)|^2 + \xi^2}\ , \tag{4}$$

where "root$_j$" is the jth root of the SW curve (2). For the case $a = 0$, the n vortices have degenerate tension:

$$T_j = 4\pi\sqrt{|\mu\Lambda|^2 + \xi^2} \ . \tag{5}$$

Therefore, we find agreement with the finding of the heterotic vortex theory: n degenerate non-BPS vacua/vortices. The physical interpretation is that the n vortices we observe in the SW low-energy effective action are nothing but the ground states of the non-Abelian heterotic vortex theory of corner (C). Something peculiar happens for the specific choice $a = \Lambda e^{\frac{i2\pi k}{n}}$. The tension of the vortices is now

$$T_j = 4\pi\sqrt{\left|2\mu\Lambda \sin\pi(j-k)\right|^2 + \xi^2} \ , \tag{6}$$

and the k-vortex is BPS-saturated, with the tension $T_k = 4\pi\xi$.

3. Perturbed $\mathcal{N} = 1$

This corner is the one where the strong dynamics is that of $\mathcal{N} = 1$ SQCD in $3 + 1$ dimensions. The classical theory has a moduli space of vacua, parametrized by gauge-invariant meson superfield $M_{ij} = \widetilde{Q}_i Q_j$ together with the two baryonic superfields $B = \epsilon_{\alpha_1 \ldots \alpha_n} Q_1^{a_1} \ldots Q_n^{\alpha_n}$ and $\widetilde{B} = \epsilon_{\alpha_1 \ldots \alpha_n} \tilde{Q}_1^{\alpha_1} \ldots \tilde{Q}_n^{\alpha_n}$. These are not independent and obey a constraint. In quantum theory, the constraint is

$$\det M_{ij} - B\widetilde{B} = \Lambda^{2n} \ . \tag{7}$$

Since we work in $\mathrm{U}(n)$ and not $\mathrm{SU}(n)$, there is also a residual $\mathrm{U}(1)_B$ gauge interaction, coupled to the baryons, which remain weakly coupled in the infrared. The presence of the FI parameter ensures that the $\mathrm{U}(1)_B$ gauge symmetry is broken, and this implies the existence of strings[7]]. But note that since it is created by the winding of the baryon field B, it has $1/n$ the charge of an Abelian $\mathrm{U}(1)_B$ vortex. It thus has the *same* flux of the non-Abelian string with respect to the $\mathrm{U}(1)_B$.

The situation we are going to face shares many features in common with the prototype vortex in the presence of flat directions.[6] We consider first the classical $\mathrm{U}(1)$ theory. The vacuum manifold is parametrized by q and $\widetilde{q}$ subject to the D-term constraint $|q|^2 - |\widetilde{q}|^2 = 2\xi$. If we sit at the base $\widetilde{q} = 0$, $|q|^2 = 2\xi$, we can have a perfectly well-defined Abelian vortex, while keeping fixed the boundary conditions at infinity. If we move away from the base of the manifold, the core of the vortex becomes a source for the massless fields that compose these extra flat directions. The outcome is a logarithmic tail, outside the core of the vortex. The logarithmic tail inevitably sets the boundary conditions to infinity.

Now back to our theory, but for the moment with $\mu \to \infty$. Vortices exist only if $\tilde{B} = 0$, and they are also BPS. The quantum mechanical analog of the base of the Higgs branch is now split into n different vacua parametrized by $M_{ij} = \mathbf{1}_{ij}\, \Lambda^2\, e^{\frac{i2\pi k}{n}}$, with $k = 1, \ldots, n$. The mesonic field is diagonal, and proportional to the nth root of the identity. Whenever we choose to move from one of these n bases, $\tilde{B} \neq 0$, any vortex solution must necessarily be non-BPS and have the log tail, typical of vortices in the presence of flat directions.

For finite μ, we must consider the effect of the operator obtained after integrating out the adjoint field Φ

$$W = -\frac{1}{\mu\sqrt{2}} \left[\mathrm{Tr}\,(Q\widetilde{Q}\; Q\widetilde{Q}) - 2a\mu \mathrm{Tr}\, Q\widetilde{Q} \right] . \tag{8}$$

This operator, in the low-energy effective theory, is a mass term that lifts the mesonic modulus M. In the $\mu \to \infty$ limit, the operator clearly vanishes, but the information we are looking for (degeneracy and susy-breaking of vortices) are all encoded in this $1/\mu$ effect. In the case $a = 0$, the effective superpotential (8) forces the vacuum to be at $\det M = 0$, and this implies $B\widetilde{B} = -\Lambda^{2n}$. The degeneracy is *not* visible from the boundary conditions at $r \to \infty$. It is encoded in the value of the M field at the core of the vortex.

We need some toy model that captures the essential properties we previously discussed. We do not know how to introduce the FI model at the effective Lagrangian level for the non-Abelian theory. What we will study corresponds to the extreme case $n = 1$. We can thus say that $B = Q$, $\widetilde{B} = \widetilde{Q}$ and the quantum deformation (7) is $M = \widetilde{Q}Q + \Lambda^2$ We also used the adiabatic approximation; that is, we impose the condition of the quantum-deformed manifold (7) over the vortex profile. That is a good approximation in the $\Lambda \gg \mu$. We can thus rewrite the superpotential (8) for the mesonic field, in terms of the quark $W = -\frac{1}{\mu\sqrt{2}} \left(\widetilde{Q}Q + \Lambda^2\right)^2$. The potential for the constituent quark q is thus

$$V = \frac{1}{2\mu^2} \left(\widetilde{q}q + \Lambda^2\right)^2 \left(|q|^2 + |\widetilde{q}|^2\right) + \frac{g^2}{2} \left(|q|^2 - |\widetilde{q}|^2 - 2\xi\right)^2 . \tag{9}$$

In Figure 1, we present numerical results for the vortex obtained with this potential and parameters $e = 1$, $\xi = .5$, $\Lambda = 2$ and $\mu = 100$. The first plot are some profile functions for q and $\widetilde{q}$; we also plot $\sqrt{\Lambda^2 + \widetilde{q}q}$. The second shows the tension density.

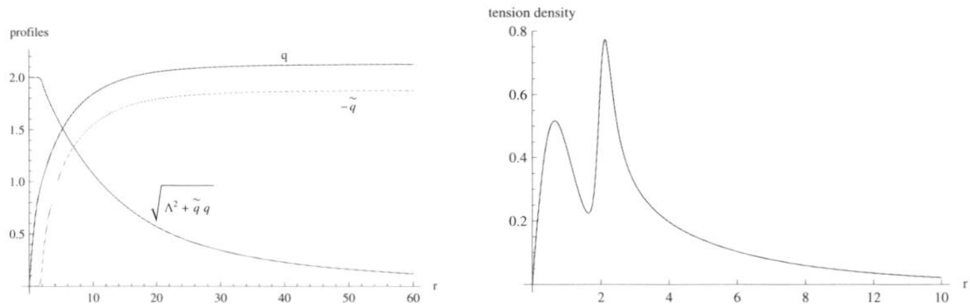

Fig. 1. Left: The profile functions for q, $\tilde{q}$. Right: The tension density, with the two peaks.

4. Heterotic Vortex Theory

Now we deal with region (C), in which non-trivial dynamics happen only on the $1+1$ worldsheet. We can study the vacuum structure and mass spectrum in the large n limit.[8]

The effective potential for the field σ (the fermionic condensate $\xi\xi$) is

$$V_{\rm eff}(\sigma) = \frac{n}{8\pi}\left\{\Lambda^2 + 2|\sigma|^2\left(\log\frac{2|\sigma|^2}{\Lambda^2} - 1\right) + 8|\sigma - a|^2\, u\right\}. \qquad (10)$$

where we introduced the dimensionless parameter u, which does not scale with n: $u = \frac{\pi|\omega|^2}{n}$ and $\omega \propto \frac{\mu}{\sqrt{\xi}}$. The minimum is given to the solution of the following equation: $2|\sigma|^2/\Lambda^2 = \exp\left[-4u(\sigma - a)/\sigma\right]$.

To have a supersymmetric solution, we need $V_{\rm eff}(\sigma_0) = 0$. There are only two possibilities for that to happen: 1) $\mu = 0$; this is the trivial $\mathcal{N} = (2,2)$ case without the superpotential; 2) $a = \frac{\Lambda}{\sqrt{2}}$; this is the non-trivial case; a particular tuning of the linear term in the superpotential preserves supersymmetry. The next part will be devoted to the last particular case, and to solve for the spectrum of the theory.

We consider the effective Lagrangian for the fields A_k, σ, and $\chi_{L,R}$. These fields are auxiliary at the classical level, but acquire a kinetic term through quantum corrections. In the large-n limit, the kinetic terms and the couplings can just be computed evaluating the 1-loop diagrams with the φ and ξ fields running inside the loops. The effective action is

$$\begin{aligned} S_{\rm eff} = \int d^2x \Big\{ & -\frac{1}{4e_\gamma^2} F_{kl}^2 + \frac{2}{e_\sigma^2}|\partial_k\sigma|^2 \\ & + \frac{1}{e_\chi^2}\,\bar{\chi}_R\, i\,\partial_L\,\chi_R + \frac{1}{e_\chi^2}\,\bar{\chi}_L\, i\,\partial_R\,\chi_L + \bar{\zeta}_L\, i\,\partial_R\,\zeta_L \\ & - V(\sigma) + \frac{n}{2\pi}\frac{{\rm Im}\,\sigma}{|\sigma|}\,\widetilde{F} - \left[\sqrt{2}\,\Gamma\,\sigma\,\bar{\chi}_R\chi_L - \omega\,\bar{\chi}_R\,\zeta_L + {\rm h.c.}\right]\Big\}, \end{aligned} \qquad (11)$$

where $V(\sigma)$ is given in Eq. (10), and $\widetilde{F}$ is the dual gauge field strength $\widetilde{F} = \frac{1}{2}\varepsilon_{kj}F_{kj}$. Here e_γ^2, e_σ^2, and e_χ^2 are the coupling constants that determine the wave function renormalization for the photon, σ, and χ fields, and Γ is the induced Yukawa coupling. These couplings are given by one-loop graphs which we have been computed in [8]. For the case in which we are interested, $D = 0$ and $a = \Lambda/\sqrt{2}$, they are all equal: $\frac{1}{e_\sigma^2} = \frac{1}{e_\gamma^2} = \frac{1}{e_\chi^2} = \frac{n}{4\pi}\frac{1}{2|\sigma|^2} = \frac{n}{4\pi}\frac{1}{\Lambda^2}$, and $\Gamma = \frac{n}{4\pi}\frac{2}{\Lambda^2}$.

We can diagonalize the fermion mass making a change of variable from χ_L, ζ_L to $\widetilde{\chi}_L$, $\widetilde{\zeta}_L$:

$$\widetilde{\chi}_L = \frac{1}{\sqrt{1+u}}\left(\chi_L - c\zeta_L\right) , \qquad \widetilde{\zeta}_L = \frac{1}{\sqrt{1+u}}\left(c^{-1}\chi_L + \zeta_L\right) \tag{12}$$

where $c = 2\omega\pi\Lambda/n$. In this new basis, $\widetilde{\zeta}_L$ is totally decoupled, and we can just ignore it. Furthermore, we want to factorize out the n factors, and also write explicitly the quadratic expansion of the potential. The outcome is

$$\begin{aligned} S_{\rm eff} = \frac{n}{4\pi\Lambda^2}\int d^2x \Big\{ & -\frac{1}{4}F_{kj}^2 + |\partial_k\sigma|^2 \\ & + \bar{\widetilde{\chi}}_L\, i\,\partial_R\,\widetilde{\chi}_L + \bar{\chi}_R\, i\,\partial_L\,\chi_R - \left[2\Lambda\sqrt{1+u}\,\bar{\chi}_R\widetilde{\chi}_L + \text{h.c.}\right] \\ & -4\Lambda^2(1+u)(\text{Re }\sigma)^2 - 4\Lambda^2 u\,(\text{Im }\sigma)^2 + 2\sqrt{2}\Lambda\,(\text{Im }\sigma)\,\widetilde{F}\Big\} \ . \end{aligned} \tag{13}$$

We thus get that the fermions χ_L and $\widetilde{\chi}_R$ get together to form a massive Dirac fermion: $m_{\chi_R} = m_{\widetilde{\chi}_L} = 2\Lambda\sqrt{1+u}$. From the quadratic term of the potential $V(\sigma)$, we calculate the mass of the real part of the σ field, $m_{\text{Re }\sigma} = 2\Lambda\sqrt{1+u}$. The anomalous $(\text{Im }\sigma)\,F^*$ mixing in (13), and the explicit potential for Im σ, gives masses to both photon and the imaginary part of σ: $m_\gamma = m_{\text{Im }\sigma} = 2\Lambda\sqrt{1+u}$.

We see that all fields from the gauge multiplets have the same mass $2\Lambda\sqrt{1+u}$ in accordance with $\mathcal{N} = (2,2)$ supersymmetry. Although in a different way, we obtain the enhancement of supersymmetry first hypothesized in [5].

5. Quantum Heterotic Vortex

The final corner is the one in which the theory is deeply $\mathcal{N} = 1$, and the Fayet-Iliopoulos breaks the gauge group in the weak coupling regime. In the extreme limit, $\mu \to \infty$, we can forget about the adjoint field. The vortex has translational and orientational zero modes, it preserves half of the supersymmetries, $\mathcal{N} = (0,2)$ on its worldsheet, and this implies that we have also the fermionic zero modes ζ_R and ξ_R^a, partners of the translational and orientational moduli. The other ones, ζ_L and ξ_L^a, become non-normalizable in the $\mu \to \infty$ limit. We are not allowed to consider them in a $1+1$ effective action. But on the other hand, a $1+1$ effective theory consisting only of the z, φ^a and their heterotic superpartners, ζ_R and ξ_R^a, is not consistent on its own. It suffers from a sigma model anomaly.[7] In the gauge formulation, this is simply the anomaly for the auxiliary gauge boson A_μ. This *is* a puzzle.

There is also another problem. The analysis of previous sections says that the dynamics of the heterotic vortex depends upon the linear term in the superpotential. This dependence upon the linear term is quite puzzling from the point of view of the region (D). Since the adjoint field Φ is completely decoupled, the two questions, about being the vortex BPS-saturated or not, and about the quantum phase of the CP moduli, should have only one answer.

We find only one plausible explanation to these issues: *The dynamical scale $\Lambda_{\rm CP}$, in the $\mu \to \infty$ limit, must go to zero.* That means that there is no running of the $1+1$ coupling constant. β, which stays frozen to the classical value β_0 from $\sqrt{\xi}$ all the way down to the infrared. So all the different phases that depend upon the linear term a coalesce into a unique phase.

For large μ but not infinite, we have a light particle with mass ξ/μ. We then have to distinguish three different regimes in the renormalization-group flow. Physics at energies above $\sqrt{\xi}$ is four-dimensional. Physics at energies below ξ/μ is $1+1$ dimensional. In the intermediate region, we are in a hybrid situation. As we shall show, there is no running of the $1+1$ coupling. The dynamical scale is thus computed starting the running only at the scale ξ/μ, with the same β_0 defined at the scale $\sqrt{\xi}$. We thus have $\Lambda_{\rm CP} \sim \Lambda_{3+1}\frac{\sqrt{\xi}}{\mu}$, and $\Lambda_{\rm CP}$ goes to zero in the $\mu \to \infty$ limit.

We need the vortex effective theory in this intermediate regime. This action has the following structure: $S_{\rm eff} = S_{3+1} + S_{1+1} + S_{\rm int}$. It is the sum of a $3+1$ action for the bulk massless moduli, plus a $1+1$ action defined on the string worldsheet that describes the orientation and translation moduli, plus an interaction term. String moduli interact with the surrounding massless moduli in the bulk. We need to take into account the interaction with the surrounding bath of $3+1$ particles. We take the expansion of the fields around some classical solution: $q(r) = q^{(0)}(r) + \delta q(r)$, $\widetilde{q}(r) = \widetilde{q}^{(0)}(r) + \delta\widetilde{q}(r)$, $A_k(r) = A_k^{(0)}(r) + \delta A_k(r)$, where the "(0)" refers to the classical vortex solution. To obtain $S_{\rm int}$ is not straightforward; we have to deal with some renormalization issues. As soon as we excite the $\widetilde{q}$ field, a logarithmic tail comes out of the vortex,[6] and makes the tension infinite.

We need to consider a fluctuation of $\widetilde{q}$ with some wavelength different from infinity. The tension of the vortex, and consequently also the interaction terms S_{1+1}, will depend on this scale. In particular, they all diverge if we try to change $\widetilde{q}$ homogeneously in all the space, at infinite wavelength. To simplify the computation at most, we can just consider the space as if it is cylindrical and compact, with a cutoff radius $1/\lambda_\perp$. If $\lambda_\perp \ll \sqrt{\xi}$, we can approximate them outside the core of the vortex with

$$\delta q(r) \sim \delta \widetilde{q}^\dagger \log (r\lambda_\perp) \ , \qquad \delta \widetilde{q}(r) \sim \delta \widetilde{q} \log (r\lambda_\perp) \ , \qquad \delta A_k(r) \sim 0 \ , \quad (14)$$

What happens inside the core is not important, as long as we consider a small enough cutoff $\lambda_\perp$. The only important thing to know is that the vortex core regularizes the log divergence in the core, and all the fields go to zero.

The term we are interested in comes out very easily from the term:

$$S = \int d^2x \left(4\pi\xi + \log\left(\frac{\sqrt{\xi}}{\lambda_\perp}\right) |\delta\widetilde{q}|^2 \right) + \dots \ . \qquad (15)$$

This is the tension of the vortex, integrated over its worldsheet. In general, it is a constant term that does not interfere with the higher-order terms containing the moduli interaction. In the case at hand, the log tail of the scalar fields modifies the vortex tension, and this introduces a mass term of the modulus $\delta\widetilde{q}$.

We need to make a distinction between parallel and orthogonal excitations. Only in the parallel case, we have the log tail typical of vortices with flat directions. We can thus extract from (15) the interaction term between the orientational modulus φ of the non-Abelian string and the bulk fluctuation $\delta\widetilde{q}$ of the Higgs branch:

$$S_{\text{int}} \sim \log\left(\frac{\sqrt{\xi}}{\lambda_\perp}\right) \frac{1}{\beta} \int d^2x \ \left|\varphi^l \ \delta\widetilde{q}_l\right|^2 \sim \frac{1}{\beta_0} \int d^2x \ \left|\varphi^l \ \delta\widetilde{q}_l\right|^2 \ . \qquad (16)$$

Where the last passage follows if we consider the renormalization of the coupling constant $\beta = \beta_0 \log(\mu_{\text{RG}} / \sqrt{\xi})$, and make the scales $\lambda_\perp \sim \mu_{\text{RG}}$.

The renormalization of the wave-function of the φ is given by the first diagram in Figure 2. It is a 2-loop diagram involving two interaction vertices with the four-dimensional field $\widetilde{q}$, and three propagators. In total, the divergence is a logarithmic one $\int d^6k \frac{1}{k^6} \sim \log \mu_{\text{RG}}$. The renormalization of the coupling constant is given by the second diagram Figure 2. It is logarithmic divergent $\int d^2k \frac{1}{k^2} \sim \log \mu_{\text{RG}}$ and gives the usual logarithmic running of the coupling β. The *physical* coupling is obtained by β divided by the wave-function renormalization of φ, and thus the theory remains conformal. In the window of energy scales below $\sqrt{\xi}$ and above ξ/μ, we nee to consider the effect of integrating out the $3+1$ degrees of freedom which interact like (16). As a consequence, the physical coupling does not run.

Fig. 2. The first is the wave-function renormalization for φ. The second graph is the one-loop renormalization of the coupling β.

6. Abelian Bound State

We shall discuss here some fundamental aspects of the SUSY breaking for the heterotic vortex. We explain why, on general grounds, the index argument cannot be used to protect the non-Abelian heterotic string from breaking supersymmetry, while instead it can be used to infer the BPS stability of the Abelian composite one. The symmetries that we need are the R symmetries in Table 1, where $\mathrm{U}(1)_{J'}$ is defined to be the sum of the two fundamental R-symmetries $\mathrm{U}(1)_J + \mathrm{U}(1)_R$. Note that the $\mathrm{U}(1)_{J'}$ charge of the mass μ is zero. The anomaly breaks it as $\mathrm{U}(1)_{J'} \to \mathrm{Z}_{2n}$.

Table 1. R symmetries and corresponding charges

	Q^1_α	Q^2_α	Φ	Q	$\widetilde{Q}$	μ	a	m	ξ	$\Lambda^{2n_c-n_f}$
$U(1)_R$	1	1	2	0	0	-2	2	2	0	$2(2n_c - n_f)$
$U(1)_J$	1	-1	0	1	1	2	0	0	0	0
$U(1)_{J'}$	2	0	2	1	1	0	2	2	0	$2(2n_c - n_f)$

In $\mathcal{N} = 2$, where $\mu = 0$, both $U(1)_R$ and $U(1)_J$ are conserved symmetries. The FI term breaks $\mathrm{SU}(2)_R \to \mathrm{U}(1)_J$, but leaves intact the eight supercharges $Q^1_\alpha = (Q^1_L, Q^1_R)$, $Q^2_\alpha = (Q^2_L, Q^2_R)$. If the vortex is straight in the positive x^3 direction, the unbroken supercharges are $(0, Q^1_R)$ and $(Q^2_L, 0)$. As a particle in $2+1$, it is a short-multiplet with four states, two bosonic and two fermionic.

$$|v\rangle\ , \quad Q^1_L|v\rangle\ , \quad Q^2_R|v\rangle\ , \quad Q^1_L Q^2_R|v\rangle\ . \tag{17}$$

For that we can compactify the x_3 direction; the theory has $\mathcal{N}=4$ in $2+1$ dimensions. Index prevents this object to break SUSY; there are not enough states to build a long multiplet.

When we add a $\mu\Phi^2/2$ superpotential, both $U(1)_R$ and $U(1)_J$ are broken, but in a way that still preserves the combination $\mathrm{U}(1)_{J'}$. The vortex is classically BPS-saturated and still has four states (the number of fermionic zero modes is not changed by μ, although the left ones are modified and not related to the others by supersymmetries). The SUSY theory on $2+1$ is now $\mathcal{N}=2$, and four states are enough for a long multiplet. So we could expect the vortex to be lifted by quantum correction and become non-BPS. But here the $\mathrm{U}(1)_{J'}$ symmetry enter into the game. The four states in (17) have different $\mathrm{U}(1)_{J'}$ charges; two have charge zero while the other two have charge two. The mixing to form a long multiplet would require a breaking $\Delta J'=\pm 2$, and thus the breaking of Z_{2n} down to Z_2.

We can thus conclude the following: *First*, if there is a dynamical mechanism that breaks the residual of $\mathrm{U}(1)_{J'}$ down to Z_2 (like the fermion condensate for the non-Abelian vortex), the mixing to form a long multiplet is allowed, and so we expect SUSY breaking to happen. *Second*, if the vortex has no internal degrees of freedom (like the Abelian composite vortex), and nothing can break the residual Z_{2n} symmetry, the mixing to form a long multiplet is forbidden, and thus the vortex remains BPS-saturated.

We address more in detail the problem using the mass deformation. This is a mass splitting for the hypermultiplet quarks. Since we will mostly focus on the $n=2$ case, that means mass m for the first quarks and mass $-m$ for the second. A single string has two ground states, one when is locked to the first flavor, the other to the second flavor. We denote these as $|\pm\rangle$.

Now assume we take two separate vortices at a certain fixed distance, and we ask how many ground states the system has. Since each can be in two different states, we have a total of four states. But the moduli space has only three disconnected components; two of the previous states belong to the same manifold (Figure 3). In particular, when the distance is zero, the total number of states is three because the previous two coalesce into a unique one (the Abelian vortex $(1,1)$ in the Figure).

The mass m explicitly breaks the $\mathrm{U}(1)_{J'}$ symmetry down to Z_2, and thus we have no reason to expect that any of these vortices, not even the Abelian one, will preserve supersymmetry. But we can nevertheless make use of an indirect argument to prove that the Abelian vortex in the $m\to 0$ limit is BPS. We need to use the $\mathrm{Z}_4\in\mathrm{U}(1)_{J'}$ and combine it with the exchange of the first and second flavor. This is a symmetry since the sign

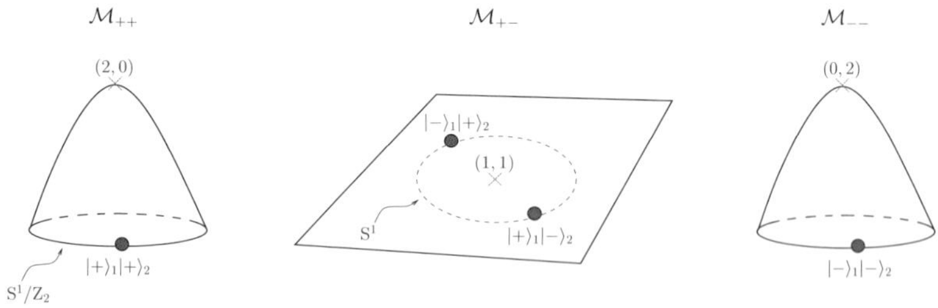

Fig. 3. The moduli space of vortices for the mass-deformed theory, for $n = 2$, has three disconnected components: $\mathcal{M}_{++}$, $\mathcal{M}_{+-}$, and $\mathcal{M}_{--}$.

of the mass m is flipped twice, once by the flavor exchange and once by the $\mathrm{U}(1)_{J'}$ element. The Abelian vortex $(1, 1)$ is invariant under this symmetry, is the only one that has this property, and will remain Z_4 invariant for all the values of m, even when $m \to 0$. When m becomes exactly zero, we restore the full flavor symmetry, and thus we can continually relate our new Z_4 symmetry with the original one that is only the subset of $\mathrm{U}(1)_{J'}$.

Acknowledgments

It is a pleasure to thank R. Auzzi, P. Bolokhov, M. Shifman, D. Tong, A. Yung, A. Vainshtein and M. Voloshin for useful discussions and in particular R. A. and M. S. for the collaboration [2] and M. S. and M. V. for the collaboration [4]. This work is supported by DOE grant DE-FG02-94ER40823.

References

1. S. Bolognesi, JHEP **0906**, 016 (2009) [arXiv:0903.3422 [hep-th]].
2. R. Auzzi, S. Bolognesi and M. Shifman, *in preparation*
3. N. Dorey, JHEP **9811**, 005 (1998) [arXiv:hep-th/9806056].
   M. Shifman and A. Yung, Phys. Rev. D **70**, 045004 (2004) [arXiv:hep-th/0403149].
4. S. Bolognesi, M. Shifman and M. B. Voloshin, Phys. Rev. D **80**, 045010 (2009) [arXiv:0905.1664 [hep-th]]; "Quantum Fusion of Domain Walls with Fluxes," arXiv:0907.3462 [hep-th].
5. M. Shifman and A. Yung, Phys. Rev. D **72**, 085017 (2005) [arXiv:hep-th/0501211].
6. A. A. Penin, V. A. Rubakov, P. G. Tinyakov and S. V. Troitsky, Phys. Lett. B **389**, 13 (1996) [arXiv:hep-ph/9609257].
7. M. Edalati and D. Tong, JHEP **0705**, 005 (2007) [arXiv:hep-th/0703045].
   D. Tong, JHEP **0709**, 022 (2007) [arXiv:hep-th/0703235].
8. M. Shifman and A. Yung, Phys. Rev. D **77**, 125016 (2008) [arXiv:0803.0158 [hep-th]]; Phys. Rev. D **77**, 125017 (2008) [arXiv:0803.0698 [hep-th]].

SECTION 3.

TOPOLOGICAL EXCITATIONS

The Innards of an Instanton

David Tong
Department of Applied Mathematics and Theoretical Physics
University of Cambridge, Cambridge, CB3 0WA, UK
E-mail: d.tong@damtp.cam.ac.uk

Instantons are composed of partons.

1. Introduction

It is both an honour and a great pleasure to deliver this talk as part of the celebrations for Misha Shifman's 60th birthday. As a graduate student, I learned about instantons and supersymmetry from Misha's beautifully written reviews. This therefore seems like the perfect place to revisit an old idea related to instantons, but with a slightly novel supersymmetric twist.[1]

The old idea that I would like to talk about is that instantons can be thought of as containing constituent objects.[2] This idea has been mooted both for instantons in Yang-Mills theories and instantons in sigma-models. In both cases, the first hint at the existence of underlying constituents comes from simply counting the number of collective coordinates:

- Instantons in Yang-Mills theories. In $SU(N)$ Yang-Mills theory, the instanton has four translational modes, a single scaling mode, and $4N-5$ orientation modes. This gives a total of $4N$ collective coordinates.
- Instantons in sigma models. In the $\mathbf{CP}^{N-1}$ sigma-model, the instanton has 2 translation modes, a single scaling mode, and $2N-3$ orientation modes. This gives a total of $2N$ collective coordinates.

For each of these solutions the collective coordinates are Goldstone modes, arising from the action of symmetries on an instanton. However, the numerology suggests that there may be a different interpretation of the collective coordinates: as the positions of N partons which comprise the instanton.

The conjecture that instantons should be thought of as containing N partonic objects is usually framed in the context of $d = 3 + 1$ dimensional Yang-Mills theories and $d = 1 + 1$ dimensional sigma-models. In both these cases, the physics is strongly coupled in the infra-red. The hope is that in the partons — which are objects localized in (Euclidean) spacetime — are somehow liberated and the vacuum is best thought of as a correlated soup of these objects. This is then invoked to explain low-energy phenomena such as confinement or chiral symmetry breaking. Interesting ideas along these lines were presented by Zhitnitsky at the Shifmania conference and a good review can be found in [3].

In this talk, I would like to examine the parton conjecture in the context of $d = 4 + 1$ dimensional Yang-Mills theories and $d = 2 + 1$ dimensional sigma-models. These theories are now weakly coupled in the infra-red, but strongly coupled in the ultra-violet. They are to be thought of as non-renormalizable effective field theories which require a UV completion. Moreover, the instanton solutions are now particle-like solitons, carrying finite energy as opposed to finite action. The parton conjecture becomes slightly better defined: it is the idea that the instanton should be thought of as a multi-particle state. Is this the right interpretation of the instanton in these models? And, if so, how can we tell?

2. Five-Dimensional Yang-Mills

It is known from the work of Seiberg and others that there exists a UV completion of certain Yang-Mills theories in $d = 4 + 1$ dimensions, at least when endowed with $\mathcal{N} = 1$ or $\mathcal{N} = 2$ supersymmetry.[4] However, we don't know a whole lot about the details of this UV theory.

The story is particularly interesting in the case of $\mathcal{N} = 2$ supersymmetry. (This means 16 supercharges). Here it is known that the UV completion of $d = 4 + 1$ dimensional $SU(N)$ Yang-Mills is really a conformal field theory in $d = 5+1$ dimensions, usually referred to as the $(2, 0)$ theory.[5] This theory is compactified on a circle of radius R, which is related to the 5d Yang-Mills coupling by $R = e^2/8\pi^2$. Little is known about the dynamics of the $(2, 0)$ theory and, in particular, there is no known Lagrangian formulation. What little knowledge we do have comes from the gauge-gravity correspondence. Most strikingly, it can be shown that the $(2, 0)$ theory has $\sim N^3$ degrees of freedom.[6] This is many more than the $\sim N^2$ degrees of freedom that are seen in the infra-red Yang-Mills theory. This means that if we're looking for the UV completion of the theory, then we need to find a lot of states! Understanding how these N^3 degrees of freedom arise is likely to tell us something important about the degrees of freedom of M-theory.

The instanton in $d = 4 + 1$ dimensional Yang-Mills plays an important role in this story: it is the Kaluza-Klein mode coming from the theory in $d = 5+1$ dimensions. This can be seen already in the classical mass formula,

$$M_{\text{inst}} = \frac{8\pi^2}{e^2} = \frac{1}{R} \tag{1}$$

The proposal that I would like to explore is the following: *The instanton in $d = 4 + 1$ dimensional Yang-Mills should be thought of as an N particle state.* Moreover, the N partons are to be thought of as the remnant of the UV degrees of freedom which comprise the $(2, 0)$ theory.

Let's look at some circumstantial evidence for this proposal. Firstly, we can compute the free energy in the infra-red and ultra-violet and ask: at what energy does the scaling change from $\sim N^2$ to $\sim N^3$? We can attack this calculation using the supergravity dual,[7] but the correct result also follows simply from assuming a second order phase transition and equating the two free energies at the critical point,

$$F \sim N^2T^5 \sim RN^3T^6 \tag{2}$$

from which we learn that the cross-over happens at the critical temperature

$$T \sim \frac{1}{NR} \sim \frac{1}{e^2N} \tag{3}$$

The factor of $1/N$ is all-important here. It tells us that, whatever the new degrees of freedom are, they come in at the energy scale below the Kaluza-Klein scale. Indeed, this had to be the case: the 5d theory becomes strongly coupled at the energy scale $E \sim 1/e^2N$ and it is here that new UV degrees of freedom are required to render the theory well-defined in the UV. This is $1/N^{\text{th}}$ the mass of an instanton. These are our conjectured partons.

There is more circumstantial evidence for the partonic nature of the instanton. There is a refinement in the count of the number of degrees of freedom that comes from looking at the anomaly coefficient. For $G = ADE$ gauge group, the coefficient is,[8,9]

$$c_2(G) \times |G| \tag{4}$$

where $c_2(G)$ is the dual Coxeter number (equal to $c_2(G) = N$ for $G = SU(N)$) while $|G|$ is the dimension of the gauge group (equal to $|G| = N(N-1)$ for $G = SU(N)$). For $G = SU(N)$, this reproduces the leading order scaling, but also gives the subleading contribution: $N^3 - N$. However, for arbitrary $G = ADE$ gauge group, this formula has an interesting interpretation in terms of the partonic nature of instantons for $G = ADE$ gauge group. This is because the dimension of the instanton moduli space is known to be $4c_2(G)$. If the anomaly coefficient is a good measure of the number degrees of freedom, it is suggesting that each parton itself transforms in the adjoint of the gauge group G. Making sense of this statement would go a long way towards understanding the parton picture of instantons.

Finally, the partonic interpretation of instantons naturally resolves a puzzle that arises upon quantization: the scaling mode of the instanton gives rise to a continuous energy spectrum above $M_{\rm inst}$. This is very peculiar behaviour for a one-particle state in quantum field theory. However, it is entirely natural for a multi-particle state.

The upshot of the above discussion is that the partonic interpretation of instantons would explain several known features of the UV behaviour of 5d Yang-Mills theories. However, there are lots of questions that remain unanswered. Most pertinent among these is: why are the partons confined to live within the instanton? This is not confinement as we know it in QCD. The existence of the instanton moduli space means that the partons are individually free to wander $\mathbf{R}^4$. Nonetheless, there are interactions between them which mean that are denied an existence on their own. What is the cause of these interactions? And how can we understand the nature of the partons given that we only have access to the low-energy physics of Yang-Mills theories.

I have not been able to answer these questions in the context of Yang-Mills. Rather, in the rest of this talk I will retreat and discuss a simple toy model which contains many of the same problems, but in a context where we can completely understand the relevant physics. In fact, the toy-model is very familiar to everyone who got stuck with QCD and looked for inspiration in something simpler: it is the $\mathbf{CP}^{N-1}$ sigma-model. Since we are interested in an analogy for Yang-Mills theory in $d = 4+1$ dimensions, we will look at the sigma-model in $d = 2+1$ dimensions.

3. Three Dimensional Sigma Models

The toy model that we consider in a supersymmetric sigma model in $d = 2+1$ dimensions. Specifically, we will consider $\mathcal{N} = 4$ supersymmetry (which means eight supercharges). While the low-energy physics will be a sigma-model, in contrast to the case of Yang-Mills theory, we will also stipulate the UV completion: it is a gauge theory first discussed by Intriligator and Seiberg.[10] We first describe this gauge theory and then explain how the sigma model emerges in the infra-red. We construct the gauge theory out of vector multiplets and hypermultiplets. These contain the following fields:

- Vector multiplet: $V = (A_\mu, \phi_i, \text{fermions})$, ϕ_i are real scalars with $i = 1, 2, 3$.
- Hypermultiplet: $Q = (q, \tilde{q}, \text{fermions})$. q and $\tilde{q}$ are both complex scalars.

The gauge theory that we consider is $U(1)^N$ with N hypermultiplets. The charges of the matter multiplets are best summarized by the following quiver diagram in which each node corresponds to a gauge group and each link to a hypermultiplet.

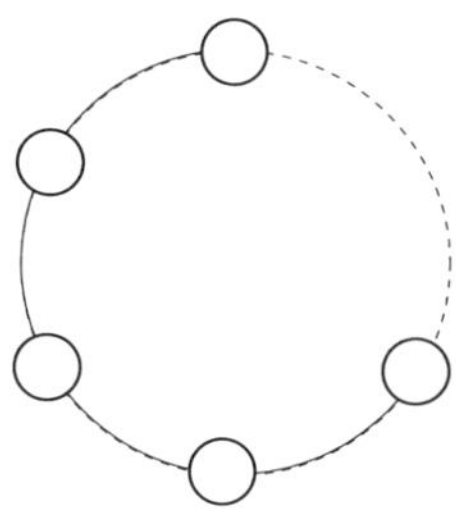

Fig. 1. The quiver diagram for the UV gauge theory.

We will need to introduce a couple of parameters for this theory. Each gauge group is assigned coupling constant g^2. Each matter multiplet is assigned bare mass m. However, the physical mass, M_a, $a = 1, \ldots, N$ of the Nth matter multiplet also depends on the expectation value of the scalars in the vector multiplets in the ath and $(a+1)^{\text{th}}$ gauge group,[a]

$$M^a = \phi^a - \phi^{a+1} + m \tag{5}$$

While the quiver gauge theory happily describes the UV physics, our real interest is in the infra-red. At low energies, we integrate out the massive hypermultiplets and seek an effective theory describing the massless vector multiplets: these are $3N$ scalar fields ϕ_i^a together with $3N$ photons, each of which can be dualised in favour of a periodic scalar field,

$$F_{\mu\nu}^a \sim g^2 \epsilon_{\mu\nu\rho}\partial_\rho \sigma^a \tag{6}$$

Integrating out the hypermultiplets induces derivative interactions between the vector multiplet fields. The resulting physics is a low-energy sigma-model on the Coulomb branch with target space [10]

$$\mathbf{R}^3 \times \mathbf{S}^1 \times T^\star \mathbf{CP}^{N-1} \tag{7}$$

The fields ϕ_i^a and σ^a provide coordinates on this space. We won't go through the derivation of this sigma-model, but it will be useful to sketch how it works in the simple example of $N = 2$. In this case, the quiver has just two nodes. The diagonal $U(1)$ gauge field has nothing charged under it and decouples (it will give rise to the $\mathbf{R}^3 \times \mathbf{S}^1$ factor of the target space). Meanwhile, the axial $U(1)$ couples to two hypermultiplets with charge $+1$ and -1. The discussion is simplest if we ignore two of the scalars in the vector multiplet (these will give rise to the cotangent bundle $T^\star$) and focus just on a single scalar ϕ and the dual photon σ. The low-energy dynamics for these fields is given by

$$\mathcal{L} = \frac{1}{g_{\text{eff}}^2}(\partial\phi)^2 + g_{\text{eff}}^2(\partial\sigma)^2 \tag{8}$$

where the effective coupling constant gets one-loop contributions from each of the hypermultiplets,

$$\frac{1}{g_{\text{eff}}^2} = \frac{1}{g^2} + \frac{1}{m-\phi} + \frac{1}{m+\phi} \tag{9}$$

The target space for $|\phi| < |m|$ is drawn in the figure below. Topologically, it is a sphere $\mathbf{CP}^1$. For finite g^2, the space has the shape of a rugby ball: it has a squashed metric with only $U(1)$ isometry. In the limit $g^2 \to \infty$,

[a] This expression actually describes a triplet of masses for each hypermultplet, corresponding to the triplet of scalars in a vector multiplet. For details of how this works, and why it's not important for the following discussion, see [1].

the target space becomes round, like a football,[b] and the isometry of the metric is enhanced to $SU(2)$.

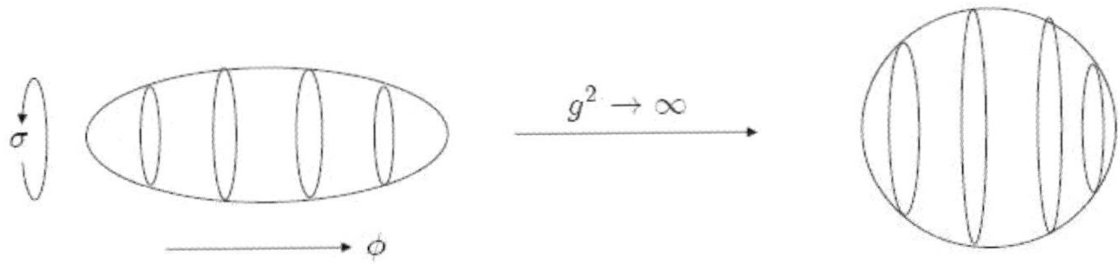

Fig. 2. The target space for the low-energy sigma model.

The Soliton

The low-energy sigma-model has a soliton. It is the sigma-model lump, a.k.a. the sigma-model instanton that we discussed in the introduction. In $d = 2 + 1$ dimensions, this is a particle-like state. For target space $\mathbf{CP}^1$ (either with a squashed or round metric), the spatial variation of the fields is governed by the first-order Bogomolnyi equation

$$\partial_\mu \phi = g^2_{\text{eff}} \epsilon_{\mu\nu} \partial_\nu \sigma \tag{10}$$

Similar first order equations hold for the soliton in $\mathbf{CP}^{N-1}$.

What does this soliton correspond to in the microscopic theory? There are a few clues. Firstly, it is a BPS state. Secondly, its mass is Nm. In fact, this is enough to identify the microscopic origin of this soliton: it is an N-particle state constructed from taking a hypermultiplet field from each link of the quiver ring [11]: $Q_1 Q_2 \ldots Q_N$. Now we see the importance of this three-dimensional example. It is an explicit model in which the partonic nature of the soliton is realised. A single soliton in the low-energy theory is indeed interpreted as a multi-particle state in the UV. Our goal now is to ask the question in reverse: given only access to the IR physics, what can we learn about the UV physics by studying the soliton? Of course, hindsight is a wonderful thing and we intend to employ it to its full extent in our study. Nonetheless, the answer is rather surprising. By studying the properties of the soliton, we will be able to reconstruct the full UV physics. In particular,

---

[b] This sporting analogy is to be taken in the European sense.

this means resolving the quantum numbers of the partons that lurk inside the soliton. In the rest of this talk, we will see how this comes about.

The Partons

Let's first ask how we can see the partons inside the soliton in this model more explicitly. As we mentioned in the introduction, the counting of collective coordinates is certainly suggestive of an interpretation in terms of the positions of N objects on the plane. However, if we simply plot the energy configuration of a single soliton, it just looks like a round blob with no hint of any internal structure. How do we reconcile these statements?

In fact, the UV theory is already telling us the right place to look. Recall that the round Fubini-Study metric on $\mathbf{CP}^{N-1}$ only arises in the limit $g^2 \to \infty$. If, instead, we study the sigma-model at finite g^2, then the metric is squashed with only a $U(1)^{N-1}$ isometry. The full $2N$ collective coordinates of a single soliton survive at finite g^2 (they are protected by index theorems), but they are no longer Goldstone modes. They now explicitly determine the positions of the partons. For example, the figure below shows the energy profile for a single soliton for a $\mathbf{CP}^1$ target space. We see that the two partons dramatically reveal themselves as the target space is squashed.

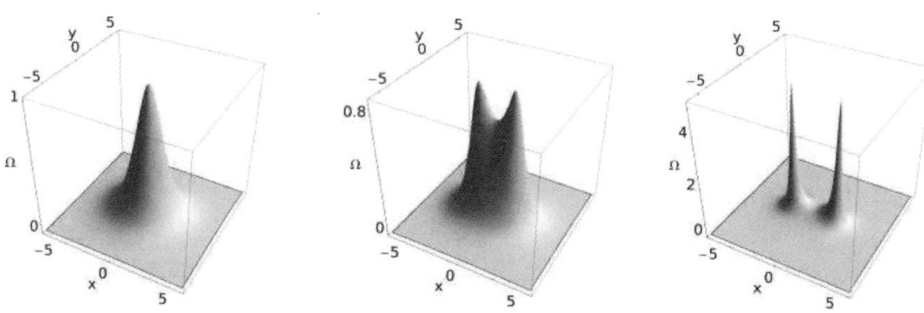

Fig. 3. The two partons inside a $\mathbf{CP}^1$ lump for $m/g^2 = 0$, 1 and 2.

A similar phenomenon happens for target space $\mathbf{CP}^{N-1}$. Here is the example of $\mathbf{CP}^2$, where the single soliton decomposes into three partons. A similar mechanism to see the partons was described in the talk at Shifmania by Ken Konishi.[12]

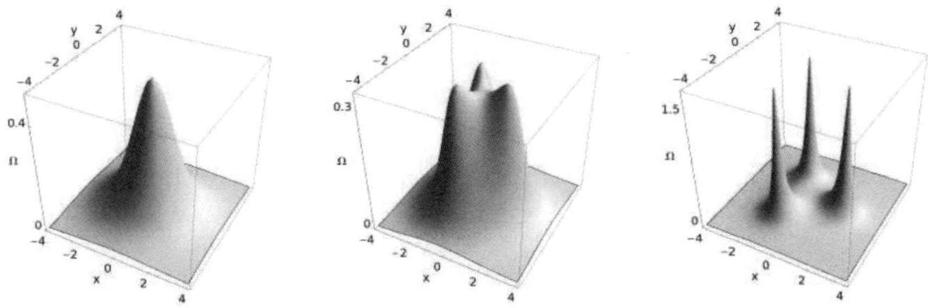

Fig. 4. The three partons inside a $\mathbf{CP}^2$ lump as the target space is squashed.

We can also see explicitly that the collective coordinates change the positions of the partons. For example, we could keep the "scale size" fixed and change the "orientation" modes of the soliton. For a round target space, this would leave the energy profile unchanged. However, with the squashed target space, the orientation modes govern the relative positions of the partons. For example, here's some plots showing the single lump in $\mathbf{CP}^2$ with different values of the collective coordinates.

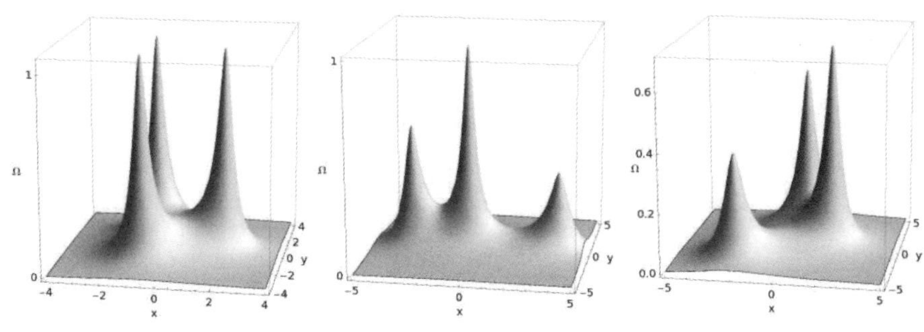

Fig. 5. Moving the partons inside a $\mathbf{CP}^2$ lump.

Let's now turn to the question of confinement. Why must the partons sit together inside a lump even though they are free to roam? The answer from the microscopic theory is simple: any electrically charged state gives rise to an electric field which asymptotically goes like $E \sim 1/r$. This gives rise to a logarithmically divergent contribution to the energy. Only gauge singlet states have finite mass, and this is the reason that the partons are bound together in the state $Q_1, \ldots Q_N$. Moreover, this log-divergence re-appears in the low-energy effective theory once we ask the partons to move: it is seen as a log-divergence in the metric on the soliton moduli space.

We have seen above that squashing the sigma-model allows us to graphically see the partons that sit inside the instanton. But how do we see their quantum numbers? Here we explain this for solitons in $\mathbf{CP}^1$. The Bogomolnyi equation for the soliton is given in (10). The k soliton has a moduli space of solutions of dimension $2k$. Using the duality transformation (6), we can rewrite this equation as

$$F_{0\mu} = \partial_\mu \phi \tag{11}$$

These are the dual Bogomolnyi equations. In contrast to (10), they have *no* smooth solutions. But this is entirely expected: after a duality transformation, solitons become fundamental excitations. These should not be associated to smooth solutions, but rather to solutions of equations with sources.

$$\partial_\mu \left(\frac{1}{g^2_{\text{eff}}} F_{0\mu} \right) = \sum_{n=1}^{k} \delta(z - z_n^+) - \delta(z - z_n^-) \tag{12}$$

One can show that solutions to (12) and (11) coincide with smooth solutions to (10). The positions of the sources $z_n^\pm$ become coordinates on the moduli space of the soliton. This provide a very simple and explicit map between fundamental excitations and solitons in a field theory.

The construction of the dual Bogomolnyi equation also works for solitons in $\mathbf{CP}^{N-1}$. It allows us to determine the quantum numbers of the partons. In this way, we can reconstruct the quiver diagram of the UV theory, details which one might have reasonably expected were lost to the winds of the renormalization group by the time we restriced ourselves to the low-energy physics. Details of this can be found in the longer paper [1].

4. Summary: Questions, not Answers

Our toy model in $d = 2 + 1$ dimensions did all that we hoped. It provides an explicit setting where the single soliton has the interpretation of an N-particle state. The partons inside the soliton are identified with the degrees of freedom necessary to form a UV completion of the sigma-model. Moreover, by studying the properties of this soliton we were able to reconstruct the quantum numbers of these partons and therefore the UV physics.

All of this is heartening. However, the real question remains: can we do the same for Yang-Mills instantons in $d = 4 + 1$ dimensions? What is the confinement mechanism that keeps these partons trapped inside an instanton and what is this telling us about the microscopic dynamics of the $(2, 0)$ theory? I don't yet have answers to these questions but I hope that further study of the Yang-Mills instanton will provide some vital clues.

References

1. B. Collie and D. Tong "*The Partonic Nature of Instantons*," arXiv:0905.2267 [hep-th].
2. A. A. Belavin, V. A. Fateev, A. S. Schwarz and Yu. S. Tyupkin, "*Quantum Fluctuations Of Multi-Instanton Solutions*," Phys. Lett. B **83**, 317 (1979).
3. A. R. Zhitnitsky, "*Confinement-deconfinement phase transition and fractional instanton quarks in dense matter*," arXiv:hep-ph/0601057.
4. N. Seiberg, "*Five dimensional SUSY field theories, non-trivial fixed points and string dynamics*," Phys. Lett. B **388**, 753 (1996) [arXiv:hep-th/9608111].
5. M. Rozali, "*Matrix theory and U-duality in seven dimensions*," Phys. Lett. B **400**, 260 (1997) [arXiv:hep-th/9702136].
6. I. R. Klebanov and A. A. Tseytlin, "*Entropy of Near-Extremal Black p-branes*," Nucl. Phys. B **475**, 164 (1996) [arXiv:hep-th/9604089].
7. N. Itzhaki, J. M. Maldacena, J. Sonnenschein and S. Yankielowicz, "*Supergravity and the large N limit of theories with sixteen supercharges*," Phys. Rev. D **58**, 046004 (1998) [arXiv:hep-th/9802042].
8. J. A. Harvey, R. Minasian and G. W. Moore, "*Non-abelian tensor-multiplet anomalies*," JHEP **9809**, 004 (1998) [arXiv:hep-th/9808060].
9. K. A. Intriligator, "*Anomaly matching and a Hopf-Wess-Zumino term in 6d, N = (2,0) field theories*," Nucl. Phys. B **581**, 257 (2000) [arXiv:hep-th/0001205].
10. K. A. Intriligator and N. Seiberg, "*Mirror symmetry in three dimensional gauge theories*," Phys. Lett. B **387**, 513 (1996) [arXiv:hep-th/9607207].
11. O. Aharony, A. Hanany, K. A. Intriligator, N. Seiberg and M. J. Strassler, "*Aspects of N = 2 supersymmetric gauge theories in three dimensions*," Nucl. Phys. B **499**, 67 (1997) [arXiv:hep-th/9703110].
12. M. Eto, T. Fujimori, S.B. Gudnason, K. Konishi, K. Ohashi, T. Nagashima, M. Nitta and W. Vinci, "*Fractional Vortices and Lumps*", arXiv:0905.3540 [hep-th].

Pandora's Box and Non-Selfdual Topological Excitations

Mithat Ünsal

SLAC and Physics Department, Stanford University, Stanford, CA 94025/94305, USA
E-mail: unsal@slac.stanford.edu

In the last few years, we have realized the existence of a new class of topological excitations, which are rather distinct from the platonic world of monopoles, monopole-instantons and instantons. All of the latter arise as solutions of the Prasad-Sommerfield type first order differential (self-duality) equations and have been extensively discussed in the context of confinement and chiral symmetry breaking for the last 30 years. However, new calculable deformations of asymptotically free chiral and vector-like gauge theories give us a new picture of these physical phenomena. Most often, the excitations which lead to confinement are not solutions to PS-type equations, they are non-selfdual and they are often bizarre. They are referred to as magnetic bions, triplets, and quintets, due to their composite nature. Bizarre as they are, combined with large-N volume independence, these novel non-self-dual excitations may also provide hope that at least some non-abelian gauge theories may be solvable.

Dedicated to Misha Shifman on the occasion of his sixtieth birthday. This write-up summarizes my interactions with him during our fruitful collaborations and important open problems in gauge theories.

The real voyage of discovery consists not in seeking new landscapes but in having new eyes. Marcel Proust.

1. Selfduality and non-selfduality: Tip and the iceberg?

A metaphor, even an imperfect one as described below, may often provide a useful insight into challenging situations. The story of Pandora's box (jar) is well-known. In its most common interpretation, Pandora opens the box due to curiosity and many evils, along with hope spread to earth. As this is an article not on moral but non-abelian gauge theories, the reader should replace the word evil with mystery or weirdness in what follows. (Although I must admit that gauge theories often have diabolical recipes for us.) Hope may remain the same, due to reasons which will become apparent.

I started to think about topological excitations in asymptotically free non-abelian gauge theories in early 2007. A distinguished member of such gauge theories, QCD, is perhaps one of the most challenging part of nature. The non-perturbative aspects of these theories have always been a source of fascination, excitement and doubled-curiosities. As described by Sasha Polyakov in his memoirs, perhaps mathematically the most important objects in addressing the non-perturbative aspects of gauge theories are instantons and monopoles. These topological excitations are solutions to first order self-duality equations, such as $F_{\mu\nu} = \star F_{\mu\nu} \equiv \frac{1}{2}\epsilon_{\mu\nu\rho\sigma}F_{\rho\sigma}$ for the BPST instanton. Similarly, the monopoles or instanton-monopoles are solutions to the first order Prasad-Sommerfield type differential equations. When I started to think about topological excitations in gauge theories, my hunch was, there would in fact be more interesting and more important topological excitations pertinent to vector-like and chiral gauge theories. The problem that we were unable to see these excitations was tied with the absence of controllable (and calculable) deformations and the right tools.

Around that time, I realized that such tools may indeed exist, if theories are compactified on a spatial (non-thermal) circle of size L. Due to the work of Gross, Pisarski and Yaffe from early 80's, I already knew that studying thermal compactifications would be beating a dead horse, so I did not care about it. My intuition was based on the sharp differences of quantum versus thermal fluctuations and I hoped that a more hospitable territory accessible via circle compactification existed.

With some optimism, I decided to examine $SU(2)$ QCD(adj) (a decidedly low-N theory) by taking advantage of a very unique and welcoming feature: its unbroken center symmetry [a] even at arbitrarily small $S^1 \times \mathbb{R}^3$. By the time I decided to do so, I did not know the index theorem for topological excitations on this geometry, but knew the famous APS-index (which applies on $\mathbb{R}^4$), and also knew lesser-known Callias (which applies on $\mathbb{R}^3$) index theorems. Combining the two, I was able to guess the form of the leading monopole operators.[b] The results were rather surprising. Although

[a] The precursor of this idea was large N volume independence of QCD(adj). A special form of volume independence is the famous Eguchi-Kawai reduction proposed in 1982. The first four dimensional working example appeared 25 years after the birth of the idea in 2007 and, as the reader may guess, it is QCD(adj) with *periodic* spin connection for fermions, a manifestly non-thermal compactification.

[b] As it should be clear, this was an educated guess. Later in 2008-CAQCD held in Minnesota, Erich Poppitz asked me a question about confinement in some gauge theories of his interest. My answer was, "if you tell me the index theorem for topological excitations, I will tell you what the answer is . . ." I will return to this later.

monopoles were there and they were the leading topological excitations in the semi-classical expansion, they were *not* responsible for the mass gap for gauge fluctuations and confinement per se, which is counter to the mid-70's dream of confinement *à la* Polyakov, 't Hooft, Mandelstam and others. I realized that I had stumbled upon a new class of topological excitations, but I did not first appreciate how general it could be. This realization was going to come from a phone call from Misha, to which I will return in the next section. First, I need to explain few aspects of these elusive topological excitations that revealed their existence in QCD(adj).

Let me briefly remind the reader the pair of relevant quantum numbers, useful in discussing topological excitations: magnetic charge and topological charge; $(\int F, \int F\widetilde{F})$. On $\mathbb{R}^3 \times S^1$, there are two types of monopoles, BPS and KK, as opposed to one (just BPS) as would be the case in the Polyakov model on $\mathbb{R}^3$. (This crucial aspect was realized in 1997 by Lee and Yi by using D-branes and Kraan and van Baal in gauge theory.) The existence of KK monopoles is due to compact topology of the holonomy. The charges of these excitations are BPS:$(1, \frac{1}{2})$, KK:$(-1, \frac{1}{2})$ with negations for the anti-monopoles: $\overline{\text{BPS}} : (-1, -\frac{1}{2})$, $\overline{\text{KK}} : (1, -\frac{1}{2})$. By using abelian duality, index theorem, and symmetries, I was able to show that the topological excitations which lead to confinement in this theory have charges $(2, 0)$ and $(-2, 0)$. Those familiar with topological excitations may find this strange for multiple-reasons: **(a)**Unlike instantons or monopoles, these new objects have vanishing topological charge. In this sense, they are indistinguishable from perturbative vacuum! **(b)** They are not solutions to Prasad-Sommerfield type equations and hence they are manifestly non-self-dual.

I referred to this topological excitation as *magnetic bion* as it is a composite, bound-state of BPS and $\overline{\text{KK}}$ monopoles, shown in Fig. 1. There is a puzzle here though: The constituents (each of which has charge $+1$) interact *repulsively* with a $1/r$ potential induced by photon exchange and there are no other apparent force carriers in the system. Perhaps, the biggest surprise that I came to grips with in this theory was that the bions were stable due to a *fermionic pairing mechanism*, induced by fermion zero-mode exchange. This exchange generates a $2n_f \log r$ attraction, which means a combined potential, $V_{\text{BPS}-\overline{\text{KK}}}(r) = \frac{1}{r} + 2n_f \log r$. This attraction implies the dynamical stability of magnetic bions.[c] The long-distance dual theory made it mani-

[c] Later, Larry Yaffe pointed out to me that similar fermion zero mode induced pairing ($\sim n_f \log r$ attraction) of topological excitations, in an instanton-anti-instanton molecule, was found earlier by Callan, Dashen and Gross in the mid 70's. Apart from this similarity, there are no others between these composites. Bions, in particular, have a net magnetic

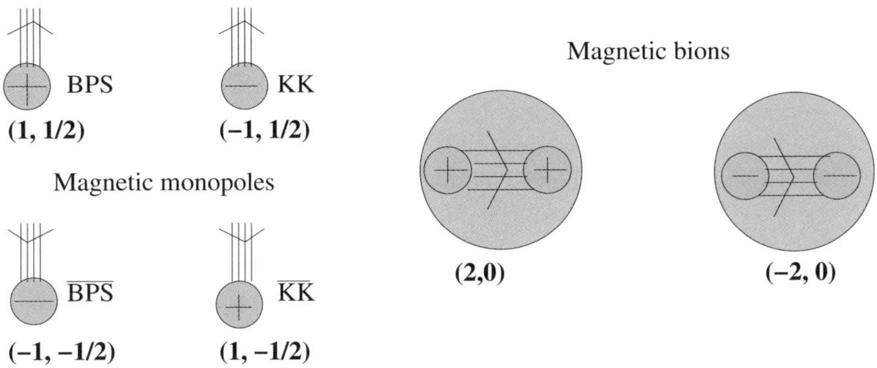

Fig. 1. Cartoon of the leading topological excitations in $SU(2)$ QCD(adj) with $n_f = 2$. The BPS and KK operators are $e^{i\sigma}\psi^4$ and $e^{-i\sigma}\psi^4$, where σ is dual photon and ψ is fermion zero mode. Flavor indices are suppressed. The product BPSKK is the usual instanton operator, ψ^8, with eight zero modes as per APS-index theorem. Because of the fermion zero modes, monopole operators cannot generate a mass gap for gauge fluctuations and confinement. Magnetic bions carry non-vanishing magnetic but zero topological charge. This corresponds to the $e^{2i\sigma} + e^{-2i\sigma}$ operator, which is the root cause of the mass gap for gauge fluctuations and confinement.

fest that the magnetic bions were the root cause of confinement and the mass gap in gauge sector of QCD(adj).

1.1. *Misha, smoothness conjecture and bions (almost) everywhere!*

The day after I posted the magnetic bion paper, I received an early morning call from Misha. Misha was pleased with the very simple fact that these excitations were *there*, but he was *not* surprised that their existence were not realized earlier. He tied the elusiveness of these objects to their indistinguishability from the perturbative vacuum in the sense of topological charge. Of course, no sensible gauge theorist with a sound mind would go out to search for topological excitations with $\int F\tilde{F} = 0$ property! To my view, we were simply lucky that these non-self-dual excitations arose so naturally in QCD(adj). Misha has his very unique perspective and deep intuition on strong coupling gauge dynamics, and he was telling me something that I may have not appreciated/pursued otherwise: His main point

charge and generate a mass gap in gauge sector. On the other hand, instanton-anti-instanton molecules are not of any particular importance for confinement.

was inspired from the large-N (planar) orbifold/orientifold equivalences. The essence of his argument was following:

> "The magnetic bions are surely there in the weak coupling—small L—domain; presumably, they must also exist in the large L domain, but in a non-dilute form. At large L, there is an exact non-perturbative equivalence between QCD(adj) and QCD(AS/S/BF).[d] Thus, the magnetic bions must also exist in these theories. How can we make them visible? "

Implicit to Misha's question was the staggering dissimilarity between QCD(adj) and QCD(AS/S/BF) in the small L domain. All of the latter theories exhibited breaking of their center at small L and were decidedly unlike QCD(adj). We needed to come-up with something which would disclose the semi-classical domain of these theories.

Incidentally, I was thinking on a proposal which makes large N volume independence work all the way down to arbitrarily small $L > 0$ for YM theory, in collaboration with Larry Yaffe. Motivated by adjoint fermion induced stabilization of center symmetry, we had the idea of using the double-trace deformations of YM theory to satisfy volume independence. This turns out to be a very interesting trick due to what I will call,

First magic of double-trace deformation: At large N, double-trace deformations of the action are an order N^2 effect and prevent center–breaking of YM theory on small S^1. On the other hand, the observables of YM theory (in the confined phase) and the ones of deformed YM theory for any $L > 0$ differ only by $O(1/N^2)$ effects in the large N limit. In a sense, the deformation, after doing the good deed, sequesters itself.[e]

One should note that this is a very peculiar deformation, as typical $O(N^2)$ deformations of the action would alter the dynamics of the theory completely. Misha wanted to know about the small N and small L incarnations of such deformations. We thought for few days and had doubts about pursuing that direction: Our concern was the apparent non-locality of the Wilson line deformation. But we realized that this was an aesthetic issue. In fact, such deformed theories have nice renormalizable UV completions which yield our deformed action upon integrating out heavy KK modes

[d] These are YM theories with n_f anti-symmetric, symmetric or bi-fundamental representation Dirac fermions. QCD(AS) is same as QCD(F) for $N = 3$, thus a natural large N limit thereof.

[e] The last sentence is coined by Gabriele Veneziano in a private conversation at GGI, Florence.

in the small circle limit, and even without such, there is no harm to view these small $S^1 \times \mathbb{R}^3$ theories as effective 3d theories. This leads to the second magic of double-trace deformations.

Smoothness conjecture: At small or finite N, a deformation could be "engineered" in such a way that a sub-class of QCD-like theories on $\mathbb{R}^4$ could be smoothly connected to the deformed-QCD (QCD*) on arbitrarily small $S^1 \times \mathbb{R}^3$.

Needless to say, this is what we were looking for despite the aesthetic issues. Regardless, this was a useful tool enabling us to move forward and identify interesting topological excitations. In the mean-time, we also learned about the existence of a very interesting lattice gauge theory work by Myers and Ogilvie. These authors have independently proposed deformations to study phases of partial center symmetry breaking and realized that, in a lattice formulation, the center symmetric phase can be continued to arbitrarily small S^1. Hence, with this non-perturbative confirmation, we decided to move on. This latter part is the main benefit of the "deformation program": It helps us to reveal the topological excitations in gauge dynamics.

Immediately, we were able to solve all interesting one-flavor QCD* theories on small $S^1 \times \mathbb{R}^3$, and analytically demonstrate the existence of a mass gap, confinement and discrete chiral symmetry breaking. This was done by pin-pointing the non-perturbative source of each effect. Non-perturbative aspects which are quite hard in the original theory become straightforward tree-level phenomena upon a duality transform. Magnetic bions are ubiquitous, bearing responsibility for the mass gap, generally speaking, in almost all QCD-like theories with 2-index representation fermions.

1.2. *One cannot overstate the importance of the index theorem*

Despite these solutions, neither Misha nor myself knew the exact index theorem for topological excitations on $S^1 \times \mathbb{R}^3$, but we were able to proceed by using intuition based on the Callias and APS index theorems and, more often than not, bions were responsible for mass gap generation. The exact form of the index theorem is crucial to reveal the origin of confinement, which depends on the matter content of the theory! (This is again an unexpected aspect that we learned recently.)

There were a number of interesting gauge theories that Misha and I were unable to address, and this was due to the absence of a useful form of index theorem. I found a mathematics paper in the Differential Geometry arXiv

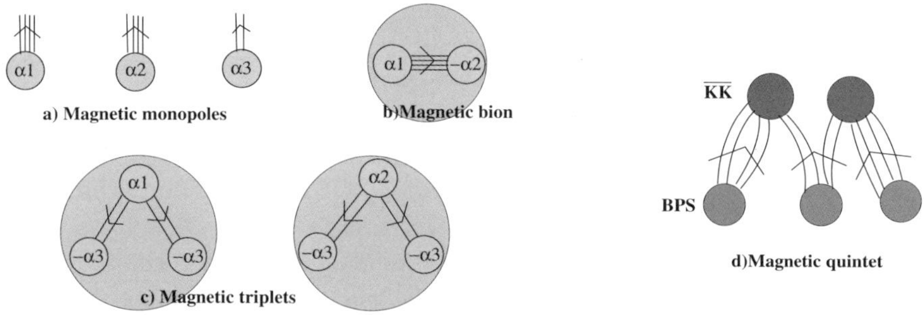

Fig. 2. $SU(3)$ QCD(S): a)monopoles with indices $(I_1, I_2, I_3) = (4, 4, 2)$ do not induce a mass gap in the gauge sector because of their zero-modes. Note that $I_{\text{inst}} = \sum_{i=1}^{3} I_i = 10$. b) magnetic bions and c) magnetic triplets generate a mass gap for gauge fluctuations. Note that all zero-modes need to be soaked up for the composite to contribute to a mass gap for gauge fluctuations. d) $SU(2)$ chiral theory with $I = 3/2$ representation fermion and a cartoon of a magnetic quintet. $(I_{\text{BPS}}, I_{\text{KK}}) = (4, 6)$. Note $I_{\text{inst.}} = \sum_{i=1}^{2} I_i = 10$. The magnetic quintet is a composite of 3 BPS and 2 $\overline{\text{KK}}$ monopoles. In the absence of index theorem on $\mathbb{R}^3 \times S^1$, it would perhaps be very hard, if not impossible, to guess these purely bosonic topological excitations, which generate confinement. The magnetic quintet is flipped from its original figure $\mathcal{W}$ to an $\mathcal{M}$(isha) for the occasion.

by Nye and Singer, which addressed an index theorem for Dirac operator on $S^1 \times \mathbb{R}^3$. Both Misha and I highly appreciated this paper and advertised this work in our paper. However, it was very hard for me to get a useful form of the index formula which we could use for concrete gauge theory applications. Eventually, Erich Poppitz and I (but mostly him) provided a new, insightful and elegant derivation of the index formula, by using axial current non-conservation—an exact operator identity valid on any four-manifold—and a center-symmetric holonomy. We were in part being influenced by E. Weinberg's work on the Callias index theorem on $\mathbb{R}^3$. The derivation also provided a useful and simple form of the index formula which is easy to use for concrete (both supersymmetric and non-supersymmetric) gauge theory applications.

In order for the reader to appreciate how weird things can get regarding confinement, and why the index formula is so important, consider two gauge theories: Vector-like $SU(3)$ QCD(S) and chiral $SU(2)$ with an $I = 3/2$ Weyl fermion. I picked these two examples, because both theories have $I_{\text{inst.}} = 10$ zero modes in the background of an instanton. The topological excitations, zero mode structures and the non-selfdual excitations which generate the mass gaps are shown in Fig. 2. This figure is the reason why knowing the index theorem is so important in our understanding of gauge theories. The actions of various topological excitations, in a center-symmetric background, relative to an instanton are given by:

$$\{S_{\rm mon},\ S_{\rm bion},\ S_{\rm triplet},\ S_{\rm quintet}\} = \left\{\frac{1}{N},\frac{2}{N},\frac{3}{N},\frac{5}{N}\right\} \times \left(S_{\rm inst} = \frac{8\pi^2}{g^2}\right).$$

Moreover, since $e^{-S_{\rm inst}} \sim e^{-N}$ and $e^{-S_{\rm mon/bion}} \sim e^{-1}$ in N-counting, the leading (composite) topological excitations, unlike instantons, are not suppressed in the large-N limit.

1.3. *Chiral gauge theories and deformation program*

Around this time, I had another phone conversation with Misha. I was thinking of pursuing the deformation program for multi-flavor QCD-like theories due to an exotic phenomena I had in mind: I wanted to show that QCD-like theories may generically possess phases of confinement without chiral symmetry breaking on $S^1 \times \mathbb{R}^3$, a phenomena first realized in QCD(adj). But this time, Misha had something different in his mind: Chiral gauge theories. On the phone, I remember him saying that "this is an uncharted territory" despite chiral theories being "familiar" for almost three decades. Lattice does not work, string theory is not useful so far and large N techniques are not useful either. The only known thing is the 't Hooft anomaly matching, while being one of the most powerful results in gauge theories, is not a dynamical framework. (In fact, there are examples of misleading anomaly matchings, that is to say, despite the matching of UV anomalies by some IR "composites", there are cases in which such composites do not describe the IR-physics.)

In a short time, Misha and I studied two main classes of chiral gauge theories by using our deformation program: Type I, chiral quiver theories and Type II, $SU(N)$ chiral theory with one AS , and $N-4$ anti-fundamental representation left-handed Weyl fermions. Currently, deformation theory is the only microscopic framework to understand the dynamics of these gauge theories.

Because of center-symmetric holonomy and abelianization, one expects topologically stable monopoles. We discovered a very strange phenomena (which never occurs in QCD-like theories). The monopole operator contributions to the non-perturbative dynamics of chiral gauge theories vanish identically![f] Upon averaging over all zero modes, in particular the global

[f] In particular, for type II theories, monopole operators are almost all fermionic and the index is in general odd(!), similar to the number of instanton zero modes in an $SU(2)$

$U(1)$ angles, all monopole operators drop out. On the other hand, one can show that new topological excitations with more involved fermion zero-mode structure (monopole-ring operators) and bions are still there. They govern, respectively, chiral symmetry realization and mass gap for gauge fluctuations.

1.4. *La Belle Époque, the supersymmetric era and today*

I think that our understanding of non-perturbative aspect of non-supersymmetric gauge theories has seen a drastic change in the last two years in an unanticipated way. It is unanticipated because earlier solutions/scenarios of confinement generally relied on self-dual topological excitations. Our solutions for vector-like and chiral theories indicate that it is often non-self-dual topological excitations which generate a mass gap in the gauge sector and confinement.

To put things into context, I would like to tell my opinions regarding past works on confinement in non-abelian gauge theories. Of course, *la belle époque* of this problem is the mid-70s to early 80s. To me, the most important solution of that era is the one of Polyakov, where he mapped the partition function of a gauge theory to a dilute gas of self-dual monopoles (3d -instantons). The dual superconductivity scenario of 't Hooft and Mandelstam was also born during that era, but had to wait almost two decades for an explicit construction, eventually being realized in the very elegant solution of Seiberg-Witten, in '94, in the supersymmetric context. In both cases, the self-dual (BPS) objects play crucial roles.

When I realized the existence of magnetic bions, I did not think that they could be this generic in gauge theories. Misha's intuition, the use of double-trace deformations and the relevant index theorem changed my opinion. Now, I think that the magnetic bions and similar non-selfdual topological excitations are very likely the most important topological configurations in typical gauge theories. These objects are rather messy, mysterious and often bizarre, as shown in Fig. 1 and Fig. 2.[g] But now, we have the correct mathematical tools to study them rather precisely, on $\mathbb{R}^3 \times S^1$, where the theory is locally four dimensional. The world of these new non-selfdual

gauge theory with an odd-number of doublets. The latter class of theories, as shown by Witten, have a global anomaly and do not exist as consistent quantum theories. In our case, there are no such global anomalies and these are fine quantum theories.

[g] It would be interesting to see whether the non-self-dual topological excitations such as magnetic bions, triplets, etc. have a counter-part in string theory, and whether they could be useful there as well.

topological excitations is almost contrary to the platonic world of self-dual monopoles, instantons, and the dreams of the past. Returning to the title of this work, I think the Pandora's box for QCD and chiral theories is now open, or at least ajar, for the first time, so that we can have a glimpse of gauge theories' inner goings-on. Many interesting things just happen to be in there and I am sure we will see more. I also think that the combination of these novel topological excitations and another unexploited concept from the past, the large-N volume independence, carry hope that we can come to a fuller understanding of most of these gauge theories.

This is, of course, an idiosyncratic point of view. Many people (including me) today hope that, gauge/string duality may provide useful insights to gauge dynamics in four dimensions. Perhaps important problems like conformality, confinement, and the mass gap in YM theories with or without fermionic matter fields will be understood by geometrization. Some prominent string theorists even argue that this is the best analytic approach at hand now, but I disagree with these claims for good reasons. When I started to think about topological excitations in gauge theories, the reason I did not pursue a stringy approach was due to my belief that one could construct new microscopic field theory techniques which could be more beneficial than what string theory can do under similar circumstances.[h] I think that a few people, including Misha and me, partially achieved this goal by revealing the existence of new non-selfdual topological excitations and their role in confinement.

1.5. *References and suggested reading*

Since this short note is written as a scientific memoir of my interactions with Misha and describes our program based on recent papers, I did not attempt

[h] Polyakov mentions in his scientific memoirs that the reason he decided to move from semi-classical techniques to the string description was the belief that the semi-classical approach cannot fully describe the strong coupling regime of QCD. I think this is most likely true. But the caveat here is that all known string theory descriptions of QCD-like theories either have non-decoupling states which prevent them from reducing to pure large-N QCD, or they are phenomenological (bottom-up) in their nature. The latter approach is not useful for the type of questions that I find interesting in gauge theories. The first may be fine modulo its caveats. In my view, the importance of semi-classical techniques is no less than any string theory approach, perhaps more. They are microscopic, can address finite N, and have taught us about the existence of new mechanisms of confinement, conformality, non-self-dual topological excitations and new scales in QCD-like theories. This is a sufficiently good reason to argue that this approach merits more attention.

to give references. Instead, I will mention a few papers which influenced my thinking about gauge theories on $\mathbb{R}^3 \times S^1$, and a few references that may be useful to guide interested readers.

Ref. [1] : This is the first use of duality in a relativistic gauge theory, and establishes the role of monopoles in confinement.

Ref. [2] : Polyakov model with adjoint fermions; and a must-read complement of Ref. [1] . The authors relies more on symmetries than dynamical aspects. Consequently, physical interpretation of some topological excitations are missing. These are explained in Ref. [6].

Ref [3] : A beautiful field theory discussion of $\mathcal{N} = 1$ SYM on $S^1 \times \mathbb{R}^3$, in the sprit of the first two, and one of the papers that I like most. The gaps regarding the interpretation of confinement in $\mathcal{N} = 1$ SYM are filled in Ref. [6] .

Ref. [4,5] : Introduced KK-monopole, a self-dual monopole-instanton, and an essential ingredient of the magnetic bion mechanism.

Ref. [6] : This is the first analytic solution of a locally four dimensional non-abelian (non-supersymmetric) gauge theory on $S^1 \times \mathbb{R}^3$. It introduced magnetic bions, a new class of non-self-dual topological excitations. Its discussion about $\mathcal{N} = 1$ SYM should be viewed as a complement to Ref. [3] .

Ref. [7] : This index theorem is the most crucial result that will shape the understanding of gauge theories on $S^1 \times \mathbb{R}^3$. However, it is hard to read and the way the results are expressed are not particularly useful for gauge theory applications. The index theorem here is far more refined than the familiar Atiyah-Singer index theorem for instantons on $\mathbb{R}^4$. This paper is so far not appreciated in the physics literature.

Ref. [8] : Gives a new derivation of the index theorem by using an exact operator identity—axial-current non-conservation—valid on any four-manifold, and by paralleling E. Weinberg's discussion of the Callias index on $\mathbb{R}^3$. It also provides a calculation of the index in specific backgrounds and a discussion of its jumps, properties which are of interest for concrete gauge theory application.

Ref. [9, 10] : The first paper is where double-trace deformations are introduced to formulate the smoothness conjecture. This construction allowed us to see that magnetic bions were pertinent to most interesting gauge theories, chiral or vector-like. The use of the deformations and the index theorem essentially puts the understanding provided in non-susy QCD or even chiral theories more or less on the same level with SQCD of Ref. [13] .

Ref. [11] : Deformed YM theory can be used to study phases of partial center symmetry breaking on lattice, and a sufficiently large deformation stabilizes the center symmetry at any value of the bare lattice coupling. See the beautiful Fig. 1 therein.

Ref. [12] : Fully center-stabilized deformations of large-N YM theories obey volume independence.

Ref. [13] : For SQCD, the smoothness conjecture is exact. The vacuum of the theory on $S^1 \times \mathbb{R}^3$ knows about the vacuum structure on $\mathbb{R}^4$. This paper has many good examples of the derivation of non-perturbative superpotentials by using holomorphy and symmetry. Using the index theorem provides a microscopic derivation of some of these superpotentials.

References

1. A. M. Polyakov, "Quark Confinement And Topology Of Gauge Groups," Nucl. Phys. B **120**, 429 (1977).
2. I. Affleck, J. A. Harvey and E. Witten, "Instantons And (Super)Symmetry Breaking In (2+1)-Dimensions," Nucl. Phys. B **206**, 413 (1982).
3. N. M. Davies, T. J. Hollowood, V. V. Khoze and M. P. Mattis, "Gluino condensate and magnetic monopoles in supersymmetric gluodynamics," Nucl. Phys. B **559**, 123 (1999) [arXiv:hep-th/9905015].
4. K. M. Lee and P. Yi, "Monopoles and instantons on partially compactified D-branes," Phys. Rev. D **56**, 3711 (1997) [arXiv:hep-th/9702107].
5. T. C. Kraan and P. van Baal, "Periodic instantons with non-trivial holonomy," Nucl. Phys. B **533**, 627 (1998) [arXiv:hep-th/9805168].
6. M. Ünsal, "Magnetic bion condensation: A new mechanism of confinement and mass gap in four dimensions," Phys. Rev. D **80**, 065001 (2009) [arXiv:0709.3269 [hep-th]].
7. T. M. W. Nye and M. A. Singer, "An L^2-Index Theorem for Dirac Operators on $S^1 \times R^3$," arXiv:math/0009144.
8. E. Poppitz and M. Ünsal, "Index theorem for topological excitations on $R^3 * S^1$ and Chern-Simons theory," JHEP **0903**, 027 (2009) [arXiv:0812.2085 [hep-th]].
9. M. Shifman and M. Ünsal, "QCD-like Theories on $R_3 \times S_1$: a Smooth Journey from Small to Large $r(S_1)$ with Double-Trace Deformations," Phys. Rev. D **78**, 065004 (2008) [arXiv:0802.1232 [hep-th]].
10. M. Shifman and M. Ünsal, "On Yang-Mills Theories with Chiral Matter at Strong Coupling," Phys. Rev. D **79**, 105010 (2009) [arXiv:0808.2485 [hep-th]].
11. J. C. Myers and M. C. Ogilvie, "New Phases of SU(3) and SU(4) at Finite Temperature," Phys. Rev. D **77**, 125030 (2008) [arXiv:0707.1869 [hep-lat]].
12. M. Ünsal and L. G. Yaffe, "Center-stabilized Yang-Mills theory: confinement and large N volume independence," Phys. Rev. D **78**, 065035 (2008) [arXiv:0803.0344 [hep-th]].
13. O. Aharony, A. Hanany, K. A. Intriligator, N. Seiberg and M. J. Strassler, "Aspects of $\mathcal{N} = 2$ supersymmetric gauge theories in three dimensions," Nucl. Phys. B **499**, 67 (1997) [arXiv:hep-th/9703110].

Solitons in Two-Dimensional Anti-de Sitter space

Tonnis ter Veldhuis*

Department of Physics and Astronomy, Macalester College,
Saint Paul, MN 55105, USA
**E-mail: terveldhuis@macalester.edu*

Soliton solutions in a scalar field theory defined on an AdS_{1+1} background space-time are investigated. An analytic soliton solution is obtained in a polynomial model, and the classical soliton mass is calculated. The fluctuation spectrum around the soliton solution is determined, and the one-loop quantum correction to the soliton mass is computed in the semi-classical approximation.

Keywords: Soliton, kink, Anti-de Sitter space

1. Introduction

Solitons (kinks) are topologically stable localized solutions to the equation of motion with finite energy. They have been extensively studied in scalar field theories in two-dimensional Minkowski space.[1] The main objective of the present work, done in collaboration with Jennifer Barnes, Daniel Schroeder, Donald Spector, and Matthew Webster, is to extend well known results for solitons obtained in flat Minkowski space to two-dimensional Anti-de Sitter (AdS) space,[2] which has constant negative curvature.

To this end a simple scalar field theory in AdS_{1+1} is investigated. An analytic classical soliton solution is determined which spontaneously breaks the $SO(2,1)$ isometry group of the AdS space-time. In contrast to the situation in Minkowski space, the soliton excitation spectrum is found to be discrete. The frequency of the lowest excitation is equal to the inverse of the AdS radius. This lowest mode is obtained by acting with a broken $SO(2,1)$ symmetry generator on the static soliton solution. It is the AdS equivalent of the generic zero mode of solitons in Minkowski space related to the spontaneously broken translation symmetry. The energy momentum tensor of the model is constructed and employed to calculate the classical soliton mass. The theory is quantized by canonical quantization. It is then rendered finite by normal ordering of the Hamiltonian in the trivial sector.

This procedure fixes the finite parts of the counter terms. A mode number cut-off is employed as an ultra-violet regulator. This regulator breaks the $SO(2,1)$ symmetry and as a consequence a counter term is generated that is not invariant. The vacuum energies in the trivial and one-soliton sector are then calculated, and the quantum corrections to the soliton mass are obtained to one loop order in the semi-classical approximation.

2. Anti de Sitter space

AdS_{1+1} space can be viewed as an $SO(2,1)$ invariant hyperboloidal hypersurface

$$(X^0)^2 - (X^1)^2 + (X^2)^2 = 1/m^2, \tag{1}$$

embedded in a three-dimensional pseudo-Euclidean space with invariant interval

$$ds^2 = (dX^0)^2 - (dX^1)^2 + (dX^2)^2. \tag{2}$$

Here m parameterizes the inverse length scale in AdS_{1+1} space. Global coordinates can be defined as

$$\begin{aligned} X^0 &= 1/m \cos(mt) \cosh(mx) \\ X^1 &= 1/m \sinh(mx) \\ X^2 &= 1/m \sin(mt) \cosh(mx). \end{aligned} \tag{3}$$

The induced metric on the hypersurface in these coordinates is

$$ds^2 = g_{\mu\nu} dx^\mu dx^\nu = \cosh^2(mx) dt^2 - dx^2. \tag{4}$$

The hyperboloid surface is covered once by $mt \in [-\pi, \pi]$ and $mx \in [-\infty, \infty]$. The space has topology S(time)$\times$ R(space). In order to avoid closed time-like curves the universal covering space is considered by unwrapping the S, and the restriction on the range of the time coordinate is lifted. Boundaries are located at $x \to \pm\infty$. Well defined time evolution requires specification of consistent boundary conditions in addition to initial conditions at a Cauchy surface.[3]

2.1. *Geodesics*

Further insight into the nature of AdS space is obtained by consideration of geodesic trajectories in this space. In terms of the proper time these geodesic trajectories take the form

$$\sinh(mx) = \pm\sqrt{l^2 - 1} \sin[m(\tau - \tau_0)], \tag{5}$$

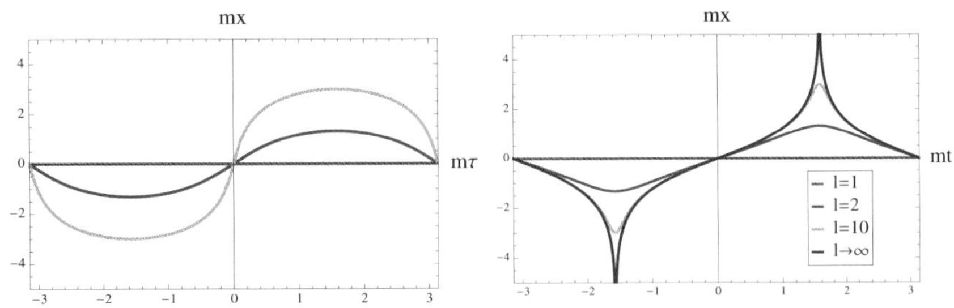

Fig. 1. Projections of geodesic trajectories in terms of proper time (left) and coordinate time (right).

while in terms of coordinate time they read

$$\sinh(mx) = \pm \frac{\sqrt{l^2 - 1}\sin[m(t - t_0)]}{\sqrt{l^2 \cos^2[m(t - t_0)] + \sin^2[m(t - t_0)]}}. \tag{6}$$

Here l and t_0 (τ_0) are integration constants. Null geodesics are obtained in the limit $l \to \infty$. Figure (1) shows how periodic geodesic trajectories start at one point, initially diverge and eventually converge again to a single point. Null geodesics are seen to reach the boundary of space-time in finite coordinate time.

3. Solitons in AdS space

In order for stable kink solutions to exist in a scalar field theory it must feature two degenerate discrete vacua. This situation naturally occurs in models with a spontaneously broken Z_2 symmetry. For simplicity, and to maintain a clear connection with well-studied solitons in Minkowski space, a "phi to the fourth" scalar field theory is considered with a negative mass term. The $SO(2,1)$ invariant Lagrangian density takes the form

$$\begin{aligned} \mathcal{L} &= \frac{1}{2} g^{\mu\nu} \partial_\mu \phi \partial_\nu \phi - V(\phi) \\ &= \frac{1}{2} \frac{1}{\cosh^2(mx)} \left(\frac{\partial \phi}{\partial t} \right)^2 - \frac{1}{2} \left(\frac{\partial \phi}{\partial x} \right)^2 - \frac{1}{2} (-\mu^2 + \lambda \phi^2)^2. \end{aligned} \tag{7}$$

In each of the two degenerate minima of the potential at $\phi_0 = \pm\mu/\sqrt{\lambda}$ the Z_2 symmetry $\phi \to -\phi$ is spontaneously broken. The maximum of the potential at $\phi = 0$ also corresponds to a perturbatively stable vacuum as long as the dimensionless ratio $\alpha = 2\lambda\mu^2/m^2$ is smaller than $1/4$. The

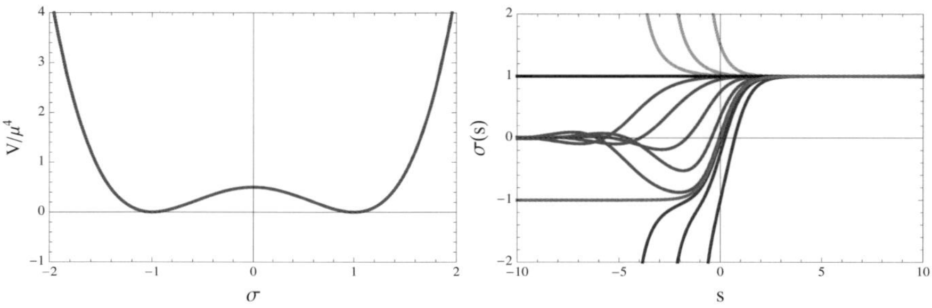

Fig. 2. The left panel shows the scalar potential with two degenerate minima for the case $\alpha = 1$. The right panel shows a family of static solutions to the equation of motion that satisfy the boundary condition $\sigma(s \to \infty) = 1$, also for $\alpha = 1$. Note that there exists only one solution which in addition satisfies the boundary condition $\sigma(s \to -\infty) = -1$.

invariant Euler-Lagrange equation of motion reads

$$\partial_\mu \left(\frac{\partial(\mathcal{L}\sqrt{-g})}{\partial(\partial_\mu \phi)} \right) - \frac{\partial(\mathcal{L}\sqrt{-g})}{\partial \phi} = 0. \tag{8}$$

In terms of the dimensionless variables

$$s \equiv xm, \qquad \tau \equiv tm \qquad \sigma \equiv \sqrt{\lambda}\frac{\phi}{\mu}, \tag{9}$$

the equation of motion takes the form

$$\frac{1}{\cosh^2(s)}\frac{\partial^2 \sigma}{\partial \tau^2} - \frac{\partial^2 \sigma}{\partial s^2} - \tanh(s)\frac{\partial \sigma}{\partial s} + \alpha\sigma(-1+\sigma^2) = 0. \tag{10}$$

The ratio α, the square of the mass scale of the field theory over the curvature of the space-time background, is identified as the physical parameter that controls the classical dynamics.

3.1. *Excitation spectrum in the trivial sector*

In order to find the excitation spectrum in the trivial (no-soliton) sector, the scalar field is expanded around one of the equivalent degenerate vacua as

$$\sigma(s, \tau) = 1 + \eta(s, \tau). \tag{11}$$

For small fluctuations only linear terms in η need to be considered. In this approximation the equation of motion becomes

$$\frac{1}{\cosh^2(s)}\frac{\partial^2 \eta}{\partial \tau^2} - \frac{\partial^2 \eta}{\partial s^2} - \tanh(s)\frac{\partial \eta}{\partial s} + 2\alpha\eta = 0. \tag{12}$$

This equation is solved by separation of variables. After making the Ansatz

$$\eta(s,\tau) = e^{i\hat{\omega}\tau} X(s), \tag{13}$$

the space dependent factor $X(s)$ of a normal mode is a solution to the equation

$$-\cosh^2(s)\frac{d^2X}{ds^2} - \sinh(s)\cosh(s)\frac{dX}{ds} + 2\alpha\cosh^2(s)X = \hat{\omega}^2 X. \tag{14}$$

This equation has the form of a time-independent Schrödinger equation, and familiar tools from quantum mechanics can be used to algebraically find the normalizable eigenfunctions and eigenvalues. In Appendix A supersymmetry and shape invariance are employed to find the spectrum of the equivalent quantum mechanical model. With the identification

$$\begin{aligned} \hat{\omega}_n^2 &= E_n^{(1)} + g_2^2 \\ g_2 &= \frac{1}{2} + \sqrt{\frac{1}{4} + 2\alpha}, \end{aligned} \tag{15}$$

where the $E_n^{(1)}$ are the energy eigenvalues of the quantum mechanical model and g_2 is a parameter that specifies its potential, the excitation spectrum of the scalar field theory in the trivial sector is determined to be

$$\hat{\omega}_n = n + \frac{1}{2} + \sqrt{\frac{1}{4} + 2\alpha}. \qquad n \in \mathbb{N}. \tag{16}$$

This equally spaced discrete spectrum is consistent with the unbroken $SO(2,1)$ symmetry.[4]

3.2. *Classical soliton solution*

A soliton at rest corresponds to a non-trivial static solution to the equation of motion. For the model defined in Eq. (7) this implies soliton solutions must satisfy the equation

$$-\frac{\partial^2\sigma}{\partial s^2} - \tanh(s)\frac{\partial\sigma}{\partial s} + \alpha\sigma(-1+\sigma^2) = 0. \tag{17}$$

Moreover, such solutions must interpolate between the two vacua at $\sigma_0 = \pm 1$ at $s \to \pm\infty$. A simple analytic solution of this type is found to exist for the specific value of the parameter $\alpha = 1$. It takes the form

$$\sigma_{\text{sol}}(s) = \tanh(s), \tag{18}$$

This solutions is depicted in the right panel of Fig.(2). Other static solutions to Eq. (17) that satisfy the boundary condition $\sigma(s \to \infty) = 1$ for $\alpha = 1$

are also shown. However, the solution specified in Eq. (18) is the unique one that interpolates between the two distinct minima of the potential for this value of α. This contrasts with the situation in Minkowski space, where one soliton solution can be translated over arbitrary distances to generate a one parameter family. The difference is a consequence of the fact that a static soliton solution is mapped onto a time dependent solution to the equation of motion under a broken $SO(2,1)$ transformation. As a result, there is no zero mode excitation for solitons in AdS space. Instead, the lowest excitation frequency of a soliton in AdS space is universally equal to m.

Soliton solutions exists for values of the parameter α other than $\alpha = 1$ as well. Simple analytic soliton solutions are found in the limits $\alpha \downarrow 0$ and $\alpha \to \infty$. Numerical soliton solutions for generic values of α can also be obtained. Such soliton solutions will be discussed in detail elsewhere. In what follows the analytic soliton solution obtained for $\alpha = 1$ and displayed in Eq. (18) will be further considered.

3.3. *Classical soliton mass*

The soliton mass is defined in terms of the conserved covariant energy momentum tensor given by

$$T_{\mu\nu} = \partial_\mu \phi \partial_\nu \phi - g_{\mu\nu}\mathcal{L}. \tag{19}$$

The classical energy functional is

$$E[\phi] = \int_{-\infty}^{\infty} dx \sqrt{-g}\mathcal{H} = \int_{-\infty}^{\infty} dx \sqrt{-g} T^0_{\ 0}. \tag{20}$$

The classical soliton mass is given by the difference in energy between the soliton and trivial vacuum field configurations as

$$\begin{aligned} M_{sol} &= \int_{-\infty}^{\infty} dx \sqrt{-g}\,[\mathcal{H}_{\text{sol}} - \mathcal{H}_0] \\ &= \int_{-\infty}^{\infty} \cosh(mx) dx \left[\frac{1}{2}\left(\frac{d\phi_{\text{sol}}}{dx}\right)^2 + V(\phi_{\text{sol}})\right]. \end{aligned} \tag{21}$$

Using this definition, the soliton mass in the special case $\alpha = 1$ is calculated to be

$$M_{\text{sol}} = \frac{3\pi}{16}\frac{m^3}{\lambda^2}. \tag{22}$$

It can be seen that the soliton mass becomes large in the limit $\lambda \downarrow 0$, making manifest its non-perturbative nature.

3.4. *Excitation spectrum in the one-soliton sector*

In order to find the excitation spectrum in the one-soliton sector, the scalar field is expanded around the classical soliton solution as

$$\sigma(s,\tau) = \tanh(s) + \eta(s,\tau).$$

The linearized equation of motion for the field η takes the form

$$\frac{1}{\cosh^2(s)}\frac{\partial^2\eta}{\partial\tau^2} - \frac{\partial^2\eta}{\partial s^2} - \tanh(s)\frac{\partial\eta}{\partial s} + 3\tanh^2(s)\eta - \eta = 0. \tag{23}$$

This equation is solved by separation of variables according to the Ansatz

$$\eta(s,\tau) = e^{i\hat{\omega}\tau}X(s).$$

The space dependent part of the normal mode equation reads

$$-\cosh^2(s)\frac{d^2X}{ds^2} - \sinh(s)\cosh(s)\frac{dX}{ds} + 2\cosh^2(s)X - 3X = \hat{\omega}^2X, \tag{24}$$

and is equivalent to a Schrödinger equation of the same type as presented in Appendix A. With the identification

$$\begin{aligned} \hat{\omega}_n^2 &= E_n^{(1)} + 1, \\ g_2 &= 2, \end{aligned} \tag{25}$$

the soliton excitation spectrum can be gleaned to be

$$\hat{\omega}_n^{\rm sol} = \sqrt{(n+2)^2-3}. \qquad n \in \mathbb{N}. \tag{26}$$

The lowest excitation frequency, corresponding to $n = 0$, in the one-soliton sector is seen to be $\hat{\omega}_0^{\rm sol} = 1$, as expected from the spontaneous $S0(2,1)$ symmetry breaking.

4. Quantization, regularization, and renormalization

The model is renormalized in the trivial sector. This procedure will also render physical observables in the one-soliton sector finite. Note that for $\alpha = 1$ the normal mode functions in the trivial and one-soliton sectors are identical since in both cases $g_2 = 2$, but the frequency spectra are distinct. According to Eq. (16) the excitation spectrum in the trivial sector for $\alpha = 1$ is given by

$$\hat{\omega}_n = n + 2, \qquad n \in \mathbb{N}, \tag{27}$$

The normal mode functions for $\alpha = 1$ can be written in terms of the Jacobi polynomials as

$$X_n(s) = B_n\frac{1}{\cosh^2(s)}P_n^{(3/2,3/2)}(\tanh(s)), \tag{28}$$

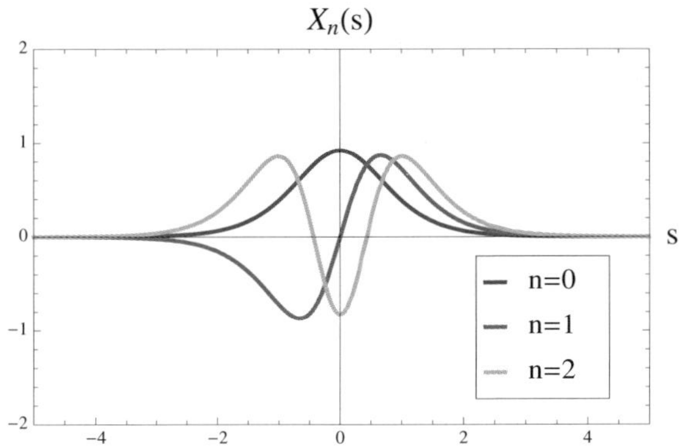

Fig. 3. The first three normal mode functions for $\alpha = 1$ describing soliton excitations.

where the normalization constant B_n is given by

$$B_n = \sqrt{\frac{(n+2)}{8}\frac{\Gamma(n+1)\Gamma(n+4)}{\Gamma(n+5/2)^2}} . \tag{29}$$

The normal mode functions satisfy the orthonormality condition

$$\int_{-\infty}^{\infty} \frac{1}{\cosh s} X_n^\dagger(s) X_m(s) ds = \delta_{nm} . \tag{30}$$

The first three normal mode functions are displayed in Eq. (3.4).

The Hamiltonian density of the model is given by

$$\mathcal{H} = \frac{1}{2}\pi^2 + \frac{1}{2}\left(\frac{\partial \phi}{\partial x}\right)^2 + V(\phi) , \tag{31}$$

where the canonical momentum π is defined as

$$\pi = \sqrt{-g}\,\frac{\partial \mathcal{L}}{\partial(\partial_0 \phi)} = \frac{1}{\cosh(mx)}\frac{\partial \phi}{\partial t} . \tag{32}$$

Both the field ϕ and the canonical momentum π are expanded in terms of normal modes according to

$$\begin{aligned} \phi &= \sum_{n=0}^{N} \frac{a_n}{\sqrt{2\omega_n}} X_n(x) e^{i\omega_n t} + \frac{a_n^\dagger}{\sqrt{2\omega_n}} X_n^\dagger(x) e^{-i\omega_n t} , \\ \pi &= \sum_{n=0}^{N} -i a_n \sqrt{\frac{\omega_n}{2}} \frac{X_n(x)}{\cosh(mx)} e^{i\omega_n t} + i a_n^\dagger \sqrt{\frac{\omega_n}{2}} \frac{X_n^\dagger(x)}{\cosh(mx)} e^{-i\omega_n t} . \end{aligned} \tag{33}$$

In order to quantize the model the commutation relation

$$\left[a_n, a_m^\dagger\right] = \delta_{nm} \tag{34}$$

is imposed.

The normalization conditions are implied by the normal ordering procedure and the finite parts of counter terms are unambiguously fixed. The relation between the normal ordered and original Hamiltonian is

$$:H: = H - \delta D - \int_{-\infty}^{\infty} \cosh(mx)dx \left[\delta\mu^2 \lambda \phi^2\right].$$

The theory is regulated by cutting off the mode sums at large mode number N. The mass counter term is logarithmically divergent,

$$\begin{aligned} \delta\mu^2 &= 3\lambda \sum_{i=0}^{N} \frac{1}{2\omega_i} X_i(x) X_i^\dagger(x) \\ &= \frac{3}{2\pi}\lambda \left[\ln(N) + \gamma - 1 + \ln(2) - \ln\left(\cosh(mx)\right)\right]. \end{aligned} \tag{35}$$

The finite part of the mass counter term is seen not to be $SO(2,1)$ invariant through its coordinate dependence. This is a consequence of the use of a regularization scheme that breaks the $SO(2,1)$ invariance explicitly. The vacuum energy counter term has a quadratic divergence and reads

$$\delta D = \sum_{i=0}^{N} \frac{1}{2}\omega_i = \frac{1}{4} m(N+4)(N+1). \tag{36}$$

4.1. *Quantum corrections to the soliton mass*

To one loop order there are two contributions to the quantum corrections of the soliton mass.[5] The first contribution is due to the mass renormalization and takes the form

$$\begin{aligned} \Delta M_1 &= -\lambda \int_{-\infty}^{\infty} cosh(mx) dx\, \delta\mu^2(x) \left[\phi_{\text{sol}}^2(x) - \phi_0^2\right] \\ &= \frac{3}{4} m \left[\ln(N) + \gamma - 1\right], \end{aligned} \tag{37}$$

while the second contribution is the difference in vacuum energy between the trivial and one soliton sectors,

$$\begin{aligned} \Delta M_2 &= \sum_{i=0}^{N} \frac{1}{2}\omega_i^{\text{sol}} - \delta D \\ &= -\frac{3}{4} m \left[\ln(N) + \gamma - 1\right] + \frac{1}{2} m C_S. \end{aligned} \tag{38}$$

Here Schroeder's number is defined as

$$C_S = \sum_{i=0}^{\infty} \left[\sqrt{(i+2)^2 - 3} - (i+2) + \frac{3}{2}\frac{1}{(i+2)}\right] \approx -0.3485.$$

The logarithmically divergent parts of the two contributions ΔM_1 and ΔM_2 cancel against each other. The finite, physical mass of the soliton including its one-loop quantum corrections is thus

$$M_{\rm sol} = M_{\rm clas} + M_1 + M_2 = \frac{3\pi}{16}\frac{m^3}{\lambda^2} + \frac{1}{2} m C_S. \tag{39}$$

This result is valid for small values of λ and $\alpha = 1$.

5. Discussion

A zero mode is generically obtained in the excitation spectrum of solitons in Minkowski space due to the spontaneously broken translation symmetry. In Anti-de-Sitter space, the corresponding spontaneously broken symmetry generator gives generically rise to a mode with $\omega = m$. The supersymmetric and shape invariant quantum mechanical models encountered in relation to the excitation spectra in the model are equivalent by a coordinate transformation to the Scarf I potentials.[6] The soliton solution in Anti-de-Sitter space was obtained as a static solution to the second order Euler-Lagrange equation of motion. Due to the explicit coordinate dependence of the Hamiltonian density it is not possible to derive a BPS bound and a first order BPS equation as is done in Minkowski space. The interpretation of the mass of the soliton in Anti-de-Sitter space as a topological charge seems unclear.

Acknowledgments

This work was supported in part by a Cottrell Award from the Research Corporation and by the NSF under grant PHY-0758073.

Appendix A.

The excitation spectra discussed in this paper are related to the Schrödinger equation with a Scarf I type potential. The spectrum of this quantum mechanical model can be algebraically determined by exploiting supersymmetry [7] and shape invariance [6] properties.

Consider Hamiltonians H_1 and H_2 which can be factorized as

$$\begin{aligned} H_1 &= A^\dagger A, \\ H_2 &= A A^\dagger, \end{aligned} \tag{A.1}$$

in terms of the differential operators A and $A^\dagger$ defined by

$$\begin{aligned} A &= +\cosh(s)\frac{d}{ds} + g_2 \sinh(s), \\ A^\dagger &= -\cosh(s)\frac{d}{ds} + g_2 \sinh(s), \end{aligned} \tag{A.2}$$

with the parameter $g_2 \geq 1$ but otherwise arbitrary. The operators

$$H = \begin{pmatrix} H_1 & 0 \\ 0 & H_2 \end{pmatrix}, \quad Q = \begin{pmatrix} 0 & 0 \\ A & 0 \end{pmatrix}, \quad Q^\dagger = \begin{pmatrix} 0 & A^\dagger \\ 0 & 0 \end{pmatrix}, \tag{A.3}$$

generate the $sl(1/1)$ superalgebra with (anti-) commutation relations

$$[H, Q] = \left[H, Q^\dagger\right] = 0, \ \ \{Q^\dagger, Q\} = H. \tag{A.4}$$

Since supersymmetry is not spontaneously broken, the ground state energy of H_1 vanishes,

$$E_0^{(1)}(g_2) = 0, \tag{A.5}$$

and the ground state wavefunction of H_1 is annihilated by the operator A,

$$AX_0^{(1)}(g_2, s) = 0. \tag{A.6}$$

Supersymmetry further relates the eigenvalues

$$E_n^{(2)}(g_2) = E_{n+1}^{(1)}(g_2) \tag{A.7}$$

and the eigenfunctions

$$\begin{aligned} X_n^{(2)}(g_2, s) &= AX_{n+1}^{(1)}(g_2, s) \\ X_{n+1}^{(1)}(g_2, s) &= A^\dagger X_n^{(2)}(g_2, s) \end{aligned} \tag{A.8}$$

of H_1 and H_2. The Hamiltonians H_1 and H_2 take the explicit forms

$$\begin{aligned} H_1 &= \cosh^2(s)\left[-\frac{d^2}{ds^2} - \tanh(s)\frac{d}{ds}\right] + (g_2^2 - g_2)\cosh^2(s) - g_2^2 \\ H_2 &= \cosh^2(s)\left[-\frac{d^2}{ds^2} - \tanh(s)\frac{dX}{ds}\right] + (g_2^2 + g_2)\cosh^2(s) - g_2^2. \end{aligned} \tag{A.9}$$

They are related by shape invariance,[6] a combination of a shift in the parameter g_2 and an additive constant, as

$$H_2(s; g_2) = H_1(s; g_2 + 1) + 2g_2 + 1. \tag{A.10}$$

The energy spectra of the Hamiltonians H_1 and H_2 are completely determined algebraically by the combination of supersymmetry and shape invariance:

$$\begin{array}{ll} E_0^{(1)}(g_1) = 0 & \\ E_1^{(1)}(g_2) = 2g_2 + 1 & E_0^{(2)}(g_2) = 2g_2 + 1 \\ E_2^{(1)}(g_2) = 4g_2 + 4 & E_1^{(2)}(g_2) = 4g_2 + 4 \\ \dots & \dots \end{array} \tag{A.11}$$

Energy eigenvalues in Eq. (A.11) are related by shape invariance in the diagonal direction, and by supersymmetry in the horizontal direction. All eigenvalues can thus be obtained starting with $E_0^{(1)}(g_1)$. The resulting spectra are graphically illustrated in Fig.(A) for $g_2 = 2$. The general expression for the energy spectrum of H_1 is

$$E_n^{(1)}(g_2) = 2ng_2 + n^2. \tag{A.12}$$

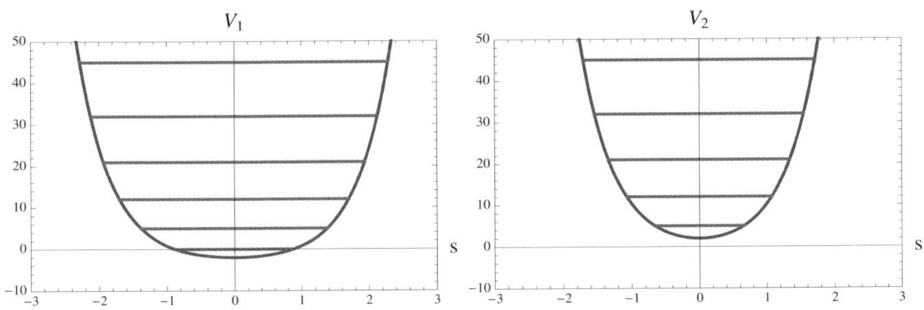

Fig. A1. Potentials and energy spectra of the Hamiltonians H_1 (left panel) and H_2 (right panel) for $g_2 = 2$.

References

1. R. Rajaraman, *"Solitons And Instantons. An Introduction To Solitons And Instantons In Quantum Field Theory,"* Amsterdam, Netherlands: North-holland (1982), and references therein.
2. S. W. Hawking and G. F. R. Ellis, *"The Large scale structure of space-time,"* Cambridge University Press, Cambridge, 1973.
3. S. J. Avis, C. J. Isham and D. Storey, Phys. Rev. D **18**, 3565 (1978).
4. D. W. Dusedau and D. Z. Freedman, Phys. Rev. D **33**, 389 (1986).
5. R. F. Dashen, B. Hasslacher and A. Neveu, Phys. Rev. D **10**, 4130 (1974).
6. F. Cooper, A. Khare and U. Sukhatme, Phys. Rept. **251**, 267 (1995) [arXiv:hep-th/9405029].
7. E. Witten, Nucl. Phys. B **188**, 513 (1981).

Effects of Extra Degrees of Freedom on a Decay Rate of Metastable Topological Objects

A. Monin

School of Physics and Astronomy, University of Minnesota, Minneapolis, MN 55455, USA
E-mail: monin@physics.umn.edu
www.physics.umn.edu.edu

M. B. Voloshin

William I. Fine Theoretical Physics Institute, University of Minnesota, Minneapolis, MN 55455, USA
and
Institute of Theoretical and Experimental Physics, Moscow, 117218, Russia
E-mail: voloshin@physics.umn.edu
www.ftpi.umn.edu

We review the recent results on a decay rate of metastable topological configurations such as strings and domain walls. The transition from a state with higher tension to a state with lower one proceeds through quantum tunneling or through thermally catalyzed quantum tunneling (at sufficiently small temperatures). It is shown that the effects of the motion in transverse direction lead to the renormalization of a mass (tension) parameter of a particle (string) associated with a boundary of a string (wall) in the semiclassical exponent. For a non-zero temperature we derive the catalysis factor for the decay rates. It is discovered that the catalysis factor is closely related to the probability (effective length) of the collision of the the Goldstone bosons, corresponding to the transverse waves on a string (wall). We find that the destruction of a string only takes place in collisions of even number of the bosons, while the destruction of the wall can occur in a collision of any number of particles.

Keywords: Nonperturbative decay; Strings; Domain walls

1. Introduction

Topological configurations of fields with the geometry of a string or a domain wall arise in various models either as solutions of classical field equations or as nonperturbative effective configurations of essentially quantum fields. String-like configurations are present in polymers, superconductors, in theoretical models of non-Abelian dynamics, as well as in models of QCD confinement, and as topological defects in models with spontaneous breaking of gauge or global symmetries. Metastable domain wall solutions arise in models with spontaneously broken approximate symmetry. The existence of such solutions was shown from different points of view, for example, in Refs. [1–3].

These configurations are classically stable, since there is no classical trajectory which takes from one topological solution into another with different topological number. But it may be the case that a classically forbidden trajectory for such a deformation exists so that the considered configurations are unstable quasiclassically and decay due to e.g. quantum tunneling effects. As an illustration of such decay process one can consider the example of a metastable domain wall investigated in Refs. [4,5] which arises in a model with a scalar field potential shown in Fig. 1. The domain wall solution corresponds to the interpolation between the same vacuum state at two spatial infinities, e.g. at $z = -\infty$ and $z = +\infty$ with the field winding around the 'peg' in the potential. Such configuration is classically stable

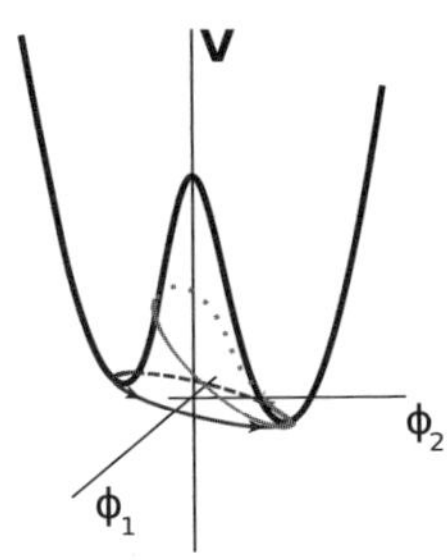

Fig. 1. Potential

in the sense that the solution with certain winding number can not evolve classically to a solution with different topological number. However, such a transition can proceed either due to a thermal fluctuation, where the path, corresponding to the solution, is lifted over the barrier or due to quantum tunneling (the motion in a classically forbidden region).

The topological configurations can be metastable with respect to either a complete breaking, or a transition to an object of lower tension emerging instead of the initial one. The former situation is relevant e.g. for a break up of a QCD string with formation of a quark - antiquark pair, or a formation of a monopole - antimonopole pair,[6] and also in a whole class of theories with spontaneous symmetry breaking.[7] The latter situation involving a phase transition between states of a string with different tension is found e.g. in Abelian Higgs models embedded in non-Abelian theories.[8,9]

2. Strings and domain walls at zero temperature.

A domain wall decay is analogous to the well-known spontaneous false vacuum decay process[10–12] in $2+1$ dimensions. Indeed (see Fig. 2), if a

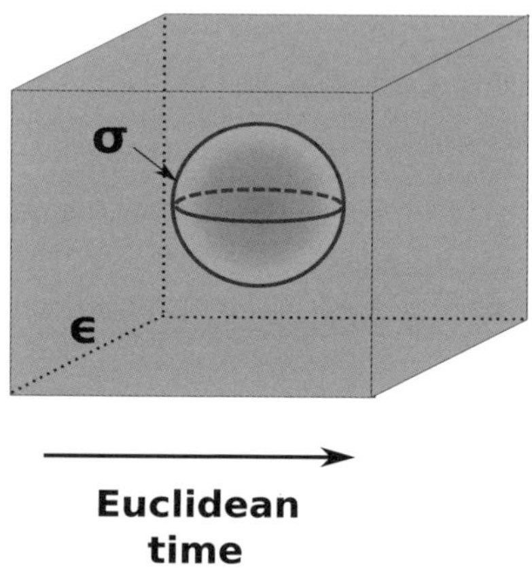

Fig. 2. Bounce configuration

hole of area $\mathcal{A}_c$ is created in a wall with tension ϵ, the gain in the energy is $\epsilon\,\mathcal{A}_c$. The barrier that inhibits the process is created by the energy $\sigma\mathcal{P}$, with $\mathcal{P}$ being the perimeter of the hole and σ being a tension associated with the interface. Thus, the 'area' energy gain exceeds the barrier energy only starting from a critical size of the hole created, i.e. starting with a round hole of radius $R_w = 2\sigma/\epsilon$. Once a critical bubble of the lower phase has nucleated due to tunneling, it expands, converting the domain wall. Therefore the probability of the transition is given by the rate of nucleation of the critical holes in the domain wall.

The same argument also applies to the string decay which is similar to a metastable vacuum decay process in $1+1$ dimensions. The bounce configuration of the string is depicted on the Fig. 3. The critical size of the gap in the initial string is then $\ell_c = 2R_s = 2\,\mu/\varepsilon$. Moreover due to the geometry of the problem such decay also bears a great similarity to the well known Schwinger process of production of pair of charged particles by

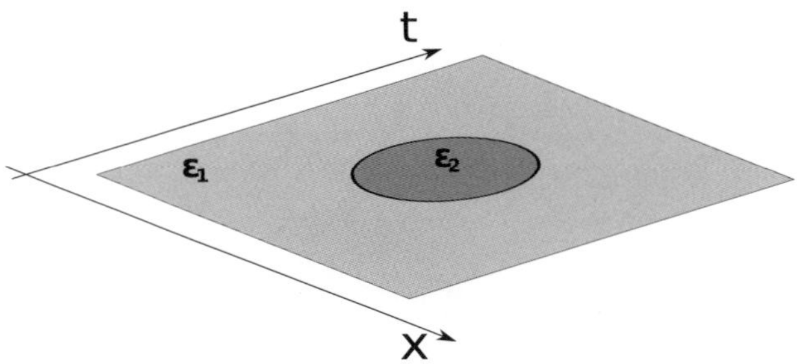

Fig. 3. The bounce configuration, describing the semiclassical tunneling trajectory.

an external electromagnetic field.[13]

Despite the similarity with the false vacuum decay the breaking of the strings and walls involves an essential difference associated with the transverse waves propagating on the latter objects, corresponding to massless bosons, which in fact are the Goldstone bosons of the spontaneously broken translational invariance. The effect of the soft modes of the field of these bosons enters the probability of breaking the topological defects at the preexponential level.[5,9]

The process of a metastable string transition from a state with tension ε_1 to a state with tension ε_2 was considered in [6–8]. Using the analogy with the false vacuum decay the decay rate was found with exponential accuracy in terms of the rate of nucleation of critical gaps per unit length of the string, $\gamma_s = d\Gamma/d\ell$, [a],

$$\gamma_s = C \, \exp\left(-\frac{\pi \, \mu^2}{\varepsilon_1 - \varepsilon_2}\right) , \tag{1}$$

where μ is the mass associated with the interface between the two phases of the string. The decay rate of an axion domain wall was considered at the level of the semiclassical exponent in Ref. [4] by adapting the expression for metastable vacuum decay in $2+1$ dimensions.

The preexponential factors for the spontaneous decay of string and walls were calculated in [5,9]. These correspond to the integration over small fluctuations around the corresponding bounce configuration. The variation

[a] It should be mentioned that in $d = 2$ there is no transverse motion for the string, and the preexponential factor can be also copied from the known result of spontaneous metastable vacuum decay

$$C_{d=2} = \frac{\varepsilon_1 - \varepsilon_2}{2\pi} .$$

of the string configuration is shown on the Fig. 4. For a string the rate of

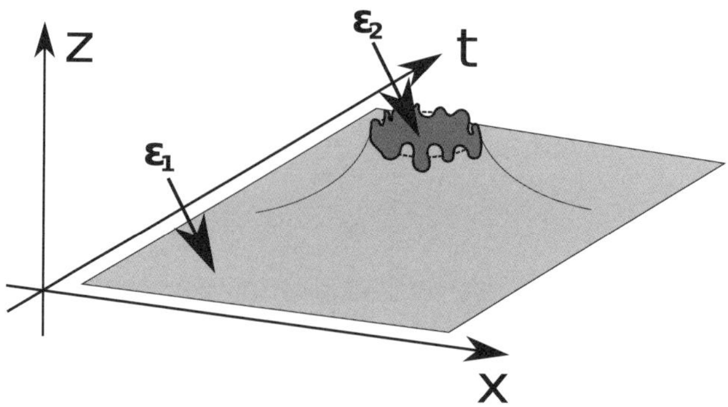

Fig. 4. Fluctuations around the bounce configuration.

the transition between the states with tension ε_1 and ε_2 was found to be

$$\gamma_s = \frac{\varepsilon_1 - \varepsilon_2}{2\pi} \left[F\left(\frac{\varepsilon_2}{\varepsilon_1}\right) \right]^{d-2} \exp\left(- \frac{\pi \, \mu_R^2}{\varepsilon_1 - \varepsilon_2} \right) , \tag{2}$$

with the dimensionless factor F given by

$$F\left(\frac{\varepsilon_2}{\varepsilon_1}\right) = \sqrt{\frac{\varepsilon_1 + \varepsilon_2}{2\,\varepsilon_1}} \, \Gamma\left(\frac{\varepsilon_1 + \varepsilon_2}{\varepsilon_1 - \varepsilon_2} + 1\right) \left(\frac{\varepsilon_1 - \varepsilon_2}{\varepsilon_1 + \varepsilon_2}\right)^{\frac{\varepsilon_1+\varepsilon_2}{\varepsilon_1-\varepsilon_2}} \times \tag{3}$$
$$\times \exp\left(\frac{\varepsilon_1 + \varepsilon_2}{\varepsilon_1 - \varepsilon_2}\right) \left(2\pi \, \frac{\varepsilon_1 + \varepsilon_2}{\varepsilon_1 - \varepsilon_2}\right)^{-1/2} ,$$

while the decay rate of a domain wall is formulated in terms of the rate γ_w of nucleation of critical holes per unit area, $\gamma_w = d\Gamma/dA$,

$$\gamma_w = \frac{\tilde{C}}{\epsilon^{7/3}} \exp\left(- \frac{16\,\pi\,\sigma_R^3}{3\,\epsilon^2} \right) , \tag{4}$$

where μ_R and σ_R are renormalized mass and tension parameters associated with the interface of the strings and domain walls respectively, and $\tilde{C}$ is a constant that does not depend on ϵ but does depend on other dimensional parameters in the underlying field theory.

3. Nonzero temperature case.

The thermal effects in the decay of metastable strings [14] further expose the difference from the false vacuum decay and the Schwinger process. Namely, as long as the temperature T is lower than the inverse of the critical length, $T < 1/\ell_c$ the thermal effects in the latter processes are exponentially suppressed as $\exp(-m/T)$ with m being the lowest scale for particle masses in the theory, and these effects are very small due to the strongly suppressed presence of massive particles in the thermal equilibrium.[15] In the case of a string, however, the transverse waves on the string are massless so that their excitation has no suppression by the mass at arbitrarily low temperature. The thermal excitations of these waves create fluctuations in the distribution of the energy in the string which catalyze the nucleation of the critical gap. The enhancement of the nucleation rate at finite temperature T can be expressed in terms of a thermal catalysis factor $\mathcal{K}_s(T) > 1$, which relates the thermal rate $\gamma_s(T)$ to that at zero temperature, γ_s, $\gamma_s(T) = \mathcal{K}_s(T)\,\gamma_s$. Clearly, at $T \ll 1/\ell_c$ the typical wavelength of the thermal waves $\sim 1/T$ is large in comparison with the critical length ℓ_c, and the thermal effect in the rate is quite small, although not exponentially small. The leading low temperature correction in the nucleation rate is given by the thermal catalysis factor[14]

$$\mathcal{K}_s = 1 + (d-2)\,\frac{\pi^8}{450}\left(\frac{\varepsilon_1 - \varepsilon_2}{3\varepsilon_1 - \varepsilon_2}\right)^2 \left(\frac{\ell_c T}{2}\right)^8 + O\left[(\ell_c T)^{12}\right] , \qquad (5)$$

while as T approaches $1/\ell_c$ (the value of temperature for which the period of the Euclidean time equals the critical size, see Fig. 5), the catalysis factor develops a singularity at $\ell_c T = 1$. At still higher temperatures the con-

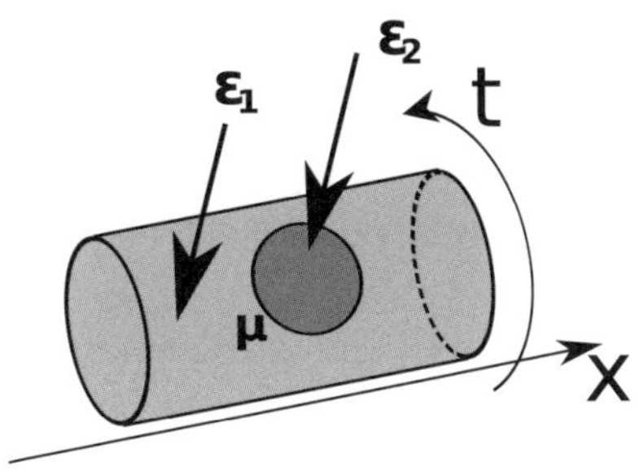

Fig. 5. Bounce configuration for nonzero temperature

sidered string transition behaves, in a sense, similarly to the false vacuum decay,[15] namely the regime of the transition changes to a different tunneling trajectory, so that the temperature dependence appears in the semiclassical exponential factor in Eq.(2), rather than in the preexponential term.

The purpose of the paper [18] was to generalize to the case of a metastable domain wall the approach used for analyzing the thermal and collision induced effects in the decay of metastable strings. We found the general expression for the thermal catalysis factor $\mathcal{K}_w$. In particular, the leading term in this factor at low temperature (in d dimensions) is given by

$$\mathcal{K}_w = 1 + 12\,(d-3)\,\zeta^2(5)\,(R_w\,T)^{10} + \ldots\,. \tag{6}$$

where R_w is the radius of the critical hole in the metastable wall, and $\zeta(x)$ is the standard notation for the Riemann ζ function, $\zeta(5) \approx 1.037$. The difference between the decay of the string and the domain wall relates to the thermal catalysis of the decay: for a string the low temperature expansion for the enhancement of the quantum tunneling is applicable everywhere where the series is convergent, i.e. up to the temperature $T = 1/\ell_c$, while for a domain wall the thermal fluctuations start dominating [15] at a lower temperature ($T = 3/(4\ell_w)$ with ℓ_w being the diameter of the critical hole in the wall), at which the thermal effects in the quantum tunneling probability are still very small numerically.

4. Decay in a collision of Goldstone bosons.

The effect of the thermally excited waves in the decay rate can in fact be considered [16] as an additional contribution to the probability of the decay due to collisions of the Goldstone bosons that are present in the thermal bath. As it turns out, it is possible to identify in the thermal catalysis factor the contribution of individual such collisions and thus calculate the probability of the destruction of metastable string by colliding particles. In particular, the first temperature dependent term in the expansion (5) is entirely due to the process of the critical gap creation in a collision of two particles (see Fig. 6) in the limit, where their center of mass energy $E = \sqrt{s}$ is much smaller than $1/\ell_c$. In order to separate the terms in the decay probability originating from collisions of different number of the bosons, the standard thermal approach to calculating the decay rate at finite temperature [14] is modified [16] by formally introducing a negative chemical potential for the bosons. This allows to find the dependence of the probability of string destruction in an n-boson collision for an arbitrary relation between ℓ_c and the energy of the bosons. One of the results of such a consideration is that the string destruction takes place only in collisions of even number of bosons and is absent at odd n. The explicit expression for the (dimensionless) probability W_2 of the critical gap creation in two-boson collisions at arbitrary values of the parameter $E\,\ell_c$ has especially simple form for the

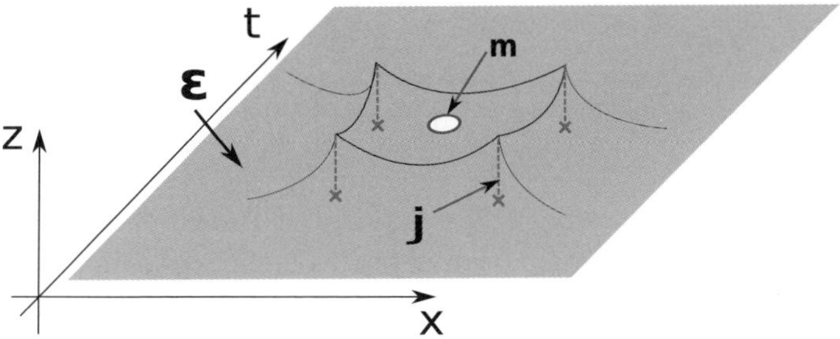

Fig. 6. Bounce configuration distorted by the sources

decay of the string into 'nothing' i.e. at $\varepsilon_2 = 0$:

$$W_2 = 2\,\pi^2\,\ell_c^2\,\gamma_s\,\left[I_3\left(\frac{E\ell_c}{2}\right)\right]^2 , \tag{7}$$

with $I_3(x)$ being the standard notation for the modified Bessel function of the third order. This expression is applicable as long as the energy E is small in comparison with the string energy scale: $E \ll \sqrt{\epsilon}$, which condition still allows for the parameter $E\,\ell_c$ to be large. At $E\,\ell_c \gg 1$ the energy dependent factor in Eq.(7) has the exponential behavior $\exp(E\ell_c)$ which matches the semiclassical approximation for the energy-dependent factor,[17] corresponding to tunneling at the energy E rather than at zero energy. In this sense the discussed energy dependence of the collision-induced probability of the decay describes the onset of the semiclassical behavior.

One can also notice that the thermal factor $\mathcal{K}_s$, is singular [14] at $\ell_c T = 1$. However the two-boson production rate does not exhibit any singularity at any value of the parameter $E\ell_c$, and a similar smooth energy behavior is also true for individual n-boson processes. Thus the 'explosion' of the thermal rate at $\ell_c T = 1$ is a result of infinite number of processes becoming important at this point, rather than due to a finite set of processes with a limited range of n developing large probability at the energy per particle of the order of $1/\ell_c$.

For the decay of the domain wall in a collision of Goldstone bosons the analog of the probability of the collision is an effective length. Although the full expression for the effective length is complicated, the two limits are rather simple.[18] The low energy limit for the effective length of destruction of the wall in a collision of two Goldstone bosons is

$$\lambda = \frac{d-3}{5}\,\pi^2\,\gamma_w\,s^3\,R_w^{10} + \ldots , \tag{8}$$

while the behavior at large energy, $R_w \sqrt{s} \gg 1$ is given by

$$\lambda \sim \exp\left(-\frac{16\,\pi\,\sigma^3}{3\,\epsilon^2} + 4\sqrt{s}\,\frac{\sigma}{\epsilon}\right), \tag{9}$$

which matches the semiclassical expression for the tunneling exponent at energy $\sqrt{s}$ [23]. At the paper [18] we aimed at providing there a systematic and self-contained presentation of the calculations leading to our final results. Such 'semi-review' format of the paper also allowed to illustrate the similarity and the difference between the string and the domain wall transitions. Although the calculations in both cases are very much similar, the results differ in some essential details. In particular the destruction of a metastable wall is induced by a collision of any number of Goldstone bosons, whereas a string can be destroyed by a collision of only even number of the bosons.

5. Summary

In this paper we presented a brief review of the results obtained in the papers considering the decays of metastable topological configurations such as string and domain wall. The summary of the results can be presented in the following form

(i) The effect on the preexponential factor in the tunneling decay rate arising from the Goldstone modes living on strings and walls is expressed in terms of boundary terms in the path integral around the bounce configuration. This effect is fully calculable within the low-energy effective action of the Nambu-Goto type, and essentially reduces to a renormalization of the tension of the interface between the two phases of the topological defect. We have found that this property is unique for the strings and walls and is generally lost in a description of decay of topological objects with more dimensions, where one would have to go beyond the effective Nambu-Goto description.

(ii) The thermal effects in the tunneling rate are also fully calculable and are described by an expansion in powers of the temperature. The power-like, rather than exponential, behavior of the thermal terms is due to the presence of the massless Goldstone bosons.

(iii) The expansion for the thermal catalysis factor converges as long as the temperature is less than the inverse diameter of the bounce, so that in the Euclidean calculation of the bounce configuration to (the imaginary part of) the free energy there is no obstruction to periodically replicating the bounce with the Matsubara period $\beta = 1/T$. For the decay of strings this range of temperatures corresponds to the dominance of the (thermally enhanced) quantum tunneling over the classical thermal effects, and at the critical temperature $1/\ell_c$ the regime of the transition changes and the classical thermal effects take over. For the domain wall decay the classical thermal activation starts dominating at a lower temperature, where the thermal enhancement of the quantum tunneling is still very small.

(iv) The thermal enhancement of the tunneling rate can be viewed as and additional contribution to the decay probability arising from collisions of the Goldstone bosons which are present in the thermal state. Such an interpretation in fact allows to determine the rate of destruction of the considered topological objects in collisions of the Goldstone bosons. In order to unambiguously identify in the thermal expansion the terms corresponding to the contribution of collisions between a given number of the bosons, we have modified the considered statistical ensemble by introducing a (fictitious) negative chemical potential ν for the Goldstone bosons. The calculated dependence on ν and T of the preexponential factor in the decay rate thus serves as a generating function for the rates of the decay induced by collisions of particles, including the dependence of those rates on the energies in the collisions.

(v) We find that a destruction of a string takes place only in collisions of even number of the Goldstone bosons, while a metastable wall can be destroyed in collisions of any number of the bosons, $n \geq 2$.

(vi) We have calculated in a closed form the energy behavior of the probability of destruction of a string in a collision of two Goldstone bosons, which is valid at an arbitrary relation between the energy $\sqrt{s}$ and the infrared scale in the problem R_s. The found expression takes an especially simple form of Eq.(7) for the decay of a string into 'nothing'

(vii) At large values of $R\sqrt{s}$ the collision-induced decay rate contains an exponential factor $\exp(2R\sqrt{s})$, which has been previously argued [17,23] within the leading semiclassical approach and corresponds to a full transfer of the energy from the colliding particles to the tunneling degrees of freedom.

One can readily notice that our calculation of the collision-induced rate through a thermal ensemble with an artificially introduced chemical potential is quite indirect. In a sense, this approach is reminiscent of the treatment in Ref. [24], where finite volume effects in the energy of a two-particle state are related to the binary scattering amplitude. The obvious difference is that we calculate the free energy of a statistical ensemble rather than of a particular quantum state. In relation to this part of our calculation it would certainly be illuminating to have a more direct, and possibly simpler method for calculating the probability of creating semiclassical objects, such as the bubbles of the stable phase, in collisions of particle. However, lacking such a method at present, we have to resort to the indirect calculation described in the present paper.

Acknowledgments

This work is supported in part by the DOE grant DE-FG02-94ER40823. The work of A.M. is also supported in part by the RFBR Grant No. 07-02-00878 and by the Scientific School Grant No. NSh-3036.2008.2.

References

1. M. A. Shifman, Prog. Part. Nucl. Phys. **39**, 1 (1997).
2. E. Witten, Phys. Rev. Lett. **81**, 2862 (1998).
3. E. Witten, JHEP **9807**, 006 (1998).
4. M. M. Forbes and A. R. Zhitnitsky, JHEP **0110**, 013 (2001).
5. A. Monin and M. B. Voloshin, Phys. Rev. D **79**, 025007 (2009).
6. A. Vilenkin, Nucl. Phys. B **196**, 240 (1982).
7. J. Preskill and A. Vilenkin, Phys. Rev. D **47**, 2324 (1993) [arXiv:hep-ph/9209210].
8. M. Shifman and A. Yung, Phys. Rev. D **66**, 045012 (2002) [arXiv:hep-th/0205025].
9. A. Monin and M. B. Voloshin, Phys. Rev. D **78**, 065048 (2008).
10. M. B. Voloshin, I. Y. Kobzarev and L. B. Okun, Sov. J. Nucl. Phys. **20**, 644 (1975) [Yad. Fiz. **20**, 1229 (1974)].
11. S. R. Coleman, Phys. Rev. D **15**, 2929 (1977) [Erratum-ibid. **16**, 1248 (1977)].
12. M. B. Voloshin, Phys. Lett. B **599**, 129 (2004).
13. J. Schwinger, Phys. Rev. **82**, 664 (1951).
14. A. Monin and M. B. Voloshin, Phys. Rev. D **78**, 125029 (2008) [arXiv: 0809.5286 [hep-th]].
15. J. Garriga, Phys. Rev. D **49**, 5497 (1994).
16. A. Monin and M. B. Voloshin, "Destruction of a metastable string by particle collisions," arXiv:0902.0407 [hep-th].
17. M. B. Voloshin and K. G. Selivanov, Sov. J. Nucl. Phys. 44, 868 (1986) [Yad. Fiz. **44**, 1336 (1986)].
18. A. Monin and M. B. Voloshin, "Spontaneous and Induced Decay of Metastable Strings and Domain Walls," arXiv:0904.1728 [hep-th].
19. C. G. Callan and S. R. Coleman, Phys. Rev. D **16**, 1762 (1977).
20. M. Stone, Phys. Rev. D **14**, 3568 (1976).
21. M. B. Voloshin, Sov. J. Nucl. Phys. **42**, 644 (1985) [Yad. Fiz. **42**, 1017 (1985)].
22. J. S. Langer, Ann. Phys. (N.Y.) **41**, 108 (1967).
23. M. B. Voloshin, Phys. Rev. D **49**, 2014 (1994).
24. M. Luscher, Commun. Math. Phys. **105**, 153 (1986).

SECTION 4.

GAUGE–STRING DUALITY

Recent Progress in AdS/CFT

John H. Schwarz*

Department of Physics, California Institute of Technology,
Pasadena, CA 91125, USA
**E-mail: jhs@theory.caltech.edu*
http://theory.caltech.edu/~jhs/

The study of AdS/CFT (or gauge/gravity) duality has been one of the most active and illuminating areas of research in string theory over the past decade. The scope of its relevance and the insights it is providing seem to be ever expanding. In this talk I briefly describe some of the attempts to explore how the duality works for maximally supersymmetric systems.

Keywords: Superstring theory; M-theory; AdS/CFT duality; Integrable models.

1. Introduction

Misha Shifman is one of the most productive (and nicest) theoretical physicists that I know. It is an honor to have been asked to speak at this gathering celebrating his 60th birthday. At this stage of my life, that age seems much younger than it used to.

Recently, in my capacity as an editor of *Reviews of Modern Physics* I persuaded Misha to coauthor a review article entitled *Supersymmetric Solitons and How They Help Us Understand Non-Abelian Gauge Theories.* I am very pleased to report that this article has now been published.[1] It is already highly cited.

I have chosen to present an overview of recent progress in AdS/CFT. An alternative, and more inclusive, name for this subject is *gauge/gravity duality.* There has been progress on various fronts. In the time available to me, I will only be able to scratch the surface of this fascinating subject. There is a great deal of impressive work that I will not mention at all. Further discussion of some of these topics, such as the Konishi multiplet, can be found in Arkady Tseytlin's contribution.[2]

2. Review of Some Basic Facts

In Maldacena's original paper [3], he proposed three maximally supersymmetric examples of AdS/CFT duality. A basic indication that the dualities (or equivalences) are plausible is that the symmetries match. In each case, there is a supergroup, which describes the isometries of the string theory or M-theory background geometry. The same supergroup appears as the superconformal symmetry group of the dual quantum field theory. Also, the string theory or M-theory solution has N units of flux threading the sphere factor in the geometry. In fact, the background configuration corresponds to the near-horizon geometry of N coincident branes, each of which contributes one unit of flux. The dual conformal field theory, which also depends on the integer N, is the low energy world-volume theory on the branes.

- **M2-brane Duality:** M-theory on $AdS_4 \times S^7$ is dual to a superconformal field theory (SCFT) in three dimensions. The supergroup is $OSp(8|4)$.
- **D3-brane Duality:** Type IIB superstring theory on $AdS_5 \times S^5$ is dual to a SCFT in four dimensions, specifically $\mathcal{N} = 4$ super Yang–Mills (SYM) theory. The supergroup is $PSU(2,2|4)$.
- **M5-brane Duality:** M theory on $AdS_7 \times S^4$ is dual to a SCFT in six dimensions. The supergroup is $OSp(6,2|4)$.

2.1. *The type IIB / $\mathcal{N} = 4$ SYM example*

This by far the most studied, and best understood, example. The N units of flux ($\int_{S^5} F_5 \approx N$) in the superstring solution correspond to the gauge group $SU(N)$ in the $\mathcal{N} = 4$ super Yang–Mills theory.[4] The gauge theory has a well-known large-N topological ('t Hooft) expansion.[5] The expansion is in powers of $1/N$ for large N at fixed λ, where the 't Hooft parameter is

$$\lambda = g_{\rm YM}^2 N. \tag{1}$$

This expansion corresponds to the loop expansion of the string theory. One also identifies

$$R^2/\alpha' \approx \sqrt{\lambda} \quad \text{and} \quad g_{\rm s} \approx \lambda/N, \tag{2}$$

where R is the radius of the S^5 and the AdS_5. $g_{\rm s}$ is the string coupling constant determined by the value of the dilaton field.

2.2. *The type IIA / ABJM example*

There has been significant progress in the last couple of years in understanding the M2-brane duality. The suggestion [6] that the three-dimensional SCFT should be Chern–Simons gauge theory was implemented for maximal supersymmetry ($\mathcal{N} = 8$) by Bagger and Lambert [7] and by Gustavsson.[8] However, their construction only works for the gauge group $SO(4)$, and it does not provide the desired dual to M-theory on $AdS_4 \times S^7$.

The correct construction was eventually obtained by Aharony, Bergman, Jafferis, and Maldacena (ABJM).[9] One key step in their work was to consider a more general problem: M-theory on $AdS_4 \times S^7/\mathbb{Z}_k$, with N units of flux. This gives 3/4 maximal supersymmetry for $k > 2$. Thus, the dual gauge theory is an $\mathcal{N} = 6$ superconformal Chern–Simons theory in three dimensions. The appropriate gauge group turns out to be $U(N)_k \times U(N)_{-k}$, where the subscripts are the levels of the Chern–Simons terms. The ABJM theory also contains bifundamental scalar and spinor fields. This theory has a topological expansion, just like the usual ones in four dimensions, with an 't Hooft parameter

$$\lambda = N/k. \tag{3}$$

The only unusual feature is that the 't Hooft parameter is rational. The extension of the supersymmetry from $\mathcal{N} = 6$ to $\mathcal{N} = 8$ for $k = 1, 2$ is a nontrivial property of the quantum theory.

The orbifold $S^7/\mathbb{Z}_k$ can be described as a circle bundle over a CP^3 base. The circle has radius R/k, where R is the S^7 radius. When $k^5 \gg N$, there is a weakly coupled type IIA superstring interpretation with string coupling constant

$$g_s \approx (N/k^5)^{1/4}. \tag{4}$$

One then obtains the correspondences

$$R^2/\alpha' \approx \sqrt{\lambda} \quad \text{and} \quad g_s \approx \lambda^{5/4}/N, \tag{5}$$

which is very similar to the previous duality, but not precisely the same.

2.3. *AdS energies and conformal dimensions*

The metric of AdS_{p+2} in global coordinates is

$$ds^2[AdS_{p+2}] = d\rho^2 - \cosh^2 \rho\, dt^2 + \sinh^2 \rho\, ds^2[S^p]. \tag{6}$$

Here, $ds^2[S^p]$ denote the metric of a unit p-dimensional sphere. Actually, AdS/CFT duality requires taking the covering space of AdS, which means that the time coordinate t runs from $-\infty$ to $+\infty$.

Witten [10] and Gubser, Klebanov, Polyakov [11] gave a prescription for relating n-point correlation functions in the gauge theory to corresponding quantities in the string theory. In the case of two-point functions, the duality relates the energy E_A of a string state $|A\rangle$ (defined with respect to the global time coordinate t)

$$H_{\text{string}}|A\rangle = E_A|A\rangle, \tag{7}$$

to the conformal dimensions Δ_A of the corresponding gauge-invariant local operator $\mathcal{O}_A$ defined by

$$\langle \mathcal{O}_A(x)\mathcal{O}_B(y)\rangle \approx \frac{\delta_{AB}}{|x-y|^{2\Delta_A}}. \tag{8}$$

Specifically, the duality requires that

$$\Delta_A(\lambda, 1/N) = E_A(R^2/\alpha', g_{\text{s}}). \tag{9}$$

The 't Hooft expansion of the dimension of $\mathcal{O}_A$ is

$$\Delta_A(\lambda, 1/N) = \Delta_A^{(0)} + \sum_{g=0}^{\infty} \frac{1}{N^{2g}} \sum_{l=1}^{\infty} \lambda^l \Delta_{l,g}. \tag{10}$$

$\Delta_A^{(0)}$ is the classical (engineering) dimension, and the rest is called the anomalous dimension.

Almost all studies have focused on the planar approximation, (genus $g = 0$), which is dual to free string theory. This restriction may make the problem fully tractable, but it is certainly not easy. After all, it would be an extraordinary achievement to solve an interacting four-dimensional quantum field theory even in the planar approximation.

2.4. *Approaches to testing the dualities*

Given that it is not possible to completely solve any of these theories, the question arises how best to test and explore the workings of AdS/CFT duality. The most obvious things—matching symmetries and the dimensions of chiral primary operators—have been done long ago. One wants to dig deeper. One approach is to match, as much as possible, energies and dimensions of fields/operators that are not protected by supersymmetry. It should be noted, however, that a complete test of the duality would also require matching three-point correlators, since a conformal field theory is completely characterized by its two-point and three-point functions. There has been much less progress on this front.

One approach that has been quite successful is the following. First, identify tractable examples of classical solutions of the string world-sheet theory. Next, examine the spectrum of small excitations about these solutions and compute their energies E_A. Finally, identify the corresponding class of operators in the dual gauge theory and compute their dimensions Δ_A in the planar approximation. Then compare to E_A. One subtlety in this analysis is that this comparison requires an extrapolation from large λ, where the classical world sheet theory is valid, to small λ, where the gauge theory can be studied perturbatively. Thus, one needs to identify examples in which this is possible. As we will see, in practice this has conjectural aspects.

A variant of the preceding procedure is to compare equations that determine E_A and Δ_A rather than the solutions. Approaches based on integrability and algebraic curves try to obtain equations of "Bethe type" on both sides and to match them. This is a very active area of research, but I will not be able to review it here. One important issue is that it is much easier to study the world-sheet theory when the range of σ is infinite (rather than a circle). In other words, the string itself is infinite, rather than a loop. In the gauge theory analysis this corresponds to the thermodynamic limit of the Bethe equations arising from a spin-chain analysis. There has been progress recently in extending the integrability techniques to the compact case.[12] However, the story is quite technical, and I don't think it is completely settled.

3. Classical String Solutions

For the reasons outlined above, we want to identify classical string solutions in the $AdS_5 \times S^5$ background that can be used to test the duality. The discussion that follows largely follows an excellent review article by Plefka.[13] Other useful reviews include [14, 15].

The bosonic part of the string world-sheet action has six cyclic coordinates:

$$(t, \varphi_1, \varphi_2; \phi_1, \phi_2, \phi_3), \tag{11}$$

where the first three coordinates pertain to AdS_5 and the second three to S^5. Specifically, we parametrize S^5 as follows:

$$ds^2(S^5) = d\gamma^2 + \cos^2\gamma\, d\phi_3^2 + \sin^2\gamma\, ds^2(S^3), \tag{12}$$

where

$$ds^2(S^3) = d\psi^2 + \cos^2\psi\, d\phi_1^2 + \sin^2\psi\, d\phi_2^2. \tag{13}$$

Associated to these cyclic coordinates one has conserved charges

$$(E, S_1, S_2; J_1, J_2, J_3). \tag{14}$$

E is the energy and the other five charges are angular momenta.

One much-studied class of string solutions involves a line up the center of AdS_5, described by $\rho = 0$ and $t = \kappa\tau$, where κ is a constant and τ is the world-sheet time coordinate. These configurations have $S_1 = S_2 = 0$.

3.1. *Point-particle solutions*

The simplest solution is a point particle (collapsed string) encircling the sphere. In addition to $\rho = 0$ and $t = \kappa\tau$, this is described by

$$\gamma = \pi/2, \quad \phi_1 = \kappa\tau, \quad \psi = 0. \tag{15}$$

This has $J_2 = J_3 = 0$.

The quantum excitations of this solution have energies that can be expanded in powers of $1/J$ for large $J = J_1$, where

$$\kappa = J/\sqrt{\lambda} \tag{16}$$

is held fixed. This is equivalent to the Berenstein, Maldacena Nastase (BMN) analysis of strings in a plane-wave background.[16] One obtains

$$E - J \approx E_2(\kappa) + \frac{1}{J}E_4(\kappa) + \ldots \tag{17}$$

The exact BMN result is

$$E_2 = \sum_{n=-\infty}^{\infty} \sqrt{n^2 + \kappa^2} N_n, \tag{18}$$

where $N_n = \sum_{i=1}^{8} \alpha_n^{i\dagger}\alpha_n^i +$ fermions is expressed in terms of ordinary oscillators

$$[\alpha_m^i, \alpha_n^{j\dagger}] = \delta^{ij}\delta_{mn}. \tag{19}$$

The level matching condition is $\sum nN_n = 0$.

The BMN paper proposed a scaling rule, known as BMN scaling, which predicts agreement with the anomalous dimensions of operators in the dual gauge theory, even though one calculation is valid for large λ and the other for small λ. In other words, their scaling hypothesis, if valid, would justify the extrapolation from small λ to large λ. In fact, it turns out that E_2 agrees perfectly, but agreement for E_4 breaks down at three loops.[17] This is not a problem for AdS/CFT duality, only for the BMN scaling conjecture.[a]

---

[a] Perhaps it would be more fair to say that the BMN scaling conjecture was made for the

3.2. *Spinning string solutions*

A class of interesting generalizations of the preceding solution describes circular or folded strings that are extended on the $S^3 \subset S^5$. These have $t = \kappa\tau$, $\rho = 0$, and $\gamma = \pi/2$, as before. But now one takes

$$\phi_1 = \omega_1 \tau, \quad \phi_2 = \omega_2 \tau, \quad \psi = \psi(\sigma). \tag{20}$$

For these choices, the string equation of motion gives

$$\psi'' + \omega_{21}^2 \sin\psi \cos\psi = 0, \tag{21}$$

where $\omega_{21}^2 = \omega_2^2 - \omega_1^2$. This is the well-known pendulum equation.

This equation has a first integral

$$\psi' = \omega_{21}\sqrt{q - \sin^2\psi}, \quad q = (\kappa^2 - \omega_1^2)/\omega_{21}^2. \tag{22}$$

The solution for $q < 1$, which involves the elliptic integrals $E(q)$ and $K(q)$, describes a folded string. It corresponds to a pendulum that oscillates back and forth. The solution for $q > 1$, which involves the elliptic integrals $E(q^{-1})$ and $K(q^{-1})$, describes a circular string. It corresponds to a pendulum that goes round and round. In the classical limit, the energy has the form

$$E = \sqrt{\lambda} F(J_1/\sqrt{\lambda}, J_2/\sqrt{\lambda}). \tag{23}$$

3.3. *Dual gauge theory analysis*

This string theory result can be extrapolated to small λ and compared to the dual gauge theory. The operators that carry J_1, J_2 charges have the form

$$\mathcal{O}_\alpha^{J_1,J_2} = \mathrm{Tr}\left(Z^{J_1} W^{J_2}\right) + \ldots \tag{24}$$

where Z and W are complex scalar fields in the adjoint of $SU(N)$. The additional terms denoted by dots involve different orderings of the Zs and Ws. Such a trace can be viewed as a ring configuration of an $S = 1/2$ quantum spin chain, where W corresponds to spin up and Z corresponds to spin down.

The conformal dimensions of operators $\mathcal{O}_\alpha^{J_1,J_2}(x)$ with these charges are eigenvalues of the dilatation operator

$$\mathcal{D}\mathcal{O}_\alpha^{J_1,J_2}(x) = \sum_\beta D_{\alpha\beta}\mathcal{O}_\beta^{J_1,J_2}(x). \tag{25}$$

plane-wave limit only, which corresponds to E_2; what fails is an attempt to generalize the scaling conjecture beyond that.

In the planar one-loop approximation the equations are precisely those of a ferromagnetic Heisenberg spin chain, which is a well-known integrable system, whose Hamiltonian is proportional to

$$\mathcal{H} = \sum_{i=1}^{J} \left(\frac{1}{4} - \vec{\sigma}_i \cdot \vec{\sigma}_{i+1} \right). \tag{26}$$

This can be solved using Bethe ansatz techniques, thereby obtaining conformal dimensions that can be compared (successfully) with energies of the corresponding string solutions. Higher-order terms, which correspond to more complicated spin-chain Hamiltonians, have also been studied.

3.4. *Strings spinning in AdS*

Another interesting class of classical string solutions are ones in which the string position is extended in the AdS space and a point moving on the sphere (rather than the other way around). The first example of this type is the straight folded string rotating in $AdS_3 \subset AdS_5$.[18] One finds that for large S

$$E = 2\Gamma(\lambda) \log S + O(S^0), \tag{27}$$

where

$$\Gamma(\lambda) = \frac{\sqrt{\lambda}}{2\pi} + O(\lambda^0) \quad \text{for} \quad \lambda \gg 1. \tag{28}$$

The dual gauge theory operators are

$$\mathrm{Tr}(D_+^{s_1} Z\, D_+^{s_2} Z) \quad s_1 + s_2 = S. \tag{29}$$

Their anomalous dimensions take the same form as the energy with

$$\Gamma(\lambda) = \frac{\lambda}{4\pi^2} + O(\lambda^2) \quad \text{for} \quad \lambda \ll 1. \tag{30}$$

In order to compare these, one needs a procedure to extrapolate between small and large λ. In fact, an exact formula for the *cusp anomalous dimension* $\Gamma(\lambda)$ has been deduced using the assumption of exact integrability.[19] It passes all tests and is likely to be correct.

The generalization of this duality to twist J operators, which have the form

$$\mathrm{Tr}(D_+^{s_1} Z\, D_+^{s_2} Z \ldots D_+^{s_J} Z) \quad \text{where} \quad \sum s_l = S, \tag{31}$$

has been explored by Dorey and Losi.[20] They computed the corresponding conformal dimensions using an $SL(2)$ spin chain model. For large S the correspond classical string solutions are *spiky strings* with J cusps. The duality predictions are verified to the extent that they have been explored.

4. Conclusion

There has been a lot of progress in testing AdS/CFT in various special cases for maximally supersymmetric theories. Much of this progress has exploited the integrability of the string world-sheet theory on the one hand and the integrability of various spin-chain models that arise in studies of the dual gauge theory in the planar approximation on the other hand.

I have described classical string solutions and the dual gauge theory operators that are relevant to the D3-brane duality. There has also been very interesting work exploring analogous constructions for the M2-brane duality following the discovery of the ABJM theory. Much less is known about the M5-brane theory, though there has been significant progress when two of the dimensions wrap a Riemann surface.[21,22]

Acknowledgments

This work was supported in part by the U.S. Dept. of Energy under Grant No. DE-FG03-92-ER40701.

References

1. M. Shifman and A. Yung, "Supersymmetric Solitons and How They Help Us Understand Non-Abelian Gauge Theories," Rev. Mod. Phys. **79**, 1139 (2007) [arXiv:hep-th/0703267].
2. A. A. Tseytlin, "Quantum Strings in $AdS_5 \times S^5$ and AdS/CFT Duality," arXiv:0907.3238 [hep-th].
3. J. M. Maldacena, "The Large N Limit of Superconformal Field Theories and Ssupergravity," Adv. Theor. Math. Phys. **2**, 231 (1998) [Int. J. Theor. Phys. **38**, 1113 (1999)] [arXiv:hep-th/9711200].
4. L. Brink, J. H. Schwarz and J. Scherk, "Supersymmetric Yang–Mills Theories," Nucl. Phys. B **121**, 77 (1977).
5. G. 't Hooft, "A Planar Diagram Theory for Strong Interactions," Nucl. Phys. B **72**, 461 (1974).
6. J. H. Schwarz, "Superconformal Chern–Simons Theories," JHEP **0411**, 078 (2004) [arXiv:hep-th/0411077].
7. J. Bagger and N. Lambert, "Gauge Symmetry and Supersymmetry of Multiple M2-Branes," Phys. Rev. D **77**, 065008 (2008) [arXiv:0711.0955 [hep-th]].
8. A. Gustavsson, "Algebraic Structures on Parallel M2-Branes," Nucl. Phys. B **811**, 66 (2009) [arXiv:0709.1260 [hep-th]].
9. O. Aharony, O. Bergman, D. L. Jafferis and J. M. Maldacena, "$\mathcal{N} = 6$ Superconformal Chern–Simons-Matter Theories, M2-Branes and Their Gravity Duals," JHEP **0810**, 091 (2008) [arXiv:0806.1218 [hep-th]].
10. E. Witten, "Anti de Sitter Space and Holography," Adv. Theor. Math. Phys. **2**, 253 (1998) [arXiv:hep-th/9802150].

11. S. S. Gubser, I. R. Klebanov and A. M. Polyakov, "Gauge Theory Correlators from Non-critical String Theory," Phys. Lett. B **428**, 105 (1998) [arXiv:hep-th/9802109].
12. N. Gromov, V. Kazakov and P. Vieira, "Exact AdS/CFT Spectrum: Konishi Dimension at any Coupling," arXiv:0906.4240 [hep-th].
13. J. Plefka, "Spinning Strings and Integrable Spin Chains in the AdS/CFT Correspondence," Living Rev. Rel. **8**, 9 (2005) [arXiv:hep-th/0507136].
14. A. A. Tseytlin, "Spinning Strings and AdS/CFT Duality," arXiv:hep-th/0311139.
15. C. Kristjansen, M. Staudacher, and A. Tseytlin, eds., *Special Issue on Integrability and the AdS/CFT Correspondence*, J. Phys. A **42**, 250301 (2009).
16. D. E. Berenstein, J. M. Maldacena and H. S. Nastase, "Strings in Flat Space and pp Waves from $N = 4$ Super Yang Mills," JHEP **0204**, 013 (2002) [arXiv:hep-th/0202021].
17. C. G. Callan, H. K. Lee, T. McLoughlin, J. H. Schwarz, I. Swanson and X. Wu, "Quantizing String Theory in $AdS_5 \times S^5$: Beyond the pp-Wave," Nucl. Phys. B **673**, 3 (2003) [arXiv:hep-th/0307032].
18. S. S. Gubser, I. R. Klebanov and A. M. Polyakov, "A Semi-Classical Limit of the Gauge/String Correspondence," Nucl. Phys. B **636**, 99 (2002) [arXiv:hep-th/0204051].
19. N. Beisert, B. Eden and M. Staudacher, "Transcendentality and Crossing," J. Stat. Mech. **0701**, P021 (2007) [arXiv:hep-th/0610251].
20. N. Dorey and M. Losi, "Spiky Strings and Spin Chains," arXiv:0812.1704 [hep-th].
21. D. Gaiotto, "$\mathcal{N} = 2$ Dualities," arXiv:0904.2715 [hep-th].
22. D. Gaiotto and J. M. Maldacena, "The Gravity Duals of $\mathcal{N} = 2$ Superconformal Field Theories," arXiv:0904.4466 [hep-th].

Quantum Strings in $AdS_5 \times S^5$ and Gauge–String Duality

A. A. Tseytlin[†] [*]

Theoretical Physics Group, Blackett Laboratory
Imperial College
London SW7 2AZ, UK
[†] *E-mail: tseytlin@imperial.ac.uk*

We review some recent progress in understanding the spectrum of energies/dimensions of strings/operators in $AdS_5 \times S^5$ – planar $\mathcal{N}=4$ super Yang-Mills correspondence. We consider leading strong coupling corrections to the energy of lightest massive string modes in $AdS_5 \times S^5$, which should be dual to members of the Konishi operator multiplet in the SYM theory. This determines the general structure of strong-coupling expansion of the anomalous dimension of the Konishi operator. We use 1-loop results for semiclassical string states to extract information about the leading coefficients in this expansion.

Keywords: AdS/CFT duality; quantum superstrings; anomalous dimensions

1. Introduction

Two important problems of modern theoretical high energy physics are to understand quantum gauge theories at any coupling (with applications to both perturbative and non-perturbative issues) and to understand string theories in non-trivial backgrounds (e.g. Ramond-Ramond ones relevant for flux compactifications). The AdS/CFT duality relates the two questions suggesting solving them together rather than separately is the best strategy. This is the modern analog of the "harmonic oscillator" or "Ising" problem – to solve two most symmetric non-trivial 4-d gauge theory and 10-d superstring theory – planar $\mathcal{N} = 4$ SYM theory and its dual – free superstring in $AdS_5 \times S^5$. They have powerful hidden symmetries allowing to solve problem "in principle" using integrability methods. The $\mathcal{N} = 4$ SYM theory has maximal supersymmetry and conformal invariance but it is a priori unclear in which sense it could be integrable. Integrability might

---

[*] Also at Lebedev Institute, Moscow.

be expected in the spectrum of anomalous dimensions (as it was previously observed in YM gluonic sector – emergence of XXX spin chain Hamiltonian as 1-loop anomalous dimension operator [1]) but hidden symmetries should play broader role as they are "inherited" via AdS/CFT from 2-d integrable QFT – string σ-model. In that sense one may hope to use 2-d integrable QFT to solve 4-d CFT.

Superstring sigma model in $AdS_5 \times S^5$ is integrable in "canonical" sense being an example of a sigma model on symmetric space. Its classical equations admit infinite number of conserved charges. In particular, it is closely related (via Pohlmeyer reduction) to (super) sine-Gordon and non-abelian Toda models.[2] For example, special motions of strings are described by integrable 1-d mechanical systems (Neumann, etc.). The integrability of string theory extends to quantum level: we have evidence of that to 2 loops in α' expansion using AdS/CFT "bootstrap" reasoning (see ref. [3] and refs. there).

Quantum integrability of string theory sigma-model should control the spectrum of string energies on $R \times S^1$ (or anomalous dimensions of 2-d primary operators = vertex operators on $R^{1,1}$). It may also have implications for correlation functions of vertex operators (or closed-string scattering amplitudes) but to which extent that might be true is unclear (for example, in flat space the string action is just a collection of free oscillators but space-time amplitudes are rather nontrivial).

The quantum integrability of string theory for all values of string tension $\frac{\sqrt{\lambda}}{2\pi}$ should then imply integrability on the gauge theory side for any value of 't Hooft coupling $\lambda = g^2_{\rm YM} N_c$. Again, this should be true for the spectrum of anomalous dimensions of single-trace gauge invariant operators, while implications of integrability for correlation functions of such operators a priori seem rather limited (though this may turn out not be so given remarkable hidden symmetries in the on-shell gluon scattering amplitudes related to cusped Wilson loops discovered recently, see, e.g., [4)].

Regarding the spectrum of states, in the last 7 years there was an impressive progress (for reviews, see, e.g. [5]). The spectrum of "long" operators or "semiclassical" string states with large quantum numbers is now understood to be described by the Asymptotic Bethe Ansatz (ABA) in its final BES form.[6] The ABA was constructed using information from perturbative gauge theory (spin chain for 1-loop anomalous dimensions, ...) and perturbative string theory (classical and 1-loop phase,...), symmetries (magnon S-matrix), and the assumption of exact integrability. The consequences of ABA were checked against all available perturbative gauge and string theory data. A key example is the ABA prediction for the cusp anomalous dimension or dimension Δ of twist 2 operator $\text{Tr}(\Phi D_+^S \Phi)$. Namely, $\Delta - 2 =_{S\to\infty} f(\lambda) \ln S$ where $f(\lambda)$ satisfies exact integral equation solution of which is known in principle to any order in small λ or large λ expansion and agrees with known perturbative results (see, e.g., [3,7,8]):

$$f(\lambda \ll 1) = \frac{\lambda}{2\pi^2}\Big[1 - \frac{\lambda}{48} + \frac{11\lambda^2}{2^8 \cdot 45} - (\frac{73}{630} + \frac{4\zeta^2(3)}{\pi^6})\frac{\lambda^3}{2^7} + ...\Big] , \quad (1)$$

$$f(\lambda \gg 1) = \frac{\sqrt{\lambda}}{\pi}\Big[1 - \frac{3\ln 2}{\sqrt{\lambda}} - \frac{K}{(\sqrt{\lambda})^2} - ...\Big] , \quad (2)$$

where K is Catalan's constant.

More recently, there was a substantial progress towards understanding the spectrum of "short" operators, i.e. the spectrum of all quantum string states. This was achieved solely on the string side by using 2-d integrable field theory methodology generalizing the ABA describing states on $R^{1,1}$ to Thermodynamic Bethe Ansatz (TBA) which should be describing states on $R \times S^1$ [9,10]. It remains to be understood why inclusion of wrapping diagram contributions to anomalous dimensions should lead to a similar modification from the gauge theory perspective. The construction of TBA is rather non-trivial due to lack of 2-d Lorentz invariance in the standard "BMN-vacuum-adapted" l.c. gauge on string theory side. In few special cases the ABA "improved" by Luscher corrections is enough: 4- and 5-loop Konishi operator dimension, and 4-loop minimal twist operator dimension were computed[11] in this way. What remains to do is to thoroughly check the TBA predictions against perturbative string and gauge theory data.

The key example of a "short" operator is the Konishi operator[12] $\text{Tr}(\bar\Phi_i \Phi_i)$ for which the weak-coupling anomalous dimension is now known up to 5 loop orders [11,13]:

$$\gamma(\lambda \ll 1) = \frac{12\lambda}{(4\pi)^2}\Big[1 - \frac{4\lambda}{(4\pi)^2} + \frac{28\lambda^2}{(4\pi)^4} - [208 - 48\zeta(3) + 120\zeta(5)]\frac{\lambda^3}{(4\pi)^6} + 8[158 + 72\zeta(3) - 54\zeta^2(3) - 90\zeta(5) + 315\zeta(7)]\frac{\lambda^4}{(4\pi)^8} + ...\Big] \quad (3)$$

The planar perturbation theory should have a finite radius of convergence; suppose we sum up this series and re-expand at strong coupling λ – what should we expect to get? The AdS/CFT correspondence suggests[15] that $\gamma(\lambda \gg 1)$ should start with $\sim \sqrt[4]{\lambda}$ term. As was argued in [14] and will be discussed below, in general, string-theory arguments imply that

$$\gamma(\lambda \gg 1) = 2\sqrt[4]{\lambda} + \mathrm{b}_0 + \frac{b_1}{\sqrt[4]{\lambda}} + \frac{b_2}{(\sqrt[4]{\lambda})^2} + \frac{b_3}{(\sqrt[4]{\lambda})^3} + \dots \,. \tag{4}$$

Here the values of b_0, b_1 should be rational, while b_3 should be transcendental. The analysis in [14] leads to

$$\mathrm{b}_0 = -4, \quad b_1 = 1 \,, \quad b_2 = 0, \tag{5}$$

while b_3 contains $\zeta(3)$. At the same time, the numerical result found from the TBA/Y-system approach [10] gives b_1 that is approximately twice as big. The reason for this disagreement remains to be understood but it is very encouraging that the numerical values of b_0, b_1 in [10] suggest that they are integer, i.e. rational as it should be according to the string-theory logic.

There are many open questions remaining: (i) which is radius of convergence of weak coupling expansion? (ii) how to find the analytic form of strong-coupling expansion from TBA/Y-system? (iii) how to carry out direct matching of short operators onto string spectrum found in near-flat-space expansion? (iv) is there "level crossing" as one increases λ? (v) is strong-coupling expansion of dimensions of short operators Borel summable or not as in the case of the cusp anomaly? (vi) are there exponential corrections $e^{-a\sqrt{\lambda}}$ in (4) like in the cusp anomaly case [8]? There are also deeper issues: how to solve string theory from first principles – which are fundamental variables? how to preserve 2-d Lorentz invariance? how to prove quantum integrability? is there a useful lattice version of string "super-coset" sigma model?

Below we shall concentrate on explaining the origin of the expansion (4) of the energy of a string state which is dual to Konishi operator following ref. [14].

2. Approaches to finding energies of quantum string states in $AdS_5 \times S^5$

According to AdS/CFT energies of quantum strings in AdS should be the same as dimensions of the corresponding gauge theory operators. The aim is to compute the leading $\alpha' \sim \frac{1}{\sqrt{\lambda}}$ corrections to the energy of the "lightest" massive string state which should be dual to the Konishi operator in SYM theory.

The members of the Konishi multiplet are operators related to the lowest canonical dimension singlet by supersymmetry. They correspond to highest-weight states of $SO(2,4) \times SO(6)$ labelled by (see, e.g., [16]) $[J_2 - J_3, J_1 -$

$J_2, J_2 + J_3]^{\Delta_0}_{(s_L,s_R)}$ with the singlet being $[0,0,0]^2_{(0,0)}$. Then full dimension is $\Delta = \Delta_0 + \gamma(\lambda)$, $\Delta_0 = 2, \frac{5}{2}, 3, ..., 10$. All operators in supermultiplet have the same anomalous dimension γ.[a] Examples of such operators are $\mathrm{Tr}(\bar{\Phi}_i \Phi_i)$, $i = 1,2,3$, with $\Delta_0 = 2$; $\mathrm{Tr}([\Phi_1, \Phi_2]^2)$ in $su(2)$ sector with $\Delta_0 = 4$; $\mathrm{Tr}(\Phi_1 D_+^2 \Phi_1)$ in $sl(2)$ sector with $\Delta_0 = 4$

Assuming no "level crossing", the Konishi operator having lowest non-trivial dimension at weak coupling should be dual to the "lightest" among massive $AdS_5 \times S^5$ string states. At large $\sqrt{\lambda} = \frac{\mathrm{R}^2}{\alpha'}$ a "small" string at the center of AdS_5 is in nearly flat space. At strong coupling we are dealing with perturbative string theory: string states are built out of "flat-string" oscillators and there is a large degeneracy of mass spectrum which is lifted once the curvature is switched on. In flat space case we have string masses $m^2 = \frac{4(n-1)}{\alpha'}$, $n = \frac{1}{2}(N + \bar{N}) = 1, 2, ...,$ $N = \bar{N}$. The $n = 1$ is the massless IIB supergravity (BPS) level (l.c. vacuum $|0>$: $(8+8)^2 = 256$ states). The $n = 2$ is the first massive level; it is highly degenerate: $[(a^i_{-1} + S^a_{-1})|0>]^2 = [(8+8)\times(8+8)]^2$. Switching on $AdS_5 \times S^5$ background fields lifts degeneracy: states with "lightest mass" are thus at first excited string level and thus should correspond to Konishi multiplet.

The string spectrum in $AdS_5 \times S^5$ is organised [16] in long multiplets of $PSU(2,2|4)$: remarkably, flat-space string spectrum can be re-organized in multiplets of $SO(2,4) \times SO(6) \subset PSU(2,2|4)$. Namely, $SO(4) \times SO(5) \subset SO(9)$ reps can be lifted to $SO(4) \times SO(6)$ reps of $SO(2,4) \times SO(6)$, etc. Then the Konishi long multiplet $\hat{T}_1 = (1 + Q + Q \wedge Q + ...)[0,0,0]_{(0,0)}$ determines the Kaluza-Klein "floor" of 1-st excited string level: $H_1 = \sum_{J=0}^{\infty} [0, J, 0]_{(0,0)} \times \hat{T}_1$. States on the first excited level with $J \neq 0$ are outside Konishi multiplet, i.e. should have higher anomalous dimension.

For scalar massive state represented by a field in AdS_5 one expects to find $(-\nabla^2 + m^2)\Phi + ... = 0$, i.e. $\Delta(\Delta - 4) = (m\mathrm{R})^2 + O(\alpha') = 4(n-1)\frac{\mathrm{R}^2}{\alpha'} + O(\alpha')$ or $\Delta = 2 + \sqrt{(m\mathrm{R})^2 + 4 + O(\alpha')}$ and thus [15]

$$\Delta(\lambda \gg 1) = \sqrt{4(n-1)\sqrt{\lambda}} + ... \tag{6}$$

Thus for the first massive level $n = 2$: $\Delta = 2\sqrt[4]{\lambda} + ...$ What about subleading corrections? In general, comparison between gauge and string theory states is non-trivial. In gauge theory for $(\lambda \ll 1)$ the operators are

---

[a] That follows from the assumption of exact conformal algebra in which the dilatation operator commutes with supersymmetry generator Q on Q itself, i.e. Q has exact conformal dimension 1/2. Then members of supermultiplet obtained from "ground state" by acting by Q will have dimensions differing by (half)integers.

built out of free fields, canonical dimension Δ_0 determines states that can mix. In string theory with ($\lambda \gg 1$) the near-flat-space string states are built out of free oscillators, and the level n determines states that can mix. One non-trivial question is then about the meaning of Δ_0 at strong coupling and the meaning of n at weak coupling. A possible strategy is to relate states with same global charges and to assume "non-intersection principle" [17]: no level crossing for states with same quantum numbers as λ changes from strong to weak coupling.

There are several approaches to computation of corrections to string energies: (i) vertex operator approach: use $AdS_5 \times S^5$ string sigma model perturbation theory to find leading terms in anomalous dimension of corresponding vertex operator [17,18]; (ii) space-time effective action approach: use near-flat-space expansion and NSR vertex operators to reconstruct $\alpha' \sim \frac{1}{\sqrt{\lambda}}$ corrections to corresponding massive string state equation of motion [19]; (iii) "light-cone" quantization approach: start with light-cone gauge $AdS_5 \times S^5$ string action and compute corrections to energy of corresponding flat-space oscillator string state [20]; (iv) semiclassical approach: identify short string state as small-spin limit of semiclassical string state – reproduce the structure of strong-coupling corrections to short operators.[14,21]

To find the spectrum one should solve the marginality (1,1) conditions on vertex operators, i.e. to diagonalize the 2-d anomalous dimension operator. For example, the scalar anomalous dimension operator $\hat{\gamma}$ acts on $T(x) = \sum c_{n...m} x^n ... x^m$ or on the coefficients $c_{n...m}$. $\hat{\gamma}$ is a differential operator in target space found from the β-function for the corresponding perturbation T

$$
\begin{aligned}
&I = \frac{1}{4\pi\alpha'} \int d^2 z \left[G_{mn}(x) \partial x^m \bar{\partial} x^n + T(x) \right] \\
&\beta_T = -2T - \tfrac{\alpha'}{2} \hat{\gamma}\, T + O(T^2) \\
&\hat{\gamma} = \Omega^{mn} D_m D_n + ... + \Omega^{m...k} D_m ... D_k + ... \\
&\Omega^{mn} = G^{mn} + O(\alpha'^3), \qquad\qquad \Omega^{....} \sim \alpha'^n R^p_{....}
\end{aligned} \tag{7}
$$

Solving $-\hat{\gamma}\, T + m^2 T = 0$ is the same as diagonalising $\hat{\gamma}$. Similar approach applies to massless (graviton, ...) and massive states: e.g., $\beta^G_{mn} = \alpha' R_{mn} + O(\alpha'^3)$ gives the Lichnerowitz operator as anomalous dimension operator

$$
(\hat{\gamma} h)_{mn} = -D^2 h_{mn} + 2R_{mknl} h^{kl} - 2R_{k(m} h^k_{n)} + O(\alpha'^3) \tag{8}
$$

For massive string states in a curved background we should get

$$
\int d^D x \sqrt{g} \left[\Phi_{...} (-D^2 + m^2 + X) \Phi_{...} + ... \right]
$$

$$m^2 = \tfrac{4}{\alpha'}(n-1) \ , \qquad X = R_{....} + O(\alpha') \tag{9}$$

In the case of $AdS_5 \times S^5$ background

$$R_{mn} - \tfrac{1}{96}(F_5 F_5)_{mn} = 0, \qquad R = 0 \ , \qquad F_5^2 = 0$$

so for a scalar operator the leading term in X should vanish. This apparently implies that the leading α' correction to the scalar string state mass should be zero:

$$\begin{aligned} &\Big[- D^2 + m^2 + \mathcal{O}(\tfrac{1}{\sqrt{\lambda}})\Big] \Phi \ = 0 \ , \\ &\Delta = 2 + \sqrt{4(n-1) + 4 + \mathcal{O}(\tfrac{1}{\sqrt{\lambda}})} \ , \\ &\Delta_{(n=2)} = 2 + 2\sqrt[4]{\lambda}\ \Big[1 + \tfrac{1}{2\sqrt{\lambda}} + \mathcal{O}(\tfrac{1}{(\sqrt{\lambda})^2})\Big]. \end{aligned} \tag{10}$$

There are possible subtleties in this argument, so one should try to rederive this result using other approaches. Also, one should compare this to predictions for non-singlet Konishi descendant states – they should have the same anomalous dimension. For example, $\mathrm{Tr}[\Phi_1, \Phi_2]^2$ corresponding to $SO(6)$ (2,2,0) state $J_1 = J_2 = 2$ should be described by a string state with a tensor wave function $\Phi_{mn;kl}$.

To find $\hat{\gamma}$ for tensor states one may use effective action approach: derive equation of motion for a massive string field in curved background from quadratic effective action reconstructed from flat-space NSR S-matrix. An example is a totally symmetric NS-NS 10-d tensor – a state on the leading Regge trajectory in flat space. It is described by a symmetric tensor $\Phi_{\mu_1 \ldots \mu_{2n}}$ in a metric+RR background $(m^2 = \frac{4(n-1)}{\alpha'})$

$$\begin{aligned} L &= R - \tfrac{1}{2\cdot 5!} F_5^2 + O(\alpha'^3) \\ &- \tfrac{1}{2}(D_\mu \Phi D^\mu \Phi + m^2 \Phi^2) + \sum_{k\geq 1} (\alpha')^{k-1} \Phi X_k(R, F_5, D)\Phi + ... \end{aligned} \tag{11}$$

The assumption is that $\alpha' n R \ll 1$, i.e. $n \ll \sqrt{\lambda}$. That corresponds to a small massive string in the middle of AdS_5 so that near-flat-space expansion should be applicable. The resulting expression is [19] (a_i are from AdS_5 and m_k are from S^5)

$$\begin{aligned} L &= \tfrac{1}{2}\Phi_{\mu_1\cdots\mu_{2n}}(-D^2 + m^2)\Phi^{\mu_1\cdots\mu_{2n}} \\ &+ \frac{n^2}{\mathrm{R}^2}\left(\Phi_{a_1 a_2 \mu_3\cdots\mu_{2n}}\Phi^{a_1 a_2 \mu_3\cdots\mu_{2n}} - \Phi_{m_1 m_2 \mu_3\cdots\mu_{2n}}\Phi^{m_1 m_2 \mu_3\cdots\mu_{2n}}\right) + ... \end{aligned} \tag{12}$$

We may consider a particular tensor with S indices in AdS_5 and K indices in S^5 and then end up with the anomalous dimension operator

$$[-D^2 + (m^2 + \tfrac{K^2-S^2}{2\mathrm{R}^2})]\Phi = 0 \ , \qquad D^2 = D^2_{AdS_5} + D^2_{S_5} \ ,$$
$$m^2 = \tfrac{4}{\alpha'}(n-1) = \tfrac{2}{\alpha'}(S+K-2) \, . \tag{13}$$

This symmetric transverse traceless tensor corresponds the highest-weight state with the Young table labels $(\Delta, S, 0; K, J, 0)$ where J is an additional KK momentum. If we extract AdS_5 radius R and set $\sqrt{\lambda} = \frac{\mathrm{R}^2}{\alpha'}$ then [19]

$$(-D^2_{AdS_5} + M^2)\Phi = 0 \ ,$$
$$M^2 = 2\sqrt{\lambda}(S+K-2) + \tfrac{1}{2}(K^2 - S^2) + J(J+4) - K \, . \tag{14}$$

Equivalently, the marginality condition for the corresponding vertex operator is

$$0 = -\sqrt{\lambda}(S+K-2)$$
$$+\tfrac{1}{2}\Big[\Delta(\Delta-4) + \tfrac{1}{2}S(S-2) - \tfrac{1}{2}K(K-2) - J(J+4)\Big] + \mathcal{O}(\tfrac{1}{\sqrt{\lambda}}) . \tag{15}$$

The lowest BPS level is $n = \frac{1}{2}(S+K) = 1$ and the first massive level is $n = \frac{1}{2}(S+K) = 2$. A state from the first massive level on leading Regge trajectory is $S = K = 2, \ J = 0$ and then

$$\Delta = 2 + \sqrt{4\sqrt{\lambda} + 4 + \mathcal{O}(\tfrac{1}{\sqrt{\lambda}})} = 2 + 2\sqrt[4]{\lambda}\Big[1 + \frac{1}{\sqrt{\lambda}} + \mathcal{O}(\tfrac{1}{(\sqrt{\lambda})^2})\Big] . \tag{16}$$

This is the same prediction as for a singlet scalar state found above in (10). The constant term here $2 = \Delta_0 - 4$ corresponds to a $\Delta_0 = 6$ operator.[14]

Let us now review how the above expressions may appear directly in the vertex operator approach. If we would like to calculate the 2d anomalous dimensions from "first principles" we should start with the $AdS_5 \times S^5$ superstring theory action [22]

$$I = \frac{\sqrt{\lambda}}{4\pi} \int d^2\sigma \Big[-\partial N_p \bar{\partial} N^p + \partial n_k \bar{\partial} n_k + \text{fermions} \Big] \tag{17}$$
$$N_p N^p \equiv N_+ N_- - N_u N_u^* - N_v N_v^* = 1 \ , \qquad n_x n_x^* + n_y n_y^* + n_z n_z^* = 1 \, ,$$
$$N_\pm = N_0 \pm i N_5, \quad N_u = N_1 + i N_2, \ ..., \quad n_x = n_1 + i n_2, ...$$

and construct marginal (1,1) operators in terms of N_p and n_k. For example, a vertex operator for a dilaton field (highest-weight state) is

$$\mathrm{V}_J = \ (N_+)^{-\Delta} \ (n_x)^J \ \Big[-\partial N_p \bar{\partial} N^p + \partial n_k \bar{\partial} n_k + \text{fermions}\Big] \tag{18}$$

where $N_+ \equiv N_0 + iN_5 = \frac{1}{z}(z^2 + x_m x_m) \sim e^{it}$, $n_x \equiv n_1 + in_2 \sim e^{i\varphi}$. The marginality condition gives

$$0 = 2 - 2 + \frac{1}{2\sqrt{\lambda}}[\Delta(\Delta-4) - J(J+4)] + \mathcal{O}(\tfrac{1}{(\sqrt{\lambda})^2}) \ , \qquad (19)$$

i.e. $\Delta = 4 + J$ as it should be for a BPS state. A vertex operator for a bosonic string state on the leading Regge trajectory in flat space is

$$\mathrm{V}_S = e^{-iEt}\big(\partial \mathrm{x} \bar{\partial} \mathrm{x}\big)^{S/2} \ , \qquad \mathrm{x} = x_1 + ix_2 \ , \qquad a' E^2 = 2(S-2) \, .$$

Candidate operators for states on leading Regge trajectory in $AdS_5 \times S^5$ are

$$\mathrm{V}_J = (N_+)^{-\Delta}\big(\partial n_x \bar{\partial} n_x\big)^{J/2} \ , \qquad n_x \equiv n_1 + in_2 \, , \qquad (20)$$

$$\mathrm{V}_S(\xi) = (N_+)^{-\Delta}\big(\partial N_u \bar{\partial} N_u\big)^{S/2} \ , \qquad N_u \equiv N_1 + iN_2 \, . \qquad (21)$$

where we ignore fermionic terms and possible $\alpha' \sim \frac{1}{\sqrt{\lambda}}$ terms from diagonalization of 2-d anomalous dimension matrix. A non-trivial question is how such operators mix with operators with same charges and dimension. In general, $\big(\partial n_x \bar{\partial} n_x\big)^{J/2}$ mixes with singlets

$$(n_x)^{2p+2q}(\partial n_x)^{J/2-2p}(\bar{\partial} n_x)^{J/2-2q}(\partial n_m \partial n_m)^p(\bar{\partial} n_k \bar{\partial} n_k)^q \, .$$

For operators like $O_{\ell,s} = f_{k_1...k_\ell m_1...m_{2s}} n_{k_1}...n_{k_\ell} \partial n_{m_1} \bar{\partial} n_{m_2} ... \partial n_{m_{2s-1}} \bar{\partial} n_{m_{2s}}$ this question was studied, e.g., in ref. [23].[b] An example of higher-level scalar operator is

$$N_+^{-\Delta}[(\partial n_k \bar{\partial} n_k)^r + ...] \ , \qquad r = 1, 2, ...$$

for which [24]

$$0 = -2(r-1) + \frac{1}{2\sqrt{\lambda}}[\Delta(\Delta-4) + 2r(r-1)] + \frac{1}{(\sqrt{\lambda})^2}[\tfrac{2}{3}r(r-1)(r-\tfrac{7}{2}) + 4r] + \dots \, . \qquad (22)$$

For $r = 2$ this is a candidate for a state on the first excited level. However, the contribution of fermions may change the value of the 2-d anomalous dimension.

---

[b] In the simplest case of the operator $f_{k_1...k_\ell} n_{k_1}...n_{k_\ell}$ with traceless $f_{k_1...k_\ell}$ it has the same anomalous dimension $\hat{\gamma}$ as its highest-weight representative $V_J = (n_x)^J$, i.e. $\hat{\gamma} = 2 - \frac{1}{2\sqrt{\lambda}} J(J+4) + \dots$. This is the same result as found for a scalar spherical harmonic that solves the Laplace equation on S^5. Similarly for the AdS_5 or $SO(2,4)$ model: replacing n_x^J and $\partial n_m \bar{\partial} n_m$ with $N_+^{-\Delta}$ and $\partial N^p \bar{\partial} N_p$, with $J = -\Delta$ and $\frac{1}{\sqrt{\lambda}} \to -\frac{1}{\sqrt{\lambda}}$. For example, the dimension of $N_+^{-\Delta} \partial N^p \bar{\partial} N_p$ is $\hat{\gamma} = \frac{1}{2\sqrt{\lambda}}\Delta(\Delta-4) + \mathcal{O}(\frac{1}{(\sqrt{\lambda})^2})$.

As an illustration, for operators with two spins $J_1 = K,\ J_2 = J$ in S^5:

$$V_{K,J} = N_+^{-\Delta} \sum_{u,v=0}^{K/2} c_{uv} M_{uv}$$
$$M_{uv} \equiv n_y^{J-u-v} n_x^{u+v} (\partial n_y)^u (\partial n_x)^{K/2-u} (\bar\partial n_y)^v (\bar\partial n_x)^{K/2-v} \tag{23}$$

the highest and lowest eigen-values of the 1-loop anomalous dimension matrix are [18]

$$2 - K + \frac{1}{2\sqrt{\lambda}}[\Delta(\Delta-4) - \tfrac{1}{2}K(K+10) - J(J+4) - 2JK] + \mathcal{O}(\tfrac{1}{(\sqrt{\lambda})^2}),$$
$$2 - K + \frac{1}{2\sqrt{\lambda}}[\Delta(\Delta-4) - \tfrac{1}{2}K(K+6) - J(J+4)] + \mathcal{O}(\tfrac{1}{(\sqrt{\lambda})^2}). \tag{24}$$

Again, the fermionic contributions may alter terms linear in K.

The main question on the way to finding energies or Δ's of the corresponding states is thus how to take the $AdS_5 \times S^5$ fermionic contributions into account? To this end we shall follow ref. [14] and employ the semiclassical approach based on the full superstring action.

3. Semiclassical approach: small-spin expansion

In semiclassical expansion for an energy of a string state one fixes the "classical" spin parameter $\mathcal{J} = \frac{J}{\sqrt{\lambda}}$ and expands in large string tension $\sim \sqrt{\lambda}$,

$$E = E(\frac{J}{\sqrt{\lambda}}, \sqrt{\lambda}) = \sqrt{\lambda}\mathcal{E}_0(\mathcal{J}) + \mathcal{E}_1(\mathcal{J}) + \frac{1}{\sqrt{\lambda}}\mathcal{E}_2(\mathcal{J}) + ... \tag{25}$$

In the "short" string limit $\mathcal{J} \ll 1$

$$\mathcal{E}_n = \sqrt{\mathcal{J}}\ (a_{0n} + a_{1n}\mathcal{J} + a_{2n}\mathcal{J}^2 + ...) \tag{26}$$

Since $\sqrt{\lambda} \gg 1$ and $\mathcal{J} = \frac{J}{\sqrt{\lambda}}$=fixed, here $J \sim \sqrt{\lambda} \gg 1$. If we knew all the terms in this expansion we could express $\mathcal{J}$ in terms of J, fix J to a finite value and re-expand in $\sqrt{\lambda}$. Then we would get

$$E = \sqrt{\sqrt{\lambda}J}\Big[a_{00} + \frac{a_{10}J + a_{01}}{\sqrt{\lambda}} + \frac{a_{20}J^2 + a_{11}J + a_{02}}{(\sqrt{\lambda})^2} + ...\Big]. \tag{27}$$

To trust the coefficient of $\frac{1}{(\sqrt{\lambda})^n}$ one would need to know coefficients of up to n-loop terms, e.g., the classical a_{10} and the 1-loop a_{01} coefficients are sufficient in order to fix the $\frac{1}{\sqrt{\lambda}}$ term.

Since here we are interested in a short string probing the near-flat-space limit, we may (i) start with classical string solutions in flat space representing states at 1-st excited string level, and (ii) then embed them into $AdS_5 \times S^5$ and compute the 1-loop correction to their energy. The two basic examples are [14]: (1) circular string with 2 spins in two orthogonal planes, and (2) folded string spinning in one plane.

The rigid circular string rotating in two planes of flat R^4 space is described by

$$t = \kappa\tau \ , \qquad \mathrm{x}_x \equiv x_1 + ix_2 = \ a\ e^{i(\tau+\sigma)} \ , \qquad \mathrm{x}_y \equiv x_3 + ix_4 = \ a\ e^{i(\tau-\sigma)} \ ,$$

$$E_{\rm flat} = \tfrac{\kappa}{\alpha'} = \sqrt{\tfrac{4}{\alpha'} J} \ , \qquad J_1 = J_2 = J = \tfrac{a^2}{\alpha'} \ . \tag{28}$$

Identifying oscillator modes that are excited on this solution we may associate it with a quantum string state created by

$$e^{-iEt}\ \Big[(\partial n_x \bar\partial \mathrm{x}_x)^{\frac{J_1}{2}} (\partial n_y \bar\partial \mathrm{x}_y)^{\frac{J_2}{2}} + ...\Big], \qquad \alpha' E^2 = 2(J_1 + J_2 - 2)\,. \tag{29}$$

In the $J_1 = J_2$ case to get the quantum-state analog of classical expression one would need to shift $J \to J - 1$, i.e.

$$E_{\rm flat} = \sqrt{\tfrac{4}{\alpha'}(J-1)} \ .$$

Then $J_1 = J_2 = 2$ corresponds to state on 1-st string level $n = 2$.

A folded string rotating in a plane in flat space is represented by

$$t = \kappa\tau \ , \qquad \mathrm{x}_1 \equiv x_1 + ix_2 = \ a \sin\sigma\ e^{i\tau} \ ,$$

$$E_{\rm flat} = \sqrt{\tfrac{2}{\alpha'} S} \ , \qquad S = \tfrac{a^2}{2\alpha'} \ . \tag{30}$$

This is a semiclassical counterpart of a quantum string state on the leading Regge trajectory created by the vertex operator

$$e^{-iEt}\ \Big[(\partial \mathrm{x}_x \bar\partial \mathrm{x}_x)^{\frac{S}{2}} + ...\Big] \ , \qquad \alpha' E^2 = 2(S-2) \ . \tag{31}$$

There are 3 obvious choices of how to embed the circular solution into $AdS_5 \times S^5$: (i) the two 2-planes may belong to S^5: $J_1 = J_2$ "small string"; (ii) the two 2-planes may belong to AdS_5: $S_1 = S_2$ "small string"; (iii) one plane in AdS_5 and the other in S^5: $S = J$ "small string". There are similar 3 choices for the folded string. We may then study each case in $AdS_5 \times S^5$, compute 1-loop correction to energy and interpolate to small values of S, J and try to match to states in the Konishi multiplet table. Getting the same expressions up to a constant shift would verify the universality of the strong-coupling expansion of the 4-d anomalous dimension of the dual gauge theory operators in the same supermultiplet.

The final result found in [14] is

$$E \equiv \Delta = 2\sqrt[4]{\lambda} + b_0 + \frac{b_1}{\sqrt[4]{\lambda}} + \frac{b_2}{\sqrt{\lambda}} + \frac{b_3}{(\sqrt[4]{\lambda})^3} ... \tag{32}$$

$$b_0 = \Delta_0 - 4 \ , \qquad\qquad b_1 = 1 \tag{33}$$

where $\Delta_0 = 4$ for the 3 circular string cases and $\Delta_0 = 6$ for the 3 folded string cases. The value of the coefficient b_2 is sensitive to the 2-loop string corrections which were not computed so far but we conjecture that that it is zero due to supersymmetry (see ref. [14]).

The simplest example is circular rotating string in S^5 with $J_1 = J_2 = J$ (see refs. [25, 14]) which should be dual to the Konishi descendant with $J_1 = J_2 = 2$: $\mathrm{Tr}([\Phi_1, \Phi_2]^2)$. The above flat solution can be directly embedded into $R_t \times S^5$ inside of $AdS_5 \times S^5$ – it represents a string on a *small* sphere inside S^5 $(X_1^2 + ... + X_6^2 = 1)$

$$X_1 + iX_2 = a\ e^{i(\tau+\sigma)}, \qquad X_3 + iX_4 = a\ e^{i(\tau-\sigma)} \ ,$$
$$X_5 + iX_6 = \sqrt{1-2a^2}, \qquad\qquad t = \kappa\tau \ ,$$
$$\mathcal{J} = \mathcal{J}_1 = \mathcal{J}_2 = a^2, \qquad \mathcal{E}^2 = \kappa^2 = 4\mathcal{J} \ . \tag{34}$$

Remarkably, the exact classical energy E_0 is just as in flat space

$$E_0 = \sqrt{\lambda}\mathcal{E} = \sqrt{4\sqrt{\lambda}J} \ , \qquad\qquad J = \sqrt{\lambda}\mathcal{J} \ . \tag{35}$$

The 1-loop quantum string correction to the energy is given by the sum of the bosonic and fermionic fluctuation frequencies

$$E_1 = \frac{1}{2\kappa}\sum_{n=-\infty}^{\infty}\Big[4\omega_n + 2n + \omega_{n+} + \omega_{n-} - 4(\omega^f_{n+} + \omega^f_{n-})\Big] \ , \tag{36}$$

where

$$AdS_5: \qquad \omega_n^2 = n^2 + 4\mathcal{J} \ ,$$
$$S^5: \qquad \omega_{n\pm}^2 = n^2 + 4(1-\mathcal{J}) \pm 2\sqrt{4(1-\mathcal{J})n^2 + 4\mathcal{J}^2} \ ,$$
$$\text{Fermions}: \quad 4\times \qquad \omega_{n\pm}^{2\,f} = n^2 + 1 + \mathcal{J} \pm \sqrt{4(1-\mathcal{J})n^2 + 4\mathcal{J}} \ .$$

Expanding in small $\mathcal{J}$ and doing the sums one finds (UV divergences cancel)

$$E_1 = \frac{1}{\sqrt{\mathcal{J}}}\Big[\mathcal{J} - \frac{3}{8}[1 + 8\zeta(3)]\mathcal{J}^2 + \dots\Big] \ ,$$
$$E = E_0 + E_1 = 2\sqrt{\sqrt{\lambda}J}\Big[1 + \frac{1}{2\sqrt{\lambda}} - \tfrac{3}{16}[1 + 8\zeta(3)]\frac{J}{(\sqrt{\lambda})^2} + ...\Big] \ . \tag{37}$$

To ensure the correct short string limit in flat space we need to shift $J \to J - 1$. F $J = J_1 = J_2 = 2$ case that would suggest that for a Konishi state $[2, 0, 2]_{(0,0)}$

$$E = 2\sqrt[4]{\lambda}\Big[1 + \tfrac{1}{2\sqrt{\lambda}} + \mathcal{O}(\tfrac{1}{(\sqrt{\lambda})^2})\Big] . \tag{38}$$

The dual state in the Konishi table has $\Delta_0 = 4$. Thus in (32) one gets $b_1 = 1$ and $b_2 = 0$ (at 1-loop order) while b_3 is transcendental (1-loop contribution to it contains $\zeta(3)$).

The same results are found for the other (2 circular and 3 folded) string solutions representing 5 other states at the 1-st massive string level dual to 5 particular operators in the Konishi multiplet table. We refer to [14] for the details.

4. Conclusions

Due to an impressive recent progress based on integrability methods we are now at the beginning of understanding the quantum string spectrum in $AdS_5 \times S^5$ or the spectrum of "short" operators in planar $\mathcal{N}$=4 SYM theory for any value of 't Hooft coupling. The predictions of the integrability approach are still to be checked against perturbative string results. This requires better understanding of perturbative quantum $AdS_5 \times S^5$ string theory.

5. Acknowledgments

I would like to thank R. Roiban for collaboration on ref. [14] and also acknowledge useful discussions with N. Gromov and R. Metsaev. I am grateful to the organizers of the Shifmania workshop for their kind invitation and hospitality.

References

1. L. N. Lipatov, "Asymptotic behavior of multicolor QCD at high energies in connection with exactly solvable spin models," JETP Lett. **59**, 596 (1994) [Pisma Zh. Eksp. Teor. Fiz. **59**, 571 (1994)].
   L. D. Faddeev and G. P. Korchemsky, "High-energy QCD as a completely integrable model," Phys. Lett. B **342**, 311 (1995) [hep-th/9404173].
2. M. Grigoriev and A. A. Tseytlin, "Pohlmeyer reduction of $AdS_5 \times S^5$ superstring sigma model," Nucl. Phys. B **800**, 450 (2008) [0711.0155 [hep-th]].
3. R. Roiban and A. A. Tseytlin, "Strong-coupling expansion of cusp anomaly from quantum superstring," JHEP **0711**, 016 (2007) [0709.0681 [hep-th]].
4. J. M. Drummond, J. Henn, G. P. Korchemsky and E. Sokatchev, "Dual

superconformal symmetry of scattering amplitudes in $\mathcal{N}=4$ super-Yang-Mills theory," 0807.1095 [hep-th].
5. A. A. Tseytlin, "Spinning strings and AdS/CFT duality," in: "From Fields to Stings: Circumnavigating Theoretical Physics", M. Shifman, A. Vainshtein, and J. Wheater, eds. (World Scientific, 2004), vol. 2, 1648-1707. [hep-th/0311139].
J. Plefka, "Spinning strings and integrable spin chains in the AdS/CFT correspondence," Living Rev. Rel. **8**, 9 (2005) [hep-th/0507136].
Special issue on integrability and AdS/CFT correspondence, C. Kristjansen, M. Staudacher and A. Tseytlin, eds. J. Phys. A **42**, 250301 (2009)
6. N. Beisert, B. Eden and M. Staudacher, "Transcendentality and crossing," J. Stat. Mech. **0701**, P021 (2007) [hep-th/0610251].
7. A. V. Kotikov, L. N. Lipatov and V. N. Velizhanin, "Anomalous dimensions of Wilson operators in N = 4 SYM theory," Phys. Lett. B **557**, 114 (2003) [hep-ph/0301021].
Z. Bern, M. Czakon, L. J. Dixon, D. A. Kosower and V. A. Smirnov, "The Four-Loop Planar Amplitude and Cusp Anomalous Dimension in Maximally Supersymmetric Yang-Mills Theory," Phys. Rev. D **75**, 085010 (2007) [hep-th/0610248].
8. B. Basso, G. P. Korchemsky and J. Kotanski, "Cusp anomalous dimension in maximally supersymmetric Yang-Mills theory at strong coupling," Phys. Rev. Lett. **100**, 091601 (2008) [0708.3933 [hep-th]].
9. J. Ambjorn, R. A. Janik and C. Kristjansen, "Wrapping interactions and a new source of corrections to the spin-chain / string duality," Nucl. Phys. B **736**, 288 (2006) [hep-th/0510171].
G. Arutyunov and S. Frolov, "On String S-matrix, Bound States and TBA," JHEP **0712**, 024 (2007) [0710.1568 [hep-th]].
G. Arutyunov and S. Frolov, "String hypothesis for the $AdS_5 \times S^5$ mirror," JHEP **0903**, 152 (2009) [0901.1417 [hep-th]].
N. Gromov, V. Kazakov and P. Vieira, "Integrability for the Full Spectrum of Planar AdS/CFT," 0901.3753 [hep-th].
D. Bombardelli, D. Fioravanti and R. Tateo, "Thermodynamic Bethe Ansatz for planar AdS/CFT: a proposal," 0902.3930 [hep-th].
N. Gromov, V. Kazakov, A. Kozak and P. Vieira, "Integrability for the Full Spectrum of Planar AdS/CFT II," 0902.4458 [hep-th].
G. Arutyunov and S. Frolov, "Thermodynamic Bethe Ansatz for the $AdS_5 \times S^5$ Mirror Model," JHEP **0905**, 068 (2009) [0903.0141 [hep-th]].
G. Arutyunov and S. Frolov, "Simplified TBA equations of the $AdS_5 \times S^5$ mirror model," 0907.2647 [hep-th].
10. N. Gromov, V. Kazakov and P. Vieira, "Exact AdS/CFT spectrum: Konishi dimension at any coupling," 0906.4240 [hep-th].
11. Z. Bajnok and R. A. Janik, "Four-loop perturbative Konishi from strings and finite size effects for multiparticle states," Nucl. Phys. B **807**, 625 (2009) [0807.0399 [hep-th]].
Z. Bajnok, R. A. Janik and T. Lukowski, "Four loop twist two, BFKL, wrapping and strings," Nucl. Phys. B **816**, 376 (2009) [0811.4448 [hep-th]].

Z. Bajnok, A. Hegedus, R. A. Janik and T. Lukowski, "Five loop Konishi from AdS/CFT," 0906.4062 [hep-th].

12. K. Konishi, "Anomalous Supersymmetry Transformation Of Some Composite Operators In SQCD," Phys. Lett. B **135**, 439 (1984).
M. Bianchi, S. Kovacs, G. Rossi and Y. S. Stanev, "Properties of the Konishi multiplet in N = 4 SYM theory," JHEP **0105**, 042 (2001) [hep-th/0104016].
13. F. Fiamberti, A. Santambrogio, C. Sieg and D. Zanon, "Anomalous dimension with wrapping at four loops in N=4 SYM," Nucl. Phys. B **805**, 231 (2008) [arXiv:0806.2095 [hep-th]].
14. R. Roiban and A. A. Tseytlin, "Quantum strings in $AdS_5 \times S^5$: strong-coupling corrections to dimension of Konishi operator," 0906.4294 [hep-th].
15. S. S. Gubser, I. R. Klebanov and A. M. Polyakov, "Gauge theory correlators from non-critical string theory," Phys. Lett. B **428**, 105 (1998) [hep-th/9802109].
16. M. Bianchi, J. F. Morales and H. Samtleben, "On stringy $AdS_5 \times S^5$ and higher spin holography," JHEP **0307**, 062 (2003) [hep-th/0305052].
17. A. M. Polyakov, "Gauge fields and space-time," Int. J. Mod. Phys. A **17S1**, 119 (2002) [hep-th/0110196].
18. A. A. Tseytlin, "On semiclassical approximation and spinning string vertex operators in $AdS_5 \times S^5$," Nucl. Phys. B **664**, 247 (2003) [hep-th/0304139].
19. B. A. Burrington and J. T. Liu, "Spinning strings in $AdS_5 \times S^5$: A worldsheet perspective," Nucl. Phys. B **742**, 230 (2006) [hep-th/0512151].
20. R. R. Metsaev, C. B. Thorn and A. A. Tseytlin, Nucl. Phys. B **596**, 151 (2001) [hep-th/0009171].
21. A. Tirziu and A. A. Tseytlin, "Quantum corrections to energy of short spinning string in AdS_5," Phys. Rev. D **78**, 066002 (2008) [0806.4758 [hep-th]].
22. R. R. Metsaev and A. A. Tseytlin, "Type IIB superstring action in $AdS_5 \times S^5$ background," Nucl. Phys. B **533**, 109 (1998) [hep-th/9805028].
23. F. Wegner, "Anomalous dimensions of high-gradient operators in the n-vector model in $2 + \epsilon$ dimensions", Z. Phys. B **78**, 33 (1990).
24. V. E. Kravtsov, I. V. Lerner and V. I. Yudson, "Anomalous Dimensions Of High Gradient Operators In The Extended Nonlinear Sigma Model And Distribution Of Mesoscopic Fluctuations," Phys. Lett. A **134**, 245 (1989).
G. E. Castilla and S. Chakravarty, "Is the phase transition in the Heisenberg model described by the $(2+\epsilon)$ expansion of the non-linear sigma-model?," Nucl. Phys. B **485**, 613 (1997) [cond-mat/9605088].
25. S. Frolov and A. A. Tseytlin, "Multi-spin string solutions in $AdS_5 \times S^5$," Nucl. Phys. B **668**, 77 (2003) [hep-th/0304255].
S. Frolov and A. A. Tseytlin, "Quantizing three-spin string solution in $AdS_5 \times S^5$," JHEP **0307**, 016 (2003) [hep-th/0306130].

M2-branes and AdS/CFT

Igor R. Klebanov

Joseph Henry Laboratories and Princeton Center for Theoretical Science
Princeton University, Princeton, NJ 08544, USA

We provide a brief introduction to the ABJM theory, the level k $U(N) \times U(N)$ superconformal Chern-Simons matter theory which has been conjectured to describe N coincident M2-branes. We discuss its dual formulation in terms of M-theory on $AdS_4 \times S^7/\mathbb{Z}_k$ and review some of the evidence in favor of the conjecture. We end with a brief discussion of the important role played by the monopole operators.

1. Introduction

It is a real pleasure to come to Minnesota to celebrate Misha Shifman's 60th birthday. I have enjoyed his friendship for over 20 years and have been greatly influenced by his many scientific accomplishments. In particular, Misha's contributions to the understanding of supersymmetric gauge dynamics have been truly seminal. The famous instanton calculation of the gluino condensate,[1] which is responsible for the chiral symmetry breaking in $\mathcal{N} = 1$ supersymmetric gluodynamics, stands out as one of the earliest exact non-perturbative effects uncovered in 4-d gauge theory. Many years later, I was happy to participate in the development of a confining model of gauge gravity duality,[2] which exhibits the same pattern of chiral symmetry breaking (for reviews see [3]).

In these notes I review another, more recent development in gauge gravity duality, which concerns the dynamics of superconformal Chern-Simons gauge theories with extended supersymmetry. Over the past two years, important progress has taken place in understanding the world volume theory of coincident supermembranes of M-theory, the M2-branes. These interacting superconformal theories are expected to be the Chern-Simons gauge theories coupled to massless matter.[4] Remarkable new theories with extended $\mathcal{N} \geq 6$ superconformal invariance have been constructed.[5–9] In particular, the ABJM theory[9] is a $U(N) \times U(N)$ Chern-Simons gauge theory with integer levels $(k, -k)$; it becomes weakly interacting for $k \gg N$. Considerable amount of evidence has accumulated that this gauge theory describes the low-energy behavior of N M2-branes placed at the $\mathcal{N} = 6$ supersymmetric orbifold $\mathbb{R}^8/\mathbb{Z}_k$. Therefore, it is conjectured to be dual, in the sense of the AdS/CFT correspondence,[10–12] to M-theory on $AdS_4 \times S^7/\mathbb{Z}_k$. This dual description becomes very useful for $N \gg k$.

Before discussing the role that M2-branes play in the AdS/CFT correspondence, let me review the correspondence for coincident D3-branes, which in flat space realize the well-known $\mathcal{N} = 4$ supersymmetric Yang-Mills theory (for reviews see for example [13–15]). The type IIB supergravity background created by a stack of N coincident D3-branes can be written as:

$$\begin{aligned} ds^2 &= h^{-1/2}(-dt^2 + dx_1^2 + dx_2^2 + dx_3^2) + h^{1/2}(dr^2 + r^2 d\Omega_5^2), \\ h(r) &= 1 + \frac{L^4}{r^4}, \qquad L^4 = 4\pi g_s N \alpha'^2, \\ g_s F_5 &= (1 + *)\, d^4x \wedge dh^{-1}(r). \end{aligned} \tag{1}$$

In the "near horizon limit," which corresponds to taking $r \ll L$, the term $h(r)$ can be approximated by L^4/r^4 and, therefore, it is easy to see that the geometry of space-time becomes that of $AdS_5 \times S^5$.

Now, let us look at this system from the point of view of the world volume theory. Considering only one D3-brane in flat space, one can see that in the low energy limit it is a free gauge theory in $3 + 1$ dimensions characterized by the lagrangian:

$$\mathcal{L} = -\frac{1}{4}F_{\mu\nu}^2 - \frac{1}{2}\sum_{i=1}^{6}(\partial_\mu \phi^i)^2 + \ldots, \tag{2}$$

where the dots indicate the fermionic terms. This lagrangian corresponds to the $\mathcal{N} = 4$ supersymmetric $U(1)$ gauge theory. Since all the fields are uncharged, this is a free theory, which implies that the moduli space is simply $\mathbb{R}^6$.

This description can be generalized to a system of N parallel D3-branes, where in the low energy limit we have the $\mathcal{N} = 4$ supersymmetric Yang-Mills theory with gauge group $U(N)$ and lagrangian:

$$\mathcal{L} = \mathrm{Tr}\left[-\frac{1}{4}F_{\mu\nu}^2 - (D_\mu \phi^i)^2 + \frac{g^2}{4}[\phi^i, \phi^j]^2\right] + \ldots, \tag{3}$$

where the scalar fields ϕ^i transform in the adjoint representation of the gauge group. The moduli space of this theory consists of diagonal matrices

$$\phi^i = \mathrm{diag}(\phi_1^i, \phi_2^i, \ldots, \phi_N^i), \tag{4}$$

with the gauge transformations acting on the fields by a permutation of the eigenvalues. From this, it can be understood that the moduli space of this theory is simply $(\mathbb{R}^6)^N/S_N$, where S_N is the permutation group of N elements.

There is a $U(1)$ subgroup under which all the fields are neutral; it decouples from the remaining $SU(N)$ gauge group. This $U(1)$ parameterizes the motion of the center of mass of the stack of N D3-branes. Thus, we can conclude that, in the low energy limit, the gauge theory living on a stack of N D3-branes is an $SU(N)$ supersymmetric Yang-Mills theory, which can be studied perturbatively in great detail. Therefore, a stack of N D3-branes in flat space can be looked at from two points of view: that of string theory which, in the near horizon limit, leads to Type IIB string theory in $AdS_5 \times S^5$ with N units of the self-dual 5-form Ramond-Ramond flux, and that of gauge theory that, in the low energy limit, leads to an $\mathcal{N} = 4$ super Yang-Mills theory with gauge group $SU(N)$. The strong version of the AdS/CFT conjecture claims that these two limits are equivalent. Therefore, the 4-d gauge fields and the 10-d strings provide two different descriptions of the same theory.

The correspondence may be generalized to the cases where a stack of parallel D3-branes is placed at the tip of a Ricci-flat cone. For this type of background, the metric has the same form as (1), but now $d\Omega_5$ refers to the base of the cone, which is an Einstein manifold often denoted as Y^5; the limit where $r \to 0$ gives the geometry of $AdS_5 \times Y^5$. When the cone is a Calabi-Yau space, the base of the cone is called a Sasaki-Einstein manifold; in these cases the supersymmetry preserved by the theory is one quarter of that of $AdS_5 \times S^5$ that we mentioned above. Hence, these are $\mathcal{N} = 1$ superconformal theories.

The example we will now consider is when the Calabi-Yau cone is the *conifold* (reviews on the conifold theory can be found in [3]). This case is of interest here because its gauge theory is related to that appearing in the AdS/CFT duality with M2-branes. Let us consider the 10 dimensional space-time $M_4 \times \mathcal{C}$, where M_4 is the usual 4 dimensional Minkowski space-time and $\mathcal{C}$ is the conifold. Introducing 4 complex variables u_i, with $i = 1, \ldots, 4$, this manifold can be defined as the locus where the following condition is satisfied:

$$u_1^2 + u_2^2 + u_3^2 + u_4^2 = 0 \tag{5}$$

Given this condition, it is clear that the conifold has 3 complex dimensions and, therefore, 6 real dimensions. If we place a D3-brane near the tip of the complex cone defined by (5), we can examine the gauge theory arising on the brane in the low energy limit. This is an $\mathcal{N} = 1$ supersymmetric Yang-Mills theory with gauge group $U(1)_1 \times U(1)_2$ and four chiral fields, that we shall denote as z_1, z_2, w_1 and w_2 [16]. They transform under the gauge groups according to Table 1.

Table 1.

	$U(1)_1$	$U(1)_2$
z_1, z_2	1	-1
w_1, w_2	-1	1

Note: Abelian charges of the chiral fields of the conifold theory under the two gauge groups.

Interestingly, all the matter fields are neutral under the diagonal gauge field[a] $A_+ = A + \hat{A}$, and are charged under the anti-diagonal gauge field $A_- = A - \hat{A}$. Also, it is important to note that this theory has no superpotential for the one D3-brane case we are considering. In order to determine the moduli space, we need to solve the D-term equations (since there's no superpotential, there are no F-term equations to impose). The D-terms for this theory can be written as:

$$\frac{D^2}{g^2} + D\left(|z_1|^2 + |z_2|^2 - |w_1|^2 - |w_2|^2\right), \tag{6}$$

where D is the usual auxiliary field. Integrating out the auxiliary field D, the D-terms can be written as (neglecting overall numerical factors):

$$g^2\left(|z_1|^2 + |z_2|^2 - |w_1|^2 - |w_2|^2\right)^2. \tag{7}$$

The moduli space of vacua is obtained by imposing the vanishing of the D-terms and dividing by the gauge group. Thus, we need to impose the condition:

$$|z_1|^2 + |z_2|^2 - |w_1|^2 - |w_2|^2 = 0, \tag{8}$$

and the identifications:

$$z_i \sim z_i e^{i\alpha}, \qquad w_i \sim w_i e^{-i\alpha}. \tag{9}$$

The moduli space of this gauge theory is precisely the conifold. This manifold can be viewed as a cone over $(SU(2) \times SU(2))/U(1)$, which in the

[a] We adopt the convention that non hatted quantities refer to gauge group 1, whereas hatted quantities refer to gauge group 2.

literature is often referred to as $T^{1,1}$. The two $SU(2)$'s can be thought of as flavour symmetries, one acting on the z_i's, which transform as a doublet, and the other acting on the w_i's, transforming as a doublet. Also, the $U(1)$ group is the interacting anti-diagonal subgroup A_- under which the chiral fields are charged.

Another way to define the conifold is by introducing four complex variables, u_{ij}, defined as:

$$u_{11} = u_1 + iu_2, \qquad u_{12} = iu_3 - u_4,$$
$$u_{21} = iu_3 + u_4, \qquad u_{22} = u_1 - iu_2, \tag{10}$$

related to the chiral fields by the relation $u_{ij} = z_i w_j$. In terms of the new variables, the defining equation for the conifold can be written simply as:

$$\det u_{ij} = 0. \tag{11}$$

This way of defining the conifold is equivalent to the one we have discussed above. Given the definition of the coordinates u_{ij} in terms of w_i and z_i, the equation (11) remains the same if we act on the chiral fields with the transformations:

$$z_i \to \lambda z_i \qquad w_i \to \lambda^{-1} w_i \qquad \lambda \in \mathbb{C}^*. \tag{12}$$

If we write $\lambda = se^{i\alpha}$, with $s \in \mathbb{R}^+$ and $\alpha \in \mathbb{R}$, the parameter s can be used to satisfy (8), while α parameterizes the gauge invariance (9).

A generalization to the case where we have a stack of N parallel D3-branes probing the tip of the cone is straightforward. In that case, we have a Yang-Mills theory with $\mathcal{N} = 1$ supersymmetry, gauge group $U(N)_1 \times U(N)_2$, and chiral matter fields which transform under these groups according to Table 2.

Table 2.

	$U(N)_1$	$U(N)_2$
z_1, z_2	$\mathbf{N}$	$\overline{\mathbf{N}}$
w_1, w_2	$\overline{\mathbf{N}}$	$\mathbf{N}$

Note: Transformations of the chiral fields under the two gauge groups. The symbol $\mathbf{N}$ indicates that a field transforms in the fundamental representation of $U(N)$, whereas $\overline{\mathbf{N}}$ indicates that a field transforms in the anti-fundamental representation of $U(N)$.

For the case with multiple branes, a superpotential needs to be added. The only superpotential quartic in the superfields and invariant under the

$SU(2) \times SU(2)$ flavor symmetry is:

$$W \propto \mathrm{Tr}\left(\epsilon^{AC}\epsilon^{BD} z_A w_B z_C w_D\right) \propto \mathrm{Tr}\left(z_1 w_1 z_2 w_2 - z_1 w_2 z_2 w_1\right) . \qquad (13)$$

Note that this superpotential vanishes in the abelian case, where all the fields commute. All the fields have R-charge equal to 1/2. Thus, at the conformal fixed point, the scaling dimension of each field is

$$\Delta = \frac{3}{2}|R| = \frac{3}{4}. \qquad (14)$$

It can be shown that the superpotential is an exactly marginal operator for this gauge theory.

The gauge group of the theory, $U(N)_1 \times U(N)_2$, contains two $U(1)$ factors. The diagonal subgroup of the gauge group, under which all the fields are neutral, decouples trivially. The anti-diagonal subgroup, commonly denoted as $U(1)_B$, becomes a global symmetry far in the IR because its gauge coupling flows to zero. Therefore, the AdS/CFT correspondence for this case states that Type IIB string theory on $AdS_5 \times T^{1,1}$ with N units of 5-form RR flux on $T^{1,1}$ is dual to $\mathcal{N} = 1$ supersymmetric Yang-Mills theory with gauge group $SU(N)_1 \times SU(N)_2$ [16]. Furthermore, after turning on M units of 3-form RR flux, and exact "warped throat" background has been found and interpreted as the gravity dual of a "cascading" $SU(N) \times SU(N+M)$ gauge theory.[2] Far in the IR this theory resembles $\mathcal{N} = 1$ supersymmetric gluodynamics and exhibits the same pattern of chiral symmetry breaking.

2. One M2-brane and Abelian Chern-Simons Theory

Having briefly introduced the AdS/CFT correspondence in the context of D3-branes, we are now ready to discuss the correspondence for M2-branes. Let us recall the formulation of 3-dimensional $\mathcal{N} = 2$ supersymmetric Chern-Simons theories coupled to charged fields.[4,17] In the superspace formalism, the Chern-Simons action for theories with a single $U(1)$ gauge group and N_f matter flavours can be written, in Wess-Zumino gauge, as:

$$S = \int \mathrm{d}^3x \int \mathrm{d}^4\theta \left(\frac{k}{4\pi}\mathcal{V}\Sigma + \sum_{i=1}^{N_f} \Phi_i e^{q_i \mathcal{V}} \Phi_i \right) , \qquad k \in \mathbb{Z} \qquad (15)$$

where Φ_i are chiral matter superfields transforming under the gauge group with charge q_i, and where

$$\begin{aligned} \mathcal{V} &= 2i\theta\bar{\theta}\sigma + 2\theta\gamma^\mu\bar{\theta}A_\mu + \sqrt{2}i\theta^2\bar{\theta}\bar{\chi} - \sqrt{2}i\bar{\theta}^2\theta\chi + \theta^2\bar{\theta}^2 D, \\ \Sigma &= \bar{D}^\alpha D_\alpha \mathcal{V}. \end{aligned} \qquad (16)$$

As we can see from above, the vector superfield $\mathcal{V}$ is composed of a gauge field A_μ, a two-component Dirac spinor χ, a scalar field σ, which comes from the A_3 component of the gauge field when we do dimensional reduction from the 3+1 dimensional theory, and another scalar field D.

The parameter k is called the Chern-Simons level: the requirement that a non-abelian theory is invariant under large gauge transformations restricts it to integer values. In components, the action can be written as:

$$S_{CS} = \frac{k}{4\pi}\int \mathrm{d}^3x\ \left(\epsilon^{\mu\nu\lambda}A_\mu\partial_\nu A_\lambda + i\bar{\chi}\chi - 2D\sigma\right),$$

$$S_{matter} = \int \mathrm{d}^3x\ \sum_{i=1}^{N_f}\Big(-\mathcal{D}_\mu\phi_i^\dagger\mathcal{D}^\mu\phi_i - i\zeta_i^\dagger\not{\mathcal{D}}\zeta_i + q_i\phi_i^\dagger D\phi_i - q_i^2\phi_i^\dagger\sigma^2\phi_i +$$
$$-q_i\zeta_i^\dagger\sigma\zeta_i + iq_i\phi_i^\dagger\bar{\chi}\zeta_i - iq_i\zeta_i^\dagger\chi\phi_i\Big), \tag{17}$$

where $\mathcal{D}_\mu$ represents the covariant derivative, and where ϕ_i and ζ_i represent, respectively, the scalar and the fermionic part of the chiral matter field Φ_i. Note that all the fields in the vector multiplet are non-dynamical, so they are all auxiliary fields. Integrating out the scalar field D, we have that:

$$\sigma = -\frac{2\pi}{k}\sum_{i=1}^{N_f} q_i\phi_i^\dagger\phi_i, \tag{18}$$

and the D-term potential can be written as:

$$V_D \propto \sum_{i=1}^{N_f} q_i^2\phi_i^\dagger\phi_i\sigma^2. \tag{19}$$

Note that, because of (18), the D-term potential is sextic in the scalar fields ϕ_i.

Now let us consider the specific gauge theory proposed in [9] as a description of a single M2-brane. This theory has gauge group $U(1)\times U(1)$, Chern-Simons levels $(k,-k)$ and four chiral superfields transforming under these groups as given in Table 1. The matter action for this theory can be written in the superspace formalism as:

$$S_{matter} = \int \mathrm{d}^3x \int \mathrm{d}^4\theta\ \left(\bar{\mathcal{Z}}_A e^{-\mathcal{V}}\mathcal{Z}^A e^{\hat{\mathcal{V}}} + \bar{\mathcal{W}}^B e^{-\hat{\mathcal{V}}}\mathcal{W}_B e^{\mathcal{V}}\right), \qquad A,B = 1,2 \tag{20}$$

where $\mathcal{Z}^A$ and $\mathcal{W}_B$ are chiral multiplets, whose lowest components are scalar fields that we denote as Z^A and W_B. Expanding the action in components, and deriving the equations of motion for the auxiliary fields, we have that:

$$\sigma = \hat{\sigma} = \frac{2\pi}{k}\left(|Z_1|^2 + |Z_2|^2 - |W_1|^2 - |W_2|^2\right). \tag{21}$$

The D-term potential for this theory is proportional to $(\sigma - \hat{\sigma})^2$; therefore, it vanishes. Since the F-term potential vanishes as well (the abelian theory has no superpotential), we could be tempted to conclude that the moduli space is simply $\mathbb{R}^8$, or $\mathbb{C}^4$. However, this is true only up to a $\mathbb{Z}_k$ identification.

Let us combine the fields as follows [9,18]:

$$Y^A = \{Z^A, W^{\dagger A}\} \qquad Y^\dagger_A = \left\{Z^\dagger_A, W_A\right\}. \tag{22}$$

The newly defined fields Y^A have the same charges as Z^A under the gauge group $U(1)_1 \times U(1)_2$.
With this definition, the bosonic part of the Chern-Simons action can be written as:

$$S_{\text{bos}} = \frac{k}{4\pi}\int \mathrm{d}^3x\ \epsilon^{\mu\nu\lambda}\left(A_\mu\partial_\nu A_\lambda - \hat{A}_\mu\partial_\nu\hat{A}_\lambda\right) + \int \mathrm{d}^3x\ \mathcal{D}_\mu Y^\dagger_A \mathcal{D}^\mu Y^A \tag{23}$$

Defining $A^\pm_\mu = A_\mu \pm \hat{A}_\mu$, the covariant derivative can be written as:

$$\mathcal{D}_\mu Y^A = \partial_\mu Y^A + i(A_\mu - \hat{A}_\mu)Y^A = \partial_\mu Y^A + iA^-_\mu Y^A. \tag{24}$$

The CS action itself can be written in a more concise form as:

$$S_{\text{CS}} = \frac{k}{4\pi}\int \mathrm{d}^3x\ \epsilon^{\mu\nu\lambda}A^-_\mu F^+_{\nu\lambda}. \tag{25}$$

If we define the theory on $\mathbb{R}^3$ and map it to $\mathbb{R}\times S^2$, then there are sectors with quantized monopole fluxes:[b]

$$\int_{S^2} F^+ = 4\pi n\ . \tag{26}$$

If we now consider a gauge transformation for which:

$$A^-_\mu \to A^-_\mu + \partial_\mu\Lambda^-, \qquad Y^A \to e^{i\Lambda^-}Y^A, \tag{27}$$

this transformation brings about a boundary term that can be written as:

$$\delta S = \frac{k}{4\pi}\Lambda^-\int_{S^2} F^+. \tag{28}$$

Since the fluxes are quantized, in order for the action to shift by 2π times an integer, we must require that:

$$\Lambda^- = \frac{2\pi l}{k}\,, \qquad l \in \mathbb{Z}\ . \tag{29}$$

---

[b] For a discussion of the extra factor of 2 appearing in the quantization condition, see [19,20].

This provides an identification on the matter fields which implies that the moduli space for this theory with Chern-Simons levels $(k, -k)$ is $\mathbb{C}^4/\mathbb{Z}_k$.

Another way to derive this result is to note that, since A^+_μ does not appear in the action, we can treat $F^+_{\mu\nu}$ as a basic variable, rather than A^+_μ itself. Thus, in order to ensure that the equation $\mathrm{d}F^+ = 0$ remains valid, we need to add a Lagrange multiplier [21,22]:

$$S_\tau = \frac{1}{4\pi}\int \mathrm{d}^3x \;\; \tau\epsilon^{\mu\nu\rho}\partial_\mu F^+_{\nu\rho}. \tag{30}$$

Having inserted this term into the action, the equations of motion of $F^+_{\nu\rho}$ can be written as:

$$A^-_\mu = \frac{1}{k}\partial_\mu\tau \; . \tag{31}$$

The quantization condition on the fluxes requires that τ be periodic with period 2π. It follows that, under the gauge transformations,

$$\tau \to \tau + k\Lambda^- \; . \tag{32}$$

We can use this gauge transformation to fix τ to be 0 but, because of its periodicity, we still have the freedom to make a transformation with $\Lambda^- = \frac{2\pi}{k}$. Examining how these "large" $\mathbb{Z}_k$ gauge transformations act on the fields Y^A, we again conclude that the moduli space of this Chern-Simons theory is $\mathbb{C}^4/\mathbb{Z}_k$.

In addition to the $\mathbb{Z}_k$ gauge symmetry, the model has a global $U(1)_b$ symmetry which corresponds to the conserved current $j_\mu \sim \epsilon_{\mu\nu\lambda}F^{+\nu\lambda}$. This symmetry acts by shifting the dual scalar τ. The monopole (anti-monopole) operators $e^{\pm i\tau}$, that are charged under the $U(1)_b$, create field configurations with magnetic flux of F^+ through the S^2 surrounding the point of insertion.

The non-abelian, $U(N) \times U(N)$ version of the model we are discussing was proposed in 2008 by Aharony, Bergman, Jafferis and Maldacena in [9], and in the literature is often referred to as the ABJM theory. It has the merit of having the conformal symmetry manifest, although the complete amount of supersymmetry is not manifest for $k = 1, 2$ (we will return to this issue in section 5). A different, older description of the gauge theory on a stack of N M2-branes is in terms of the IR limit of the gauge theory on N D2-branes, i.e. the $\mathcal{N} = 8$ supersymmetric Yang-Mills theory in $2+1$ dimensions. For example, for a single D2-brane the action is

$$S = \int \mathrm{d}^3x \left(-\frac{1}{4}F^2_{\mu\nu} - \sum_{i=1}^{7}\frac{1}{2}(\partial_\mu\phi^i)^2 \right). \tag{33}$$

Dualizing the field strength to a scalar

$$F_{\mu\nu} \sim \epsilon_{\mu\nu\lambda}\partial^{\lambda}\phi^{8}, \tag{34}$$

the $SO(8)$ symmetry acting on the 8 scalars is manifest. However, it is not known how to generalize the duality transformation to the non-abelian gauge theory on multiple D2-branes. In this approach to M2-branes, the $\mathcal{N} = 8$ supersymmetry is manifest, but the conformal invariance is not; it contains the coupling constant g_{YM} which has dimension of Energy$^{1/2}$. Presumably, the superconformal M2-brane theory is the IR sector of the theory on N D2-branes which emerges for energies much smaller than g_{YM}^2.

3. Non-Abelian Chern-Simons theory

A breakthrough in the search for a Chern-Simons matter theory with $\mathcal{N} = 8$ supersymmetry came with the work of Bagger and Lambert[5–7] and, independently, Gustavsson.[8] They constructed a theory, often referred to as the BLG model, using a so-called 3-algebra. Given a set of generators T_a, such an algebra can be defined by introducing the triple product:

$$[T^a, T^b, T^c] = f^{abc}{}_d T^d, \tag{35}$$

where f_{abcd} is a fully anti-symmetric tensor. Given this algebra, a maximally supersymmetric Chern-Simons lagrangian is:

$$\begin{aligned}
\mathcal{L}_{CS} &= \frac{1}{2}\epsilon^{\mu\nu\lambda}\left(f^{abcd}A_{\mu ab}\partial_\nu A_{\lambda cd} + \frac{2}{3}f^{cda}{}_g f^{efgb}A_{\mu ab}A_{\nu cd}A_{\lambda ef}\right)\\
\mathcal{L}_{matter} &= -\frac{1}{2}\mathcal{D}^\mu x^{aI}\mathcal{D}_\mu x^I_a + \frac{i}{2}\bar{\psi}^a\Gamma^\mu\mathcal{D}_\mu\psi_a + \frac{i}{4}\bar{\psi}_b\Gamma_{IJ}x^I_c x^J_d\psi_a f^{abcd}\\
&\quad -\frac{1}{12}\operatorname{Tr}\left([x^I,x^J,x^K][x^I,x^J,x^K]\right), \qquad I,J = 1,\ldots,8,
\end{aligned} \tag{36}$$

where A^μ_{ab} is the gauge boson, and ψ_a and $x^I = x^I_a T^a$ are matter fields[c]. If we let $a = 1,\ldots,4$, then we can obtain an $SO(4)$ gauge symmetry by choosing $f^{abcd} = f\epsilon^{abcd}$, f being a constant.[d] Not only does this seem like a natural choice, but it turns out to be the only one that gives a gauge theory with manifest unitarity and $\mathcal{N} = 8$ supersymmetry.

After this model was proposed, it was shown[23,24] that this gauge theory is equivalent to an $SU(2)\times SU(2)$ Chern-Simons gauge theory with opposite

[c] By convention, space-time indices are denoted by Greek letters $\mu, \nu, \lambda, \ldots$, gauge indices are denoted by $a, b, c, \ldots$, and SO(8) vector indices are denoted by capital letters $I, J, \ldots$.
[d] The invariance under large gauge transformations requires $f = \frac{2\pi}{k}$.

Chern-Simons levels:

$$\mathcal{L}_{CS} = \frac{k}{4\pi}\,\epsilon^{\mu\nu\lambda}\,\mathrm{Tr}\left(A_\mu\partial_\nu A_\lambda + \frac{2i}{3}A_\mu A_\nu A_\lambda - \hat{A}_\mu\partial_\nu\hat{A}_\lambda - \frac{2i}{3}\hat{A}_\mu\hat{A}_\nu\hat{A}_\lambda\right), \quad (37)$$

$$\begin{aligned}\mathcal{L}_{matter} = & -(\mathcal{D}^\mu X^I)^\dagger \mathcal{D}_\mu X^I + i\bar{\Psi}^\dagger\Gamma^\mu\mathcal{D}_\mu\Psi \\ & - \frac{4i\pi}{k}\bar{\Psi}^\dagger\Gamma^{IJ}(X^I X^{J\dagger}\Psi + X^J\Psi^\dagger X^I + \Psi X^{I\dagger}X^J) + \\ & - \frac{32\pi^2}{3k^2}\,\mathrm{Tr}\left(X^{[I}X^{\dagger J}X^{K]}X^{\dagger[K}X^J X^{\dagger I]}\right),\end{aligned} \quad (38)$$

where the covariant derivative can be written as:

$$\mathcal{D}_\mu X^I = \partial_\mu X^I + iA_\mu X^I - iX^I\hat{A}_\mu\,. \quad (39)$$

The bi-fundamental matter fields X^I are related to the ones in the $SO(4)$ notation by:

$$X^I = \frac{1}{2}(x_4^I \mathbb{I}_{2\times 2} + ix_i^I\sigma^i)\,, \quad (40)$$

where $\mathbb{I}_{2\times 2}$ is the 2×2 identity matrix and σ^i are the Pauli matrices. Also, the bi-fundamental matter fields satisfy the reality condition [24]:

$$X_a^{\hat{a}} = -\epsilon_{ab}\,X_{\hat{b}}^b\,\epsilon^{\hat{a}\hat{b}}, \qquad \epsilon^{ab} = i\sigma_2^{ab}. \quad (41)$$

An $\mathcal{N}=2$ superspace formulation of this theory was given in [9,18]. The gauge fields A and $\hat{A}$ can be thought of as components of two vector supermultiplets, $\mathcal{V}$ and $\hat{\mathcal{V}}$, whose other components are auxiliary fields. The matter fields can be combined into chiral bi-fundamental superfields, which we denote as $\mathcal{Z}^A$. The lowest component of $\mathcal{Z}^A$ is a complex scalar field Z^A, related to the X^I fields by:

$$Z^A = X^A + iX^{A+4}\,, \qquad A = 1,\ldots,4\,. \quad (42)$$

In terms of the chiral multiplets $\mathcal{Z}_a^A$ transforming under the $SO(4)$ gauge group and carrying R-charge 1/2, the superpotential can be written as:

$$W = -\frac{\pi}{24k}\epsilon_{ABCD}\epsilon^{abcd}(\mathcal{Z}_a^A\mathcal{Z}_b^B\mathcal{Z}_c^C\mathcal{Z}_d^D)\,, \quad (43)$$

The theory thus has a manifest $U(1)_R\times SU(4)$ symmetry. By virtue of (40), this superpotential can be expressed in terms of the bi-fundamental chiral superfields $\mathcal{Z}^A$ as:

$$W = \frac{\pi}{3k}\epsilon_{ABCD}(\mathcal{Z}^A\mathcal{Z}^{\ddagger B}\mathcal{Z}^C\mathcal{Z}^{\ddagger D})\,, \quad (44)$$

where we have introduced the operations:

$$\begin{aligned} Z^{\ddagger A} &= -\epsilon(Z^A)^{\mathrm{T}}\epsilon = X^{\dagger A} + iX^{\dagger A+4}, \\ \bar{Z}_A &= -\epsilon(Z^A)^*\epsilon = X^A - iX^{A+4}. \end{aligned} \quad (45)$$

Although the first of these operations might look like a hermitian conjugation, we note that there is no conjugation of the imaginary part, but only of the scalar multiplets X^I. This operation is crucial because it does not break the holomorphy of the superpotential.

It is not obvious that this $\mathcal{N} = 2$ formalism describes the BLG model. The latter has $\mathcal{N} = 8$ supersymmetry, and therefore an $SO(8)_R$ invariance, whereas (44) only has a manifest $U(1)_R \times SU(4)$ symmetry. However, expressing the action of this model in components and integrating out all the auxiliary fields, the sum of the D-term potential and of the F-term potential has the desired $SO(8)_R$ global symmetry.[18]

Since the reality condition (41) and the "double dagger" operation (45) are special to $SU(2)\times SU(2)$, it seems difficult to generalize the construction to gauge groups with a higher rank. A way to overcome this difficulty, proposed in [9], is to abandon the manifest global $SU(4)$ invariance by forming the following combinations of the bi-fundamental fields:

$$\begin{aligned} Z^1 &= X^1 + iX^5, \qquad & W^1 &= X^{3\dagger} + iX^{7\dagger} , \\ Z^2 &= X^2 + iX^6, \qquad & W^2 &= X^{4\dagger} + iX^{8\dagger} . \end{aligned} \tag{46}$$

If we promote the fields Z^A and W^A to chiral superfields, that we shall denote as $\mathcal{Z}^A$ and $\mathcal{W}^A$, the superpotential of the model can be written as [9,18]:

$$W = \frac{2\pi}{k}\epsilon_{AC}\epsilon^{BD}\,\mathrm{Tr}\left(\mathcal{Z}^A\mathcal{W}_B\mathcal{Z}^C\mathcal{W}_D\right) . \tag{47}$$

This superpotential looks exactly the same as that for the D3-brane theory on the conifold and can be easily generalized to higher rank gauge groups of the type $SU(N) \times SU(N)$. Furthermore, there is an important *caveat.* Apart from the manifest $U(1)_R \times SU(2) \times SU(2)$ global symmetry, the superpotential has also a $U(1)$ symmetry, under which the chiral multiplets transform as:

$$\mathcal{Z}^A \to e^{i\alpha}\mathcal{Z}^A, \qquad \mathcal{W}_B \to e^{-i\alpha}\mathcal{W}_B . \tag{48}$$

In the $3+1$ dimensional theory on N D3-branes at the conifold singularity, this starts out as a gauge symmetry but in the far IR becomes global and is identified with a $U(1)$ baryonic symmetry. In the $2+1$ dimensional case at hand, the dynamics is different and we have to treat this $U(1)$ as a gauge symmetry, although eventually it becomes broken to a $\mathbb{Z}_k$ subgroup (29). Using this argument, as well as type IIB brane constructions, ABJM proposed that the gauge group on N M2-branes is $U(N) \times U(N)$, and not $SU(N) \times SU(N)$. We have already checked that the moduli space of the

$N = 1$ abelian theory with Chern-Simons levels $(k, -k)$ is the $\mathbb{Z}_k$ orbifold of $\mathbb{C}^4$; thus, this theory correctly describes a single M2-brane.

An important property of the classical ABJM action is that the $U(1)_R \times SU(2) \times SU(2)$ global symmetry is enhanced to $SU(4)_R$, corresponding to $\mathcal{N} = 6$ supersymmetry. The scalar potential is made of two parts, one coming from the F-terms and the other from the D-terms. Combining the two contributions and using the notation introduced in (22), we find that the full bosonic potential is:

$$V^{\rm bos} = -\frac{4\pi^2}{3k^2}\,{\rm Tr}\Big[Y^A Y_A^\dagger Y^B Y_B^\dagger Y^C Y_C^\dagger + Y_A^\dagger Y^A Y_B^\dagger Y^B Y_C^\dagger Y^C + \\ +4Y^A Y_B^\dagger Y^C Y_A^\dagger Y^B Y_C^\dagger - 6Y^A Y_B^\dagger Y^B Y_A^\dagger Y^C Y_C^\dagger\Big]. \qquad (49)$$

The interaction terms, quadratic in the fermion fields and quartic in the scalars, possess the manifest $SU(4)_R \sim SO(6)_R$ symmetry as well.[18] Therefore, this symmetry is manifest in the classical action; it strongly suggests that, for general N and k, the ABJM theory has at least $\mathcal{N} = 6$ supersymmetry. An explicit demonstration of the $\mathcal{N} = 6$ superconformal invariance of the ABJM theory was presented in [25].

4. Gravitational description of coincident M2-branes

A stack of N coincident M2-branes creates the following extremal geometry:

$$\begin{aligned} ds_{11}^2 &= h(r)^{-2/3}\left(-dt^2 + dx_1^2 + dx_2^2\right) + h(r)^{1/3}\left(dr^2 + r^2 d\Omega_7^2\right), \\ h(r) &= 1 + \frac{L^6}{r^6}, \qquad L^6 = 32\pi^2 N l_p^6, \\ F_4 &= d^3x \wedge dh\,(r)^{-1}. \end{aligned} \qquad (50)$$

In the limit where $r \to 0$, the metric becomes that of $AdS_4 \times S^7$:

$$ds_{11}^2 = L^2\left(\frac{1}{4}\,ds_{AdS_4}^2 + ds_{S^7}^2\right). \qquad (51)$$

For the strongly coupled theory on N M2-branes the AdS/CFT correspondence predicts an interesting essential feature that has not been completely understood yet: the number of degrees of freedom scales as $N^{3/2}$ [26], not as N^2 found for the strongly coupled gauge theories on D3-branes. One way to see this is through studying the correlation functions of protected gauge invariant operators. Since their spectrum is in one-to-one correspondence with the Kaluza-Klein harmonics on S^7, we can use the gravity side of AdS/CFT to predict that all their correlation functions scale like:

$$\langle \mathcal{O}_1 \mathcal{O}_2 \ldots \mathcal{O}_m \rangle \propto (L/l_p)^9 \propto N^{3/2}, \qquad (52)$$

where we have used the relation between L and N obtained in (50). Another way is to study the thermal gauge theory.[9,14,26] The non-extremal geometry for a stack of N M2-branes is:

$$ds_{11}^2 = h(r)^{-2/3}\left(-f(r)dt^2 + dx_1^2 + dx_2^2\right) + h^{1/3}(r)\left(\frac{dr^2}{f(r)} + r^2 d\Omega_7^2\right),$$
$$h(r) = 1 + \frac{L^6}{r^6}, \qquad f(r) = 1 - \frac{r_0^6}{r^6}. \tag{53}$$

In the near-horizon region $r \ll L$ this becomes a black brane in AdS_4, and r_0 represents the Schwarzschild radius, which is related to the Hawking temperature T.

If we consider a stack of M2-branes placed at the singularity of the orbifold $\mathbb{C}^4/\mathbb{Z}_k$, which is described by the level k ABJM theory, then $d\Omega_7^2$ refers to the metric on unit $S^7/\mathbb{Z}_k$. When the 't Hooft coupling $\lambda = N/k$ is very large, so that the metric is weakly curved, the Bekenstein-Hawking entropy is written as:

$$S_{BH} = \frac{A}{4G_N} = 2^{7/2}3^{-3}\pi^2 V_2 T^2 k^{1/2} N^{3/2} + O(N^{1/2}), \tag{54}$$

where V_2 is the spatial volume of the stack of M2-branes. Thus, the AdS/CFT correspondence predicts that for large λ the number of degrees of freedom in the ABJM theory scales like $N^{3/2}k^{1/2}$. More generally, the thermal entropy of the ABJM theory can be written as:

$$\frac{S_{BH}}{V_2 T^2} \propto N^2 f(\lambda). \tag{55}$$

In the perturbative regime, $\lambda \ll 1$, we expect $f(\lambda) = 1 + O(\lambda^2)$, while the AdS/CFT correspondence predicts that

$$f(\lambda) \propto \frac{1}{\sqrt{\lambda}}, \qquad \lambda \gg 1. \tag{56}$$

This behavior is very different from the situation where a stack of D3-branes is considered. In that case the Bekenstein-Hawking entropy can be written in a similar form to (55), but when the 't Hooft coupling becomes large, the function $f(\lambda)$ approaches $3/4$ at large λ.[e]

Now, let us discuss the parity transformation. In $2+1$ dimensions, the reflection of both coordinates is simply a rotation; therefore, one has to consider the reflection of only one. Sometimes it has to be accompanied by the

[e] The situation becomes even more interesting when we consider a stack of M5-branes, because the supergravity approximation shows that the Bekenstein-Hawking entropy scales like N^3 [26].

action on other fields. For example, in $AdS_4 \times Y^7$, the parity transformation acts on one of the spatial coordinates and on the 3-form:

$$x^1 \to -x^1 , \qquad C_3 \to -C_3 \ . \tag{57}$$

If only the C_{012} component of the 3-form is turned on, this transformation preserves the background. This shows that the dual gauge theory cannot be a supersymmetric Chern-Simons theory with a single gauge group $U(N)$, where the Chern-Simons term violates the parity.[4] However, theories with two gauge groups and opposite Chern-Simons levels preserve a parity symmetry that involves the interchange of the gauge groups.[23,27] For example, in the ABJM theory, the parity transformation

$$\begin{aligned} x^1 &\to -x^1 , \qquad & A &\leftrightarrow \hat{A} , \\ Y^A &\leftrightarrow Y_A^\dagger , \qquad & \psi_\alpha^{\dagger A} &\leftrightarrow \gamma^1_{\alpha\beta} \psi^\beta_A \end{aligned} \tag{58}$$

is a symmetry of the action. This is consistent with the parity symmetry of the $AdS_4 \times S^7/\mathbb{Z}_k$ background of M-theory.

5. Supersymmetry Enhancement and Monopole Operators

Let us show that $\mathcal{N} = 6$ is the correct amount of supersymmetry for N M2-branes placed at the singularity of $\mathbb{C}^4/\mathbb{Z}_k$ for $k > 2$. The $\mathbb{Z}_k$ orbifold acts on the 4 complex coordinates of $\mathbb{C}^4$ as:

$$y^A \to e^{2\pi i/k} y^A . \tag{59}$$

Note that this preserves the $SU(4)$ symmetry that rotates the y^A; this is the R-symmetry from the gauge theory point of view. The generators of $\mathbb{Z}_k$ act on the $SO(8)$ spinors as:

$$\Psi \to e^{2\pi i (s_1 + s_2 + s_3 + s_4)/k} \Psi , \tag{60}$$

where $s_i = \pm 1/2$ are the spinor weights. By chirality projection, the sum of the s_i's has to be even, giving an 8 dimensional representation. Therefore, the spinors that survive the orbifold projection must satisfy:

$$\sum_{i=1}^{4} s_i = 0 \pmod{k} . \tag{61}$$

For $k > 2$, 6 of the 8 spinors are left invariant by the orbifold action. It follows that this supersymmetric gauge theory is expected to have 12 supercharges and, accordingly, $\mathcal{N} = 6$ supersymmetry, in agreement with that in the classical action of the ABJM theory. However, for $k = 1, 2$ all the spinors are found to be invariant, and the supersymmetry is enhanced to

$\mathcal{N} = 8$. In the following, we will investigate what causes this supersymmetry enhancement in the ABJM model.

In the ABJM model, the classical global symmetry is $U(1)_b \times SU(4)_R$. The $SU(4)$ symmetry is realized by the 15 conserved traceless currents

$$j^A_{\mu B} = i\,\mathrm{Tr}\left[Y^A \mathcal{D}_\mu Y^\dagger_B - (\mathcal{D}_\mu Y^A) Y^\dagger_B + i\psi^{\dagger A}\gamma_\mu \psi_B\right] . \tag{62}$$

The $U(1)_b$ transformation acts as

$$Y^A \to e^{i\alpha} Y^A \,, \qquad \psi_B \to e^{i\alpha}\psi_B \,. \tag{63}$$

The corresponding current is related by the A^- equation of motion (in the $U(N) \times U(N)$ theory, $A^\pm \sim \mathrm{Tr}\,A \pm \mathrm{Tr}\,\hat{A}$) to the current $j_\mu \sim \epsilon_{\mu\nu\lambda} F^{+\nu\lambda}$, which is obviously conserved. Therefore, the $U(1)_b$ charge is carried only by the field configurations that have a flux of F^+ through the 2-sphere at infinity of $\mathbb{R}^3$. Such field configurations are created by the so-called "monopole operators" that we discuss next.

In order for the supersymmetry to be enhanced to $\mathcal{N} = 8$, we need the global symmetry to be enhanced to $SO(8)_R$ which has 28 generators. Therefore, we need to find 12 conserved currents in addition to the 16 $U(1)_b \times SU(4)_R$ currents. Construction of these 12 currents is expected[9] to involve the monopole operators in the three-dimensional gauge theory.[28] Each of these operators creates a quantized flux in a $U(1)$ subgroup of the gauge group through a sphere surrounding the insertion point. In the ABJM theory, which has gauge group $U(N)_1 \times U(N)_2$, these monopole operators are labelled by the Cartan generators H and $\hat{H}$ of each of the two $U(N)$ factors of the gauge group:

$$H = \mathrm{diag}(q_1, q_2, \ldots, q_N) \,, \quad \hat{H} = \mathrm{diag}(\hat{q}_1, \hat{q}_2, \ldots, \hat{q}_N) \,, \quad q_i, \hat{q}_i \in \mathbb{N} \,, \tag{64}$$

where the condition that the entries of the generators have to be integers follows from the flux quantization condition around the S^2 surrounding the insertion point of the monopole. Also, for convenience, one can arrange the q_i's and the $\hat{q}_i$'s such that:

$$q_1 \geq q_2 \geq \ldots \geq q_N \,, \qquad \hat{q}_1 \geq \hat{q}_2 \geq \ldots \geq \hat{q}_N \,. \tag{65}$$

We will restrict our attention to monopoles with $H = \hat{H}$. We also note that, for each monopole, there is an anti-monopole.

Generically, if a gauge group has Chern-Simons level k, the monopole operators transform in the representation of the gauge group given by a Young tableaux with kq_1 boxes in the first row, kq_2 boxes in the second row, etc.[f] It follows that in the ABJM theory, the monopoles are not gauge singlets and can combine with the chiral matter to form gauge invariant operators.

Consider, for example, the $k = 1$ theory. Here, the simplest, unit-charge monopole operator corresponds to $q_1 = \hat{q}_1 = 1$. The corresponding operator, $\mathcal{M}^a_{\hat{a}}$, transforms as a fundamental under $U(N)_1$ and anti-fundamental under $U(N)_2$. There is also the anti-monopole operator that transforms in the conjugate representation, $(\mathcal{M}^{-1})^{\hat{a}}_a$. Similarly, there are doubly charged monopole operators corresponding for example to $q_1 = \hat{q}_1 = 2$, that we denote $(\mathcal{M}^2)^{ab}_{\hat{a}\hat{b}}$, which transform in the symmetric tensor representation under $U(N)_1$, and in the conjugate representation under $U(N)_2$. We will denote their anti-monopole operators by $(\mathcal{M}^{-2})^{\hat{a}\hat{b}}_{ab}$. Another type of doubly charged BPS monopole operator has $q_1 = q_2 = \hat{q}_1 = \hat{q}_2 = 1$. Such an operator transforms as an anti-symmetric tensor under $U(N)_1$, and in the conjugate representation under $U(N)_2$. As we increase the monopole charge, the variety of different monopoles increases.

Now, let us consider the current operators:

$$j^{AB}_\mu = i\left[Y^A \mathcal{D}_\mu Y^B - \mathcal{D}_\mu Y^A Y^B + i\psi^{\dagger A}\gamma_\mu \psi^{\dagger B}\right]. \tag{66}$$

These operators are not gauge invariant, but they can be combined with the monopole $(\mathcal{M}^{-2})^{\hat{a}\hat{b}}_{ab}$, which has $kq_1 = 2, k\hat{q}_1 = 2$ and $q_i = \hat{q}_i = 0$ for $i \neq 1$, to form 6 invariant currents[g] that we shall denote as J^{AB}_μ. Combining their complex conjugates $j_{AB\mu}$ with the monopole $\mathcal{M}^2$, gives another six conserved currents which are also gauge invariant. To summarize, the monopoles have provided us with 12 symmetry generators that we can add to the 16 obvious currents of $U(1)_b \times SU(4)_R$. With a total of 28 generators, the global symmetry is enhanced to $SO(8)_R$, as we expected. When $k \geq 3$, there is no way to construct the monopole operators $\mathcal{M}^{-2}$ and $\mathcal{M}^2$, which carry two indices under each $U(N)$; so the symmetry enhancement does not happen.

The importance of the monopoles goes beyond the supersymmetry enhancement for $k = 1, 2$. They are also necessary for matching the spectrum of the gauge theory with that of the dual gravity theory. In the ABJM theory with $k = 1$, the simplest gauge invariant operators have the form $Y^{\dagger\hat{a}}_{Ba}\mathcal{M}^a_{\hat{a}}$ and their conjugate. These operators, linear in the scalar fields, are expected to have scaling dimension 1/2; they can be thought of as free fields

[f] Since the gauge groups we are dealing with are $U(N)$ rather than $SU(N)$, the columns of length N must be taken into account as well.

[g] The number 6 comes from the anti-symmetry of j^{AB}_μ under the exchange of A and B.

dual to 'singleton' modes in AdS_4. The simplest scalar composite operators we can construct in the ABJM theory at level $k = 1$ can be written as:

$$\mathrm{Tr}\left[Y_A^\dagger Y^B - \frac{1}{4}\delta_A^B Y_C^\dagger Y^C\right] . \tag{67}$$

There are 15 such traceless operators of dimension 1. The AdS/CFT correspondence implies that the gauge invariant scalar operators should be in one-to-one correspondence with the Kaluza-Klein harmonics on S^7. There are 35 such harmonics that correspond to operators of dimension 1. Therefore, we need to find 20 more operators made of two scalar fields in order to match this part of spectrum with the gravity side. Once again, we proceed by forming non-gauge-invariant combinations of scalar fields and combining them with monopoles in order to make them gauge invariant. The 20 gauge invariant operators written as:

$$Y_A^\dagger Y_B^\dagger \mathcal{M}^2, \qquad Y^A Y^B \mathcal{M}^{-2}, \tag{68}$$

turn out to be what is needed to match the spectrum of the field theory with the gravity theory. These additional operators have non-vanishing $U(1)_b$ charge, which is dual to the momentum along the M-theory circle.[h]

It remains to show that the monopole operators do not alter the 'naive dimension' 1 of the scalar bilinears. Since the ABJM theory at low k is a strongly coupled gauge theory, the monopole operators are quite hard to analyse in detail as there's no perturbative approach available. However, a way to overcome this problem is to embed the ABJM theory in an $\mathcal{N} = 3$ supersymmetric Yang-Mills-Chern-Simons theory, by adding a Yang-Mills term for the gauge fields in the action [29]:

$$\begin{aligned} S_{YM} &= \frac{1}{4g^2}\int \mathrm{d}^3x \int \mathrm{d}^2\theta \;\; \mathrm{Tr}\left[\mathcal{U}^\alpha \mathcal{U}_\alpha + \hat{\mathcal{U}}^\alpha \hat{\mathcal{U}}_\alpha\right], \\ \mathcal{U}_\alpha &= \frac{1}{4}\bar{D}^2 e^{\mathcal{V}} D_\alpha e^{-\mathcal{V}}, \end{aligned} \tag{69}$$

where $\mathcal{V}$ is the vector superfield. Also, we add two dynamical adjoint superfields with kinetic term:

$$S_{\mathrm{adj}} = \frac{1}{g^2}\int \mathrm{d}^3x \int \mathrm{d}^2\theta \, \mathrm{Tr}\left[-\bar{\Phi} e^{-\mathcal{V}} \Phi e^{\mathcal{V}} - \bar{\hat{\Phi}} e^{-\hat{\mathcal{V}}} \hat{\Phi} e^{\hat{\mathcal{V}}}\right] . \tag{70}$$

The superpotential can be written as:

$$W = \mathrm{Tr}\left(\Phi \mathcal{Z}^A \mathcal{W}_A + \hat{\Phi} \mathcal{W}_A \mathcal{Z}^A\right) + \frac{k}{8\pi}\,\mathrm{Tr}\left(\Phi\Phi - \hat{\Phi}\hat{\Phi}\right) . \tag{71}$$

---

[h] If we view the S^7 as a circle fibration over CP^3, then the reduction to type IIA string theory produces an $AdS_4 \times CP^3$ background.

In the IR the gauge coupling g flows to infinity, the kinetic term (70) vanishes and the adjoint fields can be integrated out. The resulting theory is precisely the ABJM model. In contrast, in the UV the gauge coupling goes to 0, and the theory becomes a weakly coupled $\mathcal{N} = 3$ supersymmetric Yang-Mills theory that can be studied perturbatively. Therefore, we can compute quantities in the UV, where the theory is weakly coupled, and then study the flow to the IR, where the theory becomes the ABJM Chern-Simons theory. In particular, we can compute the non-abelian $SU(2)_R$ charge of the monopole operators, which is not modified by the RG flow.[29]

In the UV we may study the monopoles semi-classically. In order for a classical monopole background to be BPS (or anti-BPS), we need to turn on the vacuum expectation values of the scalar fields ϕ_i that are in the vector multiplets and that transform in the spin 1 representation of the R-symmetry group:

$$\phi_i = -\hat{\phi} = \pm\frac{H}{2}n_i(\tau) \ . \tag{72}$$

The minus sign corresponds to a BPS background, while the plus to anti-BPS. The $n_i(\tau)$ vector is a unit vector on the $SU(2)_R/U(1)_R$ 2-sphere and it may vary with the euclidean time τ adiabatically.

The vector $n_i(\tau)$ couples to the fermions; integrating them out we obtain the induced action for n_i. By solving for its quantum mechanical motion, we can determine the representations in which the monopoles transform under the $SU(2)_R$ symmetry. For the ABJM theory the singlet representation turns out to be allowed. Such monopole operators do not contribute to the R-charge, and therefore, to the scaling dimension of composite operators. Thus, the dimensions of composite operators like (68) agree with the AdS/CFT correspondence.

Acknowledgments

I thank the organizers of the Shifmania for inviting me to present a talk, and for their exceptional hospitality. I am grateful to M. Benna, T. Klose, A. Murugan, M. Smedbäck, A. Tseytlin and E. Witten for collaboration on some of the material reviewed in these notes. I also thank G. Torri for his help with preparing this manuscript, and M. Kiermaier and T. Klose for their useful comments on it. This work was supported in part by the NSF grants PHY-0756966 and PHY-0551164. I am grateful to the Kavli Institute for Theoretical Physics for hospitality during some of my work on these notes.

References

1. V. A. Novikov, M. A. Shifman, A. I. Vainshtein and V. I. Zakharov, "Supersymmetric instanton calculus: Gauge theories with matter," Nucl. Phys. B **260**, 157 (1985) [Yad. Fiz. **42**, 1499 (1985)].
2. I. R. Klebanov and M. J. Strassler, "Supergravity and a confining gauge theory: Duality cascades and chiSB-resolution of naked singularities," JHEP **0008**, 052 (2000) [arXiv:hep-th/0007191].
3. C. P. Herzog, I. R. Klebanov and P. Ouyang, "D-branes on the conifold and $\mathcal{N} = 1$ gauge / gravity dualities," arXiv:hep-th/0205100; M. K. Benna and I. R. Klebanov, "Gauge-String Dualities and Some Applications," arXiv:0803.1315 [hep-th].
4. J. H. Schwarz, "Superconformal Chern-Simons theories," JHEP **0411**, 078 (2004) [arXiv:hep-th/0411077].
5. J. Bagger and N. Lambert, "Modeling multiple M2's," Phys. Rev. D **75**, 045020 (2007) [arXiv:hep-th/0611108].
6. J. Bagger and N. Lambert, "Gauge Symmetry and Supersymmetry of Multiple M2-Branes," Phys. Rev. D **77**, 065008 (2008) [arXiv:0711.0955 [hep-th]].
7. J. Bagger and N. Lambert, "Comments On Multiple M2-branes," JHEP **0802**, 105 (2008) [arXiv:0712.3738 [hep-th]].
8. A. Gustavsson, "Algebraic structures on parallel M2-branes," Nucl. Phys. B **811**, 66 (2009) [arXiv:0709.1260 [hep-th]].
9. O. Aharony, O. Bergman, D. L. Jafferis and J. Maldacena, "$\mathcal{N}$=6 superconformal Chern-Simons-matter theories, M2-branes and their gravity duals," JHEP **0810**, 091 (2008) [arXiv:0806.1218 [hep-th]].
10. J. M. Maldacena, "The large N limit of superconformal field theories and supergravity," Adv. Theor. Math. Phys. **2** (1998) 231 [Int. J. Theor. Phys. **38** (1999) 1113] [arXiv:hep-th/9711200].
11. S. S. Gubser, I. R. Klebanov and A. M. Polyakov, "Gauge theory correlators from non-critical string theory," Phys. Lett. B **428** (1998) 105 [arXiv:hep-th/9802109].
12. E. Witten, "Anti-de Sitter space and holography," Adv. Theor. Math. Phys. **2** (1998) 253 [arXiv:hep-th/9802150].
13. O. Aharony, S. S. Gubser, J. M. Maldacena, H. Ooguri and Y. Oz, "Large N field theories, string theory and gravity," Phys. Rept. **323**, 183 (2000) [arXiv:hep-th/9905111].
14. I. R. Klebanov, "TASI lectures: Introduction to the AdS/CFT correspondence," arXiv:hep-th/0009139.
15. E. D'Hoker and D. Z. Freedman, "Supersymmetric gauge theories and the AdS/CFT correspondence," arXiv:hep-th/0201253.
16. I. R. Klebanov and E. Witten, "Superconformal field theory on threebranes at a Calabi-Yau singularity," Nucl. Phys. B **536**, 199 (1998) [arXiv:hep-th/9807080].
17. D. Gaiotto and X. Yin, "Notes on superconformal Chern-Simons-matter theories," JHEP **0708**, 056 (2007) [arXiv:0704.3740 [hep-th]].

18. M. Benna, I. Klebanov, T. Klose and M. Smedback, "Superconformal Chern-Simons Theories and AdS_4/CFT_3 Correspondence," JHEP **0809**, 072 (2008) [arXiv:0806.1519 [hep-th]].
19. Y. Imamura and K. Kimura, "On the moduli space of elliptic Maxwell-Chern-Simons theories," Prog. Theor. Phys. **120**, 509 (2008) [arXiv:0806.3727 [hep-th]].
20. D. Martelli and J. Sparks, "Moduli spaces of Chern-Simons quiver gauge theories and AdS(4)/CFT(3)," Phys. Rev. D **78**, 126005 (2008) [arXiv:0808.0912 [hep-th]].
21. J. Distler, S. Mukhi, C. Papageorgakis and M. Van Raamsdonk, "M2-branes on M-folds," JHEP **0805**, 038 (2008) [arXiv:0804.1256 [hep-th]].
22. N. Lambert and D. Tong, "Membranes on an Orbifold," Phys. Rev. Lett. **101**, 041602 (2008) [arXiv:0804.1114 [hep-th]].
23. M. A. Bandres, A. E. Lipstein and J. H. Schwarz, "N = 8 Superconformal Chern–Simons Theories," JHEP **0805**, 025 (2008) [arXiv:0803.3242 [hep-th]].
24. M. Van Raamsdonk, "Comments on the Bagger-Lambert theory and multiple M2-branes," JHEP **0805**, 105 (2008) [arXiv:0803.3803 [hep-th]].
25. M. A. Bandres, A. E. Lipstein and J. H. Schwarz, "Studies of the ABJM Theory in a Formulation with Manifest SU(4) R-Symmetry," JHEP **0809**, 027 (2008) [arXiv:0807.0880 [hep-th]].
26. I. R. Klebanov and A. A. Tseytlin, "Entropy of Near-Extremal Black p-branes," Nucl. Phys. B **475**, 164 (1996) [arXiv:hep-th/9604089].
27. S. Deser, R. Jackiw and S. Templeton, "Topologically massive gauge theories," Annals Phys. **140**, 372 (1982) [Erratum-ibid. **185**, 406.1988 APNYA,281,409 (1988 APNYA,281,409-449.2000)].
28. V. Borokhov, A. Kapustin and X. k. Wu, "Topological disorder operators in three-dimensional conformal field theory," JHEP **0211**, 049 (2002) [arXiv:hep-th/0206054].
29. M. K. Benna, I. R. Klebanov and T. Klose, "Charges of Monopole Operators in Chern-Simons Yang-Mills Theory," arXiv:0906.3008 [hep-th].

Surprising Symmetry of Scattering Amplitudes in Gauge Theories

G. P. Korchemsky

Institut de Physique Théorique, CEA Saclay, 91191 Gif-sur-Yvette Cedex, France*

I review a recent progress in computing scattering amplitudes in strongly coupled gauge theories - a fascinating subject which has been recently boosted by the formulation of the gauge/string duality in maximally supersymmetric Yang-Mills theory. In addition to the conventional symmetry of the underlying Lagrangian, the scattering amplitudes in this theory exhibit a new, dual superconformal symmetry. This symmetry is powerful enough to completely determine the scattering amplitudes for arbitrary coupling in a suitably defined limit.

1. Introduction

Scattering amplitudes in maximally supersymmetric $\mathcal{N} = 4$ Yang-Mills theory (SYM) have a number of remarkable properties both at weak and at strong coupling. Defined as matrix elements of the $S-$matrix between asymptotic on-shell states, they inherit all symmetries of the underlying gauge theory. Studying their properties, we are expecting to discover new dynamical symmetries which will eventually allow us to compute the scattering amplitudes in $\mathcal{N} = 4$ theory exactly for arbitrary coupling constant.

In a recent development, it was realized that planar scattering amplitudes in $\mathcal{N} = 4$ SYM theory have a hidden symmetry, the so-called dual superconformal symmetry.[1] It appears on top of all well-known symmetries of the scattering amplitudes (supersymmetry, conformal symmetry, etc.) and is not related (not in an obvious way, at least) to an invariance of the Lagrangian of the theory. Quite remarkably, the same symmetry also emerges at strong coupling, in the AdS/CFT description of scattering amplitudes,[2] from fermionic T-duality of the sigma model on an $\mathrm{AdS}_5 \times \mathrm{S}^5$

* Unité de Recherche Associée au CNRS URA 2306

background.[3,4] In this note, we shall review dual conformal symmetry of the scattering amplitudes in $\mathcal{N} = 4$ SYM theory.

In general, the scattering amplitudes in four-dimensional gauge theories are nontrivial functions of Mandelstam kinematical invariants and gauge coupling constant. At weak coupling, they can be computed order by order in the coupling by employing a conventional Feynman diagram approach.[5–7] The properties of the scattering amplitudes are quite different at tree and loop levels.

Tree amplitudes can be computed from the classical Lagrangian directly in four dimensions while loop amplitudes require regularization due to appearance of infrared divergences. The spectrum of on-shell scattering states and, as a consequence, the full set of tree amplitudes, depends on the theory under consideration. For instance, in pure Yang-Mills theory the on-shell states are transversely polarized gluons while in the maximally supersymmetric $\mathcal{N} = 4$ theory there are in addition four real scalars and four gluinos, all in the adjoint representation of the gauge group. Among all tree amplitudes, the gluon amplitudes play a special rôle since they are the same in all gauge theories.

The situation changes however at loop level. Since Feynman diagrams involve all particles propagating inside loops, perturbative corrections to the amplitudes are not universal and depend on the gauge theory under consideration. In addition, the corresponding Feynman integrals suffer from both ultraviolet and infrared (IR) divergences. The former divergences come from integration over large loop momenta, or equivalently, due to interaction at short distances. As a consequence, ultraviolet divergences can be eliminated at once by expressing the scattering amplitudes in terms of the renormalized coupling. At the same time, IR divergences are associated with propagation of massless particles (necessarily present in the spectrum of any gauge theory) with vanishing momenta, or equivalently, due to interaction at large distances. As a result, all-loop scattering amplitudes suffer from IR divergences and their definition requires IR regularization, for example, through the dimensional reduction of Feynman integrals from $D = 4$ to $D = 4 - 2\epsilon$ dimensions (with $\epsilon < 0$).

The fact that the all-loop scattering amplitudes are necessarily defined away from four dimensions immediately implies that some of the classical symmetries of tree amplitudes (like dilatations and conformal boosts) are inevitably broken at loop level and, therefore, they loose their predictive power. It is therefore remarkable that loop amplitudes reveal some 'hidden' symmetry. It is not manifest at the level of classical Lagrangian and leads to the additional nontrivial relations for all-loop amplitudes. In this note, we shall address two main questions:

- What is the hidden symmetry of tree amplitudes?
- What happens to this symmetry at loop level?

Answering these questions, we shall start with $\mathcal{N} = 4$ SYM theory and, then, extend consideration to gauge theories with less supersymmetry including QCD.

To simplify analysis, let us consider the special subclass of $n-$gluon scattering amplitudes in $\mathcal{N} = 4$ SYM theory. Each scattered gluon is characterized by the set of quantum numbers $|i\rangle = |p_i, h_i, a_i\rangle$ consisting of its on-shell momentum p_i^μ (with $p_i^2 = 0$), helicity $h_i = \pm 1$ and color $a_i = 1, \ldots, N^2 - 1$. The $n-$gluon scattering amplitude is a function of quantum numbers of all particles $\mathcal{A}_n = \mathcal{A}_n(\{p_i, h_i, a_i\})$ and it admits an expansion in powers of the coupling constant $g^2_{\rm YM}$. As was already mentioned, at lowest order in the coupling, the gluon amplitudes are the same as in QCD but this is not the case anymore at loop level.

Additional simplification occurs in the planar limit, $a = g^2_{\rm YM} N/(8\pi^2) =$ fixed as $N \to \infty$, in which case $\mathcal{A}_n$ can be expanded over color-ordered planar partial amplitudes [7]

$$\mathcal{A}_n = \mathrm{tr}\left[t^{a_1} t^{a_2} \ldots t^{a_n}\right] A_n^{h_1, h_2, \ldots, h_n}(p_1, p_2, \ldots, p_n) + \ldots . \tag{1}$$

Here t^a are generators of the $SU(N)$ group in the fundamental representation and ellipses denote nonplanar $O(1/N^2)$ corrections as well as terms needed to ensure Bose symmetry of $\mathcal{A}_n$ under exchange of any pair of gluons.

It is convenient to classify the partial amplitudes according to the total helicity $h = h_1 + \ldots + h_n$ with $h_i = \pm 1$. The supersymmetry implies that the gluon amplitudes with the total helicity $h = n$ and $h = (n-2)$ vanish to all orders in 't Hooft coupling [8]

$$A_n^{++\ldots+} = A_n^{-+\ldots+} = 0 . \tag{2}$$

The same is true for complex conjugated amplitudes with $h = -n$ and $h = -(n-2)$. The simplest nontrivial gluon amplitudes like $A_n^{--+\ldots+}$, $A_n^{-+-\ldots+}$, $\ldots$ have $h = n - 4$ and they are known as maximal helicity violating amplitudes (MHV). In similar manner, for $h = n - 4 - 4p$ the corresponding gluon amplitudes are of N^{p}MHV type. It is easy to see that four- and five-gluon amplitudes are all MHV like whereas NMHV amplitudes only arise starting from $n = 6$ particles.

In general, color-ordered planar partial amplitudes have a complicated dependence on the helicities of external gluons at loop level. A distinguished feature of MHV amplitudes in $\mathcal{N}=4$ SYM theory is that all-loop corrections to various MHV amplitudes are described by a universal, helicity independent function of 't Hooft coupling and Mandelstam kinematical invariants s_{ij} [8]

$$A_n^{\rm MHV}(1^+ ... a^- ... b^- ... n^+) = A_n^{\rm MHV;0}(1^+ ... a^- ... b^- ... n^+)\, M_n(\{s_{ij}\}, a)\,, \tag{3}$$

where $A_n^{\rm MHV;0}$ stands for tree-level $n-$gluon MHV amplitude.

As first hint for hidden symmetry, let us examine the explicit one-loop expression for the function $M_{n=4}$ describing perturbative corrections to four-gluon amplitude (3). It turns out that in $\mathcal{N}=4$ SYM theory it is given by scalar box Feynman integral [9]

$$M_{n=4} = 1 - \frac{1}{2} a I_\epsilon^{\square} + O(a^2)\,, \tag{4}$$

$$I_\epsilon^{\square} = \frac{i}{\pi^2}\int \frac{d^{4-2\epsilon}k\,(p_1+p_2)^2(p_2+p_3)^2}{k^2(k-p_1)^2(k-p_1-p_2)^2(k+p_4)^2}\,,$$

where integration goes over loop momentum k. Here four external gluon momenta are considered to be incoming so that $\sum_{i=1}^4 p_i = 0$ and Mandelstam invariants are $s=(p_1+p_2)^2$ and $t=(p_2+p_3)^2$. The scalar box integral in (4) has IR divergences and they appear as a double pole in ϵ.

To reveal a new interesting symmetry of (4) let us ignore for a moment IR divergences and put formally $\epsilon=0$ in (4). Then, we change variables to go from five momenta $k, p_1, \dots, p_4$ to *dual momenta* x_i (with $i=0,\dots,4$)

$$p_i = x_i - x_{i+1}\,, \qquad k = x_1 - x_0\,, \tag{5}$$

with $x_5 \equiv x_1$, and obtain

$$I_{\epsilon=0}^{\square} = \frac{i}{\pi^2}\int \frac{d^4x_0\, x_{13}^2 x_{24}^2}{x_{10}^2 x_{20}^2 x_{30}^2 x_{40}^2}\,. \tag{6}$$

This integral is obviously invariant under translations, dilatations and rotations in the dual $x-$space. It is less trivial to verify that $I_{\epsilon=0}^{\square}$ is also (formally) invariant under conformal inversions $x_i^\mu \to x_i^\mu/x_i^2$ (with $i=0,\dots,4$). Thus, the integral (6) is invariant under conformal $SO(2,4)$ transformations in the dual $x-$space.[10] That is the reason why the corresponding symmetry is called *dual conformal symmetry*.

It is important to realize that this symmetry is not related to conventional conformal symmetry of $\mathcal{N}=4$ SYM theory. Indeed, the latter symmetry acts locally on the coordinates of particles and, after Fourier transform, it is realized nonlinearly when acting on momenta of particles. The dual $x-$variable introduced in (5) have the meaning of the momenta and the dual conformal symmetry acts on the momenta x_i locally as if they were coordinates.

The dual conformal symmetry arises as a property of one-loop correction to $M_{n=4}$ and one may wonder whether this symmetry is also present at high loops. This can be checked using explicit three-loop expression for $M_{n=4}$. It is given by the sum over scalar three-loop Feynman diagrams of special topology,[11,12] both planar and nonplanar. It turns out [13] that, when expressed in terms of dual $x-$coordinates, the planar diagrams are invariant under conformal transformation while for nonplanar diagrams this is not the case. This suggests that the dual conformal invariance is the property of planar scattering amplitudes only.

Notice that the integral (6) suffers from IR divergences and, therefore, it is not well-defined in four dimensions. Introducing the dimensional regularization, $d^4x_0 \to d^{4-2\epsilon}x_0$ (with $\epsilon < 0$), we regularize the scalar box integral and simultaneously break its dual conformal invariance. In other words, the dual conformal symmetry of planar amplitudes is broken by IR divergences. As a result, in order to make use of this symmetry we should be able to control its breaking. We shall return to this issue in section 3 (see Eq. (30) below).

2. Dual conformal symmetry of tree amplitudes

Dual conformal symmetry was identified in the previous section as a particular feature of one-loop correction to four-gluon MHV amplitude. It is natural to ask whether this symmetry has any manifestation at tree level.

Let us start with tree-level $n-$particle MHV amplitudes. By definition, these amplitudes are characterized by the property that their total helicity equals $n - 4$. We recall that the on-shell states in $\mathcal{N} = 4$ SYM theory are gluons of helicity (± 1) that we shall denote as $G^{\pm}$, helicity$\pm\frac{1}{2}$ gluinos $\Gamma_A, \bar{\Gamma}^A$ and helicity 0 scalar S_{AB} (with $A, B = 1, \dots, 4$). Examples of MHV amplitudes include pure gluonic amplitudes like $A(G_1^- G_2^- G_3^+ \dots G_n^+)$ as well as amplitudes involving gluinos and scalars, like $A(G_1^- S_2 S_3 \dots G_n^+)$ and $A(G_1^- \Gamma_2 \bar{\Gamma}_3 \dots G_n^+)$.

The light-cone superspace formalism provides a convenient framework for discussing symmetries of scattering amplitudes in $\mathcal{N} = 4$ SYM theory. A unique feature of this theory is that all on-shell states can be combined into a single CPT self-conjugated superstate

$$\Phi(p,\eta) = G^+(p) + \eta^A\Gamma_A(p) + \frac{1}{2}\eta^A\eta^B S_{AB}(p) + \frac{1}{3!}\eta^A\eta^B\eta^C\epsilon_{ABCD}\bar{\Gamma}^D(p) + \frac{1}{4!}\eta^A\eta^B\eta^C\eta^D\epsilon_{ABCD}G^-(p)\,. \qquad (7)$$

Here expansion runs in powers of four Grassman valued variables η^A ($A = 1,\ldots,4$) with the expansion coefficients given by on-shell states of various particles. In this way, instead of considering the scattering of different particles we may consider the scattering of superstates $\Phi(p_i,\eta_i)$. The resulting superamplitude $\mathcal{A}_n \equiv \mathcal{A}(\Phi_1,\ldots,\Phi_n)$ can be thought of a generating functional for all possible scattering amplitudes in $\mathcal{N}=4$ theory. In particular, the MHV amplitudes mentioned in the beginning of this section correspond to two terms in the expansion of the superamplitude in η's

$$\mathcal{A}_n = (\eta_1)^4(\eta_2)^4 \times A(G_1^- G_2^- G_3^+ \ldots G_n^+) + (\eta_1)^4(\eta_2)^2(\eta_3)^3 \times A(G_1^- \bar{S}_2\, S_3 \ldots G_n^+) + \ldots\,. \qquad (8)$$

Due to invariance of the superamplitude under SU(4) rotations of η's, this expansion runs in powers of η's multiple of four. The MHV amplitudes are described by homogenous polynomials in η's of degree 8. Moreover, the supersymmetric relations (2) imply that the superamplitude $\mathcal{A}_n$ does not contain terms in η's of degree 0 and 4 since the corresponding amplitudes vanish.

Additional simplification occurs after we employ the spinor helicity formalism to resolve the on-shell conditions for light-like momenta of particles in terms of two commuting spinors λ^α and $\tilde{\lambda}^{\dot\alpha}$, carrying the helicities $+\frac{1}{2}$ and $-\frac{1}{2}$, respectively,

$$p_i^2 = 0 \quad \Longrightarrow \quad p_i^{\alpha\dot\alpha} \equiv p_i^\mu(\sigma_\mu)^{\alpha\dot\alpha} = \lambda_i^\alpha\tilde{\lambda}_i^{\dot\alpha}\,, \qquad (9)$$

with $\sigma^\mu = (1, i\vec{\sigma})$ expressed in terms of Pauli matrices. Then, $n-$particle superamplitude becomes a function of n copies of kinematical (super)variables λ_i, $\tilde{\lambda}_i$ and η_i^A. It is remarkable that all tree-level MHV amplitudes in $\mathcal{N}=4$ SYM theory are combined into a single superamplitude [14,15]

$$\mathcal{A}_n^{\text{MHV};0} = i(2\pi)^4\frac{\delta^{(4)}(\sum_1^n \lambda_i\tilde{\lambda}_i)\delta^{(8)}(\sum_1^n \lambda_i\eta_i)}{\langle 12\rangle\langle 23\rangle\ldots\langle n1\rangle}\,, \qquad (10)$$

where the notation is introduced for contraction of spinors, $\langle ij\rangle = \lambda_i^\alpha\lambda_{j\,\alpha}$.

As was already mentioned, the tree amplitudes should respect conventional superconformal symmetry of $\mathcal{N}=4$ theory. Indeed, two delta functions in (10) ensure that the amplitude carries vanishing total momentum $p^{\alpha\dot\alpha} = \sum_1^n \lambda_i^\alpha \tilde\lambda_i^{\dot\alpha}$ and zero supercharges, $q^{\alpha A} = \sum_1^n \lambda_i^\alpha \eta_i^A$ and $\bar q_{\dot\alpha}^A = \sum_1^n \eta_i^A \partial/\partial\tilde\lambda_i^{\dot\alpha}$

$$p^{\alpha\dot\alpha} \mathcal{A}_n^{\text{MHV};0} = q^{\alpha\, A} \mathcal{A}_n^{\text{MHV};0} = \bar q^{\dot\alpha\, A} \mathcal{A}_n^{\text{MHV};0} = 0\,. \tag{11}$$

It is much less trivial to verify [15] that the tree MHV superamplitude is also annihilated by conformal boost generator

$$k_{\alpha\dot\alpha} \mathcal{A}_n^{\text{MHV};0} = \sum_1^n \frac{\partial^2}{\partial\lambda_i^\alpha \partial\lambda_i^{\dot\alpha}} \mathcal{A}_n^{\text{MHV};0} = 0\,, \tag{12}$$

and its supersymmetric counterparts, s and $\bar s$.

It turns out that, in addition to conventional superconformal symmetry, the tree MHV superamplitude (10) possesses a much bigger, dual superconformal symmetry.[1] In distinction with the former symmetry, it operates locally on the dual (super) momenta $x_i^{\alpha\dot\alpha}$ and $\theta_{i\alpha}^A$ defined as

$$p_i^{\alpha\dot\alpha} = (x_i - x_{i+1})^{\alpha\dot\alpha}\,, \qquad \lambda_i^\alpha \eta^A = \theta_i^\alpha - \theta_{i+1}^\alpha \tag{13}$$

subject to periodicity condition, $x_{n+1} \equiv x_1$ and $\theta_{n+1} \equiv \theta_1$. The dual superconformal symmetry acts on the dual $x-$ and $\theta-$variables as if they were coordinates in some fictitious configuration space. Then, one can use the explicit expression for tree MHV superamplitude (10) to verify that it transforms covariantly under the dual superconformal transformations. In particular, we find that under conformal inversion

$$x_i \to \frac{x_i}{x_i^2}\,, \qquad \theta_i \to \theta_i \frac{x_i}{x_i^2}\,, \qquad \mathcal{A}_n^{\text{MHV};0} \to [x_1^2 x_2^2 \dots x_n^2] \mathcal{A}_n^{\text{MHV};0}\,. \tag{14}$$

Quite remarkably, the dual conformal symmetry is a property of all (MHV, NMHV, N^2MHV, ...) tree-level superamplitudes in $\mathcal{N} = 4$ SYM theory.[1,16]

3. Dual conformal symmetry at loop level

Going from tree to loop level amplitudes, we should deal with infrared divergences. In the dimensional regularization with $D = 4 - 2\epsilon$, they appear as poles ϵ in such a way that to $\ell-$loop order the leading IR singularity is $1/\epsilon^{2\ell}$.

Infrared divergences of scattering amplitudes have been intensively studied in past and their properties are well known.[17] Namely, all-loop planar amplitudes can be factorized into a product of IR divergent factor that resums all poles in ϵ and infrared finite factor. For a logarithm of the planar amplitude in $\mathcal{N} = 4$ SYM, the same property can be expressed as

$$\ln \mathcal{A}_n = \text{Div}(1/\epsilon) + \text{Finite} + O(\epsilon)\,. \tag{15}$$

Here the finite part depends on the type of the superamplitude (MHV, NMHV, N^{2}MHV,...) while the divergent part has a universal form

$$\text{Div}(1/\epsilon) = -\frac{1}{4}\sum_{l=1}^{\infty} a^l \left(\frac{\Gamma^{(l)}_{\text{cusp}}}{(l\epsilon)^2} + \frac{G^{(l)}}{l\epsilon}\right)\sum_{i=1}^{n}(-s_{i,i+1}/\mu^2)^{-l\epsilon}\,, \tag{16}$$

where $s_{i,i+1} = (p_i + p_{i+1})^2$ is the two-particle invariant mass and the scale μ^2 has the meaning of IR cut-off. Notice that the right-hand side of (16) only contains double and simple poles in ϵ. The corresponding residues can be identified as coefficients in the perturbative expansion of two functions

$$\Gamma_{\text{cusp}}(a) = \sum_{l\geq 1} a^l\, \Gamma^{(l)}_{\text{cusp}}\,, \qquad G(a) = \sum_{l\geq 1} a^l\, G^{(l)}\,, \tag{17}$$

known as the cusp anomalous dimension and collinear anomalous dimension, respectively. By definition, the cusp anomalous dimension describes the dependence of light-like Wilson loop on ultraviolet cut-off and it seems to have nothing to do with IR divergences of the scattering amplitudes. Nevertheless, as was shown in,[18] infrared divergences of the scattering amplitudes are in one-to-one correspondence with ultraviolet divergences of cusped Wilson loops.

One of the immediate consequences of (15) and (16) is that IR divergences break both conformal and dual conformal symmetries. Indeed, applying the conformal boost operator (12) to both sides of (15) one can verify that

$$k_{\alpha\dot{\alpha}}\mathcal{A}_n^{(\text{all loops})} \neq 0\,. \tag{18}$$

Similar relation also holds for the generator of dual conformal symmetry. Being combined together, the conformal and dual conformal symmetries prove to be very powerful in determining the tree superamplitudes in $\mathcal{N}=4$ theory. To make use of these symmetries at loop level, we should be able to control the corresponding anomalies. At present, not much is known about the conformal $k-$anomaly, partly due to a relatively complicated form of the corresponding generator (12). At the same time, as we will explain in a moment, the dual conformal anomaly can be computed to all loops using the conjectured duality between MHV amplitudes and light-like Wilson loops.

We recall that perturbative corrections to all MHV amplitudes are described by a single scalar function of Mandelstam variables and 't Hooft coupling constant, Eq. (3). In terms of superamplitudes, this means that all-loop $n-$particle MHV superamplitude takes the following form in $\mathcal{N}=4$

theory

$$\mathcal{A}_n^{\text{MHV;(all loops)}} = \mathcal{A}_n^{\text{MHV;0}} M_n(\{s_{ij}\}; a) , \tag{19}$$

with tree-level superamplitude $\mathcal{A}_n^{\text{MHV;0}}$ given by (10).

To lowest order in the coupling, $M_n(\{s_{ij}\}; a)$ is given by the sum of scalar Feynman diagrams of a special topology [12] such that the corresponding Feynman integrals are (formally) dual conformal invariant.[10,13] As we will see in a moment, this property can be understood from the conjectured duality between MHV amplitudes and light-like Wilson loops [2,13,19]

$$\ln M_n(\{s_{ij}\}; a) = \ln W(C_n) + O(1/N^2) , \tag{20}$$

where $W(C_n)$ stands for Wilson loop vacuum expectation value

$$W(C_n) = \frac{1}{N} \langle 0| \operatorname{tr} P \exp\Big(\oint_{C_n} dx \cdot A(x)\Big) |0\rangle . \tag{21}$$

Here the gauge field is integrated over the closed n–gon contour C_n in Minkowski space-time

$$C_n = [x_1, x_2] \cup [x_2, x_3] \cup \ldots \cup [x_n, x_1] \tag{22}$$

with the vertices located at x_i^μ and separated by light-like intervals, $(x_i - x_{i+1})^2 = 0$.

The duality relation (20) is expected to hold in planar $\mathcal{N} = 4$ theory for arbitrary 't Hooft coupling. At first glance, relation (20) looks unnatural since the quantities entering its both sides are defined in two different representations – M_n depends on the momenta of particles while $W(C_n)$ is a nonlocal functional of the contour C_n in the coordinate space. The relation between the two representations is established through identification of light-like edges of C_n with on-shell momenta of scattered particles

$$p_i = x_i - x_{i+1} . \tag{23}$$

Notice that we already encountered the same relation when we discussed dual conformal properties of one-loop scalar box integral, Eq. (5).

Let us examine the duality relation (20) for $n = 4$ at one loop, by comparing corrections to four-gluon scattering amplitude and four-edged (rectangular) Wilson loop. According to (21), $W(C_4)$ is given at weak coupling by

$$\ln W(C_4) = \frac{1}{2}(ig)^2 C_F \oint_{C_4} dx_\mu \oint_{C_4} dx'_\nu \, D^{\mu\nu}(x - x') + O(g^4) , \tag{24}$$

where $C_F = (N^2 - 1)/(2N)$ is the quadratic Casimir of the $SU(N)$ group and $D^{\mu\nu}(x - x')$ is the gauge field propagator in the configuration space.

Here, the end-points of the propagator are integrated over the contour C_4. A close examination of (24) reveals that the contour integral has divergences originating from integration over x and/or x' close the cusp points x_i^μ. In the dimensional regularization with $D = 4-2\epsilon$ one finds [13] (with $x_{ij} = x_i - x_j$)

$$\ln W(C_4) = a\left\{-\frac{1}{\epsilon^2}\left[\left(-x_{13}^2\mu^2\right)^\epsilon + \left(-x_{24}^2\mu^2\right)^\epsilon\right] + \frac{1}{2}\ln^2\left(\frac{x_{13}^2}{x_{24}^2}\right) + \text{const}\right\} + O(a^2). \tag{25}$$

Let us compare this relation with the one-loop correction to the function M_4. It is given by (dimensionally regularized) scalar box diagram (4) and reads

$$\ln M_4(s,t) = a\left\{-\frac{1}{\epsilon^2}\left[\left(-s/\mu^2\right)^{-\epsilon} + \left(-t/\mu^2\right)^{-\epsilon}\right] + \frac{1}{2}\ln^2\left(\frac{s}{t}\right) + \text{const}\right\} + O(a^2). \tag{26}$$

Then, we apply (23) and identify two diagonals of the rectangle C_4 as Mandelstam invariants of four gluon amplitude

$$x_{13}^2 := (p_1+p_2)^2 = s\,, \qquad x_{24}^2 := (p_2+p_3)^2 = t\,. \tag{27}$$

In this way, we find that, after appropriate identification of regularization parameters, $\epsilon \to -\epsilon$ and $\mu \to 1/\mu$, ultraviolet divergences of light-like Wilson loop (25) match infrared divergences of four-gluon amplitude (26). Most importantly, the finite parts of the two quantities coincide, in a perfect agreement with (20).

Recently, the conjecture was put forward by Bern, Dixon and Smirnov [12] (BDS) about possible form of M_4 to all loops

$$\ln M_4(s,t) = \text{Div}(1/\epsilon) + \frac{1}{4}\Gamma_{\text{cusp}}(a)\ln^2\left(\frac{s}{t}\right) + \text{const}\,, \tag{28}$$

where the divergent part $\text{Div}(1/\epsilon)$ has the well-known form (16) and, in comparison with one-loop result (26), the coupling constant a in front of $\ln^2(s/t)$ is replaced by $\Gamma_{\text{cusp}}(a)$. We recall that the cusp anomalous dimension $\Gamma_{\text{cusp}}(a)$ controls the leading infrared divergence of the amplitude. A unusual feature of (28) calling for additional explanation is that the cusp anomalous dimension also determines the finite part. As we will see shortly, this property can be derived from MHV amplitude/Wilson loop duality (20).

At present, the duality relation (20) has been verified both at weak and strong coupling. In particular, using stringy description of gluon scattering amplitudes in AdS/CFT, Alday and Maldacena [2] reproduced the BDS ansatz (28) at strong coupling. At weak coupling, the light-like Wilson loop

$W(C_n)$ was computed to two loops for arbitrary $n \geq 4$ in [13,19–23]. It was found that the resulting expression for the Wilson loop $W(C_n)$ matches the BDS ansatz for $n = 4$ and $n = 5$ points, in agreement with (20), but the two quantities differ from each other for $n \geq 6$ starting from two loops. This means that for $n \geq 6$ the BDS ansatz (28) and the duality relation (20) can not hold simultaneously. We shall return to this point at the end of the section.

Let us assume that the duality relation (20) holds to all loops and examine its consequences. Notice that the Wilson loop $W(C_n)$ in (20) is computed in a conformal invariant $\mathcal{N} = 4$ SYM theory. Therefore, $W(C_n)$ should be invariant under conformal transformations acting on four-dimensional vectors x_i^μ defining the position of vertices of the contour C_n in Minkowski space-time. Then, the duality relation (20) implies that the function $M_n(\{s_{ij}\}; a)$ expressed in terms of dual $x-$variables (23) should be invariant under conformal transformations acting on x's. This is exactly the dual conformal symmetry of the scattering amplitudes described in Sect. 1.

We recall that the dual conformal symmetry of the amplitudes is broken by infrared divergences. In the similar manner, conformal symmetry of Wilson loop $W(C_n)$ is broken by ultraviolet (cusp) divergences. If the duality relation (20) survives the dimensional regularization, then the anomaly of the dual conformal symmetry for the amplitudes should match the anomaly of conventional conformal symmetry for light-like Wilson loops. The latter one can be derived by the standard methods.[21] Namely, separating the Wilson loop into divergent and finite parts as

$$\ln W(C_n) = F_n^{(\mathrm{WL})} + \mathrm{Div}(1/\epsilon) + O(\epsilon) \tag{29}$$

we find that the finite part $F_n^{(\mathrm{WL})}$ satisfies the following all-loop anomalous conformal Ward identity[21]

$$K^\mu F_n \equiv \sum_{i=1}^{n} \left[2x_i^\mu (x_i \cdot \partial_{x_i}) - x_i^2 \partial_{x_i}^\mu\right] F_n = \frac{1}{2}\Gamma_{\mathrm{cusp}}(a) \sum_{i=1}^{n} x_{i,i+1}^\mu \ln\Big(\frac{x_{i,i+2}^2}{x_{i-1,i+1}^2}\Big) . \tag{30}$$

The duality relation (20) allows us to identify F_n with the finite part of $n-$particle MHV amplitude and, then, to translate (30) into the relation for the finite part of all-loop MHV amplitude.

Solving the differential equation (30) we find that its general solution is defined up to an arbitrary function of conformal invariants in the form of cross-ratios

$$F_n \to F_n + f_n(\{u_{ijkl}\}), \qquad u_{ijkl} = \frac{x_{il}^2 x_{jk}^2}{x_{ik}^2 x_{jl}^2}. \tag{31}$$

Since dual $x-$variables are related to on-shell momenta of particles as $p_i = x_i - x_{i+1}$, the conformal ratios u_{ijkl} are given by ratio of Mandelstam variables. Examining the corresponding expressions for different number of particles n, we find that $n = 4$ and $n = 5$ are special since there are no (nontrivial) conformal ratios due to light-like condition $x_{i,i+1}^2 = 0$. As a result, for $n = 4$ and $n = 5$ the anomalous Ward identity (30) has a *unique solution* (up to an additive constant) [21]

$$F_4 = \frac{1}{4}\Gamma_{\text{cusp}}(a)\ln^2\Big(\frac{x_{13}^2}{x_{24}^2}\Big) + \text{ const },$$

$$F_5 = -\frac{1}{8}\Gamma_{\text{cusp}}(a)\sum_{i=1}^{5}\ln\Big(\frac{x_{i,i+2}^2}{x_{i,i+3}^2}\Big)\ln\Big(\frac{x_{i+1,i+3}^2}{x_{i+2,i+4}^2}\Big) + \text{ const }. \tag{32}$$

For $n = 4$, we verify that F_4 coincides with the finite part of the BDS ansatz [12] for four-gluon amplitude (28). In the similar manner, one can verify that F_5 matches the BDS ansatz for five-point MHV amplitudes.

Starting from $n = 6$ there are nontrivial cross-ratios (31) and, therefore, the anomalous Ward identity (30) does not fix F_n uniquely. As a consequence, the finite part of $n-$particle amplitude, $F_n^{(\text{MHV})}$, predicted by the BDS ansatz, $F_n^{(\text{BDS})}$, and by the light-like Wilson loop, $F_n^{(\text{WL})}$, may differ from each other for $n \geq 6$. To verify this the hexagon Wilson loop $F_6^{(\text{WL})}$ has been been computed to two loops.[22,24] In a parallel development, six-gluon MHV amplitude $F_6^{(\text{MHV})}$ has been constructed.[25] Comparing the explicit two-loop expressions with the BDS ansatz for $n = 6$ we find that

$$F_6^{(\text{WL})} = F_6^{(\text{MHV})} \neq F_6^{(\text{BDS})}. \tag{33}$$

We conclude from this relation that the Wilson loop/MHV scattering amplitude duality holds at $n = 6$ to two loops while the BDS ansatz needs to be modified.

4. Dual conformal symmetry beyond MHV

We have seen in the previous section that the MHV superamplitudes have a number remarkable properties in planar $\mathcal{N} = 4$ SYM theory. But what about non-MHV superamplitudes, do they share the same properties? As compared to MHV amplitudes, their calculation is more involved and explicit one-loop expressions have a much more complicated structure.

To begin with, we consider next-to-MHV amplitudes. To one loop, $n-$gluon NMHV amplitudes have been computed in [26]. Similar to MHV case, all NMHV amplitudes can be combined into a single NMHV superamplitude.[1] It takes the following form when expressed in terms of dual (super)coordinates (13)

$$\mathcal{A}_n^{\rm NMHV}(x_i,\lambda_i,\theta_i^A) = \mathcal{A}_n^{\rm MHV} \times \left[R_n^{\rm NMHV}(x_i,\lambda_i,\theta_i^A) + O(\epsilon)\right] , \qquad (34)$$

where $\mathcal{A}_n^{\rm MHV}$ is all-loop MHV superamplitude (19) and the 'ratio function' $R_n^{\rm NMHV}$ is a homogenous polynomial in θ's of degree 4.

The explicit one-loop expression for $R_n^{\rm NMHV}$ was found in [1,27]. Quite remarkably, it turned out to be dual conformal invariant. This property is by no means obvious from (34) since both superamplitudes, $\mathcal{A}_n^{\rm MHV}$ and $\mathcal{A}_n^{\rm NMHV}$, suffer from IR divergences and their dual conformal symmetry is broken. However, IR divergences of both amplitudes have universal form (16) and, as a consequence, they cancel in their ratio $R_n^{\rm NMHV}$. The fact that $R_n^{\rm NMHV}$ is invariant under dual conformal transformations implies that $\mathcal{A}_n^{\rm MHV}$ and $\mathcal{A}_n^{\rm NMHV}$ have the same dual conformal anomaly. For MHV superamplitude this anomaly was found in (30) by making use of the duality with Wilson loops. Dual conformal symmetry of $R_n^{\rm NMHV}$ suggests that the finite part of NMHV superamplitude satisfies the same relation (30).

Taking into account the obtained relations for MHV and NMHV superamplitudes, we are now ready to formulate the DHKS conjecture[1,27] for the unifying superamplitude in $\mathcal{N}=4$ SYM theory

$$\mathcal{A}_n(x_i,\lambda_i,\theta_i^A) = \mathcal{A}_n^{\rm MHV} + \mathcal{A}_n^{\rm NMHV} + \mathcal{A}_n^{\rm N^2MHV} + \ldots + \mathcal{A}_n^{\overline{\rm MHV}} , \qquad (35)$$

where $p^{\rm th}$ term in this expansion describes N^{p}MHV superamplitude:

- At tree level, the superamplitude $\mathcal{A}_n(x_i,\lambda_i,\theta_i^A)$ is free from IR divergences and, as a consequence, it should respect dual conformal symmetry. More precisely, the tree superamplitude $\mathcal{A}_n^{\rm (tree)}$ should transform under dual superconformal transformations in the same way as the MHV superamplitude, Eq. (14);
- At loop level, this symmetry becomes anomalous due to IR divergences. These divergences cancel however in the 'ratio' of superamplitudes

$$\mathcal{A}_n(x_i,\lambda_i,\theta_i^A) = \mathcal{A}_n^{\rm MHV} \times \left[R_n(x_i,\lambda_i,\theta_i^A) + O(\epsilon)\right] \qquad (36)$$

where IR finite ratio function takes the form

$$R_n = 1 + R_n^{\rm NMHV} + R_n^{\rm N^2MHV} + \ldots \qquad (37)$$

with $R_n^{\mathrm{N^pMHV}}$ being homogenous polynomial in θ's of degree $4p$. The dual conformal symmetry should be restored in the 'ratio' of superamplitudes and, therefore, R_n should be dual conformal invariant to all loops.

At present, these properties has been firmly established at one loop only [27–29] but the complete all-loop proof is still missing.

5. Scattering amplitudes/Wilson loop duality in QCD

So far our consideration was restricted to maximally supersymmetric Yang-Mills theory. What happens with remarkable symmetries of scattering amplitudes described above after one reduces the amount of supersymmetry and arrives eventually at QCD?

We recall that gluon tree amplitudes look alike in all gauge theory and, therefore, QCD tree amplitudes inherit the dual conformal symmetry of $\mathcal{N} = 4$ theory. This is not true anymore at loop level. For instance, the simplest four-gluon amplitude in QCD is given to two loops by a lengthy expression [30] which is way more complicated compared to $\mathcal{N} = 4$ result (28). This seems to indicate that in QCD there is no obvious relation between four-gluon amplitude $A_4^{(\mathrm{QCD})}(s,t)$ and the corresponding light-like Wilson loop $W^{(\mathrm{QCD})}(C_4)$.

Let us however examine $A_4^{(\mathrm{QCD})}(s,t)$ in the Regge limit, $s \gg -t$. It is well-known that in this limit the planar amplitudes exhibit power-like (Regge) asymptotic behavior [31,32]

$$A_4^{(\mathrm{QCD})}(s,t) \stackrel{s\gg -t}{\sim} (s/(-t))^{\omega_R(t)} , \tag{38}$$

dominated by the contribution of reggeized gluon trajectory $\alpha_R(t) = 1 + \omega_R(t)$ whose explicit two-loop expression has been found in [33]. How does the corresponding rectangular Wilson loop $W^{(\mathrm{QCD})}(C_4)$ behave in the same limit? The edges of the integration contour C_4 are just four light-like momenta of gluons, Eq. (23), and, in the Regge limit, one of the diagonals of the rectangular C_4 is much bigger than another one, $|x_{13}^2| \gg |x_{24}^2|$. In this limit, the Wilson loop is known to have the following asymptotic behavior [34]

$$W^{(\mathrm{QCD})}(C_4) \stackrel{|x_{13}^2|\gg|x_{24}^2|}{\sim} \left(x_{13}^2/(-x_{24}^2)\right)^{\omega_{\mathrm{W}}(x_{24}^2)} , \tag{39}$$

where the exponent $\omega_{\mathrm{W}}(x_{24}^2)$ is given to all loops by

$$\omega_{\mathrm{W}}(x_{24}^2) = \frac{1}{2}\int_{(-x_{24}^2)}^{1/\mu^2} \frac{dk_\perp^2}{k_\perp^2} \Gamma_{\mathrm{cusp}}(a(k_\perp^2)) + \Gamma_{\mathrm{W}}(a(-x_{24}^2)) . \tag{40}$$

Here $\Gamma_{\rm cusp}$ stands for the cusp anomalous dimension in QCD and $\Gamma_{\rm W}$ is analogous to the collinear anomalous dimension. Notice that the QCD coupling constant depends on the normalization scale.

Identifying the scales as $x_{13}^2 := s$ and $x_{24}^2 := t$ we find that $A_4^{(\rm QCD)}(s,t)$ and $W^{(\rm QCD)}(C_4)$ have similar asymptotic behavior in the Regge limit, Eqs. (38) and (40), respectively. It is controlled by two (a priori unrelated) functions $\omega_R(t)$ and $\omega_{\rm W}(t)$. However, examining their explicit two-loop expressions we find that they coincide upon an appropriate identification of renormalization parameters [34]

$$\omega_R(t) = \omega_{\rm W}(t)\,. \tag{41}$$

Assuming that this relation holds at all loops we conclude that the duality between scattering amplitudes and Wilson loops also holds in QCD but in the Regge limit only.[34] This should be compared with $\mathcal{N}=4$ SYM theory in which case the duality is the exact statement for arbitrary s and t.[13]

6. Concluding remarks

Scattering amplitudes in planar $\mathcal{N}=4$ SYM theory exhibit a surprising dual conformal symmetry. It appears on top of all known symmetries of this theory and imposes severe constraints on the possible form of the planar amplitudes.

In addition, MHV scattering amplitudes in planar $\mathcal{N}=4$ theory are related to the Wilson loops evaluated along a closed polygon-like contour in Minkowski space-time built from light-like segments defined by on-shell gluon momenta. As a consequence, the dual conformal symmetry of MHV amplitudes follows from the conventional conformal symmetry of light-like Wilson loops. For non-MHV amplitudes the situation is quite different. These amplitudes also possess the dual superconformal symmetry but the corresponding dual object analogous to Wilson loop has not been identified yet. Despite of this, non-MHV amplitudes satisfy the same anomalous all-loop Ward identity as light-like Wilson loop. This suggests that $n-$gon Wilson loop should be one of the building blocks in the dual description of non-MHV amplitudes.

The planar scattering amplitudes in $\mathcal{N}=4$ SYM theory should have the dual conformal symmetry for arbitrary coupling. At strong coupling, this symmetry is believed to be related through the AdS/CFT correspondence to invariance of the stringy sigma model on $\mathrm{AdS}_5\times \mathrm{S}^5$ background under a combined bosonic [35] and fermionic T duality.[3,4] It is even more intriguing that the dual symmetry is also present in QCD but in the Regge limit only.

It worth mentioning that it is in the Regge limit that QCD is expected to admit a dual description in terms of (still mysterious) hadronic string. In a close analogy with $\mathcal{N} = 4$ theory, we expect that the dual conformal symmetry provides yet another glimpse into QCD/string duality.

References

1. J. M. Drummond, J. Henn, G. P. Korchemsky and E. Sokatchev, "Dual superconformal symmetry of scattering amplitudes in $\mathcal{N}$=4 super-Yang-Mills theory," arXiv:0807.1095 [hep-th].
2. L. F. Alday and J. M. Maldacena, "Gluon scattering amplitudes at strong coupling," JHEP **0706**, 064 (2007) [arXiv:0705.0303 [hep-th]];
L. F. Alday and J. Maldacena, "Comments on gluon scattering amplitudes via AdS/CFT," JHEP **0711**, 068 (2007) [arXiv:0710.1060 [hep-th]].
3. N. Berkovits and J. Maldacena, "Fermionic T-Duality, Dual Superconformal Symmetry, and the Amplitude/Wilson Loop Connection," JHEP **0809**, 062 (2008) [arXiv:0807.3196 [hep-th]].
4. N. Beisert, R. Ricci, A. A. Tseytlin and M. Wolf, "Dual Superconformal Symmetry from AdS5 x S5 Superstring Integrability," Phys. Rev. D **78**, 126004 (2008) [arXiv:0807.3228 [hep-th]].
5. S. J. Parke and T. R. Taylor, "An Amplitude for n Gluon Scattering," Phys. Rev. Lett. **56**, 2459 (1986).
6. F. A. Berends and W. T. Giele, "Recursive Calculations for Processes with n Gluons," Nucl. Phys. B **306**, 759 (1988).
7. M. L. Mangano and S. J. Parke, "Multi-Parton Amplitudes in Gauge Theories," Phys. Rept. **200**, 301 (1991) [arXiv:hep-th/0509223].
8. M. T. Grisaru, H. N. Pendleton and P. van Nieuwenhuizen, "Supergravity And The S Matrix," Phys. Rev. D **15**, 996 (1977);
M. T. Grisaru and H. N. Pendleton, "Some Properties Of Scattering Amplitudes In Supersymmetric Theories," Nucl. Phys. B **124**, 81 (1977).
9. M. B. Green, J. H. Schwarz and L. Brink, "N=4 Yang-Mills And N=8 Supergravity As Limits Of String Theories," Nucl. Phys. B **198**, 474 (1982).
10. D. J. Broadhurst, "Summation of an infinite series of ladder diagrams," Phys. Lett. B **307**, 132 (1993);
J. M. Drummond, J. Henn, V. A. Smirnov and E. Sokatchev, "Magic identities for conformal four-point integrals," JHEP **0701**, 064 (2007) [arXiv:hep-th/0607160].
11. C. Anastasiou, Z. Bern, L. J. Dixon and D. A. Kosower, "Planar amplitudes in maximally supersymmetric Yang-Mills theory," Phys. Rev. Lett. **91**, 251602 (2003) [arXiv:hep-th/0309040].
12. Z. Bern, L. J. Dixon and V. A. Smirnov, "Iteration of planar amplitudes in maximally supersymmetric Yang-Mills theory at three loops and beyond," Phys. Rev. D **72**, 085001 (2005) [arXiv:hep-th/0505205].
13. J. M. Drummond, G. P. Korchemsky and E. Sokatchev, "Conformal properties of four-gluon planar amplitudes and Wilson loops," Nucl. Phys. B **795**, 385 (2008) [arXiv:0707.0243 [hep-th]].

14. V. P. Nair, "A current algebra for some gauge theory amplitudes," Phys. Lett. B **214**, 215 (1988).
15. E. Witten, "Perturbative gauge theory as a string theory in twistor space," Commun. Math. Phys. **252**, 189 (2004) [arXiv:hep-th/0312171].
16. A. Brandhuber, P. Heslop and G. Travaglini, "A note on dual superconformal symmetry of the N=4 super Yang-Mills S-matrix," Phys. Rev. D **78**, 125005 (2008) [arXiv:0807.4097 [hep-th]].
17. A. H. Mueller, "On The Asymptotic Behavior Of The Sudakov Form-Factor," Phys. Rev. D **20**, 2037 (1979);
J. C. Collins, "Algorithm To Compute Corrections To The Sudakov Form-Factor," Phys. Rev. D **22**, 1478 (1980);
A. Sen, "Asymptotic Behavior Of The Sudakov Form-Factor In QCD," Phys. Rev. D **24**, 3281 (1981);
G. P. Korchemsky, "Double logarithmic asymptotics in QCD," Phys. Lett. B **217**, 330 (1989);
L. Magnea and G. Sterman, "Analytic continuation of the Sudakov form-factor in QCD," Phys. Rev. D **42**, 4222 (1990).
18. G. P. Korchemsky and A. V. Radyushkin, "Loop Space Formalism And Renormalization Group For The Infrared Asymptotics Phys. Lett. B **171**, 459 (1986);
S. V. Ivanov, G. P. Korchemsky and A. V. Radyushkin, "Infrared Asymptotics Of Perturbative QCD: Contour Gauges," Yad. Fiz. **44**, 230 (1986) [Sov. J. Nucl. Phys. **44**, 145 (1986)].
19. A. Brandhuber, P. Heslop and G. Travaglini, "MHV Amplitudes in N=4 Super Yang-Mills and Wilson Loops," Nucl. Phys. B **794**, 231 (2008) [arXiv:0707.1153 [hep-th]].
20. J. M. Drummond, J. Henn, G. P. Korchemsky and E. Sokatchev, "On planar gluon amplitudes/Wilson loops duality," Nucl. Phys. B **795**, 52 (2008) [arXiv:0709.2368 [hep-th]].
21. J. M. Drummond, J. Henn, G. P. Korchemsky and E. Sokatchev, "Conformal Ward identities for Wilson loops and a test of the duality with gluon amplitudes," arXiv:0712.1223 [hep-th].
22. J. M. Drummond, J. Henn, G. P. Korchemsky and E. Sokatchev, "The hexagon Wilson loop and the BDS ansatz for the six-gluon amplitude," Phys. Lett. B **662**, 456 (2008) [arXiv:0712.4138 [hep-th]].
23. C. Anastasiou, A. Brandhuber, P. Heslop, V. V. Khoze, B. Spence and G. Travaglini, "Two-Loop Polygon Wilson Loops in N=4 SYM," JHEP **0905**, 115 (2009) [arXiv:0902.2245 [hep-th]].
24. J. M. Drummond, J. Henn, G. P. Korchemsky and E. Sokatchev, "Hexagon Wilson loop = six-gluon MHV amplitude," Nucl. Phys. B **815**, 142 (2009) [arXiv:0803.1466 [hep-th]].
25. Z. Bern, L. J. Dixon, D. A. Kosower, R. Roiban, M. Spradlin, C. Vergu and A. Volovich, "The Two-Loop Six-Gluon MHV Amplitude in Maximally Supersymmetric Yang-Mills Theory," Phys. Rev. D **78**, 045007 (2008) [arXiv:0803.1465 [hep-th]].
26. Z. Bern, V. Del Duca, L. J. Dixon and D. A. Kosower, "All non-maximally-

helicity-violating one-loop seven-gluon amplitudes in N = 4 super-Yang-Mills theory," Phys. Rev. D **71**, 045006 (2005) [arXiv:hep-th/0410224];
Z. Bern, L. J. Dixon and D. A. Kosower, "All next-to-maximally helicity-violating one-loop gluon amplitudes in N=4 super-Yang-Mills theory," Phys. Rev. D **72**, 045014 (2005) [arXiv:hep-th/0412210].

27. J. M. Drummond, J. Henn, G. P. Korchemsky and E. Sokatchev, "Generalized unitarity for N=4 super-amplitudes," arXiv:0808.0491 [hep-th].
28. A. Brandhuber, P. Heslop and G. Travaglini, "One-Loop Amplitudes in N=4 Super Yang-Mills and Anomalous Dual Conformal Symmetry," arXiv:0905.4377 [hep-th].
29. H. Elvang, D. Z. Freedman and M. Kiermaier, "Dual conformal symmetry of 1-loop NMHV amplitudes in N=4 SYM theory," arXiv:0905.4379 [hep-th].
30. E. W. N. Glover, C. Oleari and M. E. Tejeda-Yeomans, "Two-loop QCD corrections to gluon gluon scattering," Nucl. Phys. B **605**, 467 (2001) [arXiv:hep-ph/0102201].
31. S. Mandelstam, "Non-Regge Terms in the Vector-Spinor Theory", Phys. Rev. B **137**, 949 (1965);
M. T. Grisaru, H. J. Schnitzer and H. S. Tsao, "Reggeization of Yang-Mills gauge mesons in theories with a spontaneously broken symmetry," Phys. Rev. Lett. **30**, 811 (1973);
M. T. Grisaru, H. J. Schnitzer and H. S. Tsao, "Reggeization of elementary particles in renormalizable gauge theories - vectors and spinors," Phys. Rev. D **8**, 4498 (1973);
M. T. Grisaru, H. J. Schnitzer and H. S. Tsao, "The Reggeization of elementary particles in renormalizable gauge theories: scalars," Phys. Rev. D **9**, 2864 (1974).
32. E. A. Kuraev, L. N. Lipatov and V. S. Fadin, "Multi - Reggeon Processes In The Yang-Mills Theory," Sov. Phys. JETP **44**, 443 (1976) [Zh. Eksp. Teor. Fiz. **71**, 840 (1976)].
33. V. S. Fadin, R. Fiore and M. I. Kotsky, "Gluon Regge trajectory in the two-loop approximation," Phys. Lett. B **387**, 593 (1996) [arXiv:hep-ph/9605357].
34. I. A. Korchemskaya and G. P. Korchemsky, "Evolution equation for gluon Regge trajectory," Phys. Lett. B **387**, 346 (1996) [arXiv:hep-ph/9607229].
35. R. Kallosh and A. A. Tseytlin, "Simplifying superstring action on AdS5xS5," JHEP **9810**, 016 (1998) [arXiv:hep-th/9808088].

SECTION 5.

SUPERSYMMETRIC FIELD THEORIES

Still in Light-Cone Superspace

P. Ramond*

Physics Department,
Institute for Fundamental Theory,
University of Florida,
Gainesville, FL 32611, USA
E-mail: ramond@phys.ufl.edu

The recently formulated Bagger-Lambert-Gustavsson (BLG) theory in three dimensions is described in terms of a constrained chiral superfield in light-cone superspace. We discuss the use of Superconformal symmetry to determine the form of its interactions, in complete analogy with $N = 4$ SuperYang-Mills in four dimensions.

Keywords: Superspace; Light-cone; Superconformal Theories

1. Introduction

Maximally supersymmetric theories live in two different superspaces. The first with eight complex Grassmann variables is used to describe $N = 1$ supergravity in $d = 11$, $N = 8$ supergravity in $d = 4$, $N = 16$ Supergravity in $d = 3$, and so-on. With a dimensionful coupling, these theories are not superconformal. Instead they contain non-compact and non-linear symmetries, $E_{7(7)}$ and $E_{8(8)}$ in $d = 4$ and 3, respectively.

The second superspace with only four complex Grassmann variables is equally rich. It houses theories with Superconformal symmetry in $d = 6, 5, 4$ and 3 dimensions. The latter theory has been recently formulated covariantly,[1,2] and on the light-cone.[3]

It has already been shown[4] how the fully interacting $N = 4$ SuperYang-Mills theory[5] in $d = 4$ can be determined by requiring $PSU(2, 2\,|\,4)$ Superconformal symmetry on a constrained chiral superfield in light-cone superspace with four complex Grassmann variables.

The following is a progress report on using the same technique, now ap-

*In collaboration with D. Belyaev, L. Brink, and S-S. Kim.

plied to $OSp(2,2\,|\,8)$ Superconformal symmetry on the same chiral superfield. On the light-cone, supersymmetries split into kinematical and dynamical supersymmetries. Kinematical supersymmetries are linearly realized on the chiral superfield. The dynamical ones also contain a linear term (free theory), but also terms non-linear in the (super)fields, which, in superconformal theories, suffice to *completely* determine the theory. Our technique has been to use algebraic consistency to find its expression.

Consistency with the kinematical constraints yields two possible expressions for the dynamical supersymmetries, each determined in terms of four integers, and with an unknown four-index tensor f^{abcd}, where the indices label the superfields. The values of these integers are determined by requiring that the light-cone Hamiltonian and boosts commute with one another. At Shifmania (this proceeding), we reported an unexpected solution, with fractional light-cone derivatives acting on the chiral superfield, without assuming any symmetry among the indices of f^{abcd}. Since then, we have found[6] that by requiring antisymmetry in three of its indices, $b \leftrightarrow c$, $b \leftrightarrow d$ and $c \leftrightarrow d$, the BLG solution emerges from these algebraic constraints, apparently uniquely.

2. Superconformal Theories

In 1978, W. Nahm[7] catalogued all relativistic field theories which extended the Poincaré symmetry to Superconformal symmetry. We only list those in spacetime dimensions $d = 6, 5, 4$, and 3:

$$\begin{aligned}
&d = 6 \quad OSp(2n\,|\,6,2) \supset SO(6,2) \times Sp(2n)_R \,, \\
&d = 5 \quad F[4] \supset SO(5,2) \times SU(2)_R \,, \\
&d = 4 \quad SSU(2,2\,|\,n) \supset SO(4,2) \times SU(n)_R \times U(1)_R \,, \\
&d = 4 \quad PSU(2,2\,|\,4) \supset SO(4,2) \times SU(4)_R \,, \\
&d = 3 \quad OSp(2,2\,|\,n) \supset SO(3,2) \times SO(n)_R \,,
\end{aligned}$$

using Kac's notation for the superalgebras. The conformal group in d spacetime dimensions is $SO(d,2)$; for $d = 4$, it is a non-compact form of $SU(2,2)$, and for $d = 3$ it is isomorphic to $Sp(2,2)$. These theories have large global R-symmetries. The theories with special number of R-symmetries, $n = 2$ in $d = 6$, $n = 4$ in $d = 4$, and $n = 8$ in $d = 3$, can be described in terms of constrained chiral superfields in light-cone superspace. Since then, it has been realized that many superconformal theories are seminal, not only in quantum field theory but also in Superstrings and M-theory.[8]

3. $N = 4$ Light-Cone Superspace

We introduce the usual light-cone variables

$$x^{\pm} = \frac{1}{\sqrt{2}}(x^0 \pm x^3) , \quad \partial^{\pm} = \frac{1}{\sqrt{2}}(\partial^0 \pm \partial^3) , \tag{1}$$

and denote the transverse variables by $x_1 ... x_{d-2}$. The relevant superspace contains four complex Grassmann variables, θ^m and $\bar{\theta}_m$, in terms of which we define the chiral derivatives

$$d^m = -\frac{\partial}{\partial \bar{\theta}_m} - \frac{i}{\sqrt{2}}\, \theta^m\, \partial^+ ; \qquad \bar{d}_n = \frac{\partial}{\partial \theta_n} + \frac{i}{\sqrt{2}}\, \bar{\theta}_n\, \partial^+ ;$$

they satisfy

$$\{\, d^m\, ,\, \bar{d}_n\, \} = -i\,\sqrt{2}\,\delta^m_n\,\partial^+ . \tag{2}$$

The chiral superfields

$$\varphi^a(y) = \frac{1}{\partial^+}\, A^a(y) + \frac{i}{\sqrt{2}}\,\theta^m\,\theta^n\,\overline{C}^{\,a}_{mn}(y) + \frac{1}{12}\,\theta^m\,\theta^n\,\theta^p\,\theta^q\,\epsilon_{mnpq}\,\partial^+\,\bar{A}^a(y)$$
$$+\ \frac{i}{\partial^+}\,\theta^m\,\bar{\chi}^a_m(y) + \frac{\sqrt{2}}{6}\theta^m\,\theta^n\,\theta^p\,\epsilon_{mnpq}\,\chi^{q\,a}(y)$$

where a is a taxonomic index, are chiral by construction,

$$d^m\,\varphi^a(y) = 0 , \tag{3}$$

where the component fields depend on the chiral coordinates

$$y = (\, x_1, ..., x_{d-2}\, ,\, x^- - i\,\frac{\theta^m \bar{\theta}_m}{\sqrt{2}}\,). \tag{4}$$

The parameter x^+ is set to zero without loss of generality. The chiral superfields obey the "inside-out" constraint

$$\bar{d}_m\,\bar{d}_n\,\varphi^a = \frac{1}{2}\,\epsilon_{mnpq}\,d^p\,d^q\,\overline{\varphi}^a . \tag{5}$$

In $d = 4$ SuperYang-Mills, this important constraint allowed us[4] to write its light-cone interacting Hamiltonian as a *positive definite* quadratic form.

The component fields of each chiral superfield represent sixteen physical degrees of freedom, eight bosons and eight fermions. They are organized in terms of an $SO(8)$ R-symmetry, with the bosons transforming as a vector, the fermions as a spinor.

Introduce the operators

$$q^m = -\frac{\partial}{\partial \bar{\theta}_m} + \frac{i}{\sqrt{2}}\,\theta^m\,\partial^+ ; \quad \bar{q}_n = \frac{\partial}{\partial \theta^n} - \frac{i}{\sqrt{2}}\,\bar{\theta}_n\,\partial^+ , \tag{6}$$

which satisfy

$$\{ q^m \, , \, \bar{q}_n \} \; = \; i\,\sqrt{2}\,\delta^m_n\,\partial^+ \; , \tag{7}$$

and do not alter chirality, since they anticommute with the chiral derivatives

$$\{ q^m \, , \, \bar{d}_n \} \; = \; \{ q^m \, , \, d^n \} \; = \; 0 \; . \tag{8}$$

The $SO(8)$ transformations are written in terms of those of its $SO(6) \times U(1) \sim SU(4) \times U(1)$ subgroup, with parameters $\omega^m{}_n$, and ω:

$$\delta_{SO(6)}\,\varphi^a \; = \; \omega^m{}_n\,\frac{i}{\sqrt{2}}\left(q^n\,\bar{q}_m - \frac{1}{4}\delta^n_m\,q^l\,\bar{q}_l\right)\frac{1}{\partial^+}\,\varphi^a \; ; \tag{9}$$

$$\delta_{U(1)}\,\varphi^a \; = \; \omega\,\frac{i}{4\sqrt{2}}\,(q^m\,\bar{q}_m - \bar{q}_m\,q^m\,)\,\frac{1}{\partial^+}\,\varphi^a \; ; \tag{10}$$

and the coset parameters ω^{mn}, and $\overline{\omega}_{mn}$,

$$\delta_{\overline{coset}}\,\varphi^a \; = \; \omega^{mn}\,\frac{i}{\sqrt{2}}\,\bar{q}_m\,\bar{q}_n\,\frac{1}{\partial^+}\,\varphi^a \; ; \quad \delta_{coset}\,\varphi^a \; = \; \omega_{mn}\,\frac{i}{\sqrt{2}}\,q^m\,q^n\,\frac{1}{\partial^+}\,\varphi^a \; . \tag{11}$$

This chiral superfield can be used to define theories in different dimensions, with the only modifications of increase the number of transverse coordinates of its component fields:

- $d = 10$
  The superfield describes $N = 1$ in $d = 10$ dimensions. This theory is not superconformal, as it is the zero slope limit of an open superstring. The $SO(8)$ transformations are interpreted as the "spin" part of the transverse little group, the orbital part being supplied by the appropriate number of transverse coordinates. There are no modifications to the chiral superfield, except for the dependence of its components on the six extra transverse coordinates.

- $d = 6$
  The superconformal group is $OSp(\,4\,|\,6,2\,)$. The transverse light-cone little group is $SO(4) \sim SU(2) \times SU(2)$. The first $SU(2)$ has only an orbital part, whereas the spin part of the second $SU(2)$ is to be found in the decomposition

$$SO(8) \; \supset \; SU(2) \times Sp(4)_R \; , \tag{12}$$

  where the physical fields decompose as

$$\mathbf{8}_b \; = \; (\,\mathbf{3}\,,\,\mathbf{1}\,) + (\,\mathbf{1}\,,\,\mathbf{5}\,) \; , \qquad \mathbf{8}_f \; = \; (\,\mathbf{2}\,,\,\mathbf{4}\,) \; . \tag{13}$$

The bosons split into an R-quintet of scalar fields and an R-singlet tensor, a second rank antisymmetric tensor with self-dual three-form field strength.

- $d = 5$
  The superconformal symmetry group is $F[4]$. The transverse little group is $SO(3) \sim SU(2)$, and its spin part is to be found in the decomposition

$$SO(8) \supset SU(2) \times SU(2)_R \; . \tag{14}$$

  The R-symmetry reduces to $SU(2)$. This decomposition is similar to that in $d = 6$, with the anomalous embedding of $SU(2)$ in $Sp(4)$ with

$$Sp(4) \supset SU(2) \; , \qquad \mathbf{5} = \mathbf{5} \; , \quad \mathbf{4} = \mathbf{4} \; , \tag{15}$$

  so that the scalar bosons split into one R-singlet vector, and five scalars with R-spin 2 and the fermions R-spin 3/2.

- $d = 4$
  The little group is now just $SO(2)$ whose spin part is found in

$$SO(8) \supset SO(2) \times SO(6)_R \sim U(1) \times SU(4)_R \; . \tag{16}$$

  This leads to the well-known $N = 4$ SuperYang-Mills theory, symmetric under $PSU(2,2\,|\,4)$, with one vector and six scalars.

- $d = 3$
  There is no light-cone little group, and the R-symmetry is the full $SO(8)$. The chiral superfield describes the degrees of freedom in the Nahm theory with $n = 8$, and symmetry $OSp(2,2\,|\,8)$. The bosons (fermions) form an R-symmetry vector (spinor) octet.
  The light-cone formulation of this theory will occupy the rest of this paper, using algebraic techniques previously developed for the $N = 4$ theory in four dimensions.

4. $OSp(2,2\,|\,8)$ Generators

We begin with

$$OSp(2,2\,|\,8) \supset Sp(2,2) \times SO(8) \; ,$$

where the first factor group is the conformal group in three dimensions, and the second factor group is the R-symmetry.

In light-cone coordinates, the space-time generators are either kinematical or dynamical. The kinematical generators operate within the initial surface ($x^+ = 0$), while the dynamical generators, called hamiltonians by Dirac, act transversely to the initial surface. The kinematical operators are the same in free and interacting theories, and are linear in the (super)fields. The dynamical operators also contain a part linear in the (super)fields for the free theory, but in the interacting theory, they develop non-linear dependence on the (super)fields.

The ten generators of the conformal group in three dimensions are given by

$$\text{Conformal Group} \quad \begin{cases} \text{Lorentz Group}: & J^{+-}\,, J^{+}\;; \quad \mathcal{J}^{-} \\ \text{Translations}: & P\,, P^{+}\;; \quad \mathcal{P}^{-} \\ \text{Dilatation}: & D \\ \text{Conformal}: & K\,, K^{+}\;; \quad \mathcal{K}^{-} \end{cases} ,$$

with the dynamical generators written in capital calligraphic letters. Note that J^+ and K^+ can be viewed as kinematical as long as we set the parameter x^+ to zero.

The supersymmetry and superconformal generators

$$\text{Supers} \quad \begin{cases} \text{Supersymmetry}: & q\,, \bar{q}\;; \quad \mathcal{Q}\,, \overline{\mathcal{Q}} \\ \text{Superconformal}: & s\,, \bar{s}\;; \quad \mathcal{S}\,, \overline{\mathcal{S}} \end{cases} ,$$

also split into kinematical and dynamical operators. All R-symmetry generators are kinematical, and given by Eqs.(9-11).

4.1. *Kinematical Transformations*

They are expressed in terms of

$$\mathcal{N} = \theta^m \frac{\partial}{\partial \theta^m} + \bar{\theta}_m \frac{\partial}{\partial \bar{\theta}_m}\;; \qquad \mathcal{A} \equiv x^-\,\partial^+ - \frac{x}{2}\,\partial - \frac{1}{2}\mathcal{N} + \frac{1}{2}\,. \tag{17}$$

The kinematical Poincaré transformations are

$$\delta_{P^+}\,\varphi^a = -i\,\partial^+\,\varphi^a\;; \qquad \delta_P\,\varphi^a = -i\,\partial\,\varphi^a\;; \tag{18}$$

$$\delta_{J^+}\,\varphi^a = ix\,\partial^+\,\varphi^a\;; \qquad \delta_{J^{+-}}\,\varphi^a = i(\,x^-\,\partial^+ - \frac{1}{2}\mathcal{N} + 1\,)\,\varphi^a\;; \tag{19}$$

$$\delta_{P^+}\,\varphi^a = -\,i\partial^+\,\varphi^a\;; \qquad \delta_P\,\varphi^a = -\,i\partial\,\varphi^a\,, \tag{20}$$

followed by the kinematical conformal symmetries

$$\delta_D\,\varphi^a = i\,(\,x^-\partial^+ \;-\; x\,\partial - \frac{1}{2}\mathcal{N} + \frac{1}{2}\,)\,\varphi^a\ ; \tag{21}$$

$$\delta_K\,\varphi^a = 2i\,x\,\mathcal{A}\,\varphi^a\ ; \qquad \delta_{K^+}\,\varphi^a \;= i\,x^2\partial^+\,\varphi^a\ . \tag{22}$$

Similarly, the kinematical (spectrum generating) supersymmetries, with parameters ε^m and $\bar{\varepsilon}_m$, are

$$\delta^{kin}_{\varepsilon\bar{q}}\,\varphi^a \;=\; \varepsilon^m\bar{q}_m\,\varphi^a\ ; \qquad \delta^{kin}_{\bar{\varepsilon}q}\,\varphi^a \;=\; \bar{\varepsilon}_m q^m\,\varphi^a\ , \tag{23}$$

and finally kinematical superconformal transformations with parameters α^m and $\bar{\alpha}_m$

$$\delta_{\alpha\bar{s}}\,\varphi^a \;= -ix\alpha^m\,\bar{q}_m\,\varphi^a\ ; \qquad \delta_{\bar{\alpha}s}\,\varphi^a \;=\; ix\,\bar{\alpha}_m\,q^m\,\varphi^a\ . \tag{24}$$

4.2. *Free Dynamical Transformations*

In superconformal theories, **all** dynamical generators are determined by the algebra from the dynamical supersymmetry transformations, because the algebra is simple. To see how this works, we start from the *free* dynamical supersymmetry transformations (written in bold), which are given by

$$\boldsymbol{\delta}^{free}_{\varepsilon\overline{\mathcal{Q}}}\,\varphi^a \;=\; \frac{1}{\sqrt{2}}\varepsilon^m\bar{q}_m\,\frac{\partial}{\partial^+}\,\varphi^a\ , \qquad \boldsymbol{\delta}^{free}_{\bar{\varepsilon}\mathcal{Q}}\,\varphi^a \;=\; \frac{1}{\sqrt{2}}\bar{\varepsilon}_m q^m\,\frac{\partial}{\partial^+}\,\varphi^a\ . \tag{25}$$

We then use the commutators

$$[\,\boldsymbol{\delta}_{\varepsilon\overline{\mathcal{Q}}}\,,\,\boldsymbol{\delta}_{\bar{\varepsilon}\mathcal{Q}}\,]\,\varphi^a \;=\; \sqrt{2}\,\bar{\varepsilon}_m\varepsilon^m\,\boldsymbol{\delta}_{\mathcal{P}^-}\,\varphi^a \qquad \rightarrow \qquad \boldsymbol{\delta}_{\mathcal{P}^-}\,\varphi^a\ , \tag{26}$$

$$[\,\delta_K\,,\,\boldsymbol{\delta}_{\mathcal{P}^-}\,]\,\varphi^a \;=\; 2i\,\boldsymbol{\delta}_{\mathcal{J}^-}\,\varphi^a \qquad \rightarrow \qquad \boldsymbol{\delta}_{\mathcal{J}^-}\,\varphi^a\ , \tag{27}$$

$$[\,\delta_K\,,\,\boldsymbol{\delta}_{\mathcal{J}^-}\,]\,\varphi^a \;=\; -\,i\,\boldsymbol{\delta}_{\mathcal{K}^-}\,\varphi^a \qquad \rightarrow \qquad \boldsymbol{\delta}_{\mathcal{K}^-}\,\varphi^a\ , \tag{28}$$

$$[\,\delta_K\,,\,\boldsymbol{\delta}_{\varepsilon\overline{\mathcal{Q}}}\,]\,\varphi^a \;=\; \sqrt{2}\,\boldsymbol{\delta}_{\varepsilon\overline{\mathcal{S}}}\,\varphi^a \qquad \rightarrow \qquad \boldsymbol{\delta}_{\varepsilon\overline{\mathcal{S}}}\,\varphi^a\ , \tag{29}$$

to compute the remaining dynamical transformations. Evaluation of the commutators yields

$$\text{Time Translation}:\quad \boldsymbol{\delta}^{free}_{\mathcal{P}^-}\,\varphi^a \;= -i\,\frac{\partial^2}{2\,\partial^+}\,\varphi^a\ ,$$

$$\text{Lorentz Boost}:\quad \boldsymbol{\delta}^{free}_{\mathcal{J}^-}\,\varphi^a \;=\; -\,i\frac{\partial}{\partial^+}\mathcal{A}\,\varphi^a\ ,$$

$$\text{Conformal Boost}:\quad \boldsymbol{\delta}^{free}_{\mathcal{K}^-}\,\varphi^a \;=\; 2i\,\frac{1}{\partial^+}\,\mathcal{A}\,(\mathcal{A} - \frac{1}{2})\,\varphi^a\ ,$$

$$\text{SuperConformal}:\quad \boldsymbol{\delta}^{free}_{\alpha\overline{\mathcal{S}}}\,\varphi^a \;=\; i\,\alpha^m\bar{q}_m\,\frac{1}{\partial^+}\,\mathcal{A}\,\varphi^a\ ,$$

$$\boldsymbol{\delta}^{free}_{\bar{\alpha}\mathcal{S}}\,\varphi^a \;= -i\,\bar{\alpha}_m q^m\,\frac{1}{\partial^+}\,\mathcal{A}\,\varphi^a\ .$$

These are valid in the free theory, and need to be altered in the interacting theory.

4.3. *Interacting Dynamical Supersymmetries*

Just as in the free case, it suffices to determine the form of the dynamical supersymmetry transformations. We write

$$\boldsymbol{\delta}_{\varepsilon\overline{Q}}\,\varphi^a \;=\; \boldsymbol{\delta}^{free}_{\varepsilon\overline{Q}}\,\varphi^a + \boldsymbol{\delta}^{int}_{\varepsilon\overline{Q}}\,\varphi^a\;, \qquad \boldsymbol{\delta}_{\overline{\varepsilon}Q}\,\varphi^a \;=\; \boldsymbol{\delta}^{free}_{\overline{\varepsilon}Q}\,\varphi^a + \boldsymbol{\delta}^{int}_{\overline{\varepsilon}Q}\,\varphi^a\;. \tag{30}$$

The expressions $\boldsymbol{\delta}^{int}_{\varepsilon\overline{Q}}\,\varphi^a$ and $\boldsymbol{\delta}^{int}_{\overline{\varepsilon}Q}\,\varphi^a$are highly restricted, by the following ten constraints:

(1) Chirality

$$d^m\,\boldsymbol{\delta}^{int}_{\varepsilon\overline{Q}}\,\varphi^a \;=\; d^m\,\boldsymbol{\delta}^{int}_{\overline{\varepsilon}Q}\,\varphi^a \;=\; 0\,. \tag{31}$$

(2) Both $\boldsymbol{\delta}^{int}_{\varepsilon\overline{Q}}\,\varphi^a$ and $\boldsymbol{\delta}^{int}_{\overline{\varepsilon}Q}\,\varphi^a$ are cubic in the superfields.
In three dimensions, canonical Bose fields have mass dimension of one-half, so that the chiral superfield has half-odd integer canonical dimension itself, assuming integer power of derivatives. Since we are looking for a conformal theory with no dimensionful parameters, $\boldsymbol{\delta}_{\varepsilon\overline{Q}}\,\varphi^a$ and $\boldsymbol{\delta}_{\overline{\varepsilon}Q}\,\varphi^a$ must then both be an odd power of superfields. Also, conformal invariance requires a Hamiltonian with a local sixth-order interaction in the superfields: the non-linear part of the dynamical supersymmetry transformation must be *cubic* in the superfields[9]: the theory must have a tensor with four indices.[a]

(3) Both are independent of x^-, using

$$[\,\delta_{P+}\,,\,\boldsymbol{\delta}_{\varepsilon\overline{Q}}\,]\,\varphi^a \;=\; [\,\delta_{P+}\,,\,\boldsymbol{\delta}_{\overline{\varepsilon}Q}\,]\,\varphi^a \;=\; 0\,. \tag{32}$$

(4) $\boldsymbol{\delta}^{int}_{\varepsilon\overline{Q}}\,\varphi^a$ is independent of x, since

$$[\,\delta_{P}\,,\,\boldsymbol{\delta}_{\varepsilon\overline{Q}}\,]\,\varphi^a \;=\; 0\,. \tag{33}$$

(5) Neither have transverse derivatives ∂: from

$$[\,\delta_{J+}\,,\,\boldsymbol{\delta}_{\varepsilon\overline{Q}}\,]\,\varphi^a \;=-\frac{i}{2}\,\boldsymbol{\delta}_{\varepsilon\bar{q}}\,\varphi^a\;,$$

it follows that

$$[\,\delta_{J+}\,,\,\boldsymbol{\delta}^{int}_{\varepsilon\overline{Q}}\,]\,\varphi^a \;=\; 0\,. \tag{34}$$

---

[a] In $d = 4$, similar considerations suggested a tensor with three indices, f^{abc}, which turned out to be the structure functions of the gauge algebra.

(6) From

$$[\,\delta_{\bar\varepsilon q}\,,\,\boldsymbol{\delta}_{\varepsilon\overline{Q}}\,]\,\varphi^a \;=-\bar\varepsilon_m\varepsilon^m\,\delta_P\,\varphi^a\;, \qquad [\,\delta_{\varepsilon\bar q}\,,\,\boldsymbol{\delta}_{\bar\varepsilon Q}\,]\,\varphi^a \;=\; \bar\varepsilon_m\varepsilon^m\,\delta_P\,\varphi^a\;,$$

we deduce that

$$[\,\delta_{\bar\varepsilon q}\,,\,\boldsymbol{\delta}^{int}_{\varepsilon\overline{Q}}\,]\,\varphi^a \;=\; [\,\delta_{\bar\varepsilon q}\,,\,\boldsymbol{\delta}^{int}_{\bar\varepsilon Q}\,]\,\varphi^a \;=\; 0\;. \tag{35}$$

(7) Proper transformation under J^{+-}

$$[\,\delta_{J^{+-}}\,,\,\boldsymbol{\delta}^{int}_{\varepsilon\overline{Q}}\,]\,\varphi^a \;=-\frac{i}{2}\,\boldsymbol{\delta}^{int}_{\varepsilon\overline{Q}}\,\varphi^a\;, \qquad [\,\delta_{J^{+-}}\,,\,\boldsymbol{\delta}^{int}_{\bar\varepsilon Q}\,]\,\varphi^a \;=-\frac{i}{2}\,\boldsymbol{\delta}^{int}_{\bar\varepsilon Q}\,\varphi^a \tag{36}$$

(8) Dimension analysis requires

$$[\,\delta_D\,,\,\boldsymbol{\delta}^{int}_{\varepsilon\overline{Q}}\,]\,\varphi^a \;=\; \frac{i}{2}\,\boldsymbol{\delta}^{int}_{\varepsilon\overline{Q}}\,\varphi^a\;, \qquad [\,\delta_D\,,\,\boldsymbol{\delta}^{int}_{\bar\varepsilon Q}\,]\,\varphi^a \;=\; \frac{i}{2}\,\boldsymbol{\delta}^{int}_{\bar\varepsilon Q}\,\varphi^a\;. \tag{37}$$

(9) They have opposite $U(1)$ R-charge,

$$[\,\delta_J\,,\,\boldsymbol{\delta}^{int}_{\varepsilon\overline{Q}}\,]\,\varphi^a \;=\; \frac{1}{2}\,\boldsymbol{\delta}^{int}_{\varepsilon\overline{Q}}\,\varphi^a\;, \qquad [\,\delta_J\,,\,\boldsymbol{\delta}^{int}_{\bar\varepsilon Q}\,]\,\varphi^a \;=-\frac{1}{2}\,\boldsymbol{\delta}^{int}_{\bar\varepsilon Q}\,\varphi^a\;. \tag{38}$$

(10) The eight interacting supersymmetries must also transform as an $SO(8)$ vector: with $\bar\varepsilon'_m = \omega_{mn}\varepsilon^n$,

$$[\,\delta_{coset}\,,\,\boldsymbol{\delta}_{\varepsilon\overline{Q}}\,]\,\varphi^a \;=\; \boldsymbol{\delta}_{\bar\varepsilon' Q}\,\varphi^a\;, \tag{39}$$

$$[\,\delta_{\overline{coset}}\,,\,\boldsymbol{\delta}_{\varepsilon\overline{Q}}\,]\,\varphi^a \;=\; 0\;. \tag{40}$$

Similarly, with $\varepsilon'^m = \omega^{mn}\bar\varepsilon_n$,

$$[\,\delta_{\overline{coset}}\,,\,\boldsymbol{\delta}_{\bar\varepsilon Q}\,]\,\varphi^a \;=\; \boldsymbol{\delta}_{\varepsilon'\overline{Q}}\,\varphi^a\;, \tag{41}$$

$$[\,\delta_{coset}\,,\,\boldsymbol{\delta}_{\bar\varepsilon Q}\,]\,\varphi^a \;=\; 0\;. \tag{42}$$

These ten requirements limit the possible forms of the dynamical supersymmetries.

5. Solving the Kinematical Restrictions

In order to satisfy the first two requirements, we must construct chiral cubic polynomials in the superfields, which requires a bit of algebraic technology.

5.1. *Chiral Engineering*

Introduce the coherent state operators

$$E_\eta \;=\; e^{\,\eta\cdot\widehat{\bar{d}}}\ , \tag{43}$$

where the hat denotes division by ∂^+, and η^m are arbitrary Grassmann parameters. Since

$$d^m\,\left(E_\eta\,\varphi^a\right) \;=\; i\sqrt{2}\,\eta^m\,\left(E_\eta\,\varphi^a\right)\ , \tag{44}$$

$E_\eta\,\varphi^a$ are eigenstates of the chiral derivatives. It follows that the quadratic combination

$$Z^{bc(\eta)} \;=\; (E_\eta\partial^{+B}\,\varphi^b)\,(E_{-\eta}\partial^{+C}\,\varphi^c)\ , \tag{45}$$

is manifestly chiral,

$$d^m\,Z^{bc(\eta)} \;=\; 0\ . \tag{46}$$

Chiral cubic polynomials in the superfields are then constructed in *nested form*,

$$\mathcal{C}^{bcd\,(\eta,\zeta)} \;=\; (E_\eta\partial^{+B}\,\varphi^b\,)\,E_{-\eta}\frac{1}{\partial^{+M}}\,(\,(E_\zeta\partial^{+C}\,\varphi^c\,)(\,E_{-\zeta}\partial^{+D}\,\varphi^d\,)\,)\ , \tag{47}$$

which is manifestly chiral

$$d^m\,\mathcal{C}^{bcd(\eta,\zeta)} \;=\; 0\ , \tag{48}$$

and serves as a generating function where the chiral cubic polynomials in the superfields appear as the coefficients in the series expansion in the independent Grassmann variables η and ζ.

5.2. *Dynamical Supersymmetry*

To find it, we introduce the supersymmetry parameters in the nested Ansatz through the combinations

$$E_\varepsilon \;=\; e^{\,\varepsilon\cdot\widehat{\bar{q}}}\ ,\qquad E_{\bar\varepsilon} \;=\; e^{\,\bar\varepsilon\cdot\widehat{q}}\ , \tag{49}$$

which allows us to keep track of requirement (6), without affecting chirality. This leads to the nested ansätze of the form

$$\begin{aligned}\boldsymbol{\delta}_{\varepsilon\overline{\mathcal{Q}}}\,\varphi^a = & \frac{f^{abcd}}{\partial^{+A_\alpha}}\Big[(E_\varepsilon E_\eta\partial^{+B_\alpha}\varphi^b)E_{-\varepsilon}E_{-\eta}\frac{1}{\partial^{+M_\alpha}}\left((E_\zeta\partial^{+C_\alpha}\varphi^c)(E_{-\zeta}\partial^{+D_\alpha}\varphi^d)\right)\Big]\\ & \equiv \mathcal{K}^{a\,(\varepsilon,\eta,\zeta)}_\alpha\ ,\end{aligned} \tag{50}$$

keeping only the first order in the supersymmetry parameters ε^m. The f^{abcd} are unknown coefficients, and the exponents A_α, B_α, M_α, C_α, D_α have yet to be determined. In this form, many of the ten requirements are manifestly satisfied:

- Chirality is manifest since the $\bar{q}_n$ anticommute with the chiral derivatives.
- Requirements (3), (4), (5), and (6) are clearly satisfied.
- The proper transformation under J^{+-}, (7), restricts the power of the ∂^+ derivatives so that

$$A_\alpha + M_\alpha - B_\alpha - C_\alpha - D_\alpha + 4 \;=\; 0\;, \tag{51}$$

which also satisfies the dimension requirement (8).
- The correct $U(1)$ R-charge, requirement (9), demands after some computation

$$\left(\eta^m\frac{\partial}{\partial\eta^m}+\zeta^m\frac{\partial}{\partial\zeta^m}\;-\;4\right)\mathcal{K}^{a\,(\varepsilon,\eta,\zeta)}_\alpha\;=\;0\;. \tag{52}$$

- The tenth requirement, that the eight supersymmetries transform as an $SO(8)$ vector, is the hardest to satisfy. Computation of the commutator yields

$$\begin{aligned}[\,\delta_{coset}\,,\,\boldsymbol{\delta}_{\varepsilon\overline{\mathcal{Q}}}\,]\,\varphi^a &= \boldsymbol{\delta}_{\overline{\varepsilon}'\mathcal{Q}}\,\varphi^a \\ &\quad + \omega_{mn}\big(\,\eta^m\eta^n(\widehat{U}_1+\widehat{U}_2)+\zeta^m\zeta^n(\widehat{U}_3+\widehat{U}_4)\,\big)\mathcal{K}^{a\,(\epsilon,\eta,\zeta)}_\alpha.\end{aligned} \tag{53}$$

Here, $\widehat{U}_i$ means insertion of $\frac{1}{\partial^+}$ in the ith position; for instance

$$\widehat{U}_2\mathcal{K}^{a\,(0,\eta,\zeta)}_\alpha = \frac{f^{abcd}}{\partial^{+A_\alpha}}\left((E_\eta\partial^{+B_\alpha}\varphi^b)E_{-\eta}\frac{1}{\partial^{+(M_\alpha+1)}}\left((E_\zeta\partial^{+C_\alpha}\varphi^c)(E_{-\zeta}\partial^{+D_\alpha}\varphi^d\,)\right)\right)\;,$$

and so on. Hence the tenth kinematical requirement of $SO(8)$ covariance is achieved as long as

$$\Big(\,\eta^m\eta^n(\widehat{U}_1+\widehat{U}_2)+\zeta^m\zeta^n(\widehat{U}_3+\widehat{U}_4)\Big)\,\mathcal{K}^{a\,(\epsilon,\eta,\zeta)}_\alpha\;=\;0\;. \tag{54}$$

After some algebra, we find two solutions to this equation.

The *odd solution*

$$\delta^{\text{int odd}}_{\varepsilon\overline{\mathcal{Q}}}\,\varphi^a\;=\;\sum_{odd}\mathcal{K}^{a\,(\varepsilon,\eta,\zeta)}_\alpha\Big|_{\eta=\zeta=0}\;, \tag{55}$$

where the sum stands for

$$\sum_{odd}\;\equiv\;\sum_{\alpha=\pm\frac{1}{2}}(-1)^{\alpha+\frac{1}{2}}\,\frac{\partial}{\partial\eta^{[2-2\alpha]}}\,\frac{\partial}{\partial\zeta^{[2+2\alpha]}}\;, \tag{56}$$

with

$$\frac{\partial}{\partial\eta^{[2-2\alpha]}}\frac{\partial}{\partial\zeta^{[2+2\alpha]}} \equiv \frac{\epsilon^{i_1\cdots i_{2-2\alpha}\cdots i_4}}{(2+2\alpha)!(2-2\alpha)!}\frac{\partial}{\partial\eta^{i_1\cdots i_{2-2\alpha}}}\frac{\partial}{\partial\zeta^{i_{3-2\alpha}\cdots i_4}} . \tag{57}$$

The second is the *even solution*

$$\delta^{\text{int }even}_{\varepsilon\overline{\mathcal{Q}}}\,\varphi^a = \sum_{even}\mathcal{K}^{a\,(\varepsilon,\eta,\zeta)}_{\alpha}\Big|_{\eta=\zeta=0} , \tag{58}$$

with

$$\sum_{even} \equiv \sum_{\alpha=0,\pm1}(-1)^{\alpha}\frac{\partial}{\partial\eta^{[2-2\alpha]}}\frac{\partial}{\partial\zeta^{[2+2\alpha]}} . \tag{59}$$

In both cases, the powers of the ∂^+ derivatives are related by

$$A_{\alpha-1} = A_\alpha + 1 , \quad B_{\alpha-1} = B_\alpha + 1 , \quad M_{\alpha-1} = M_\alpha - 2 , \tag{60}$$

as well as

$$C_{\alpha-1} = C_\alpha - 1 , \quad D_{\alpha-1} = D_\alpha - 1 . \tag{61}$$

Both even and odd solutions are seen to satisfy Eq.(40). Their forms suggest that $SO(8)$ triality is at work, with α denoting the $U(1)$ charges in its vector and spinor representations.
Both solutions are conveniently written in the form

$$\delta^{\,int\,odd(even)}_{\varepsilon\overline{\mathcal{Q}}}\,\varphi^a \equiv \big[\,A_\alpha, B_\alpha, M_\alpha, C_\alpha, D_\alpha\,\big]_{\,odd(even)} , \tag{62}$$

with $\alpha = -1/2(-1)$ in the odd(even) case.

It can be checked that these two solutions satisfy the correct commutations with the kinematical conformal supersymmetries

$$[\,\delta_{\bar{\alpha}s}\,,\,\boldsymbol{\delta}_{\varepsilon\overline{\mathcal{Q}}}\,]\,\varphi^a = \left(i\delta_D\,\varphi^a - i\delta_{J^{+-}}\,\varphi^a + \frac{1}{2}\,\delta_J\,\varphi^a\right) + \frac{1}{\sqrt{2}}\,\delta_{SO(6)}\,\varphi^a , \tag{63}$$

as well as

$$[\,\delta_{\alpha\bar{s}}\,,\,\boldsymbol{\delta}_{\varepsilon\overline{\mathcal{Q}}}\,]\,\varphi^a = \delta_{\overline{coset}}\,\varphi^a . \tag{64}$$

Finally, we note that the conjugate supersymmetries are obtained by simply changing E_ε into $E_{\bar{\varepsilon}}$.

5.3. *Hamiltonian and Boost*

In the previous section, the form of the dynamical supersymmetry transformations have been narrowed down to two solutions with yet undetermined powers of the light-cone derivatives. In the $d = 4$ SuperYang-Mills case, their values were determined from the vanishing of the commutator between the light-cone boost and Hamiltonian. We expect the same to hold in the $d = 3$ theory.

The light-cone Hamiltonian is computed from

$$[\,\boldsymbol{\delta}^{free}_{\bar{\varepsilon}\mathcal{Q}} + \boldsymbol{\delta}^{int}_{\bar{\varepsilon}\mathcal{Q}}\,,\,\boldsymbol{\delta}^{free}_{\varepsilon\overline{\mathcal{Q}}} + \boldsymbol{\delta}^{int}_{\varepsilon\overline{\mathcal{Q}}}\,]\varphi^a \;=\; \sqrt{2}\,\bar{\varepsilon}_m\varepsilon^m\,\boldsymbol{\delta}_{\mathcal{P}^-}\,\varphi^a\,. \tag{65}$$

The commutator yields terms linear and quadratic in f^{abcd}. The first order stems from

$$[\,\boldsymbol{\delta}^{free}_{\bar{\varepsilon}\mathcal{Q}}\,,\,\boldsymbol{\delta}^{int}_{\varepsilon\overline{\mathcal{Q}}}\,]\varphi^a + [\,\boldsymbol{\delta}^{int}_{\bar{\varepsilon}\mathcal{Q}}\,,\,\boldsymbol{\delta}^{free}_{\varepsilon\overline{\mathcal{Q}}}\,]\varphi^a \tag{66}$$

The results of the computation are expressed in terms of

$$\begin{aligned} K^{a\,[r,1]}_{\alpha} &\equiv (E_r\,U_1)(E_{-r}\,U_2)\,\mathcal{K}^{a\,(0,\eta,\zeta)}_{\alpha} \\ &= \frac{f^{abcd}}{\partial^{+A_\alpha}}\Big[(E_rE_\eta\partial^{+B_\alpha}\varphi^b)E_{-r}E_{-\eta}\frac{1}{\partial^{+M_\alpha}}\,\big((E_\zeta\partial^{+C_\alpha}\varphi^c)(E_{-\zeta}\partial^{+D_\alpha}\varphi^d)\big)\Big]\,, \end{aligned}$$

and

$$\begin{aligned} K^{a\,[1,r]}_{\alpha} &\equiv (E_r\,U_3)(E_{-r}\,U_4)\,\mathcal{K}^{a\,(0,\eta,\zeta)}_{\alpha} \\ &= \frac{f^{abcd}}{\partial^{+A_\alpha}}\Big[(E_\eta\partial^{+B_\alpha}\varphi^b)E_{-\eta}\frac{1}{\partial^{+M_\alpha}}\,\big((E_rE_\zeta\partial^{+C_\alpha}\varphi^c)(E_{-r}E_{-\zeta}\partial^{+D_\alpha}\varphi^d)\big)\Big]\,, \end{aligned}$$

where the transverse derivative is introduced through

$$E_r \;=\; e^{\,r\,\widehat{\partial}}\,, \tag{67}$$

and r is a dimensionless parameter.

The computation of these commutators yields, for the odd case,

$$\boldsymbol{\delta}^{int\,odd}_{\mathcal{P}^-}\,\varphi^a \;=\; \frac{\partial}{\partial r}\left(\sum_{even}K^{a\,[1,r]}_{\alpha} + \sum_{odd}K^{a\,[r,1]}_{\alpha+\frac{1}{2}}\right)_{r=0}\,, \tag{68}$$

for the linear part in f^{abcd}.

A similar expression is found in the even case,

$$\boldsymbol{\delta}^{int\,even}_{\mathcal{P}^-}\,\varphi^a \;=\; \frac{\partial}{\partial r}\left(\sum_{odd}K^{a\,[1,r]}_{\alpha} - \sum_{even}K^{a\,[r,1]}_{\alpha+\frac{1}{2}}\right)_{r=0}\,. \tag{69}$$

The boost transformation is computed from the commutator

$$\boldsymbol{\delta}_{\mathcal{J}^-}\,\varphi^a \;=\; -\frac{i}{2}\,[\,\delta_K\,,\,\boldsymbol{\delta}_{\mathcal{P}^-}\,]\,\varphi^a\,. \tag{70}$$

for both odd and even cases. Its expression is not particularly enlightening, and will be published elsewhere.[6]

5.4. *Dynamical Constraints*

The next step is to require

$$[\,\delta_{\mathcal{P}^-}\,,\,\boldsymbol{\delta}_{\mathcal{J}^-}\,]\,\varphi^a \;=\; 0\;. \tag{71}$$

This condition, as in the Yang-Mills case, is expected to fix the unknown exponents, and the interactions. After a lengthy calculation, keeping only the terms linear in f^{abcd}, the result can be written in the form

$$[\,\overset{odd}{\delta_{\mathcal{P}^-}}\,,\,\overset{odd}{\boldsymbol{\delta}_{\mathcal{J}^-}}\,]\,\varphi^a \;=\; \mathcal{S}\,\frac{\partial^2}{\partial r\partial r'}\,\left(F\,\mathcal{O}_1^{odd}+G\,\mathcal{O}_2^{odd}\right)_{r=r'=0}\,, \tag{72}$$

where $\mathcal{S}$ is a shift operator

$$\mathcal{S}:\quad A_\alpha\to A_\alpha+1\,,\quad B_\alpha\to B_\alpha+1\,,\quad M_\alpha\to M_\alpha-1\,. \tag{73}$$

In addition,

$$F\equiv(B_{-\frac{1}{2}}-3)\widehat{U}_1+(M_{-\frac{1}{2}}-C_{-\frac{1}{2}}-D_{-\frac{1}{2}}+3)\widehat{U}_2\,, \tag{74}$$

$$G\equiv 2(C_{-\frac{1}{2}}-\frac{3}{2})\widehat{U}_3-2(D_{-\frac{1}{2}}-\frac{3}{2})\widehat{U}_4\,, \tag{75}$$

and

$$\mathcal{O}_1^{odd}=\sum_{odd}(K_\alpha^{[rr',1]}-K_{\alpha+1}^{[rr',1]})+2\sum_{even}K_{\alpha+\frac{1}{2}}^{[r,r']}\,, \tag{76}$$

$$\mathcal{O}_2^{odd}=\sum_{odd}K_{\alpha+1}^{[r,r']}-\frac{1}{2}\sum_{even}(K_{\alpha+\frac{1}{2}}^{[r,r']}-K_{\alpha+\frac{3}{2}}^{[1,rr']})\,. \tag{77}$$

A similar equation obtains in the even case.

We can envisage two types of solutions to the vanishing of this commutator.

- The "trivial" solution is when $F=G=0$, which determines the values of all the exponents, and leads to

$$\delta_{\varepsilon\mathcal{Q}}^{\,\mathrm{int}\;odd}\,\varphi^a \;=\; \left[\,2,3,0,\frac{3}{2},\frac{3}{2}\,\right]_{odd}\,, \tag{78}$$

which exists only if $f^{abcd}=-f^{abdc}$.

The even solution, given by,

$$\delta_{\varepsilon\mathcal{Q}}^{\,\mathrm{int}\;even}\,\varphi^a \;=\; \left[\,\frac{5}{2},\frac{7}{2},-1,1,1\,\right]_{even}\,, \tag{79}$$

requires $f^{abcd}=+f^{abdc}$.

In both cases, there are no further symmetry requirements on f^{abcd}. Both solutions require fractional powers of ∂^+, which have to be further interpreted.
We have not checked the validity of this solution any further: there remains to check the vanishing of the commutators

$$[\,\delta_K\,,\,\boldsymbol{\delta}_{\mathcal{K}^-}\,]\,\varphi^a\;, \qquad [\,\boldsymbol{\delta}_{\mathcal{P}^-}\,,\,\boldsymbol{\delta}_{\mathcal{K}^-}\,]\,\varphi^a\;, \qquad [\,\delta_{\mathcal{J}^-}\,,\,\boldsymbol{\delta}_{\mathcal{K}^-}\,]\,\varphi^a\;. \tag{80}$$

In the Yang-Mills case, these did not put any further restrictions on the solution. The vanishing of the second one may be explained by the Jacobi identity:

$$[\,\boldsymbol{\delta}_{\mathcal{P}^-}\,,\,[\,\delta_K\,,\,\boldsymbol{\delta}_{\mathcal{J}^-}\,]\,]+[\,\delta_K\,,\,[\,\boldsymbol{\delta}_{\mathcal{J}^-}\,,\,\boldsymbol{\delta}_{\mathcal{P}^-}\,]\,]+[\,\boldsymbol{\delta}_{\mathcal{J}^-}\,,\,[\,\boldsymbol{\delta}_{\mathcal{P}^-}\,,\,\delta_K\,]\,]\;=\;0\;. \tag{81}$$

Since

$$[\,\delta_K\,,\,\boldsymbol{\delta}_{\mathcal{J}^-}\,]\,\varphi^a\;=\;-i\,\boldsymbol{\delta}_{\mathcal{K}^-}\,\varphi^a\;, \tag{82}$$

it follows that

$$[\,\boldsymbol{\delta}_{\mathcal{P}^-}\,,\,\boldsymbol{\delta}_{\mathcal{J}^-}\,]\,\varphi^a\;=\;0\;. \tag{83}$$

The algebraic validity of the fractional power solution hinges on the first commutator

$$[\,\delta_K\,,\,\boldsymbol{\delta}_{\mathcal{K}^-}\,]\,\varphi^a\;=\;0\;, \tag{84}$$

which we have not yet checked. Through the Jacobi identity, it would ensure that

$$[\,\boldsymbol{\delta}_{\mathcal{J}^-}\,,\,\boldsymbol{\delta}_{\mathcal{K}^-}\,]\,\varphi^a\;=\;0\;. \tag{85}$$

- The less trivial solution(s) relies on the symmetries of f^{abcd} under the interchange of *three* of its indices. It appears to lead uniquely to the BLG solution; since at the time of Shifmania, we had not obtained it, its details will appear elsewhere.[6]

Much remains to be done. For one, we have not derived the quadratic term in f^{abcd} in the Hamiltonian. In the Yang-Mills case, this led to the Jacobi identity of the f^{abc} and identified them as structure functions.

Acknowledgments

It has been an honor to be invited to speak at Misha's celebration, as well as a pleasure to attend Shifmania, where Misha's superb skills and influence were so deservedly lauded, and the wonderful Russian hospitality was much in evidence.

This research is partially supported by the Department of Energy Grant No. DE-FG02-97ER41029.

References

1. J. Bagger and N. Lambert, Phys. Rev. D **75**, 045020 (2007); Phys. Rev. D **77**, 065008 (2008).
2. A. Gustavsson, Nucl. Phys. B **811**, 66 (2009).
3. B. E. W. Nilsson, Class. Quant. Grav. **26**, 175001 (2009).
4. S. Ananth, L. Brink, S-S. Kim and P. Ramond, Nucl. Phys. B **722**, 166 (2005).
5. L. Brink, J. H. Schwarz and J. Scherk, Nucl. Phys. B **121**, 77 (1977); F. Gliozzi, J. Scherk and D. I. Olive, Nucl. Phys. B **122**, 253 (1977).
6. D. Belyaev, L. Brink, S-S. Kim and P. Ramond, in preparation.
7. W. Nahm, Nucl. Phys. B **135**, 149 (1978).
8. E. Witten, "*Some comments on string dynamics*," arXiv:hep-th/9507121; M. R. Douglas and G. W. Moore, *"D-branes, quivers, and ALE instantons"*, arXiv:hep-th/9603167; J. Distler and A. Hanany, Nucl. Phys. B **490**, 75(1997); J. M. Maldacena, Adv. Theor. Math. Phys. **2**, 231 (1998), and Int. J. Theor. Phys. **38**, 1113 (1999); O. Aharony, A. Hanany and B. Kol, JHEP **9801**, 002 (1998); C. Grojean and J. Mourad, Class. Quant. Grav. **15**, 3397 (1998); P. Arvidsson, E. Flink and M. Henningson, JHEP **0405**, 048 (2004); J. H. Schwarz, JHEP **0411**, 078 (2004).
9. A. Basu and J. A. Harvey, Nucl. Phys. B **713**,136 (2005).

Brane Tilings: The NSVZ Beta Function

Amihay Hanany

Theoretical Physics Group, The Blackett Laboratory,
Imperial College London, Prince Consort Road
London, SW7 2AZ, UK
E-mail: a.hanany@imperial.ac.uk

In honor of Misha Shifman's 60th birthday this note will start with one of Misha's many contributions, the famous NSVZ beta function, and study how it affects the construction of Brane Tilings. The main features of Brane Tilings for 3+1 dimensions SCFT's are reviewed and the generalization to 2+1 dimensional SCFT's is discussed.

Keywords: Brane Tilings; AdS/CFT.

Imperial/TP/09/AH/05

1. Conditions for Conformal Invariance

One of Misha's well known contributions is the computation of the beta function of a supersymmetric gauge theory to all orders in perturbation theory.[1] The present contribution to the proceedings of this conference will discuss one of the many applications that resulted from the NSVZ work. It will start with the NSVZ beta function and set several conditions and assumptions that will eventually lead to the construction of Brane Tilings. To date, the latter form the largest known class of superconformal field theories that have a known AdS dual.

Let us start with a discussion along the lines presented in Section 2 of [2]. The Beta function of a $\mathcal{N} = 1$ supersymmetric gauge theory in 3+1 dimensions with gauge group $SU(N)$ is given by the renowned NSVZ formula [1]

$$\beta_{\frac{1}{g^2}} = \frac{1}{8\pi^2}\frac{3N - \sum_M \mu[\mathcal{R}_M](1-\gamma_M(g))}{1-\frac{g^2 N}{8\pi^2}}, \tag{1}$$

where the sum is over all matter fields that transform under the representation $\mathcal{R}_M$, $\mu[\mathcal{R}]$ is the Dynkin index of the representation $\mathcal{R}$ defined by

$\mathrm{Tr}_{\mathcal{R}}(T^a T^b) = \mu[\mathcal{R}]\delta^{ab}$, with the convention that $\mu = \frac{1}{2}$ for the fundamental representation implying that $\mu = N$ for the adjoint representation, and γ_M is the anomalous dimension for the matter field M.

This formula is general for any collection of matter fields but for the purposes of this paper we will make some simplifications which will ease up the calculation.

We will be interested in superconformal field theories and hence require that the beta functions vanish, hence only the numerator of Eq. (1) is going to be relevant. Another property of superconformal field theories is the relation between the scaling dimension Δ and the R charge of a chiral superfield,

$$\Delta = 1 + \frac{\gamma}{2} = \frac{3}{2} r\,. \tag{2}$$

Using Eq. (2), the condition for a vanishing beta function for a single gauge group becomes

$$0 = N - \sum_M \mu[\mathcal{R}_M](1 - r_M) \tag{3}$$

Our motivation comes from string theory and gauge theories that live on the world volume of D branes. There can be several types of D branes and therefore the gauge group will consists of a product of several $U(N)$ factors, one per each type of D brane, say $\prod_{a=1}^{G} U(N_a)$ for some choice of ranks, N_a. Perturbatively, matter fields arise from strings ending on D branes and since the string can only have two ends we will restrict to matter fields that are either in the bi-fundamental representation, in cases that the string stretches between one type of D brane to another, or in the adjoint representation in cases that the string stretches from a given type of D brane to the same type.

Further restricting Eq. (3) to several gauge groups with adjoints and bi-fundamentals gives

$$0 = N_a - \sum_{A \in \mathrm{adj}][a]} N_a(1 - r_A) - \frac{1}{2} \sum_{B \in \mathrm{bifund[a,b]}} N_b(1 - r_{a,b})\,, \tag{4}$$

where N_a is the rank of the a-th gauge group, r_A is the R charge of the adjoint field of gauge group a, if there is such a matter field, and $r_{a,b}$ is the R charge of the matter field in the bi-fundamental representation of gauge groups a and b. We further use the property that the adjoint representation can be viewed as a bifundamental representation from a gauge group to itself, denoting the corresponding R charge by $r_{a,a}$, and arrive at the condition for a vanishing beta function for any gauge group

$$\sum_b N_b(1 - r_{a,b}) = 2N_a\,, \qquad \text{each } a\,. \tag{5}$$

This formula begins to look like a geometric condition but before we see this let us use another assumption that all the gauge groups have the same rank, $N_a = N$. In this case, the condition further loses the N dependence,

$$\sum_b (1 - r_{a,b}) = 2\,, \qquad \text{each } a\,. \tag{6}$$

This form of the condition for scale invariance is very nice and in fact looks very similar to the condition for the superpotential. If W is a polynomial function of the chiral superfields then a necessary condition for conformal invariance is that every monomial will have an R charge 2,

$$\sum_{(a,b)\in\text{monomial}} r_{a,b} = 2\,, \qquad \text{each monomial.} \tag{7}$$

where a and b are the indices of the gauge groups under which the chiral superfields in the given monomial are charged under. Thus we summarize that the necessary conditions for conformal invariance for this class of gauge theories is[3]

$$\begin{aligned} &\sum_b (1 - r_{a,b}) = 2\,, \qquad \text{each gauge group } a\,, \\ &\sum_{(a,b)\in\text{monomial}} r_{a,b} = 2\,. \qquad \text{each monomial in } W\,. \end{aligned} \tag{8}$$

We can further use these conditions and sum Eq. (6) over all possible gauge groups, assuming there are G such gauge groups and a total of E matter fields,

$$2E - \sum_{a,b} r_{a,b} = 2G\,. \tag{9}$$

The first term comes from the fact that the sum over 1 is all over matter fields and that each matter field appears precisely twice in this sum, once for its fundamental representation and once for its anti-fundamental representation. Soon we will see that in a quiver representation it is due to the fact that each arrow has precisely one start and precisely one end. The right hand side is a sum over all gauge groups. To further evaluate this expression, we can sum over all monomial terms in the superpotential.

$$\sum_{a,b} r_{a,b} = 2N_T\,, \tag{10}$$

where N_T is the number of monomial terms in the superpotential. This relation can be substituted into Eq. (9) only if each bifundamental field appears precisely twice in the superpotential. Otherwise the sum over a and b in Eq. (10) does not coincide with the sum over a and b in Eq. (9).

Thus we are led to assume the next condition, called the "Toric condition" [4]: Each chiral multiplet appears precisely twice in W, once with a plus sign and once with a minus sign. The toric condition in fact asserts the second point about the signs which are not important for the current discussion. These signs will show up below as rather crucial. Given the toric condition, we can write

$$G - E + N_T = 0\,. \tag{11}$$

With these results in mind let us look for some graphical representation of these equations, something which will help in getting the most general solution to these conditions. Let us recall in Fig. 1 a typical quiver that can describe a gauge theory of the form which is discussed here. Further,

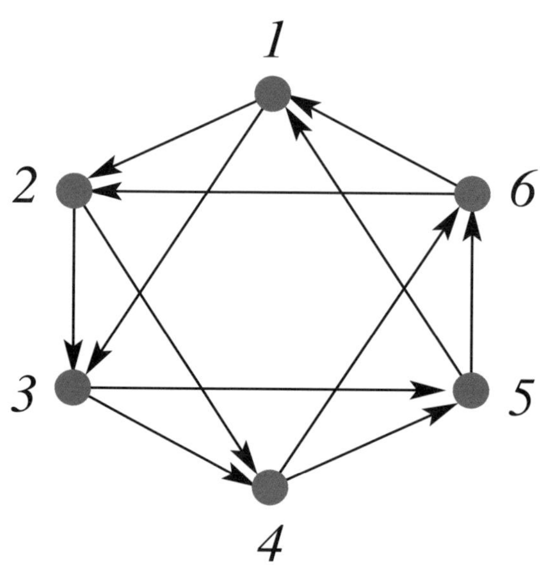

Fig. 1. A typical quiver for the class of gauge theories discussed in this note. The gauge theory has 6 gauge groups and 12 matter fields. The quiver is chosen due to its shape in the form of the star of David, acknowledging Misha's great love and care for Jewish culture.

we recall that a Lagrangian for a $\mathcal{N} = 1$ supersymmetric theory in 3+1 dimensions needs 3 pieces of data:

- Gauge fields in vector multiplets
- Matter fields in chiral multiplets
- Interaction terms in the superpotential – a sum of gauge invariant products of chiral multiplets

The typical quiver of Fig. 1 contains information about the different vector multiplets of the theory (the nodes) and information on the chiral multiplets and their charges under the gauge groups (the arrows). This form is very suitable for gauge theories that arises on D branes since every arrow has a start and an end, corresponding to open strings that start and end on D branes. This form, however, does not contain any information on the superpotential. As such it lacks the third property for constructing the Lagrangian and therefore we are in need of a better graphical representation. Another property which can be seen on the quiver is the fact that each node satisfies the condition that the number of incoming arrows is equal to the number of outgoing arrows. For the corresponding gauge theory this is the chiral anomaly condition which requires that the number of fundamental multiplets should be the same as the number of antifundamental multiplets for each gauge group. The quiver does satisfy it but not in any special way. There is no specific graphical way that implies that this condition is satisfied. It is present as an external constraint, without an explicit way which guaranties this by construction. This is another drawback of the quiver that we wish to address and below we see that this can be done with success.

One key feature which can give a clue is the relation in Eq. (11). This looks really suggestive like an Euler condition where each of the 3 objects corresponds to a given simplex and the different simplexes are tiling a torus. There are 3 simplexes that are tiling a torus:

- Faces (Tiles)
- Edges
- Nodes

We can now match the two lists above, pick a tiling of the two dimensional torus, and identify faces with vector multiplets, edges with chiral multiplets, and nodes with interaction terms in the superpotential. There is a possible ambiguity in this mapping as in fact the nodes and faces can be exchanged without violating any of the previous assumptions. So we could equivalently identify nodes with vector multiplets, edges with chiral multiplets, and faces with interaction terms in the superpotential. The first mapping will be called "Brane Tilings" [6] while the second mapping will be called the "Periodic Quiver" [6]. The reason for the first name is more natural from a D brane perspective of embedding branes into space time and looking at the gauge theory which arises on their world volume, where every tile corresponds to a D brane and a corresponding $U(N)$ gauge group that lives on its world volume. It is a natural continuation to the notions of Brane Intervals,[7] where the superconformal field theories form a one dimensional tiling, and of Brane Box Models,[8] and Brane Diamonds [9] which are earlier versions of the two dimensional construction. The reason for the second name is by the traditional use of a quiver gauge theory that maps nodes to gauge fields and arrows to matter fields. The novelty is by the periodic nature of the quiver, as it wraps the torus. An earlier appearance of this notion is for example Figure 3 of [10]. Furthermore, this periodicity allows the encoding of superpotential terms as the faces of the periodic quiver.

The association of the node in the brane tiling or equivalently the face in the periodic quiver with a term in the superpotential requires further look. Each term in the superpotential needs to be gauge invariant with respect to all gauge groups. Such gauge invariance is achieved by forming closed paths in the periodic quiver. This requires the face of the periodic quiver to be oriented such that all arrows are going in one direction. Equivalently, this puts an orientation around the nodes of the brane tiling. There can be two such orientations - one clockwise and the other counterclockwise. Since an arrow in the periodic quiver is at the boundary of two faces, then one of faces must have a sequence of clockwise arrows and the other must have a sequence of counterclockwise arrows. We therefore get two important consequences: The faces in the periodic quiver alternate in a clockwise and counterclockwise direction. Periodicity requires the number of faces in the fundamental domain of the torus to be even such that the number of clockwise faces is equal to the number of counterclockwise faces. Equivalently, there are two types of nodes in the brane tiling. Let us distinguish these two types of nodes by giving a color (black) white to the node of (counter)clockwise orientation. By convention we can assign an arrow in the periodic quiver to an edge in the tiling by setting the direction of the arrow perpendicular to the edge, and pointing it such that the white node is always to the right. See Fig. 2 for an example of turning the star of David quiver together with a specific superpotential into a periodic quiver.[6]

Thus we are led to the following formulation of Brane Tilings: Pick a

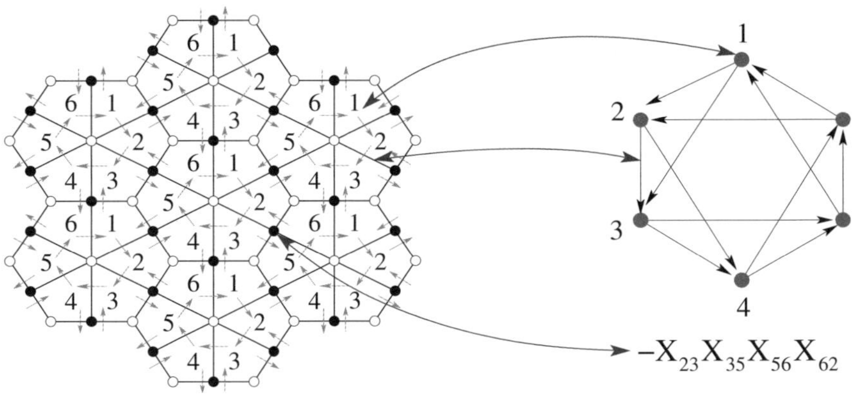

Fig. 2. The brane tiling (left) for the star of David quiver (right) with a specific superpotential. Each red node in the quiver maps to a face in the tiling. The green arrows form the periodic quiver and are mapped to the black arrows in the quiver. Each node in the fundamental domain of the tiling corresponds to a term in the superpotential. There are 6 such terms.

periodic bi-bi-partite tiling of the plane and a positive integer N and write down a Lagrangian according to the following rules:

(1) For each face (tile) assign a $U(N)$ gauge group.
(2) For each edge of the tiling assign a chiral multiplet. Such an edge is the boundary of two faces and therefore charged under both in the bi-fundamental representation.
(3) For each node of the tiling assign a term in the superpotential by taking product of chiral multiplets associated with the arrows going around it. A white node comes with a + sign and a black node comes with a − sign.

These rules define uniquely a Lagrangian in 3+1 dimensions. There is a notion of consistent tiling which is introduced in [3]. All such theories are claimed to form an infinite class of SCFT's in 3+1 dimensions with a known AdS dual. The moduli space of vacua for a theory defined by a tiling is a non-compact singular toric CY cone of dimension 3. This CY cone has a Sasaki-Einstein base that enters into the AdS dual background. A simple criterion for testing if a tiling is consistent or not is proposed in [11]. This conditions suggests that if twice the area of the toric diagram, as computed by the Kasteleyn matrix, is equal to the number of tiles, then the corresponding brane tiling is consistent.

In fact, Eq. (8) receive a new interpretation as the conditions for locally flat tiles and locally flat nodes, respectively.[3] In addition, the R charges are interpreted as angles in the tiling.[3]

1.1. *Seiberg Duality*

A brane realization for a theory in 3+1 dimensions with 4 supercharges must come equipped with a natural action of Seiberg Duality.[13] Indeed such an action exists for quadrilateral tiles.[6] The corresponding transformation and action on the brane tiling is depicted in Fig. 3.

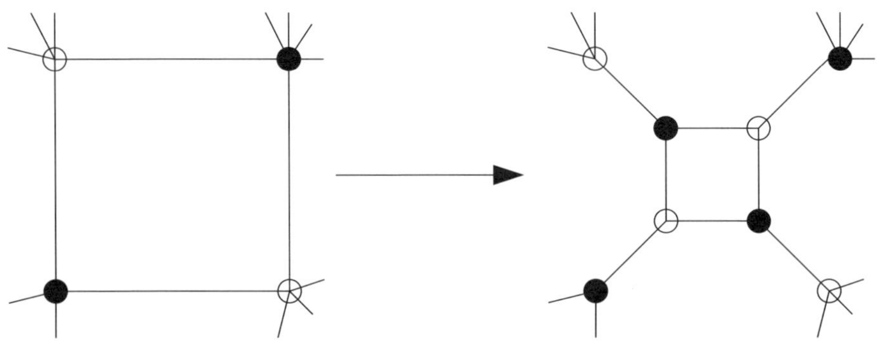

Fig. 3. A quadrilateral tile (left) which undergoes Seiberg duality. The original edges (quarks) are erased and get replaced by dual quarks, represented by the horizontal and vertical edges. The corresponding edges have their orientation inverted. Each pair of quarks gives rise to Seiberg mesons, represented by the diagonal edges. Finally, for each meson there is a corresponding term in W, given by the 4 new nodes in the tiling. If the resulting tiling contains nodes of valence 2, then they need to be integrated out [6].

1.2. *Integrating Out Massive Fields*

Nodes of valence two give rise to a mass term in the superpotential. In such cases, as the theory flows to the IR, the corresponding chiral fields are frozen and integrated out. This process of integrating out amounts in the tiling to the collapse of the node of valence two, together with the two nodes connected to it, to a single node. This is depicted in figure Fig. 4.

2. Brane Tilings for 2+1 dimensional theories

In a recent paper [12] it was realized that the above rules can be modified slightly and generate a large class of SCFT's in 2+1 dimensions with 4 supercharges, corresponding to $\mathcal{N} = 2$ supersymmetry in 2+1 dimensions:

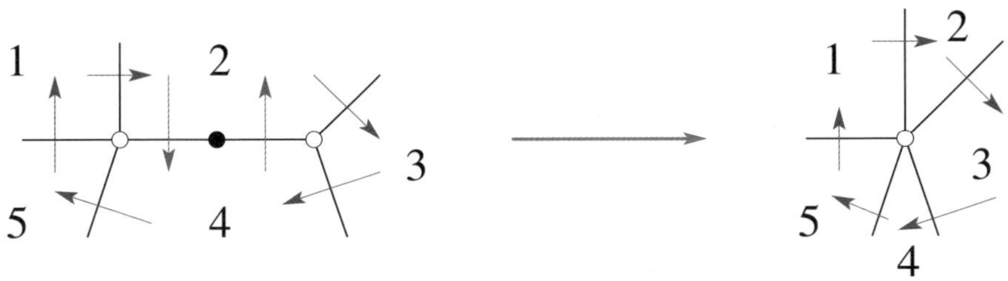

Fig. 4. A node of valence 2 and its neighbors are replaced by a single node in the tiling. This is equivalent to integrating out massive fields.[6]

(1′) For each face (tile) assign a $U(N)$ gauge group with a CS term of coefficient k_a but no gauge kinetic term

Thus, with these rules we form an infinite class of SCFT's in 2+1 dimensions. There has been a lot of progress in this direction recently and the present contribution can not cover all aspects. We will just mention some features which are relevant to the discussion on Brane Tilings. The set of CS levels can admit an alternative formulation by setting the CS levels on edges. Assign an integer n_i to each edge such that the CS of a given gauge group is given by

$$k_a = d_{ai} n_i \,, \qquad d_{ai} = \begin{cases} 1 & \text{edge } i \text{ starts at node } a \\ -1 & \text{edge } i \text{ ends at node } a \\ 0 & \text{otherwise} \end{cases} \tag{12}$$

where d_{ai} is the $G \times E$ incidence matrix that encodes the quiver data. The moduli space of vacua for a theory defined by the tiling in 2+1 dimensions is a non-compact singular toric CY cone of dimension 4. As above, this CY cone has a Sasaki-Einstein base that enters into the AdS dual background.

3. The Kasteleyn matrix

As stated above, the moduli space of a brane tiling in 3+1 (2+1) dimensions is a singular toric CY3 (CY4). The simplest way to represent this moduli space is by specifying its 2d (3d) toric diagram. The most efficient way of computing the toric diagram is by using the Kasteleyn matrix. Methods for computing the Kasteleyn matrix and its determinant are discussed in detail in, for example [15,16].

4. R Charges

An important ingredient in solving the scaling dimensions of a SCFT is in computing the spectrum of R charges for BPS operators. There has been a lot of progress in this direction and to date there is a collection of impressive set of results and predictions for scaling dimensions of various SCFT's. Let us summarize the main properties. The R charges in 3+1 dimensions are computed by using the a maximization.[17] Explicit expression for the central charge in the special case that the moduli space is a singular toric CY3 is given by [18]

$$a = \sum_{i<j<k} |\det\{v_i, v_j, v_k\}| R_i R_j R_k \, , \tag{13}$$

where $v_i, i = 1 \ldots n$ are the vertices of the 3d toric diagram of the CY3.[a] n is the number of external nodes of the toric diagram, and R_i is the R charge assigned to the external vertex of the toric diagram such that $\sum_{i=1}^{n} R_i = 2$. Alternatively, a geometric method to compute the R charges is by using Z minimization.[19] An equivalence of the a maximization and Z minimization is demonstrated in [20].

5. Classification of Brane Tilings

Since brane tilings bare important properties on SCFT's in 3+1 and in 2+1 dimensions, it is important to look for a classification of all such objects. It is indeed a challenge to classify all brane tilings, consistent or not and only partial can be reported. In a recent publication [14] the first steps in this direction were performed using an algorithm which can be used to classify all brane tilings with up to $N_T = 8$ terms in the superpotential. The classification is arranged by two order parameters, N_T and G, with a range of $N_T/2 \leq G \leq N_T$. The number of consistent brane tilings turns out to be 2, 4, 10, 25 for $N_T = 2, 4, 6, 8$, respectively.[14]

References

1. V. A. Novikov, M. A. Shifman, A. I. Vainshtein and V. I. Zakharov, "Exact Gell-Mann-Low Function Of Supersymmetric Yang-Mills Theories From Instanton Calculus," Nucl. Phys. B **229**, 381 (1983).

[a] Since the toric diagram is embedded in a 2d sub-lattice, it is customary to embed the 2d toric diagram into the $x_3 = 1$ plane, where x_3 is the 3rd coordinate of the vertices, v_i.

2. S. Benvenuti and A. Hanany, "New results on superconformal quivers," JHEP **0604**, 032 (2006) [arXiv:hep-th/0411262].
3. A. Hanany and D. Vegh, "Quivers, tilings, branes and rhombi," JHEP **0710**, 029 (2007) [arXiv:hep-th/0511063].
4. B. Feng, S. Franco, A. Hanany and Y. H. He, "Symmetries of toric duality," JHEP **0212**, 076 (2002) [arXiv:hep-th/0205144].
5. A. Hanany and K. D. Kennaway, "Dimer models and toric diagrams," arXiv:hep-th/0503149.
6. S. Franco, A. Hanany, K. D. Kennaway, D. Vegh and B. Wecht, "Brane Dimers and Quiver Gauge Theories," JHEP **0601**, 096 (2006) [arXiv:hep-th/0504110].
7. A. Hanany and E. Witten, "Type IIB superstrings, BPS monopoles, and three-dimensional gauge dynamics," Nucl. Phys. B **492**, 152 (1997) [arXiv:hep-th/9611230].
8. A. Hanany and A. Zaffaroni, "On the realization of chiral four-dimensional gauge theories using branes," JHEP **9805**, 001 (1998) [arXiv:hep-th/9801134].
9. M. Aganagic, A. Karch, D. Lust and A. Miemiec, "Mirror symmetries for brane configurations and branes at singularities," Nucl. Phys. B **569**, 277 (2000) [arXiv:hep-th/9903093].
10. A. Hanany, M. J. Strassler and A. M. Uranga, "Finite theories and marginal operators on the brane," JHEP **9806**, 011 (1998) [arXiv:hep-th/9803086].
11. D. R. Gulotta, "Properly ordered dimers, R-charges, and an efficient inverse algorithm," JHEP **0810**, 014 (2008) [arXiv:0807.3012 [hep-th]].
12. A. Hanany and A. Zaffaroni, "Tilings, Chern-Simons Theories and M2 Branes," JHEP **0810**, 111 (2008) [arXiv:0808.1244 [hep-th]].
13. N. Seiberg, "Electric - magnetic duality in supersymmetric nonAbelian gauge theories," Nucl. Phys. B **435**, 129 (1995) [arXiv:hep-th/9411149].
14. J. Davey, A. Hanany and J. Pasukonis, "On the Classification of Brane Tilings," arXiv:0909.2868 [hep-th].
15. A. Hanany and Y. H. He, "M2-Branes and Quiver Chern-Simons: A Taxonomic Study," arXiv:0811.4044 [hep-th].
16. J. Davey, A. Hanany, N. Mekareeya and G. Torri, "Phases of M2-brane Theories," JHEP **0906**, 025 (2009) [arXiv:0903.3234 [hep-th]].
17. K. A. Intriligator and B. Wecht, "The exact superconformal R-symmetry maximizes a," Nucl. Phys. B **667**, 183 (2003) [arXiv:hep-th/0304128].
18. A. Kato, "Zonotopes and four-dimensional superconformal field theories," JHEP **0706**, 037 (2007) [arXiv:hep-th/0610266].
19. D. Martelli, J. Sparks and S. T. Yau, "The geometric dual of a-maximisation for toric Sasaki-Einstein manifolds," Commun. Math. Phys. **268**, 39 (2006) [arXiv:hep-th/0503183].
20. A. Butti and A. Zaffaroni, "R-charges from toric diagrams and the equivalence of a-maximization and Z-minimization," JHEP **0511**, 019 (2005) [arXiv:hep-th/0506232].

Aspects of Supersymmetric Field Theories

Kenneth Intriligator

Department of Physics, UC San Diego
La Jolla, CA 92039 USA

We discuss the constraints on SUSY field theories and their IR phases, including bounds on the ratio a/c. Proceeding of talk at Shifmania, in honor of Misha Shifman's 60th birthday.

1. Review of the supersymmetric multiplet of anomalies

Supersymmetric theories (specifically, global $\mathcal{N} = 1$ in 4d) have a stress-energy tensor supermultiplet, $\mathcal{T}_{\alpha\dot{\alpha}}$, which contains the stress-energy tensor $T_{\mu\nu}$, supercurrents $S_{\alpha\mu}$ and $\bar{S}_{\dot{\alpha}\mu}$, and a global $U(1)_R$ current j_μ^R: $\mathcal{T}_\mu = j_\mu^R + \theta^\alpha S_{\alpha\mu} + \cdots + (\bar{\theta}\theta)^\nu T_{\mu\nu} + \ldots$. Conservation of the energy tensor and supercurrents is written in superspace as

$$\bar{D}^{\dot{\alpha}}\mathcal{T}_{\alpha\dot{\alpha}} = D_\alpha X\,, \tag{1}$$

where X is a chiral superfield. The F-component of X is related to the lack of scale invariance, and the lack of conservation of the $U(1)_R$ current:

$$\frac{2}{3}T^\mu_\mu + i\partial^\mu j_\mu^R = X|_{\theta^2}\,. \tag{2}$$

Holomorphy thus links (non)conservation of the dilatation current to that of the $U(1)_R$ current. There can be non-conservation contributions both at tree-level, and from anomalies. Matching the ABJ anomaly to the imaginary part of the LHS of Eqn. (2) gives the anomaly contribution

$$X \supset -\frac{1}{16\pi^2}\,\mathrm{Tr}(RG^2)\mathrm{Tr}\mathrm{W}^2\,, \tag{3}$$

which is 1-loop exact by the Adler-Bardeen theorem. Here $\mathrm{Tr}(RG^2)$ is the coefficient of the ABJ triangle diagram with j_μ^R at one vertex and G gauge fields at the other two. Using holomorphy to link the real and imaginary parts of Eqns. (2) and (3) gives the multiplet of anomalies

$$\beta(g^{-2}) = \frac{3}{2}\frac{1}{16\pi^2}\,\mathrm{Tr}(RG^2)\,, \tag{4}$$

which has been an important puzzle, as the gauge beta function is not one-loop exact. This puzzle was discussed in many papers, with many proposed resolutions. The most fruitful and inspirational was that of Shifman and Vainshtein (see [1] and references cited therein) which ties in with the Novikov, Shifman, Vainshtein, Zakharov (NSVZ) exact beta function.[2]

The holomorphic relation between the dilatation and R-current implies a corresponding relation for the charges of chiral superfields:

$$\Delta(Q_i) \equiv 1 + \frac{1}{2}\gamma_i(g) = \frac{3}{2}R(Q_i)\,, \tag{5}$$

where we'll interpret the RHS as renormalization group running R-charges when the theory is running between RG fixed points. (We do not need to worry about the fact that the Q_i are not gauge invariant: the relation Eqn. (5) remains useful once we assemble gauge invariant chiral operators $\mathcal{O}$, because their R-charges, and hence also their operator dimensions, are simply additive for chiral operators.) So there is a hidden g dependence in Eqns. (3) and (4), via the charges $R_i = R_i(g)$. Indeed, using the R-charges of Eqn. (5) in Eqn. (4) gives the numerator of the NSVZ exact beta function. Similarly, the holomorphic relation Eqn. (2) implies that superpotential couplings to a general operator $\mathcal{O}$, e.g. Yukawa couplings, have exact beta functions related to the term's explicit R-violation,

$$W = h\mathcal{O} \qquad \text{has} \qquad \beta_h = \frac{3}{2}h(R(\mathcal{O}) - 2)\,. \tag{6}$$

2. RG flows and IR fixed point SCFTs

Let's consider why the R-charges change along RG flows, $R_i = R_i(g)$. See e.g. Ref. [3] for a related discussion. Along the RG flow, the j_μ^R current in the $\mathcal{T}_\mu$ supermultiplet is a linear combination of a conserved part, together with a violated part, $j_\mu^R = j_\mu^{R,cons} + j_\mu^{R,viol}$. The violated part, $j_\mu^{R,viol}$ generally includes both tree-level breaking, as in Eqn. (6), and anomalous breaking, as in Eqn. (4). The R-charges evolve along the flow because the violated part $j_\mu^{R,viol}$ picks up a non-zero, positive anomalous dimension, and becomes irrelevant in the IR, $j_\mu^{R,viol} \to 0$; this is the holomorphic partner of the fact that the couplings are approaching a zero of the beta functions. Finally, at the IR, scale invariant, RG fixed point, $j_\mu^R \to j_\mu^{R,cons}$, and the exact chiral operator dimensions are given by Eqn. (5), evaluated for the superconformal R-charges, R_*, associated with $j_\mu^{R,cons}$.

In some theories, like the classic examples of SQCD and their Seiberg duals,[4] there is a unique conserved R_*, determined by the symmetries. More generally, there can be a multi-parameter family of conserved $j_\mu^{R,cons}$, since a conserved R-symmetry can mix with all conserved, non-R, global $U(1)$ flavor symmetries, with charges given by $R = R_0 + \sum_I s_I F_I$. Here R_0 is an arbitrary choice of a conserved R-symmetry, the I label the independent $U(1)$ flavor symmetries, with charge F_I, and the s_I are parameters labeling the most general, conserved, R-symmetry. Along the RG flow to the IR, the conserved part $j_\mu^{R,cons}$ can then also evolve, in the s_I parameter space. Finally, at the IR fixed point, the s_I take some special values s_I^*, and the superconformal R-charges, which determine the exact dimensions of chiral operators in the IR SCFT, is given by $R_* = R_0 + \sum_I s_I^* F_I$.

The correct R_* can be simply, and exactly, determined by a-maximization Ref. [5]. A quick derivation is to consider Eqn. (2) on a fixed curved background, where there is an additional contribution to X coming from the curvature

$$X \supset \frac{c}{32\pi^2}\mathcal{W}^2 - \frac{a}{32\pi^2}\Xi\,, \tag{7}$$

where the chiral superfield $\mathcal{W}^2$'s first component is the gravitino squared, and its F-component contains the Weyl curvature squared, whereas Ξ is a topological term, whose F-component contains the Euler density. The imaginary part of the holomorphic F-components also contain linear combinations of curvature $R\widetilde{R}$ and background $U(1)_{R*}$ gauge field $F_{R*}\widetilde{F}_{R^*}$. See Ref. [6] and references therein. Considering the real part of Eqn. (2) at a RG fixed point shows that the coefficients a and c in Eqn. (7) are the two types of conformal anomalies in 4d CFTs. Considering the imaginary part of Eqn. (7) relates a and c to the 't Hooft anomalies of the superconformal R-symmetry, as found in Ref. [6]:

$$a_* = \frac{3}{32}\left(3\mathrm{Tr}R_*^3 - \mathrm{Tr}R_*\right), \qquad c_* = \frac{1}{32}\left(9\mathrm{Tr}R^3 - 5\mathrm{Tr}R\right), \tag{8}$$

where the $*$ means that the quantities are evaluated at the RG fixed point.

Now suppose that there is a multi-parameter family of conserved R-symmetries $R = R_0 + \sum_I s_I F_I$. The result of a-maximization [5] is that the superconformal R-symmetry, $R_* = R_0 + \sum_I s_I^* F_I$ is the unique local maximum of the function

$$a(R) = \frac{3}{32}\left(3\mathrm{Tr}R^3 - \mathrm{Tr}R\right) \tag{9}$$

among all possible R-symmetries. To see why, consider varying the j_μ^R R-symmetry in the parameter space of conserved R-symmetries, $R \to R + \epsilon F$, (where F is an arbitrary linear combination of the F_I). In terms of currents, we vary $j_\mu^R \subset \mathcal{T}_\mu$ as $j_\mu^R \to j_\mu^R + \epsilon j_\mu^F$, where j_μ^R is a non-R flavor current, which resides in a real superfield J_F. Then Eqn. (1) says that we also vary

$X \to X + \epsilon \bar{D}^2 J_F$, where $\bar{D}^2 J_F$ is the violation of the current J_F associated with the 't Hooft anomalies of a curved background, and background gauge fields coupled to the global currents. These anomalies are given by the superspace relation $\bar{D}^2 J_F = k_{FFF} W_F^2 + k_F \mathcal{W}^2$, where $k_F = \mathrm{Tr} F$ and and $k_{FFF} = \mathrm{Tr} F^3$. Note that there is no Ξ term in $\bar{D}^2 J_F$, so the variation $X \to X + \epsilon \bar{D}^2 J_F$ in Eqn. (7) leaves a stationary, $a \to a$. Thus the R-symmetry in the $\mathcal{T}_\mu$ locally extremizes the combination of 't Hooft anomalies in Eqn. (9) (unitarity implies that the extremum is a local maximum).

a-maximization determines R_* at the RG fixed point, but the above variation argument is not limited to RG fixed points – it is a statement about the RG flowing R-charges of the $U(1)_R$ current in the $\mathcal{T}_\mu$ supermultiplet. This fits with the conjecture of Ref. [7] that a-maximization can be extended to apply along the entire RG flow, to define an a-function on coupling constant space, whose gradients are related to the beta functions. See Refs. [8,9] for additional discussion and checks.

Maximizing the function Eqn. (9) determines the exact R-charges at RG fixed points as the roots of quadratic equations, with integer coefficients. The coefficients are integers because the flavor symmetries have rational charges (they're compact). a-maximization thus implies that the exact R-charges R_{i*}, and also the conformal anomalies a and c of SCFTs are necessarily *algebraic numbers*. A consequence of this is that the superconformal R-charges and a and c can't vary continuously, so they all must be constant along any RG line of fixed points.

3. IR free or interacting? A proposed diagnostic

Some UV theories RG flow to interacting scale invariant RG fixed points, whereas others are instead IR free. We also know of examples containing both in the IR: an interacting sector, and also a decoupled IR free sector. In the context of non-susy models, technicolor model building provided motivation for many studies of how to determine the IR phase, and the conformal window, using a variety of approximate methods and conjectures. See e.g. Ref. [10] for a recent proposal, and for references to some of the literature with other proposals.

Here, I'll briefly mention the conjecture of Ref. [11]: that the correct IR phase is that with the larger value of the conformal anomaly a_{IR}. The intuition is that the shorter IR flow is correct, and that Δa measures the length of the RG flow. For susy theories, this conjecture implies that only chiral operators X with superconformal R-charge $R(X) < 5/3$ can become IR free. This conjecture works in all understood examples.

Let's mention one example which is not yet definitively understood: $\mathcal{N} = 1$ supersymmetric $SU(2)$ gauge theory, with a matter field Q in the $I = 3/2$ rep of $SU(2)$. This example was discussed in Ref. [12] as an example of a simple model of dynamical supersymmetry breaking, if the theory is IR free. Alternatively, the theory might instead flow to an interacting SCFT, in which case the theory is less interesting, in that susy is not broken. The IR free phase possibility is supported by a 't Hooft anomaly matching which looks too non-trivial to be a fluke (though there are other examples of non-trivial 't Hooft anomaly matching which are known to be misleading[13]).

If the theory is superconformal in the IR (without accidental symmetries), there is a uniquely determined superconformal $U(1)_R$ symmetry: the anomaly free R-symmetry with $R(Q) = 3/5$. The gauge invariant chiral superfield $X = Q^4$ thus has $R(X) = 12/5$. If the theory is IR free, there would be an accidental symmetry that allows $R_{accid}(X) = 2/3$. But the interacting phase has a larger value of the conformal anomaly: $a_{IR,interact} > a_{IR,free}$, as follows immediately from the fact that $R(X) > 5/3$. So, according to the conjecture of [11], this theory is interacting in the IR, rather than IR free. Recent work of Poppitz and Ünsal Ref. [14], presented at this workshop, also suggests that the $SU(2)$ with $I = 3/2$ theory is likely in the interacting CFT phase, rather than IR free.

4. The status of Cardy's conjecture?

The intuition that RG flows to the IR course grain away degrees of freedom suggests that there should be a way to quantify the number of degrees of freedom of a QFT, with a quantity that is always reduced in (unitary) RG flows to the IR. In 2d, this was made precise by the Zamolodchikov c-function. Cardy's conjecture Ref. is that the conformal anomaly a is the analogous quantity in 4d CFTs, with $a_{IR} < a_{UV}$ for all RG flows. The conjecture was analyzed (and verified) in perturbation theory, and a proposal was given for a monotonically decreasing a-function on the RG flow was given in Ref.

In the context of supersymmetric gauge theories, Cardy's conjecture survived many, many very non-trivial tests, including some that initially seemed to violate the conjecture. See e.g. [15] for a large class of examples of RG flows, all non-trivially satisfying $a_{IR} < a_{UV}$. Indeed, a-maximization can be used to "almost prove" Cardy's conjecture, modulo the major caveat that there is a loophole with general, possible accidental symmetries. The loophole can be closed[8] for those accidental symmetries associated with gauge invariant chiral operators hitting the unitarity bound $R(X) \geq 2/3$).

In a very interesting and surprising work last year, Shapere and Tachikawa[16] presented a claimed counterexample to Cardy's conjecture. The example is based on strongly coupled $\mathcal{N}=2$ SCFTs. It doesn't contradict a-maximization since the superconformal symmetry, including the superconformal R-symmetry, is completely accidental. The values of the superconformal anomalies a and c are computed using TQFT results, as explained in [17]. The theory in the UV is a $\mathcal{N}=2$ supersymmetric $SU(N+1)$ gauge theory, with $N_f=2N$ flavors. The flavors are given a mass, and the vacuum is on the Coulomb branch of the moduli space, at an Argyres-Douglas[18] point. There the theory RG flows to an interacting SCFT, with conformal anomaly computed in [16,17] to be $a_{AD}=7N^2/36$. An added relevant deformation then drives the theory farther in the IR to $SU(N)$ with $N_f=2N$ flavors, with conformal anomaly $a_{IR}=7N^2/24$, which is larger than the value computed for a_{AD}; hence the counterexample [16] to $a_{IR}<a_{UV}$ for all RG flows. Given the surprise of the result, the work [16] was careful to address and rule out loopholes in their analysis. It is an interesting question for the future to see if there might still be some as yet unseen loophole or subtlety in their argument, or if there is perhaps some modified version, $\widetilde{a}$, of the appropriate d.o.f. counting quantity for these theories.

5. Hofman-Maldacena inequalities for a/c

In Ref. [19], Hofman and Maldacena (HM) consider correlation functions of energy flux operators, $\mathcal{E}(\widehat{n})=\int dtr^2n^iT^0_i(t,r\vec{n})|_{r\to\infty}$, and conjecture that their correlation functions are always non-negative. For the case of $\mathcal{N}=1$ supersymmetric theories, they find (where $\widehat{\epsilon}$ and $\widehat{n}$ are unit vectors)

$$\langle J_R\cdot\epsilon|\mathcal{E}(\widehat{n})|J_R\cdot\epsilon\rangle=1+3\frac{c-a}{c}(\widehat{\epsilon}\cdot\widehat{n}^2-\frac{1}{3}),\tag{10}$$

$$\langle T\cdot\epsilon|\mathcal{E}(\widehat{n})|T\cdot\epsilon\rangle=1+6\frac{c-a}{c}(\frac{\epsilon^*_{ij}\epsilon_{il}n_jn_l}{\epsilon^*_{ij}\epsilon_{ij}}-\frac{1}{3}),\tag{11}$$

where J_R is the superconformal $U(1)_R$ symmetry, T is the stress-energy tensor. The conjectured non-negativity of the RHSs then implies inequalities for the ratio a/c: for $\mathcal{N}=1$ theories, it's

$$\frac{3}{2}\geq\frac{a}{c}\geq\frac{1}{2}\tag{12}$$

(similarly, for $\mathcal{N}=0$ theories it's $\frac{31}{18}\geq\frac{a}{c}\geq\frac{1}{3}$, and for $\mathcal{N}=2$ theories, it's $\frac{5}{4}\geq\frac{a}{c}\geq\frac{1}{2}$; we'll focus on the $\mathcal{N}=1$ case). The upper limit is always saturated by a free vector field, and the lower limit is saturated by a free matter field. The inequalities have been verified to be satisfied (comfortably, as we'll discuss next) in every example checked.

We'll discuss the sign of $c - a$, and note that typical examples have $c \geq a$. Expressions show that the energy flux is peaked in the direction $\hat{n}$ parallel to the state's polarization ϵ if $c - a > 0$. On the other hand, if $c - a < 0$ the energy flux would be peaked in the orthogonal direction. This latter situation sounds more exotic, perhaps fitting with the fact that typical examples have $c \geq a$.

6. Tighter a/c inequalities for fully interacting theories?

The upper and lower bounds of the conjectured a/c inequalities are saturated by free gauge fields and free matter fields, respectively. But all interacting 4d $\mathcal{N} = 1$ SCFTs are obtained from an appropriate blend of both gauge fields and matter. There are no non-trivial interacting 4d $\mathcal{N} = 1$ RG fixed points near theories with only gauge fields, or near theories with no gauge fields. (Similar remarks apply for $\mathcal{N} = 0$ and $\mathcal{N} = 2$). This suggests that there should be tighter bounds on a/c for fully interacting SCFTs. We do not yet have a specific statement or conjecture for what the tighter bounds might be – just that they should exist. As an example, in the interacting conformal window of SQCD [4], $3N_c > N_f > \frac{3}{2}N_c$, the a/c ratio is bounded between $\frac{5}{6} > \frac{a}{c} > \frac{2}{3}$, where the upper limit is saturated for $N_f = 3N_c$, and a/c monotonically decreases with N_f, hitting $a/c = 2/3$ at $N_f = \frac{3}{2}N_c$.

The qualifier "fully interacting" is because there are many known examples of interacting SCFTs with IR free subsectors. An example of when this happens is when some operators hit the unitary bound, e.g. $\Delta(\mathcal{O}) \geq 1$ for scalar operators. If an operator hits the bound, it becomes a decoupled IR free field, and decouples from the rest of the theory. This happens, for example, for the meson chiral superfield M in SQCD for $N_f \leq \frac{3}{2}N_c$, where in that case the entire theory is IR free, with also the IR free magnetic gauge fields.[4] More generally, part of the theory can remain interacting even if some operators become IR free, see e.g. the example in [20]. There are also examples where operators become IR free even if not required by unitary bounds (see e.g. the discussion in [11] for a proposed necessary condition). Such partially interacting theories can of course have a/c outside of the expected tighter bounds for fully interacting theories (so apparent violations of expected tighter bounds could become a way to show that there must be an IR free subsector).

Let's first consider $\mathcal{N} = 1$ Banks-Zaks type SCFTs, with large N_c, with zeros of the beta function at weak coupling, $\beta(g_*) = 0$ for $g_* \ll 1$. This is arranged by taking the matter representation r of the group G ($G = SU(N_c)$, or $SO(N_c)$, or $Sp(N_c)$) to have $T_2(r) \approx 3T_2(G)$, so the gauge coupling is just barely asymptotically free. In this limit, we compute using Eqn. (8)

$$\frac{a}{c} \approx 1 - \frac{|r| - 3|G|}{6|G| - 2|r|}.$$

The asymptotically free, large N_c reps, with $T_2(r) \approx 3T_3(G)$, all have $|r| \geq 3|G|$ (of the the adjoint representation minimizes $|r|/|G|$ for fixed $T_2(r)/T_2(G)$). So large N_c Banks-Zaks RG fixed points all have $a/c \leq 1$. For example, for SQCD with $N_f \approx 3N_c$, $a/c \approx 5/6$. The inequality $a/c \leq 1$ for weakly coupled SCFTs is saturated for $\mathcal{N} = 4$ theories (3 $\mathcal{N} = 1$ adjoints and appropriate superpotential).

Incidentally, the sign of $a - c$ has also been recently discussed in the AdS/CFT literature. The Einstein-Hilbert action in AdS_5 leads to $a = c$ for the 4d boundary CFT [21]. Adding higher curvature terms to the bulk action, starting with R^2 terms, leads to $a \neq c$. Such higher curvature terms also affect the viscosity/entropy ratio of the KSS conjecture,[22] $\eta/s \geq \frac{1}{4\pi}$. As discussed in [23–25 the relation is

$$\frac{\eta}{s} = \frac{1}{4\pi}\left(1 - \frac{c-a}{c} + \ldots\right),$$

where the ... are contributions from higher curvature terms, which can be small if $|(c-a)/c| \ll 1$. The KSS conjectured bound then might suggest that a must be bigger than c, at least if $|(c-a)/c| \ll 1$; however, this is not the case. (As shown in [26] microcausality is equivalent to the Hofman-Maldacena lower bound, $a/c \geq \frac{1}{3}$, and not $a/c \geq 1$.) In fact, the opposite, $\frac{a}{c} \leq 1$, is typical for interacting theories! The above Banks-Zaks examples illustrate $a < c$, as also noted in [25]. Moreover, as we'll discuss below, $\frac{a}{c} \leq 1$ is rather generic in known $N = 1$ SCFTs.

For $\mathcal{N} = 1$ susy theories, we can use (8) and (9) to express the HM bounds in terms of the 't Hooft anomalies for the superconformal $U(1)_R$:

$$\frac{3}{2} \geq \frac{a}{c} \qquad \leftrightarrow \qquad \mathrm{Tr}R^3 \geq \mathrm{Tr}R, \tag{13}$$

$$\frac{a}{c} \geq \frac{1}{2} \qquad \leftrightarrow \qquad \mathrm{Tr}R^3 \geq \frac{1}{9}\mathrm{Tr}R. \tag{14}$$

Note also that

$$a - c = \frac{1}{16}\,\mathrm{Tr}R. \tag{15}$$

The a/c upper bound inequality Eqn. (13) is more-or-less clearly satisfied, the reason is that the gauge field contribution to this difference of 't Hooft anomalies cancels, because the gauginos have $R = 1$(indeed, the inequality is saturated for a free vector multiplet). So this combination of 't Hooft anomalies only gets contributions from the matter chiral superfields:

$$\mathrm{Tr}\,R^3 - \mathrm{Tr}\,R = \sum_{\text{matter } f} \left((R_f - 1)^3 - (R_f - 1)\right).$$

Typically each term in the sum is positive, as chiral superfields contribute only contribute negatively to the sum if they have either $R_f < 0$, or if $1 < R_f < 2$, whereas typically fields have $0 < R_f < \frac{2}{3}$. There are cases where some fields contribute negatively, and then more work is required to show that the net sum is positive. (E.g. this happens for SQCD at the upper end of the conformal window, $N_f \approx 3N_c$. In the electric description, it's clear that the sum is positive, since all electric quarks contribute positively. But, in the magnetic dual description, the mesons have $R(M) \approx 4/3$, so they contribute negatively to the sum; the sum remains positive, in any case, because of the larger contribution of the magnetic quarks.)

On the other hand, the a/c lower bound inequality, $\mathrm{Tr}R^3 - \frac{1}{9}\mathrm{Tr}R \geq 0$, is more subtle, since both gauge fields and matter contribute (only free matter, with $R = 2/3$, makes a vanishing contribution).

Now, if $a > c$, i.e. if $\mathrm{Tr}R < 0$, the more subtle bound, Eqn. (14), is in fact the stronger of the two, which is perhaps reasonable. On the other hand, if $a < c$ then $\mathrm{Tr}R > 0$ and Eqn. (13) becomes stronger, and Eqn. (14) is automatically satisfied (because $c > 0$). This latter situation is curious, so one might wonder if perhaps fully interacting CFTs can not have $a < c$. Indeed, in many checked examples, it is the case that $a < c$. For example, there is a large family of SCFTs based on SQCD with added adjoints. These theories have an ADE classification, and RG flows among them were analyzed in [15] (and it was verified, in particular, that all RG flows have $a_{IR} < a_{UV}$). The ratio a/c was recently computed for all of these theories in Ref. [27], and not only did all satisfy the HM bounds, but all in fact have $a/c < 1$.

So we ask: can we find any (fully) interacting (S)CFT examples with $a/c > 1$? We have found some examples, but they're examples where we're not 100% certain about the nature of the IR fixed point. For example, consider $SO(N)$ gauge theory with a 2-index symmetric tensor S. As discussed in [13], this theory has a highly non-trivial 't Hooft anomaly matching, which misleadingly suggests confinement in terms of operators $\mathcal{O}_n = \mathrm{Tr} S^n$, even though the theory is evidently interacting (it is certainly not IR free in the $\mathcal{O}_n$.) Assuming the minimal accidental symmetries, i.e. that only those $\mathcal{O}_n$ below the unitarity bound $\Delta \geq 1$ are IR free, then we find (in the large N limit) $a/c = 381/362 = 1.05249$. If we ignored the effect of the accidental symmetries, we would have obtained $a/c = 1.05$, so the difference is just in the last 3 decimal places. Note that the effect of accidental symmetries is to modify the 't Hooft anomaly

$$\mathrm{Tr}\, R \to \mathrm{Tr}\, R + \sum_{\text{free } \mathcal{O}} (2/3 - R_{naive}(\mathcal{O})),$$

so those minimal accidental symmetries associated with chiral operators having hit the unitarity bound, i.e. having $R_{naive}(\mathcal{O}) < 2/3$, will contribute positively to $\mathrm{Tr}R$ and thus increase a/c. Unseen accidental symmetries, from IR free operators with $R_{naive}(\mathcal{O}) > 2/3$, on the other hand, can decrease $a - c$. So we can not be completely sure if $a > c$ in these examples, because there could be some unseen accidental symmetries.

Finally, we mention again the $SU(2)$ with $Q_{I=3/2}$ example of Ref. [12]. The 't Hooft anomalies are $\mathrm{Tr}R = 7/5$ and $\mathrm{Tr}R = (7/5)^3$. Again, a non-trivial anomaly matching suggests that this theory is IR free, in which case $a/c = 1/2$ (the lower limit of the HM bound). On other hand, as mentioned above, more recent studies suggest that the theory is actually an interacting CFT. If we suppose that it is fully interacting, without accidental symmetries, then we find $a/c = 183/158 = 1.16$. This theory holds the record for the largest value of a/c that we've found in a fully interacting CFT... if it is indeed actually fully interacting...

Acknowledgements

I would like to thank Arkady and Misha for the invitation to this special workshop, and to wish Misha a happy 60th birthday, and many good future years! Work supported in part by DOE grant DOE-FG03-97ER40546.

References

1. M. A. Shifman, Int. J. Mod. Phys. A **14**, 5017 (1999) [arXiv:hep-th/9906049].
2. V. A. Novikov, M. A. Shifman, A. I. Vainshtein and V. I. Zakharov, Nucl. Phys. B **229**, 381 (1983).
3. I. I. Kogan, M. A. Shifman and A. I. Vainshtein, Phys. Rev. D **53**, 4526 (1996) [Erratum-ibid. D **59**, 109903 (1999)] [arXiv:hep-th/9507170].

4. N. Seiberg, Nucl. Phys. B **435**, 129 (1995) [arXiv:hep-th/9411149].
5. K. A. Intriligator and B. Wecht, Nucl. Phys. B **667**, 183 (2003) [arXiv:hep-th/0304128].
6. D. Anselmi, D. Z. Freedman, M. T. Grisaru and A. A. Johansen, Nucl. Phys. B **526**, 543 (1998) [arXiv:hep-th/9708042].
7. D. Kutasov, "New results on the 'a-theorem' in four dimensional supersymmetric field theory," arXiv:hep-th/0312098.
8. E. Barnes, K. A. Intriligator, B. Wecht and J. Wright, Nucl. Phys. B **702**, 131 (2004) [arXiv:hep-th/0408156].
9. D. Kutasov and A. Schwimmer, Nucl. Phys. B **702**, 369 (2004) [arXiv:hep-th/0409029].
10. E. Poppitz and M. Ünsal, JHEP **0909**, 050 (2009) [arXiv:0906.5156 [hep-th]].
11. K. A. Intriligator, Nucl. Phys. B **730**, 239 (2005) [arXiv:hep-th/0509085].
12. K. A. Intriligator, N. Seiberg and S. H. Shenker, Phys. Lett. B **342**, 152 (1995) [arXiv:hep-ph/9410203].
13. J. H. Brodie, P. L. Cho and K. A. Intriligator, Phys. Lett. B **429**, 319 (1998) [arXiv:hep-th/9802092].
14. E. Poppitz and M. Ünsal, JHEP **0907**, 060 (2009) [arXiv:0905.0634 [hep-th]].
15. K. A. Intriligator and B. Wecht, Nucl. Phys. B **677**, 223 (2004) [arXiv:hep-th/0309201].
16. A. D. Shapere and Y. Tachikawa, JHEP **0812**, 020 (2008) [arXiv:0809.3238 [hep-th]].
17. A. D. Shapere and Y. Tachikawa, JHEP **0809**, 109 (2008) [arXiv:0804.1957 [hep-th]].
18. P. C. Argyres and M. R. Douglas, Nucl. Phys. B **448**, 93 (1995) [arXiv:hep-th/9505062].
19. D. M. Hofman and J. Maldacena, JHEP **0805**, 012 (2008) [arXiv:0803.1467 [hep-th]].
20. D. Kutasov, A. Parnachev and D. A. Sahakyan, JHEP **0311**, 013 (2003) [arXiv:hep-th/0308071].
21. M. Henningson and K. Skenderis, JHEP **9807**, 023 (1998) [arXiv:hep-th/9806087].
22. P. Kovtun, D. T. Son and A. O. Starinets, Phys. Rev. Lett. **94**, 111601 (2005) [arXiv:hep-th/0405231].
23. Y. Kats and P. Petrov, JHEP **0901**, 044 (2009) [arXiv:0712.0743 [hep-th]].
24. M. Brigante, H. Liu, R. C. Myers, S. Shenker and S. Yaida, Phys. Rev. D **77**, 126006 (2008) [arXiv:0712.0805 [hep-th]].
25. A. Buchel, R. C. Myers and A. Sinha, JHEP **0903**, 084 (2009) [arXiv:0812.2521 [hep-th]].
26. M. Brigante, H. Liu, R. C. Myers, S. Shenker and S. Yaida, Phys. Rev. Lett. **100**, 191601 (2008) [arXiv:0802.3318 [hep-th]].
27. A. Parnachev and S. S. Razamat, JHEP **0907**, 010 (2009) [arXiv:0812.0781 [hep-th]].

High-Energy Amplitudes in $\mathcal{N}$=4 SYM in the Next-to-Leading Order

Ian Balitsky

Physics Dept., ODU, Norfolk VA 23529,
and
Theory Group, Jlab, 12000 Jefferson Ave, Newport News, VA 23606
E-mail: balitsky@jlab.org

The high-energy behavior of the of $\mathcal{N}$=4 SYM amplitudes in the Regge limit can be calculated order by order in perturbation theory using the high-energy operator expansion in Wilson lines. At large N_c, a typical four-point amplitude is determined by a single BFKL pomeron. The conformal structure of the four-point amplitude is fixed in terms of two functions: pomeron intercept and the coefficient function in front of the pomeron (the product of two residues). The pomeron intercept is universal while the coefficient function depends on the correlator in question. The intercept is known in first two orders in coupling constant : LO BFKL intercept and NLO BFKL calculated in Ref. [1]. As an example of using the Wilson-line OPE, we calculate the coefficient function in front of the pomeron for the correlator of four Z^2 currents in the leading and next-to-leading order.

Keywords: High energy, Conformal invariance, Wilson lines.

1. Introduction

The high-energy scattering in a gauge theory can be described in terms of Wilson lines - infinite gauge factors ordered along the straight lines (see e.g. the review [2]). Indeed, the fast particle moves along its straight-line classical trajectory and the only quantum effect is the eikonal phase factor acquired along this propagation path. For a fast particle scattering off some target, this eikonal phase factor is a Wilson line - an infinite gauge link ordered along the straight line collinear to particle's velocity n^μ:

$$U^Y(x_\perp) = \mathrm{Pexp}\Big\{ig\int_{-\infty}^{\infty} du\; n_\mu\; A^\mu(un + x_\perp)\Big\}\,. \qquad (1)$$

Here A_μ is the gluon field of the target, $x_\perp$ is the transverse position of the particle which remains unchanged throughout the collision, and the index

Y labels the rapidity of the particle.

The high-energy behavior of the amplitudes can be studied in the framework of the rapidity evolution of Wilson-line operators forming color dipoles.[3,4] The main idea[5,6] is factorization in rapidity: we separate a typical functional integral into (**i**) the integral over the gluon (and gluino) fields with rapidity close to the rapidity of the spectator Y_A, (**ii**) the integral over the gluons with rapidity close to the rapidity of the target Y_B, and (**iii**) the integral over the intermediate region of rapidities $Y_A > Y > Y_B$. The result of the first integration is a certain coefficient function ("impact factor") times color dipole (ordered in the direction of spectator's velocity) with rapidities up to Y_A. Similarly, the result of second integration is again the impact factor times color dipole ordered in the direction of target's velocity with rapidities greater than Y_B. The result of last integration is the correlation function of two dipoles which can be calculated using the evolution equation for color dipoles which is known in the leading and next-to-leading order. As an example of practical use of this factorization scheme in the NLO approximation, in present paper we calculate the high-energy behavior of the "scattering amplitude of scalar particles" (the four-point correlation function of scalar currents).

The high-energy (Regge) limit of a four-point amplitude $A(x, y; x', y')$ in the coordinate space can be achieved as

$$\begin{aligned} x &= \lambda x_* p_1 + x_\perp, \quad & y &= \lambda y_* p_2 + y_\perp, \\ x' &= \rho x'_\bullet p_2 + x'_\perp, \quad & y' &= \rho y'_\bullet p_2 + y'_\perp \end{aligned} \tag{2}$$

with $\lambda, \rho \to \infty$ and $x_* > 0 > y_*$, $x'_\bullet > 0 > y'_\bullet$. Hereafter we use the notations $x_\bullet = p_1^\mu x_\mu$, $x_* = p_2^\mu x_\mu$ where p_1 and p_2 are light-like vectors normalized by $2(p_1, p_2) = s$. These "Sudakov variables" are related to the usual light-cone coordinates $x^\pm = \frac{1}{\sqrt{2}}(x^0 \pm x^3)$ by $x_* = x^+\sqrt{s/2}$, $x_\bullet = x^-\sqrt{s/2}$ so $x = \frac{2}{s}x_* p_1 + \frac{2}{s}x_\bullet p_2 + x_\perp$. We use the (1,-1,-1,-1) metric so $x^2 = \frac{4}{s}x_\bullet x_* - \vec{x}_\perp^2$. In the Regge limit (2) the full conformal group reduces to Möbius subgroup SL(2,C) leaving the transverse plane $(0, 0, z_\perp)$ invariant.

For simplicity, let us consider correlation function of four scalar currents

$$(x-y)^4(x'-y')^4\langle \mathcal{O}(x)\mathcal{O}^\dagger(y)\mathcal{O}(x')\mathcal{O}^\dagger(y')\rangle \tag{3}$$

where $\mathcal{O} \equiv \frac{4\pi^2\sqrt{2}}{\sqrt{N_c^2-1}}\mathrm{Tr}\{Z^2\}$ $(Z = \frac{1}{\sqrt{2}}(\phi_1 + i\phi_2))$ is a renorm-invariant chiral primary operator.

In a conformal theory this four-point amplitude $A(x, y; x', y')$ depends on two conformal ratios which can be chosen as

$$R = \frac{(x-x')(y-y')^2}{(x-y)^2(x'-y')^2},$$
$$r = R\Big[1 - \frac{(x-y')^2(y-x')^2}{(x-x')^2(y-y')^2} + \frac{1}{R}\Big]^2. \tag{4}$$

In the Regge limit (2) the conformal ratio R scales as $\lambda^2\rho^2$ while r does not depend on λ or ρ.

As demonstrated in Ref. [7], the pomeron contribution in a conformal theory can be represented as an integral over one real variable ν

$$(x-y)^4(x'-y')^4\langle\mathcal{O}(x)\mathcal{O}^\dagger(y)\mathcal{O}(x')\mathcal{O}^\dagger(y')\rangle \tag{5}$$
$$= \frac{i}{2}\int d\nu\ \tilde{f}_+(\nu)\frac{\tanh\pi\nu}{\nu}F(\nu)\Omega(r,\nu)R^{\frac{1}{2}\omega(\nu)}.$$

Here $\omega(\nu) \equiv \omega(0,\nu)$ is the pomeron intercept, $\tilde{f}_+(\nu) \equiv \tilde{f}_+(\omega(\nu))$ where $\tilde{f}_+(\omega) = (e^{i\pi\omega}-1)/\sin\pi\omega$ is the signature factor in the coordinate space, and $F(\nu)$ is the "pomeron residue" (strictly speaking, the product of two pomeron residues). The conformal function $\Omega(r,\nu)$ is given by a hypergeometric function (see Ref. [10]) but for our purposes it is convenient to use the representation in terms of the two-dimensional integral

$$\Omega(r,\nu) = \frac{\nu^2}{\pi^3}\int d^2z\Big(\frac{\kappa^2}{(2\kappa\cdot\zeta)^2}\Big)^{\frac{1}{2}+i\nu}\Big(\frac{\kappa'^2}{(2\kappa'\cdot\zeta)^2}\Big)^{\frac{1}{2}-i\nu} \tag{6}$$

where $\zeta \equiv p_1 + \frac{z_\perp^2}{s}p_2 + z_\perp$ and

$$\kappa = \frac{\sqrt{s}}{2x_*}(p_1 - \frac{x^2}{s}p_2 + x_\perp) - \frac{\sqrt{s}}{2y_*}(p_1 - \frac{y^2}{s}p_2 + y_\perp) \tag{7}$$
$$\kappa' = \frac{\sqrt{s}}{2x'_\bullet}(p_1 - \frac{x'^2}{s}p_2 + x'_\perp) - \frac{\sqrt{s}}{2y'_\bullet}(p_1 - \frac{y'^2}{s}p_2 + y'_\perp)$$

are two SL(2,C)-invariant vectors[10] such that

$$\kappa^2\kappa'^2 = \frac{1}{R} \qquad \text{and} \qquad 4(\kappa\cdot\kappa')^2 = \frac{r}{R}. \tag{8}$$

Here $x^2 = -x_\perp^2$, $x'^2 = -x'^2_\perp$ and similarly for y. Note that all the dependence on large energy ($\equiv$ large λ,ρ) is contained in $R^{\frac{1}{2}\omega(\nu)}$.

The dynamical information about the conformal theory is encoded in two functions: pomeron intercept and pomeron residue. The pomeron intercept is known both in the small and large α_s limit. At small α_s [8]

$$\omega(n,\nu) = \frac{\alpha_s}{\pi}N_c\Big[\chi(n,\frac{1}{2}+i\nu)+\frac{\alpha_s N_c}{4\pi}\delta(n,\frac{1}{2}+i\nu)\Big],$$
$$\delta(n,\gamma) = 6\zeta(3)-\frac{\pi^2}{3}\chi(n,\gamma)-\chi\text{''}(n,\gamma)-2\Phi(n,\gamma)-2\Phi(n,1-\gamma) \quad (9)$$

where $\chi(n,\gamma)=2\psi(1)-\psi(\gamma+\frac{n}{2})-\psi(1-\gamma+\frac{n}{2})$ and[1]

$$\Phi(n,\gamma) = \int_0^1\frac{dt}{1+t}\,t^{\gamma-1+\frac{n}{2}}\Big\{\frac{\pi^2}{12}-\frac{1}{2}\psi'\Big(\frac{n+1}{2}\Big)$$
$$-\mathrm{Li}_2(t)-\mathrm{Li}_2(-t)-\Big(\psi(n+1)-\psi(1)+\ln(1+t)$$
$$+\sum_{k=1}^{\infty}\frac{(-t)^k}{k+n}\Big)\ln t-\sum_{k=1}^{\infty}\frac{t^k}{(k+n)^2}[1-(-1)^k]\Big\}. \quad (10)$$

Our main goal is the description of the amplitude in the next-to-leading order in perturbation theory, but it is worth noting that the pomeron intercept is known also in the limit of large λ

$$\omega(\nu) = 2-\frac{\nu^2+4}{2\sqrt{\lambda}} \quad (11)$$

where 2 is the graviton spin and the first correction was calculated in Ref. [9].

The pomeron residue $F(\nu)$ is known in the leading order both at small[10,11] and large[7] 't Hooft coupling

$$F(\nu)\stackrel{\lambda\to 0}{\to}\lambda^2\frac{\pi\sin\pi\nu}{4\nu\cos^3\pi\nu},\qquad F(\nu)\stackrel{\lambda\to\infty}{\to}\frac{\pi^3\nu^2(1+\nu)^2}{\sinh^2\pi\nu} \quad (12)$$

To find the amplitude in the next-to-leading order, we must also calculate the "pomeron residue" $F(\nu)$. In the rest of the paper we will do this (and confirm $\omega(\nu)$ as well) using the high-energy operator product expansion in Wilson lines.[5]

2. Operator expansion in conformal dipoles

We consider the light-like dipoles (in the p_1 direction) and impose the cutoff on the maximal α emitted by any gluon from the Wilson lines so

$$U^\eta_x = \mathrm{Pexp}\Big[ig\int_{-\infty}^{\infty}du\,p_1^\mu A^\eta_\mu(up_1+x_\perp)\Big],$$
$$A^\eta_\mu(x) = \int\frac{d^4k}{16\pi^4}\theta(e^\eta-|\alpha_k|)e^{-ik\cdot x}A_\mu(k). \quad (13)$$

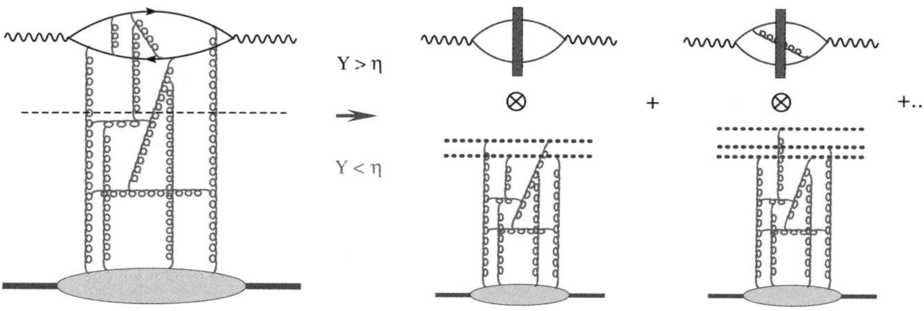

Fig. 1. High-enery operator expansion in Wilson lines

Formally, the light-like Wilson lines are Möbius invariant. Unfortunately, the light-like Wilson lines are divergent in the longitudinal direction and moreover, it is exactly the evolution equation with respect to this longitudinal cutoff which governs the high-energy behavior of amplitudes. At present, it is not known how to find the conformally invariant cutoff in the longitudinal direction. When we use the non-invariant cutoff we expect, as usual, the invariance to hold in the leading order but to be violated in higher orders in perturbation theory. In our calculation we restrict the longitudinal momentum of the gluons composing Wilson lines, and with this non-invariant cutoff the NLO evolution equation in QCD has extra non-conformal parts not related to the running of coupling constant. Similarly, there will be non-conformal parts coming from the longitudinal cutoff of Wilson lines in the $\mathcal{N} = 4$ SYM equation. We will demonstrate below that it is possible to construct the "composite conformal dipole operator" (order by order in perturbation theory) which mimics the conformal cutoff in the longitudinal direction so the corresponding evolution equation has no extra non-conformal parts. This is similar to the construction of the composite renormalized local operator in the case when the UV cutoff does not respect the symmetries of the bare operator - in this case the symmetry of the UV-regularized operator is preserved order by order in perturbation theory by subtraction of the symmetry-restoring counterterms. Following Ref. [12] we choose he conformal composite operator in the form

$$[\mathrm{Tr}\{\hat{U}_{z_1}\hat{U}^\dagger_{z_2}\}]^{\mathrm{conf}}_a = \mathrm{Tr}\{\hat{U}^\eta_{z_1}\hat{U}^{\dagger\eta}_{z_2}\} + \frac{\alpha_s}{2\pi^2}\int d^2z_3\,\frac{z_{12}^2}{z_{13}^2 z_{23}^2} \tag{14}$$
$$\times\,[\mathrm{Tr}\{T^n\hat{U}^\eta_{z_1}\hat{U}^{\dagger\eta}_{z_3}T^n\hat{U}^\eta_{z_3}\hat{U}^{\dagger\eta}_{z_2}\} - N_c\mathrm{Tr}\{\hat{U}^\eta_{z_1}\hat{U}^{\dagger\eta}_{z_2}\}]\ln\frac{az_{12}^2}{z_{13}^2 z_{23}^2} + O(\alpha_s^2)$$

where a is an arbitrary constant. It is convenient to choose the rapidity-dependent constant $a = a_0 e^{-2\eta}$ so that the $[\mathrm{Tr}\{\hat U^\eta_{z_1}\hat U^{\dagger\eta}_{z_2}\}]^a_{\rm conf}$ does not depend on η and all the rapidity dependence is encoded into a-dependence. The operator expansion in conformal dipoles has the form[12]

$$\begin{aligned}(x-y)^4 T\{\hat O(x)\hat O(y)\} &= \frac{1}{\pi^2(N_c^2-1)}\int\frac{d^2z_1d^2z_2}{z_{12}^4}\mathcal{R}^2\,\Big\{[\mathrm{Tr}\{\hat U^\eta_{z_1}\hat U^{\dagger\eta}_{z_2}\}]^{\rm conf}\\ &-\frac{\alpha_s}{2\pi^2}\int d^2z_3\frac{z_{12}^2}{z_{13}^2z_{23}^2}\Big(\ln\frac{x_*y_*z_{12}^2e^{2\eta}}{16(x-y)_\perp^2z_{13}^2z_{23}^2}\mathcal{Z}_3^2 - i\pi+2C\Big)\\ &\times\,[\mathrm{Tr}\{T^n\hat U^\eta_{z_1}\hat U^{\dagger\eta}_{z_3}T^n\hat U^\eta_{z_3}\hat U^{\dagger\eta}_{z_2}\} - N_c\mathrm{Tr}\{\hat U^\eta_{z_1}\hat U^{\dagger\eta}_{z_2}\}]\Big\}\end{aligned} \tag{15}$$

where the conformal dipole $[\mathrm{Tr}\{\hat U^\eta_{z_1}\hat U^{\dagger\eta}_{z_2}\}]^{\rm conf}$ is given by Eq. (14) with $a = \frac{x_*y_*}{s^2(x-y)^2}$. (This choice of a follows from the requirement that the impact factor in the r.h.s of Eq. (15) should not change under the rescaling (2)). Hereafter we use the notations $\mathcal{Z}_i \equiv \frac{(x-z_i)^2}{x_*} - \frac{(y-z_i)^2}{y_*}$ and

$$\mathcal{R} = \frac{(x-y)^2z_{12}^2}{x_*y_*\mathcal{Z}_1\mathcal{Z}_2}\,. \tag{16}$$

Now it is evident that the impact factor in the r.h.s. of this equation is Möbius invariant and does not scale with λ so Eq. (14) gives conformally invariant operator up to α_s^2 order. In higher orders, one should expect the correction terms with more Wilson lines. This procedure of finding the dipole with conformally regularized rapidity divergence is analogous to the construction of the composite renormalized local operator by adding the appropriate counterterms order by order in perturbation theory.

To find the amplitude (3) we it is sufficient to take into account only two gluons in the t-channel (BKFL approximation). In this case we need to expand our Wilson lines up to two-gluon accuracy. It is convenient to define the "color dipole in the adjoint representation"

$$\hat{\mathcal{U}}^\eta(x,y) = 1-\frac{1}{N_c^2-1}\mathrm{Tr}\{\hat U^\eta_x\hat U^{\dagger\eta}_y\}\,. \tag{17}$$

With this two-gluon accuracy

$$\begin{aligned}&[\frac{1}{N_c}\mathrm{Tr}\{T^n\hat U^\eta_{z_1}\hat U^{\dagger\eta}_{z_3}T^n\hat U^\eta_{z_3}\hat U^{\dagger\eta}_{z_2}\} - \mathrm{Tr}\{\hat U^\eta_{z_1}\hat U^{\dagger\eta}_{z_2}\}] \simeq\\ &= -\frac{1}{2}(N_c^2-1)\big[\hat{\mathcal{U}}^\eta_{\rm conf}(z_1,z_3)+\hat{\mathcal{U}}^\eta_{\rm conf}(z_2,z_3)-\hat{\mathcal{U}}^\eta_{\rm conf}(z_1,z_2)\big]\,.\end{aligned}$$

The corresponding conformal dipole operator in the BFKL approximation has the form

$$\hat{\mathcal{U}}^a_{\rm conf}(z_1,z_2) = \hat{\mathcal{U}}^\eta(z_1,z_2) \tag{18}$$
$$+ \frac{\alpha_s N_c}{4\pi^2}\int d^2z\frac{z_{12}^2}{z_{13}^2z_{23}^2}\ln\frac{az_{12}^2}{z_{13}^2z_{23}^2}[\hat{\mathcal{U}}^\eta(z_1,z_3)+\hat{\mathcal{U}}^\eta(z_2,z_3)-\hat{\mathcal{U}}^\eta(z_1,z_2)]\,.$$

With the two-gluon accuracy one more integration in the r.h.s. of Eq. (15) can be performed and the result is

$$(x-y)^4T\{\hat{\mathcal{O}}(x)\hat{\mathcal{O}}(y)\} \tag{19}$$
$$= -\frac{1}{\pi^2}\int\frac{dz_1dz_2}{z_{12}^4}\,\hat{\mathcal{U}}^{a_0}_{\rm conf}(z_1,z_2)\mathcal{R}^2\Big(1-\frac{\alpha_sN_c}{2\pi^2}\Big[-\frac{\pi^2}{3}$$
$$+ 2\mathrm{Li}_2(1-\mathcal{R})+\ln^2\mathcal{R}-\frac{\ln\mathcal{R}}{\mathcal{R}}+\big[i\pi+4\ln 2-2C\big]\big(\ln\mathcal{R}-\frac{1}{\mathcal{R}}+2\big)\Big]\Big)\,.$$

We need the projection of the T-product in the l.h.s. of this equation onto the conformal eigenfunctions of the BFKL equation with conformal spin 0. Using the decomposition of the product of the transverse δ-functions in conformal 3-point functions

$$\delta^{(2)}(z_1-w_1)\delta^{(2)}(z_2-w_2) \tag{20}$$
$$= \sum_{-\infty}^{\infty}\int\frac{d\nu}{\pi^4}\frac{\nu^2+\frac{n^2}{4}}{z_{12}^2w_{12}^2}\int d^2\rho\; E^*_{\nu,n}(w_1-\rho,w_2-\rho)E_{\nu,n}(z_1-\rho,z_2-\rho)$$

where

$$E_{\nu,n}(z_{10},z_{20}) = \Big[\frac{\tilde{z}_{12}}{\tilde{z}_{10}\tilde{z}_{20}}\Big]^{\frac12+i\nu+\frac{n}{2}}\Big[\frac{\bar{z}_{12}}{\bar{z}_{10}\bar{z}_{20}}\Big]^{\frac12+i\nu-\frac{n}{2}} \tag{21}$$

($\tilde{z}=z_x+iz_y, \bar{z}=z_x-iz_y$, $z_{10}\equiv z_1-z_0$ etc.) we obtain

$$(x-y)^4T\{\hat{\mathcal{O}}(x)\hat{\mathcal{O}}(y)\} = -\frac{1}{\pi^2}\int d\nu\int d^2z_0\,\frac{1+4\nu^2}{8\pi}\frac{\Gamma^2(\frac12-i\nu)}{\Gamma(1-2i\nu)}$$
$$\times\Big(\frac{\kappa^2}{4(\kappa\cdot\zeta_0)^2}\Big)^{\frac12+i\nu}I_0(\nu)\big[1-\frac{\alpha_sN_c}{4\pi}\Phi_1(\nu)\big]\hat{\mathcal{U}}^a_{\rm conf}(\nu,z_0)\,. \tag{22}$$

Here $\zeta_0\equiv p_1+\frac{z_{0\perp}^2}{s}p_2+z_{0\perp}$, $I_0(\nu)=\frac{2\pi^2\alpha_s}{\cosh\pi\nu}N_c/\sqrt{N_c^2-1}$,

$$\hat{\mathcal{U}}^a_{\rm conf}(\nu,z_0) = \int\frac{d^2z_1d^2z_2}{z_{12}^4}\Big(\frac{z_{12}^2}{z_{10}^2z_{20}^2}\Big)^{\frac12+i\nu}\hat{\mathcal{U}}^a_{\rm conf}(z_1,z_2) \tag{23}$$

is a conformal dipole in the z_0,ν representation, and ($\gamma=\frac12+i\nu$)

$$\Phi_1(\nu) = 4\Big(\psi'(\gamma)+\psi'(1-\gamma)+\frac{1-\chi(\gamma)}{\gamma(1-\gamma)}+\chi(\gamma)(-\frac{i\pi}{2}+C)-\frac{\pi^2}{3}\Big)\,. \tag{24}$$

Similarly, one can write down the expansion of the bottom part of the diagram in color dipoles (we change $\nu \to -\nu'$ for future use):

$$(x'-y')^4 T\{\hat{\mathcal{O}}(x')\hat{\mathcal{O}}(y')\} = -\frac{1}{\pi^2}\int d\nu' \int d^2 z_0' \, \frac{1+4\nu'^2}{8\pi} \frac{\Gamma^2(\frac{1}{2}+i\nu')}{\Gamma(1+2i\nu')}$$
$$\times \Big(\frac{\kappa'^2}{4(\kappa'\cdot\zeta_0')^2}\Big)^{\frac{1}{2}-i\nu'} I_0(\nu)\big[1-\frac{\alpha_s N_c}{4\pi}\Phi_1(-\nu')\big]\hat{\mathcal{V}}^{b_0\,*}_{\rm conf}(\nu',z_0')\,. \quad (25)$$

Here $\zeta_0' \equiv p_1 + \frac{z'^2_{0\perp}}{s}p_2 + z'_{0\perp}$,

$$\hat{\mathcal{V}}^{b\,*}_{\rm conf}(\nu',z_0') = \int \frac{d^2 z_1 d^2 z_2}{z_{12}^4}\Big(\frac{z_{12}^2}{z_{10}^2 z_{20}^2}\Big)^{\frac{1}{2}-i\nu'} \hat{\mathcal{V}}^b_{\rm conf}(z_1,z_2) \quad (26)$$

and the conformal operator

$$\hat{\mathcal{V}}^b_{\rm conf}(z_1,z_2) = \hat{\mathcal{V}}^\eta(z_1,z_2) \quad (27)$$
$$+\frac{\alpha_s N_c}{4\pi^2}\int d^2 z \frac{z_{12}^2}{z_{13}^2 z_{23}^2}\ln\frac{b z_{12}^2}{z_{13}^2 z_{23}^2}[\hat{\mathcal{V}}^\eta(z_1,z_3)+\hat{\mathcal{V}}^\eta(z_2,z_3)-\hat{\mathcal{V}}^\eta(z_1,z_2)]$$

is made from the dipoles $\hat{\mathcal{V}}^\eta(x,y) = 1-\frac{1}{N_c^2-1}{\rm Tr}\{\hat{V}^\eta_x \hat{V}^{\dagger\eta}_y\}$ (cf. Eq. (17) ordered along the straight line $\parallel p_2$ with the rapidity restriction:

$$V^\eta_x = {\rm Pexp}\Big[ig\int_{-\infty}^{\infty} du p_1^\mu A^\eta_\mu(up_2+x_\perp)\Big]\,,$$
$$A^\eta_\mu(x) = \int \mathit{đ}^4 k\theta(\beta_k - e^\eta)e^{-ik\cdot x}A_\mu(k)\,. \quad (28)$$

The correlation function (3) takes the form

$$(x-y)^4(x'-y')^4\langle T\{\hat{\mathcal{O}}(x)\hat{\mathcal{O}}(y)\hat{\mathcal{O}}(x')\hat{\mathcal{O}}(y')\}\rangle$$
$$= -\frac{1}{\pi^4}\int d\nu d\nu' \int d^2 z_0 d^2 z_0' \, \frac{1+4\nu^2}{8\pi}\frac{\Gamma^2(\frac{1}{2}-i\nu)}{\Gamma(1-2i\nu)}\Big(\frac{\kappa^2}{4(\kappa\cdot\zeta_0)^2}\Big)^{\frac{1}{2}+i\nu} I_0(\nu)$$
$$\times \big[1-\frac{\alpha_s N_c}{4\pi}\Phi_1(\nu)\big]\frac{1+4\nu'^2}{8\pi}\frac{\Gamma^2(\frac{1}{2}+i\nu')}{\Gamma(1+2i\nu')}\Big(\frac{\kappa'^2}{4(\kappa'\cdot\zeta_0')^2}\Big)^{\frac{1}{2}-i\nu'}$$
$$\times I_0(\nu')\big[1-\frac{\alpha_s N_c}{4\pi}\Phi_1(\nu')\big]\langle \hat{\mathcal{U}}^{a_0}_{\rm conf}(\nu,z_0)\hat{\mathcal{V}}^{b_0\,*}_{\rm conf}(\nu',z_0')\rangle\,. \quad (29)$$

3. NLO scattering of conformal dipoles

The last step is to find the NLO amplitude of the scattering of conformal dipoles $\hat{\mathcal{U}}^a_{\rm conf}(z_0,\nu)$ and $\hat{\mathcal{V}}^b_{\rm conf}(z_0',\nu')$. First we need to write down the NLO BFKL evolution (as we discussed above the rapidity dependence is now encoded in a-evolution):

$$2a\frac{d}{da}\hat{\mathcal{V}}^a_{\rm conf}(z_1,z_2) = \int d^2 z_3 d^2 z_4 K(z_1,z_2;z_3,z_4)\,\hat{\mathcal{V}}^a_{\rm conf}(z_3,z_4) \quad (30)$$

where the kernel $K(z_1, z_2; z_3, z_4)$ in the first two orders has the form[13]

$$\begin{aligned} K_{\rm LO}(z_1, z_2; z_3, z_4) &= \frac{\alpha_s N_c}{2\pi^2}\Big[\frac{z_{12}^2\delta^2(z_{13})}{z_{14}^2 z_{24}^2} + \frac{z_{12}^2\delta^2(z_{24})}{z_{13}^2 z_{23}^2} \\ &- \delta^2(z_{13})\delta^2(z_{24})\int d^2z\,\frac{z_{12}^2}{(z_1-z)^2(z_2-z)^2}\Big], \end{aligned} \tag{31}$$

$$\begin{aligned} K_{\rm NLO}(z_1, z_2; z_3, z_4) &= -\frac{\alpha_s N_c}{4\pi}\frac{\pi^2}{3}K_{\rm LO}(z_1, z_2; z_3, z_4) \\ &+ \frac{\alpha_s^2 N_c^2}{8\pi^4 z_{34}^4}\left[\frac{z_{12}^2 z_{34}^2}{z_{13}^2 z_{24}^2}\left\{\left(1 + \frac{z_{12}^2 z_{34}^2}{z_{13}^2 z_{24}^2 - z_{14}^2 z_{23}^2}\right)\ln\frac{z_{13}^2 z_{24}^2}{z_{14}^2 z_{23}^2}\right.\right. \\ &\left.\left. + 2\ln\frac{z_{12}^2 z_{34}^2}{z_{14}^2 z_{23}^2}\right\} + 12\pi^2\zeta(3) z_{34}^4\delta(z_{13})\delta(z_{24})\right]. \end{aligned} \tag{32}$$

The eigenfunctions of the kernel K are the given by Eq. (21) and the eigenvalues by the pomeron intercept (9),

$$\int d^2z_3 d^2z_4 K(z_1, z_2; z_3, z_4)E_{\nu,n}(z_{30}, z_{40}) = \omega(n,\nu)E_{\nu,n}(z_{10}, z_{20}). \tag{33}$$

For the conformal operators with definite conformal spin (23) the evolution equation (30) takes the simple form

$$2a\frac{d}{da}\hat{\mathcal{U}}^a_{\rm conf}(\nu, z_0) = \omega(\nu)\hat{\mathcal{U}}^a_{\rm conf}(\nu, z_0). \tag{34}$$

(Since Eq. (3) is a correlation functions of scalar currents, we need only the $n = 0$ projection of this evolution).

The result of the evolution is

$$\hat{\mathcal{U}}^a_{\rm conf}(\nu, z_0) = (a/\tilde{a})^{\frac{1}{2}\omega(\nu)}\hat{\mathcal{U}}^{a_0}_{\rm conf}(\nu, z_0) \tag{35}$$

where the endpoint of the evolution $\tilde{a}$ should be taken from the requirement that the amplitude of scattering of conformal dipoles with "normalization points" $\tilde{a}$ and b should not contain large logarithms of energy so it will serve as the initial point of the evolution. (This is similar to taking μ^2 around 1 GeV for the initial point of the DGLAP evolution).

The amplitude of scattering of two conformal dipoles in the first two orders of perturbation theory can be taken from the amplitude of scattering of two dipoles calculated in Ref. [14]. As we discussed above, we need only the projection of this amplitude on the eigenvalues with conformal spin 0.

It has the form:

$$\int \frac{d^2z'_1 d^2z'_2}{z'^2_{12}} \langle \mathcal{U}^a_{\rm conf}(z_1,z_2)\mathcal{V}^b_{\rm conf}(z'_1,z'_2)\rangle \Big(\frac{z^2_{1'2'}}{z'^2_{10}z'^2_{20}}\Big)^{1-\gamma}$$
$$= \; -2\pi^4\alpha_s^2\frac{N_c^2}{N_c^2-1}\frac{\Gamma(-\gamma)\Gamma(\gamma-1)}{\Gamma(1+\gamma)\Gamma(2-\gamma)}\Big(\frac{z^2_{12}}{z^2_{10}z^2_{20}}\Big)^{1-\gamma}$$
$$\times \Big\{1+\frac{\alpha_s N_c}{2\pi}\Big(\chi(\gamma)\big[\ln\frac{s^2}{16}\tilde{a}b - 4C - \frac{2}{\gamma(1-\gamma)}\big] - \frac{\pi^2}{3}\Big)\Big\} \tag{36}$$

where as usual $\gamma = \frac{1}{2}+i\nu$. Using this equation as an initial condition for the evolution (35) we get the amplitude of scattering of two conformal dipoles with definite conformal spin in the form:

$$\langle \hat{\mathcal{U}}^a_{\rm conf}(\nu,z_0)\hat{\mathcal{V}}^b_{\rm conf}(\nu',z'_0)\rangle \tag{37}$$
$$= \delta(z_0-z'_0)\delta(\nu-\nu')\Big(\frac{ab}{16s^2}\Big)^{\frac{\omega(\nu)}{2}}\Big[1-\frac{\alpha_s N_c}{2\pi}\Big(\chi(\gamma)\big[4C+\frac{2}{\gamma(1-\gamma)}\big]+\frac{\pi^2}{3}\Big)\Big].$$

Finally,

$$(x-y)^4(x'-y')^4\langle T\{\hat{O}(x)\hat{O}(y)\hat{O}(x')\hat{O}(y')\}\rangle \tag{38}$$
$$= \; -\frac{1}{\pi^4}\int d\nu\int d^2z_0\;\frac{1+4\nu^2}{8\pi}\frac{\Gamma^2(\frac{1}{2}-i\nu)}{\Gamma(1-2i\nu)}\Big(\frac{\kappa^2}{4(\kappa\cdot\zeta_0)^2}\Big)^{\frac{1}{2}+i\nu}\Big(\frac{a_0b_0}{16s^2}\Big)^{\frac{1}{2}\omega(\nu)}$$
$$\times\; I_0(\nu)\big[1-\frac{\alpha_s N_c}{4\pi}\Phi_1(\nu)\big]\frac{1+4\nu'^2}{8\pi}\frac{\Gamma^2(\frac{1}{2}+i\nu')}{\Gamma(1+2i\nu')}\Big(\frac{\kappa'^2}{4(\kappa'\cdot\zeta'_0)^2}\Big)^{\frac{1}{2}-i\nu'}$$
$$\times\; I_0(\nu)\big[1-\frac{\alpha_s N_c}{4\pi}\Phi_1(\nu)\big]\Big[1-\frac{\alpha_s N_c}{2\pi}\Big(\chi(\gamma)\big[4C+\frac{2}{\gamma(1-\gamma)}\big]+\frac{\pi^2}{3}\Big)\Big]$$

which coincides with Eq. (5) with

$$F(\nu) \;=\; \frac{N_c^2}{N_c^2-1}\frac{16\pi^4\alpha_s^2}{\cosh^2\pi\nu}\big[1-\frac{\alpha_s N_c}{4\pi}\Phi_1(\nu)\big]\big[1-\frac{\alpha_s N_c}{4\pi}\Phi_1(\nu)\big]$$
$$\times\Big[1-\frac{\alpha_s N_c}{2\pi}\Big(\chi(\gamma)\Big\{4C+\frac{2}{\gamma(1-\gamma)}\Big\}+\frac{\pi^2}{3}\Big)\Big]+O(\alpha_s^2) \tag{39}$$

which gives the pomeron residue in the next-to-leading order.

4. Conclusions

The main result of the paper is that the rapidity factorization and high-energy operator expansion in color dipoles works at the NLO level. There are many examples of the factorization which are fine at the leading order but fail at the NLO level. We believe that the high-energy OPE has the same status as usual light-cone expansion in light-ray operators so one can calculate the amplitudes level by level in perturbation theory. As an outlook we intend to apply the NLO high-energy operator expansion for the description of QCD amplitudes. Although our composite dipole (14) is no longer conformal in QCD, we believe that the that the effects due to the running coupling calculated in Refs. [15,16] can be incorporated in some sort of structure resembling the formula (5) for $\mathcal{N} = 4$ SYM. As an application of the machinery developed here we intend to calculate the photon impact factor for the structure function$F_2(x)$ of deep inelastic scattering which will compete the calculation of small-x structure functions at the NLO level. The study is in progress.

The authors are grateful to L.N. Lipatov and J. Penedones for valuable discussions. This work was supported by contract DE-AC05-06OR23177 under which the Jefferson Science Associates, LLC operate the Thomas Jefferson National Accelerator Facility. G.A.C. work was supported by DOE grant DE-FG02-97ER41028 and by the Jefferson Lab Graduate Fellowship.

References

1. A. V. Kotikov and L. N. Lipatov, Nucl. Phys. B **582**, 19 (2000); Nucl. Phys. B **661**, 19 (2003). Erratum-ibid., **685**, 405 (2004).
2. I. Balitsky, *"High-Energy QCD and Wilson Lines,"* In *Shifman, M. (ed.): At the frontier of particle physics, vol. 2*, p. 1237-1342 (World Scientific, Singapore,2001) [hep-ph/0101042]
3. A. H. Mueller, Nucl. Phys. B **415**, 373 (1994); A. H. Mueller and B. Patel, Nucl. Phys. B **425**, 471 (1994).
4. N. N. Nikolaev and B. G. Zakharov, Phys. Lett. B **332**, 184 (1994); Z. Phys. C **64**, 631 (1994); N. N. Nikolaev B. G. Zakharov, and V. R. Zoller, JETP Letters **59**, 6 (1994).
5. I. Balitsky, Nucl. Phys. B **463**, 99 (1996); *"Operator expansion for diffractive high-energy scattering,"* hep-ph/9706411.
6. I. Balitsky, Phys. Rev. D **60**, 014020 (1999).
7. L. Cornalba, *"Eikonal methods in AdS/CFT: Regge theory and multi-reggeon exchange,"* arXiv:0710.5480 [hep-th].
8. V. S. Fadin, E. A. Kuraev and L. N. Lipatov, Phys. Lett. B **60**, 50 (1975); I. Balitsky and L. N. Lipatov, Sov. Journ. Nucl. Phys. **28**, 822 (1978).
9. R. C. Brower, J. Polchinski, M. J. Strassler and C. I. Tan, JHEP **0712**, 005 (2007) [arXiv:hep-th/0603115].
10. L. Cornalba, M. S. Costa and J. Penedones, JHEP **048**, 0806 (2008).
11. J. Penedones, *"High Energy Scattering in the AdS/CFT Correspondence,"* arXiv:0712.0802 [hep-th].
12. I. Balitsky and G.A. Chirilli, Nucl. Phys. B **822**, 45 (2009).
13. I. Balitsky and G.A. Chirilli, Phys. Rev. D **79**, 031502 (2009).
14. A. Babansky and I. Balitsky, Phys. Rev. D **67**, 054026 (2003).
15. I. Balitsky, Phys. Rev. D **75**, 014001 (2007).
16. Yu. V. Kovchegov and H. Weigert, Nucl. Phys. A **784**, 188 (2007),

SECTION 6.

HOLOGRAPHIC QCD

How Well Does AdS/QCD Describe QCD?

J. Erlich

Physics Department, College of William and Mary
Williamsburg, VA, 23185, USA
E-mail: erlich@physics.wm.edu

AdS/QCD is an extra-dimensional approach to modeling hadronic physics, motivated by the AdS/CFT correspondence in string theory. AdS/QCD models are often more accurate than would have been expected at energies below a few GeV. We address the question of why these models are so successful, and respond to some of the criticisms that have been waged against these models.

Keywords: AdS/QCD, holographic QCD

1. What is AdS/QCD?

The AdS/CFT correspondence[1] provides a powerful tool for extra-dimensional model-building. Qualitative features of electroweak symmetry breaking models with warped extra dimensions can often be predicted by analogy with the AdS/CFT correspondence.[2] Even quantitatively, simple extra-dimensional models of QCD motivated by the AdS/CFT correspondence have proven successful at reproducing low-energy hadronic data like meson masses, decay constants, and coefficients of the chiral Lagrangian.[3–5] These models fall into two classes: top-down models based on brane constructions in string theory, and bottom-up models which are more phenomenological. Both top-down and bottom-up models in this framework are referred to as AdS/QCD models or holographic QCD.

AdS/QCD models are related to a number of earlier ideas which are useful for understanding properties of QCD at low energies: chiral symmetry breaking, hidden local symmetry,[6] vector meson dominance,[7] large N,[8] the Weinberg and Shifman-Vainshtein-Zakharov (SVZ) sum rules,[9,10] Skyrmions,[11] and matching of the low energy and high energy regimes.[12] Aside from the pattern of chiral symmetry breaking, which is an input of some AdS/QCD models, aspects of all of the features listed above are a natural consequence of the extra-dimensional nature of these models. Hidden local symmetry is related to the gauging of the global symmetry as suggested by the AdS/CFT correspondence [13]; large N is required by the classical limit of the AdS/CFT correspondence [1] and is reflected in poles in correlation functions at real values of squared momenta; dominance of the lightest rho meson in hadronic couplings is due to oscillations of the heavier Kaluza-Klein modes in the extra dimensions [13,14]; sum rules become exact relations for sums over Kaluza-Klein modes [4,5,15,16]; solitonic solutions describing baryons in these models are closely related to the Skyrme model [17–19]; resonance masses and decay constants conspire to reproduce the high energy behavior of correlators [20,21].

The AdS/CFT correspondence relates certain strongly coupled large-N gauge theories to weakly coupled theories containing gravity.[1] The prototypical example is $\mathcal{N}=4$ supersymmetric SU(N) gauge theory in the limit $N \to \infty$ and 't Hooft coupling $g^2N \gg 1$, which is identified with Type IIB supergravity in an $AdS_5 \times S^5$ spacetime background with nonvanishing 5-form flux. The five-sphere S^5 plays an important role in matching supergravity fields to operators in the $\mathcal{N}=4$ gauge theory, but will not play an important role in our discussion of QCD-like models. Anti-de Sitter space in 4+1 dimensions, AdS_5, is described by the metric

$$ds^2 = \frac{R^2}{z^2}\left(\eta_{\mu\nu}\, dx^\mu\, dx^\nu - dz^2\right), \tag{1}$$

where $\mu, \nu \in (0, 1, 2, 3)$ and $\eta_{\mu\nu}$ is the Minkowski tensor in 3+1 dimensions (4D). The coordinate z is sometimes referred to as the radial coordinate. The region $z > 0$ covers half (a Poincaré patch) of the full AdS spacetime, which is enough for our purposes. Although Anti-de Sitter space is a maximally symmetric spacetime, the radial coordinate plays a distinguished role in the metric Eq. (1). An interesting relationship between the radial coordinate and light-front dynamics has been observed by Brodsky and De Teramond (see, for example, Refs. [22,23]).

1.1. *The top-down approach*

In top-down AdS/QCD models, a brane construction in string theory is engineered which at low energies describes a gauge theory with features similar to QCD.[17,24–27] The top-down model most similar to QCD is the Sakai-Sugimoto model,[17,28] in which N D4-branes in Type IIA string theory are wrapped on a circle, and are intersected by N_f D8-branes and N_f $\overline{\text{D8}}$-branes, as in Fig. 1. Open strings with both ends on the D4-branes describe a SU(N) gauge theory with sixteen supercharges. However, the

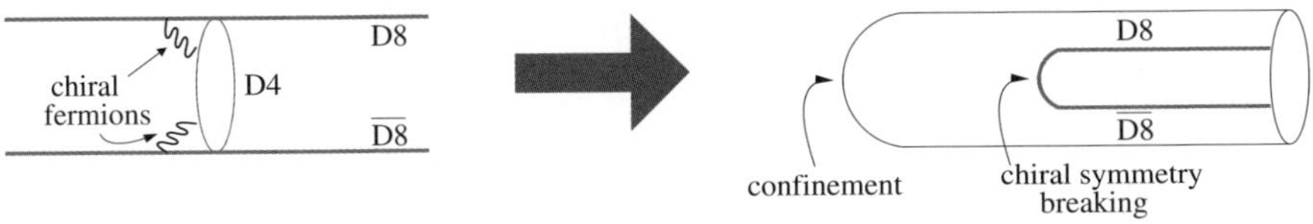

Fig. 1. The Sakai-Sugimoto model.

supersymmetry is broken if fermions are given antiperiodic boundary conditions around the circle, and at low energies the system is described by a pure SU(N) gauge theory with fermions and scalar fields whose masses are determined by the size of the circle.[29] (Unfortunately, the mass gap is comparable to the masses of the extraneous fermions and scalars in this model.) The D8 and $\overline{\text{D8}}$-branes also break the supersymmetry. A string with one end on a D4-brane and one end on a D8-brane ($\overline{\text{D8}}$-brane) describes a left-handed (right-handed) fermion in the effective 4D theory.

The separation of left and right-handed fermions makes chiral symmetry manifest in this theory, and makes the Sakai-Sugimoto model the AdS/CFT model of choice for describing QCD. The gravitational backreaction of the D4-branes generates a horizon, which is related to confinement as the horizon creates a mass gap in gravitational fluctuations which are interpreted as glueballs.[29,30] Chiral symmetry breaking in the Sakai-Sugimoto model is due to the fact that the D8-branes and $\overline{\text{D8}}$-branes are joined together as a result of the curved spacetime geometry, as in Fig. 1. In this model the location at which the D8 and $\overline{\text{D8}}$-branes meet is an adjustable parameter and controls the ratio of confining to chiral symmetry breaking scales.[31–33]

1.2. *The bottom-up approach*

In bottom-up AdS/QCD models, we specify an extra-dimensional spacetime geometry and the fields that propagate in them based on the properties of QCD which we would like to be incorporated. Towers of Kaluza-Klein modes are identified with towers of radial excitations of QCD states. Boundary conditions on gauge fields break the higher-dimensional gauge invariance, while the corresponding global symmetry remains in the effective 3+1 dimensional theory. For example, an SU(2)×SU(2) gauge theory in 4+1 dimensions (5D) can represent the approximate chiral symmetry of QCD. A set of scalar fields transforming in the fundamental representation of both SU(2) gauge groups can spontaneously break the symmetry to a diagonal subgroup which may be identified with isospin. This is the basic scenario in some basic bottom-up models.[4,5] Rather than use scalar fields to break the chiral symmetry, modified spacetime geometry and boundary conditions can do the same.[15,34] A more general approach includes the scalar fields with modified boundary conditions.[35]

For simplicity the spacetime geometry in bottom-up models is often chosen to be a slice of AdS_5 with metric given by Eq. (1) between $z = \epsilon$ and $z = z_m$, where $\epsilon/z_m \ll 1$. Such models are called hard-wall models because of the sharp boundary at $z = z_m$.[4,5,36–38] The AdS_5 spacetime has an SO(2,4) isometry which is broken only by the boundaries of the spacetime. The consequence is that the effective 3+1 dimensional theory has an SO(2,4) symmetry at high energies which is identified with the conformal invariance of QCD at high energies. The conformal symmetry ensures that correlation functions of operators as derived in these models will have the form required by asymptotic freedom. However, other aspects of these models are not expected to be valid at energies above a few GeV where stringy effects become important in QCD, so the matching of certain predictions of AdS/QCD models to the UV is somewhat *ad hoc*.

In summary, a basic bottom-up model is described by an SU(2)×SU(2) gauge theory in a slice of AdS_5 with bifundamental scalar field X and action

$$\begin{aligned} S = \int d^4x \int_\epsilon^{z_m} dz \sqrt{|g|} \left[-1/(2g_5^2) \,\mathrm{Tr}\left(L_{MN}L^{MN} + R_{MN}R^{MN} \right) \right. \\ \left. + \mathrm{Tr}\left(|D_M X|^2 - m_X^2 |X|^2 \right) \right], \end{aligned} \tag{2}$$

where Lorentz indices M and N run from 0 to 4, and L_{MN} and R_{MN} are field strengths of the two sets of SU(2) gauge fields. The fields X_{ij}, $i, j \in (1, 2)$, have the quantum numbers of the scalar quark bilinear $\overline{q}_{Li} q_{Rj}$, where the subscripts L and R refer to left and right-handed chirality, respectively, and q_1 and q_2 are the up and down quark fields, respectively. According to the AdS/CFT correspondence we would identify the mass m_X of the field dual to an operator of scaling dimension $\Delta = 3$, via[39,40]

$$m_X^2 R^2 = \Delta(\Delta - 4) = -3\,, \tag{3}$$

where R is the AdS curvature scale of Eq. (1). We may make this choice for definiteness, although the AdS/CFT correspondence is not strictly valid for this model and we would be ignoring renormalization effects which modify operator scaling dimensions away from the high energy regime.

The solutions to the classical equations of motion with vanishing gauge fields and X field of the form $X_0(z)$ are

$$X_0(z) = m_q\, z^{4-\Delta} + \frac{\sigma}{4\Delta - 8}\, z^{\Delta}\,. \tag{4}$$

For $2 < \Delta < 4$, the term proportional to the parameter m_q is non-normalizable in the sense that the contribution to the action from the integral over z diverges if the UV cutoff length $\epsilon \to 0$. On the other hand, the term proportional to σ is normalizable. We will allow both normalizable and non-normalizable backgrounds for the field X. The factor of $1/(4\Delta - 8)$ in Eq. (4) is motivated by the AdS/CFT identification of σ as the (isospin-preserving) chiral condensate $\langle \overline{u}_L u_R \rangle = \langle \overline{d}_L d_R \rangle$ [41].

The predictions of the model are insensitive to ϵ as long as $\epsilon/z_m \ll 1$. Hence, the free parameters in the model are m_X, g_5, m_q, σ, and z_{IR}. If we fix $m_X^2 = -3/R^2$ then there are four parameters in addition to the choice of boundary conditions on the gauge fields. We can also fix g_5 by extending AdS/QCD predictions to the UV and matching with perturbative QCD, leaving only three model parameters.[4,5] However, this model is especially simplistic. For example, there could be a nontrivial potential for X, as in J. Kapusta's talk,[42] and the geometry does not need to be Anti-de Sitter space.[34,43]

Meson masses and decay constants are determined by the effective 3+1 dimensional theory. The Kaluza-Klein modes of the gauge fields have the quantum numbers of vector and axial-vector mesons. The masses of the Kaluza-Klein modes are determined by the eigenvalues of the equations of motion together with the prescribed boundary conditions. The decay constants are determined by computing the mixing between the Kaluza-Klein modes and the zero modes for the gauge fields.[17,28] Alternatively, the AdS/CFT correspondence can be used to calculate the correlation function of products of currents, and the the decay constants are the residues of poles at the location of the Kaluza-Klein masses.[4,5] Some of the early predictions of this model, with particular values of the parameters, appear in Table 1. Some of the coefficients of the chiral Lagrangian are related to these observables, and were calculated in the same model with a different choice of parameters in Ref. [5].

Numerous additional results in AdS/QCD models have been calculated. In a hard-wall AdS/QCD model including a strange quark mass parameter, predictions for masses and decay constants of light mesons containing strange quarks mostly agree with experimental central values at the few percent level.[45] The q^2 dependence of form factors and moments of generalized parton distributions in various AdS/QCD models are in relatively good agreement with experiment up to a few GeV (see, for example, Refs. [46–48]). From the behavior of form factors at small q^2, these models predict that mesons have a larger size as determined by their charge

Table 1. Predictions of a hard-wall model. For more details see Refs. [4,44].

Observable	Measured (MeV, central values)	Model (MeV)
m_π	139.6	141
m_ρ	775.8	832
m_{a_1}	1230	1220
f_π	92.4	84.0
$F_\rho^{1/2}$	345	353
$F_{a_1}^{1/2}$	433	440
$g_{\rho\pi\pi}$	6.03	5.29
m_{f_2}	1275	1240

distribution than by their momentum distribution.[48]

2. How well does AdS/QCD work?

The summary of our discussion thus far is that AdS/QCD models seem to be reasonably reliable for predicting observables at below a few GeV, but tend to make poor predictions at higher energies. Until we better understand why certain models work especially well, we should at best trust only those predictions which are independent of model details such as the choice of spacetime geometry or boundary conditions.

According to the AdS/CFT correspondence, finite temperature physics can be studied by introducing a black hole into the higher-dimensional spacetime. From general properties of black hole horizons, a universal prediction of the ratio of shear viscosity η to entropy density s was discovered by Kovtun, Son and Starinets [49]: $\eta/s = 1/(4\pi)$. Other universal relations have subsequently been found, such as a relation between electrical conductivity to charge susceptibility.[50] The zero-temperature observables quoted earlier are not universal, but we can test their dependence on model parameters. In the hard-wall model defined in the previous section, allowing the effective dimension Δ of the operator $\overline{q}q$ to vary from its UV value $\Delta_{UV} = 3$ is tantamount to varying the mass of the field X via Eq. (3). The mass varies from the Breitenlohner-Freedman bound [51] $m_X^2 = -4/R^2$ for $\Delta = 2$, to $m_X^2 = 0$ at $\Delta = 4$. The dependence of some AdS/QCD predictions as a function of Δ over part of this range is shown in Fig. 2, holding fixed m_π, f_π and m_ρ [35]. Note the insensitivity of hard-wall AdS/QCD predictions to the parameter Δ. However, there is more sensitivity if the boundary conditions on the gauge fields in the IR are modified.[35] If the geometry is allowed to vary in a certain class of AdS/QCD models, it seems that the

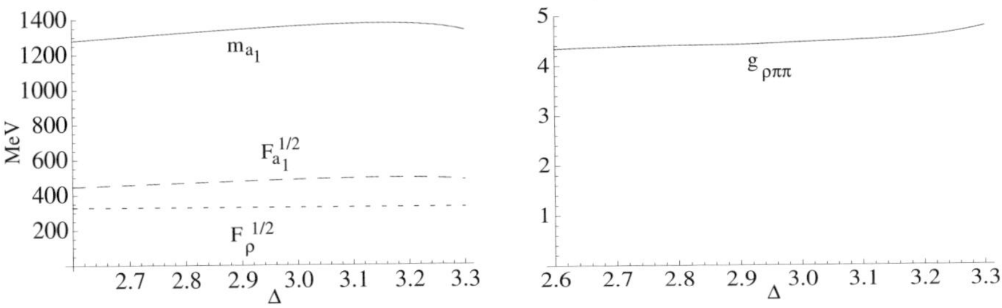

Fig. 2. AdS/QCD predictions are not sensitive to varying $\bar{q}q$ scaling dimension Δ around its UV value $\Delta_{UV} = 3$, from Ref. [35].

AdS geometry provides an especially good fit to data, but certain predictions are relatively insensitive to the details of the choice of geometry.[43] The hard-wall model has a root-mean-squared error with seven observables and three parameters (as defined in Ref. [4]) of under 10%. In general, AdS/QCD models are consistent with one another at around the 20-25% level, and I am hopeful that we will better understand the unreasonable success of specific models.

3. Challenges for AdS/QCD

I would like to address some of the criticisms of the AdS/QCD approach that have been raised, both in the literature and in private (and not-so-private) communication. I contend that most of these objections are red herrings if the scope of AdS/QCD models is properly understood, but AdS/QCD still faces some phenomenological challenges.

- *AdS/QCD is not AdS/CFT.*

The $\mathcal{N}$=4 Yang-Mills theory which serves as the prototypical example of the AdS/CFT correspondence is supersymmetric, conformal, and has a large number of colors and large 't Hooft coupling for calculability. This is all quite different from QCD, so why should we rely on the AdS/CFT correspondence to model QCD?

First of all, it is important to note that the existence of gravity duals is not tied to conformal invariance or supersymmetry. The Sakai-Sugimoto model is an example without either feature. The large-N limit is required for a classical description with decoupled stringy physics, and the narrow resonances predicted by large N are indeed features of classical AdS/QCD calculations. It would be interesting to pursue $1/N$ corrections from quantum corrections in the 5D models, and as far as I know little work has been done in this regard (although there are close analogies to 5D models of electroweak symmetry breaking for which quantum corrections have been studied). For example, in AdS/QCD models the $\rho\pi\pi$ coupling has been calculated and compared with the decay rate of the rho meson. Such couplings can be used to self-consistently modify corresponding spectral functions.

The analogy with the AdS/CFT correspondence is useful for inferring qualitative features of AdS/QCD models, but these models stand on their own as 5D models independent of the AdS/CFT correspondence.

- *The spectrum is not Regge-like.*

Misha Shifman noted that hard-wall AdS/QCD models have spectra which depend on radial excitation number n like $m_n^2 \sim n^2$ as opposed to the Regge-like behavior $m_n^2 \sim n$ [20]. The soft-wall model [52] obtains the correct Regge-like spectrum by introducing a dilaton background which modifies the equations of motion appropriately. At low energies the predictions of the soft-wall and hard-wall models are comparable.

- *The model is matched to the UV, where QCD is weakly coupled.*

In three places the hard-wall model described above is matched to the UV: in the choice of AdS_5 geometry, in specifying the 5D gauge coupling g_5 by comparison with UV current correlators, and in assigning the squared mass $m_X^2 = -3/R^2$ to the scalar field by analogy with the AdS/CFT correspondence and the UV dimension of the operator $\overline{q}q$. Matching to the UV was done for definiteness, but one can allow parameters to vary, as in Ref. [35]. The conclusion is that matching to experiment prefers g_5 fixed to around the value obtained by matching to the UV, while there is less sensitivity to variation of m_X.

On a related note, Misha Shifman has pointed out that AdS/QCD predictions for correlators of isospin currents agree with Migdal's Padé approximation for the UV behavior of current correlators.[20] The interesting thing about this approach is that one can systematically observe the radial direction of Anti-de Sitter space open up as the number of poles included in the Padé approximation increases.[21]

- *Identification of QCD parameters depends on N.*

The pattern of chiral symmetry breaking is built into the AdS/QCD models we have been discussing. If the AdS/CFT correspondence were strictly valid for this model, then m_q and σ of Eq. (4) would play the role of the source and expectation value of the operator $\overline{q}_L q_R$ dual to the scalar field X, up to an N-dependent rescaling.[53] The N dependence is not included in Eq. (4), so we should allow the possibility of rescaling m_q and σ appropriately. On the other hand, if we treat m_q and σ simply as model parameters, then the interpretation of those parameters as quark mass and chiral condensate is understood to be imprecise from the outset.

- *AdS/QCD has incorrect high energy behavior.*

Even though Anti-de Sitter space guarantees correct conformal behavior at high energies and the soft-wall model produces Regge-like spectra, there are many things AdS/QCD continues to get wrong at high energies. The density of states does not agree with the spectrum expected from string theory, and high-energy scattering amplitudes are not correctly predicted.[54–56]

We are forced at the moment to understand these models as low-energy descriptions valid only below some cutoff scale which may differ between models but is generally a few GeV. It is still valuable to try to match certain aspects of these models to the UV, but at the same time we have to recognize the limitations of these models at high energies. Perhaps we can reconcile AdS/QCD predictions with high-energy scattering by including progressively more 5D fields into the model as the energy scale of interest increases, although so doing would introduce more model parameters and a decrease in predictivity.

- *Important operators are not accounted for.*

The glueball condensate is important for certain observables but is not included in some of these models. Simple AdS/QCD models like the hard-wall and soft-wall models also do not include fields which would be dual to the operators $\overline{q}[\gamma^i, \gamma^0]T^a q$, which have significant overlap between vacuum and one-rho meson states.[57] The basic response to such objections is that there is a natural way to include fields into the 5D theory which would describe corresponding operators and condensates in the effective 4D theory. The glueball condensate is related to the dilaton profile. New 5D fields can be included with quantum numbers conjugate to the operators in question, and one should study the effects of those fields. Any 5D field charged under the gauged isospin symmetry will lead to new couplings involving the rho meson which can be compared with experiment. At higher energies, more 5D fields are expected to be required in order to describe higher-dimension 4D operators that become important at those scales.

- *AdS/QCD is an uncontrolled expansion.*

If we understand AdS/QCD as a low-energy effective theory, then we should expect quantum corrections to generate higher-dimension operators in the 5D action suppressed by some scale. In AdS/QCD the relevant scale is the confining scale (or the chiral symmetry breaking scale, which is comparable). However, we are interested in physics above that scale. This is an honest difficulty for AdS/QCD models, and deserves further exploration.

Higher-dimension operators are generated by integrating out fields above the scale of interest. In the large-N limit with large 't Hooft coupling, high-spin states decouple from the theory.[a] Perhaps the effectiveness of the large-N expansion in this limit is also related to the effectiveness of the AdS/QCD approach at low energies, summing only over "supergravity states" with spin less than or equal to two.

4. Conclusions

In summary, AdS/QCD models are extra-dimensional models of QCD resonances, and are generally accurate below a few GeV at the 10-25% level depending on the details of the model. The benefit of an extra-dimensional approach is that several features of QCD are immediate consequences of extra dimensions. QCD sum rules, vector meson dominance, and hidden local symmetry are all natural features of extra-dimensional models. A number of criticisms have been raised which challenge the AdS/QCD approach, based on the comparison of AdS/QCD models with the AdS/CFT correspondence in string theory and based on flawed phenomenology of these models, especially at high energies. Most of these objections disappear if we understand AdS/QCD as a class of effective low-energy models independent of the AdS/CFT correspondence.

It is an honor and a pleasure to have been part of this wonderful conference celebrating the vast achievements of Misha Shifman, who shows no sign of slowing down. Misha has played an important role in identifying challenges for AdS/QCD and has influenced the development of this field over the past few years. Thanks, and happy birthday Misha.

References

1. J. M. Maldacena, Adv. Theor. Math. Phys. **2**, 231 (1998) [Int. J. Theor. Phys. **38**, 1113 (1999)] [arXiv:hep-th/9711200].
2. N. Arkani-Hamed, M. Porrati and L. Randall, JHEP **0108**, 017 (2001) [arXiv:hep-th/0012148].

[a] I thank Kaustubh Agashe for discussion on this point.

3. G. F. de Teramond and S. J. Brodsky, Phys. Rev. Lett. **94**, 201601 (2005) [arXiv:hep-th/0501022].
4. J. Erlich, E. Katz, D. T. Son and M. A. Stephanov, Phys. Rev. Lett. **95**, 261602 (2005) [arXiv:hep-ph/0501128].
5. L. Da Rold and A. Pomarol, Nucl. Phys. B **721**, 79 (2005) [arXiv:hep-ph/0501218].
6. M. Bando, T. Kugo, S. Uehara, K. Yamawaki and T. Yanagida, Phys. Rev. Lett. **54**, 1215 (1985).
7. J. J. Sakurai, Phys. Rev. Lett. **22**, 981 (1969).
8. G. 't Hooft, Nucl. Phys. B **72**, 461 (1974).
9. S. Weinberg, Phys. Rev. Lett. **18**, 507 (1967).
10. M. A. Shifman, A. I. Vainshtein and V. I. Zakharov, Nucl. Phys. B **147**, 385 (1979).
11. T. H. R. Skyrme, Nucl. Phys. **31**, 556 (1962).
12. A. A. Migdal, Annals Phys. **109**, 365 (1977).
13. D. T. Son and M. A. Stephanov, Phys. Rev. D **69**, 065020 (2004) [arXiv:hep-ph/0304182].
14. S. Hong, S. Yoon and M. J. Strassler, arXiv:hep-ph/0501197.
15. J. Hirn and V. Sanz, JHEP **0512**, 030 (2005) [arXiv:hep-ph/0507049].
16. J. Hirn and V. Sanz, Phys. Rev. D **76**, 044022 (2007) [arXiv:hep-ph/0702005].
17. T. Sakai and S. Sugimoto, Prog. Theor. Phys. **113**, 843 (2005) [arXiv:hep-th/0412141].
18. K. Nawa, H. Suganuma and T. Kojo, Phys. Rev. D **75**, 086003 (2007) [arXiv:hep-th/0612187].
19. A. Pomarol and A. Wulzer, arXiv:0904.2272 [hep-ph].
20. M. Shifman, arXiv:hep-ph/0507246.
21. J. Erlich, G. D. Kribs and I. Low, Phys. Rev. D **73**, 096001 (2006) [arXiv:hep-th/0602110].
22. S. J. Brodsky and G. F. de Teramond, Phys. Lett. B **582**, 211 (2004) [arXiv:hep-th/0310227].
23. G. F. de Teramond and S. J. Brodsky, Phys. Rev. Lett. **102**, 081601 (2009) [arXiv:0809.4899 [hep-ph]].
24. I. R. Klebanov and M. J. Strassler, JHEP **0008**, 052 (2000) [arXiv:hep-th/0007191].
25. M. Kruczenski, D. Mateos, R. C. Myers and D. J. Winters, JHEP **0405**, 041 (2004) [arXiv:hep-th/0311270].
26. J. Babington, J. Erdmenger, N. J. Evans, Z. Guralnik and I. Kirsch, Phys. Rev. D **69**, 066007 (2004) [arXiv:hep-th/0306018].
27. E. Antonyan, J. A. Harvey, S. Jensen and D. Kutasov, arXiv:hep-th/0604017.
28. T. Sakai and S. Sugimoto, Prog. Theor. Phys. **114**, 1083 (2005) [arXiv:hep-th/0507073].
29. E. Witten, Adv. Theor. Math. Phys. **2**, 505 (1998) [arXiv:hep-th/9803131].
30. C. Csaki, H. Ooguri, Y. Oz and J. Terning, JHEP **9901**, 017 (1999) [arXiv:hep-th/9806021].
31. E. Antonyan, J. A. Harvey and D. Kutasov, Nucl. Phys. B **784**, 1 (2007) [arXiv:hep-th/0608177].

32. O. Aharony, J. Sonnenschein and S. Yankielowicz, Annals Phys. **322**, 1420 (2007) [arXiv:hep-th/0604161].
33. C. D. Carone, J. Erlich and M. Sher, Phys. Rev. D **76**, 015015 (2007) [arXiv:0704.3084 [hep-th]].
34. J. Hirn, N. Rius and V. Sanz, Phys. Rev. D **73**, 085005 (2006) [arXiv:hep-ph/0512240].
35. J. Erlich and C. Westenberger, Phys. Rev. D **79**, 066014 (2009) [arXiv:0812.5105 [hep-ph]].
36. J. Polchinski and M. J. Strassler, arXiv:hep-th/0003136.
37. A. Karch, E. Katz and N. Weiner, Phys. Rev. Lett. **90**, 091601 (2003) [arXiv:hep-th/0211107].
38. S. Hong, S. Yoon and M. J. Strassler, JHEP **0604**, 003 (2006) [arXiv:hep-th/0409118].
39. E. Witten, Adv. Theor. Math. Phys. **2**, 253 (1998) [arXiv:hep-th/9802150].
40. S. S. Gubser, I. R. Klebanov and A. M. Polyakov, Phys. Lett. B **428**, 105 (1998) [arXiv:hep-th/9802109].
41. I. R. Klebanov and E. Witten, Nucl. Phys. B **556**, 89 (1999) [arXiv:hep-th/9905104].
42. T. Gherghetta, J. I. Kapusta and T. M. Kelley, Phys. Rev. D **79**, 076003 (2009) [arXiv:0902.1998 [hep-ph]].
43. D. Becciolini, M. Redi and A. Wulzer, arXiv:0906.4562 [hep-ph].
44. E. Katz, A. Lewandowski and M. D. Schwartz, Phys. Rev. D **74**, 086004 (2006) [arXiv:hep-ph/0510388].
45. E. Katz, plenary talk at Lattice 2008 conference. Talk posted at http://conferences.jlab.org/lattice2008/plenary.html
46. H. J. Kwee and R. F. Lebed, Phys. Rev. D **77**, 115007 (2008) [arXiv:0712.1811 [hep-ph]].
47. H. R. Grigoryan and A. V. Radyushkin, Phys. Lett. B **650**, 421 (2007) [arXiv:hep-ph/0703069].
48. Z. Abidin and C. E. Carlson, Phys. Rev. D **77**, 095007 (2008) [arXiv:0801.3839 [hep-ph]].
49. P. Kovtun, D. T. Son and A. O. Starinets, Phys. Rev. Lett. **94**, 111601 (2005) [arXiv:hep-th/0405231].
50. P. Kovtun and A. Ritz, Phys. Rev. D **78**, 066009 (2008) [arXiv:0806.0110 [hep-th]].
51. P. Breitenlohner and D. Z. Freedman, Annals Phys. **144**, 249 (1982).
52. A. Karch, E. Katz, D. T. Son and M. A. Stephanov, Phys. Rev. D **74**, 015005 (2006) [arXiv:hep-ph/0602229].
53. A. Cherman, T. D. Cohen and E. S. Werbos, arXiv:0804.1096 [hep-ph].
54. D. M. Hofman and J. Maldacena, JHEP **0805**, 012 (2008) [arXiv:0803.1467 [hep-th]].
55. C. Csaki, M. Reece and J. Terning, JHEP **0905**, 067 (2009) [arXiv:0811.3001 [hep-ph]].
56. B. Pire, C. Roiesnel, L. Szymanowski and S. Wallon, Phys. Lett. B **670**, 84 (2008) [arXiv:0805.4346 [hep-ph]].
57. L. Y. Glozman, arXiv:0903.3923 [hep-ph].

Conformality Lost

D. B. Kaplan, J.-W. Lee and D. T. Son*

Institute for Nuclear Theory, University of Washington,
Seattle, WA 98195-1550, USA
**E-mail: son@phys.washington.edu*

M. A. Stephanov

Department of Physics, University of Illinois,
Chicago, Illinois 60607-7059, USA

We consider zero-temperature transitions from conformal to non-conformal phases in quantum theories. We argue that there are three generic mechanisms for the loss of conformality: (i) fixed point goes to zero coupling, (ii) fixed point runs off to infinite coupling, or (iii) an IR fixed point annihilates with a UV fixed point and they both disappear into the complex plane. We give examples of the last case and show that the critical behavior of the mass gap is similar to that of the inverse correlation length in the finite temperature Berezinskii-Kosterlitz-Thouless (BKT) phase transition, $\xi \sim \exp(c/|T - T_c|^{1/2})$. We speculate that the chiral phase transition in QCD at large number of fermion flavors belongs to this universality class, and attempt to identify the UV fixed point that annihilates with the Banks-Zaks fixed point at the lower end of the conformal window.

Keywords: Conformal window; phase transitions

1. Introduction

The renormalization group (RG) underlies our understanding of second-order phase transitions, with critical points being identified with fixed points of the appropriate RG equation.[1] Near the phase transition the characteristic energy or momentum scale m (the inverse correlation length) goes to zero as $m \sim |\alpha - \alpha_*|^{\nu}$, where α is a parameter that can vary continuously, and $\alpha = \alpha_*$ is the location of the critical point.

In this paper, we argue that there is wide class of phase transitions in which the correlation length behaves very differently, vanishing exponentially on one side of the phase transition, while being strictly zero on the other side

$$m \sim \Lambda_{\rm UV}\, \theta(\alpha_* - \alpha) \exp\left(-\frac{c}{\sqrt{\alpha_* - \alpha}}\right), \qquad c > 0 \ . \tag{1}$$

This peculiar behavior has been observed before in the Berezinskii-Kosterlitz-Thouless (BKT) phase transition;[2] therefore we will refer to eq. (1) as "BKT scaling." The BKT transition is a classical phase transition in two dimensions that can be described in terms of vortex condensation. While this transition is peculiar to two dimensions, we will show that the mechanism underlying BKT scaling from an RG point of view is far more general, and is one of three generic behaviors that can occur when a system in any dimension makes a transition from a conformal to a non-conformal phase. In particular, as we will show, it follows when an IR fixed point of the system merges with a UV fixed point. In this language it is easy to see why BKT scaling can be found in a wide variety of systems.

The basic mechanism can be illustrated with a simple model with a dimensionless coupling g depending on an external parameter α, for which the β-function takes the form (Fig. 1(a))

$$\beta(g;\alpha) = \frac{\partial g}{\partial t} = (\alpha - \alpha_*) - (g - g_*)^2 \ , \tag{2}$$

where $t = \ln\mu$, μ being the renormalization scale. For $(\alpha - \alpha_*) > 0$, the fixed points for this system (zeros of β) are given by

$$g_\pm = g_* \pm \sqrt{\alpha - \alpha_*} \ , \tag{3}$$

where g_-, g_+ correspond to IR and UV fixed points respectively, each describing a conformal phase of the theory. As α decreases, these two fixed points approach each other until they merge at $g_\pm = g_*$ for $\alpha = \alpha_*$. For $\alpha < \alpha_*$ the solutions to $\beta = 0$ are complex, and the theory no longer has a conformal phase.

To see that fixed point merger generically gives rise to BKT scaling, consider the case where α is slightly below α_*, and that at a UV scale $\Lambda_{\rm UV}$ the coupling takes an initial value $g_{\rm UV} < g_*$. On scaling to the IR, the coupling then flows to larger values, lingering near $g = g_*$ where the β-function is small, and then blowing up quickly, defining an intrinsic IR scale $\Lambda_{\rm IR}$, which is insensitive to the initial value $g_{\rm UV}$. This behavior is displayed in Fig. 1(b). The scale $\Lambda_{\rm IR}$ will characterize the longest correlation lengths in this theory, and can be computed by integrating eq. (2):

$$\frac{\Lambda_{\rm IR}}{\Lambda_{\rm UV}} = \exp\left[t_{\rm IR} - t_{\rm UV}\right] = \exp\left[\int_{g_{\rm UV}}^{g_{\rm IR}} \frac{dg}{\beta(g;\alpha)}\right] \simeq e^{-\pi/\sqrt{(\alpha_* - \alpha)}} \ , \tag{4}$$

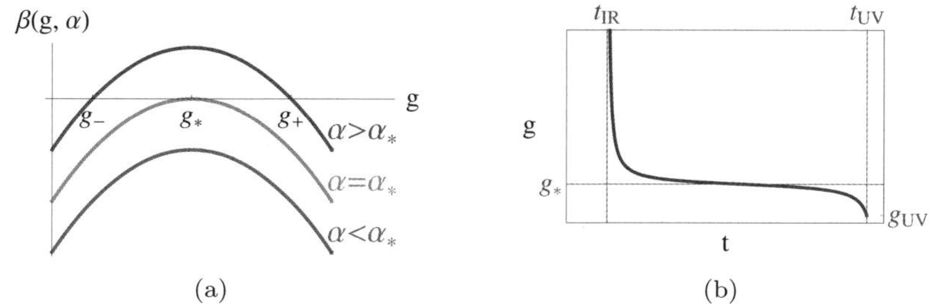

Fig. 1. (a) A toy β-function. For $\alpha > \alpha_*$ there are fixed points at $g_\pm$ which are UV- and IR-stable respectively; these fixed points merge at g_* for $\alpha = \alpha_*$, and disappear for $\alpha < \alpha_*$; (b) The RG flow of the coupling g as a function of $t = \ln \mu$ in the non-conformal phase, with $(t_{\rm UV} - t_{\rm IR}) \propto 1/\sqrt{\alpha_* - \alpha}$.

where we have assumed $|g_{\rm IR,UV} - g_*| \gg |\alpha - \alpha_*|$.

How general is this mechanism of fixed point annihilation? Suppose a system has a nontrivial IR fixed point at $g = \bar{g}(\alpha)$ whose location depends continuously on a parameter α, and that at a critical value $\alpha = \alpha_*$ there is a phase transition where conformality is lost. At this phase transition, the β-function must somehow lose a zero. This can come about in three ways:

(A) $\bar{g}$ can decrease until it merges with the trivial fixed point at $g = 0$, giving rise to a trivial, asymptotically unfree theory;
(B) $\bar{g}$ can run off to infinite coupling and disappear;
(C) $\bar{g}$ can merge with a UV fixed point, as in our toy model, giving rise to BKT scaling.

Examples of scenarios (A) and (B) are afforded by supersymmetric QCD (SQCD). At large number of colors N_c, the parameter $x \equiv N_f/N_c$ may be treated as continuous, where N_f is the number of quark flavors. It has been shown by Seiberg [3,4] that SQCD is conformal in the window $3/2 \le x \le 3$. For x just below 3, the theory has a Banks-Zaks fixed point at weak coupling [5]; approaching $x = 3$ from below, this fixed point merges with the trivial fixed point at $g = 0$, and for $x > 3$ the theory is in the asymptotically unfree "free electric phase." This is an example of mechanism "A" above. In contrast, at the lower end of the conformal window at $x = 3/2$, SQCD goes from a strongly coupled conformal theory when $x \gtrsim 3/2$ to a "free magnetic phase" when $x \lesssim 3/2$. In the free magnetic phase, the Coulomb force between charges takes the form $e^2 \ln(\Lambda r)/r^2$ where Λ is associated with the Landau pole of the dual magnetic theory. The log behavior of the coupling can be explained by a β-function which is negative and approaches zero as $\beta \sim -1/g$ for large g. Thus it appears that conformality in the electric description is lost via mechanism (B).

In this paper we analyze the quantum mechanical example of a $1/r^2$ potential in d dimensions, which can be solved nonperturbatively and which exhibits the phenomenon of fixed point merger. This analysis has many parallels in the AdS/CFT correspondence,[6–8] and that loss of conformality via fixed point merger is analogous (if not holographically dual) to the instability of AdS space at the Breitenlohner-Freedman (BF) bound.[9]

One of the motivations for this paper is to understand the chiral phase transition that happens in (nonsupersymmetric) large-N_c QCD when the number of flavors of massless fermions N_f varies, and so to answer a question that Misha Shifman and collaborators posed in hep-th/0403071. As with SQCD, we know there exists a conformal window for QCD in the parameter $x = N_f/N_c$ where the upper end occurs at $x_* = 11/2$, near which the Banks-Zaks calculation is perturbative and reliable. For decreasing x conformality must eventually be lost, since for small x chiral symmetry breaking is expected. We speculate that the phase transition at this lower boundary of the conformal window occurs due to fixed point merger. If this picture is correct, then near the transition the chiral condensate must exhibit BKT scaling. Incidentally, this exponential behavior is also typically found when one solves the gap equation obtained from (an unsystematic) truncation of the Schwinger-Dyson hierarchy (see, e.g., Ref. [10–14]). A priori, the relationship between the RG picture of merging fixed points and the gap equation is not obvious; however our analysis of a relativistic defect fermion model yields the same result in the regime where both approaches are reliable.

If QCD does indeed exhibit BKT scaling, then our arguments suggest that within the conformal window there exists another theory, QCD*, which is defined at the UV fixed point. We conclude with speculations about this theory.

2. A nonrelativistic example: quantum mechanics in $1/r^2$ potential

It is well known that the solutions for a quantum particle in a potential

$$V(r) = \alpha/r^2 \tag{5}$$

possess conformal symmetry when the potential is repulsive or weakly attractive ($\alpha > \alpha_*$), but that for sufficiently attractive potential ($\alpha < \alpha_*$), conformality is lost and the potential has discrete bound states. For a range of α, the zero-energy, s-wave solution to the Schrödinger equation for two particles with mass $m = 1$ in d dimensions interacting via the potential $V(r)$ is given by

$$\psi = c_- r^{\nu_-} + c_+ r^{\nu_+}\,, \quad \nu_\pm = -\frac{d-2}{2} \pm \sqrt{\alpha - \alpha_*}\,, \quad \alpha_* \equiv -\frac{(d-2)^2}{4}\,. \tag{6}$$

This solution is valid for α in the range

$$\alpha_* \leq \alpha \leq \alpha_* + 1\,; \tag{7}$$

for $\alpha < \alpha_*$ the above solution becomes complex, and the Hamiltonian does not have ground state, while for $\alpha > \alpha_* + 1$ then $\nu_- < -d/2$ and the r^{ν_-} solution is not normalizable near $r \sim 0$. Within the range eq. (7), if either c_+ or c_- vanish, then the solution is scale-invariant. Solutions for which both c_+ and c_- are nonzero define an intrinsic length scale, $L \equiv (c_+/c_-)^{1/(\nu_- - \nu_+)}$ and therefore do not exhibit conformal invariance; however in this case the solution always approaches $c_+ r^{\nu_+}$ for large r (since $\nu_+ \geq \nu_-$) and so we can identify the $c_- = 0$ solution with an IR attractive fixed point and the $c_+ = 0$ solution with a UV fixed point, in a manner we can make precise. Arranging to have one of these solutions or the other requires different boundary conditions at the origin, so we see that the theory is actually not well defined with the potential eq. (5), but that it must be augmented by a δ function at the origin which controls the boundary condition at $r = 0$:

$$V(r) = \alpha/r^2 - g\delta^d(r)\,. \tag{8}$$

We will show that the coupling g obeys an RG equation analogous to our toy model eq. (2), and that the two conformal solutions $c_- = 0$ and $c_+ = 0$ will correspond to two different fixed points of the coupling g. As α approaches α_* from above, we will show that the two fixed points merge, $g_\pm = g_*$, at a value for g_* which we will compute. For $\alpha < \alpha_*$ a UV cutoff must be imposed on the theory in order to have a ground state and an IR scale emerges which is related to the UV cutoff through the BKT scaling formula eq. (4). We show this by performing a nonperturbative quantum-mechanical analysis.

2.1. *A nonperturbative calculation*

2.1.1. *The exact wavefunction and energy*

To solve the Schrödinger equation exactly for two particle scattering via a $1/r^2$ potential in d dimensions we need to regulate the singularity at $r = 0$.

We choose to do so by considering the potential

$$V(r) = \begin{cases} \alpha/r^2, & r > r_0, \\ -g/r_0^2, & r < r_0, \end{cases} \tag{9}$$

where r_0^{-1} will serve as the cutoff $\Lambda_{\rm UV}$.

At low energy, there is a region $r_0 < r \ll 1/\sqrt{E}$ where the $E\psi$ term in the Schrödinger equation can be neglected, and for $\alpha > \alpha_*$ we find the solution eq. (6)

$$\psi = c_- r^{\nu_-} + c_+ r^{\nu_+} , \tag{10}$$

with the ratio c_+/c_- given in terms of Bessel functions as

$$\frac{c_+}{c_-} = -r_0^{(\nu_- - \nu_+)} \frac{\gamma + \nu_-}{\gamma + \nu_+} , \qquad \gamma \equiv \left[\frac{\sqrt{g} J_{\frac{d}{2}}\left(\sqrt{g}\right)}{J_{\frac{d-2}{2}}\left(\sqrt{g}\right)} \right] . \tag{11}$$

The quantity (c_+/c_-) is a dimensionful quantity characterizing this solution; by requiring that it does not change as we change the UV cutoff r_0, we arrive at the exact β-function for γ (defining RG time $t = -\ln r_0$):

$$\beta_\gamma = \frac{\partial \gamma}{\partial t} = -(\gamma + \nu_+)(\gamma + \nu_-) = (\alpha - \alpha_*) - (\gamma - \gamma_*)^2 , \tag{12}$$

with

$$\alpha_* = -\frac{(d-2)^2}{4} , \quad \gamma_* = \frac{d-2}{2} , \quad \gamma_\pm = -\nu_\mp = \frac{d-2}{2} \pm \sqrt{\alpha - \alpha_*} . \tag{13}$$

We recognize this to be the same β-function as our toy model eq. (2) with fixed points at $\gamma_\pm$; referring to eq. (11) we see that the IR fixed point corresponds to $\gamma = \gamma_-$ and $c_- = 0$, while the UV fixed point is associated with $\gamma = \gamma_+$ and $c_+ = 0$.

For general d and $\alpha < \alpha_*$, scaling solutions do not exist; physical quantities, such as the bound state energy, depend on the UV cutoff. Motivated by the discussion of coupling constant flow in our toy model, we know that physical quantities will be insensitive to the value of γ (the UV coupling) so long as $\gamma < \gamma_*$, as seen in Fig. 1(b). So we take $\gamma \to -\infty$, which is reached in the limit of a hardcore repulsive potential for $r < r_0$, $g \to -\infty$. The ground state wavefunction is then described by the Bessel function $\psi(r) = r^{-(d-2)/2} K_{i\eta}(kr)$ with $\eta = \sqrt{\alpha_* - \alpha}$ and the boundary condition $\psi(r_0) = 0$. For small real η we can solve for k and find the binding energy

$$B = k^2 = \frac{1}{r_0^2} \exp\left(-\frac{2\pi}{\sqrt{\alpha_* - \alpha}} + O(1) \right) , \tag{14}$$

Note that this scale for the binding energy is easily attained from the RG analysis as

$$B \simeq \Lambda_{\rm IR}^2 = \left(\frac{1}{r_0} e^{\int_{-\infty}^{\infty} d\gamma/\beta_\gamma}\right)^2 = \frac{1}{r_0^2} e^{-2\pi/\sqrt{\alpha_* - \alpha}} \,. \tag{15}$$

2.1.2. *Operator anomalous dimensions at the IR and UV fixed points.*

We can gain insight about the two fixed points by looking at the dimension of the operators. Let us consider the two-particle operator $\psi\psi$. According to the operator/state correspondence developed in Ref. [15], one can find dimensions of this operator by putting two particles in a harmonic potential. The Hamiltonian of the system is given by

$$H = -\frac{1}{2}\nabla_1^2 - \frac{1}{2}\nabla_2^2 + V(|\mathbf{r}_1 - \mathbf{r}_2|) + \frac{1}{2}\omega^2(r_1^2 + r_2^2) \,. \tag{16}$$

In terms of the center of mass coordinate $\mathbf{R}$ and relative coordinate $\mathbf{r}$, the Hamiltonian can be rewritten as $H = H_R + H_r$ where the ground state energy of H_R equals $d\omega/2$ and

$$H_r = -\nabla_r^2 + V(r) + \frac{1}{4}\omega^2 r^2 , \tag{17}$$

where the potential is given in eq. (9). The ground state wavefunctions and energies for this Hamiltonian for g tuned to one of the fixed points $g_\pm$ is easily seen to equal

$$\psi_\pm = e^{-\omega r^2/4} r^{\nu_\pm} \,, \qquad E_r^\pm = \left(\frac{d}{2} + \nu_\pm\right)\omega \tag{18}$$

in the limit $r_0 \to 0$. (Recall that the fixed points $g_\pm$ correspond to solutions $\psi = r^{\nu_\pm}$ in the absence of the harmonic potential). Therefore the total ground state energy is

$$E^\pm = (d + \nu_\pm)\,\omega \,, \tag{19}$$

and so the scaling dimensions of the two-particle operator $\psi\psi$ are

$$\Delta_\pm = (d + \nu_\pm) = \frac{d+2}{2} \pm \sqrt{\alpha - \alpha_*} \,, \tag{20}$$

where Δ_+ and Δ_- are the operator dimension at the IR and UV fixed points respectively. We emphasize the fact that

$$(\Delta_+ + \Delta_-) = d + 2 \tag{21}$$

in any spatial dimension d and for any in the range $\alpha_* \le \alpha < (\alpha_* + 1)$ in eq. (7), with $\Delta_+ = \Delta_- = (d+2)/2$ at $\alpha = \alpha_*$. Note that $(d+2)$ is the scaling dimension of a nonrelativistic Lagrange density, since time has twice the scaling dimension as space; this is what one expects from AdS/CFT correspondence.

3. QCD at large N_c and N_f.

We now turn our attention to the most interesting, and most difficult, example; the chiral phase transition in QCD with large N_c and N_f. Denote $x = N_f/N_c$. We consider the Veneziano limit $N_c \to \infty$, $N_f \to \infty$, x fixed, and denote the (rescaled) 't Hooft coupling as

$$a_s = \frac{g^2 N_c}{(4\pi)^2}\,. \tag{22}$$

The beta function of QCD in this regime is

$$\beta(a_s) = -\frac{2}{3}\left[(11-2x)a_s^2 + (34-13x)a_s^3 + \cdots\right]. \tag{23}$$

For $x < 11/2$ the theory is asymptotically free. If x is slightly below $11/2$ the beta function has a nontrivial zero which is still in the perturbative regime:

$$a_{s*} = \frac{2}{75}\,(11-2x)\,. \tag{24}$$

This is the Banks-Zaks (BZ) fixed point.[5] In the IR, the theory is an interacting CFT. This fixed point moves to strong coupling as one makes $11/2 - x \sim 1$. For small x, $x \ll 1$, we believe that the theory has chiral symmetry breaking and a confinement scale. It is natural to assume that there is a critical N_f/N_c ratio $x_{\rm crit}$ at which the chiral condensate goes to zero.

If the picture emerging from the previous examples also holds for QCD, then conformality is lost when the BZ fixed point annihilates with another UV fixed point. Therefore, we predict that when x is slightly larger than $x_{\rm crit}$, QCD has an UV fixed point, in addition to the IR fixed point (and the free UV fixed point).

The situation is illustrated in Fig. 2. The UV fixed point called QCD*, is a different CFT compared to the usual IR fixed point; for example, the dimension of the operator $\bar\psi\psi$ should be different between the two fixed points.

What is the nature of QCD*? It could be that the β-function for the QCD gauge coupling simply has an unstable fixed point at strong coupling.

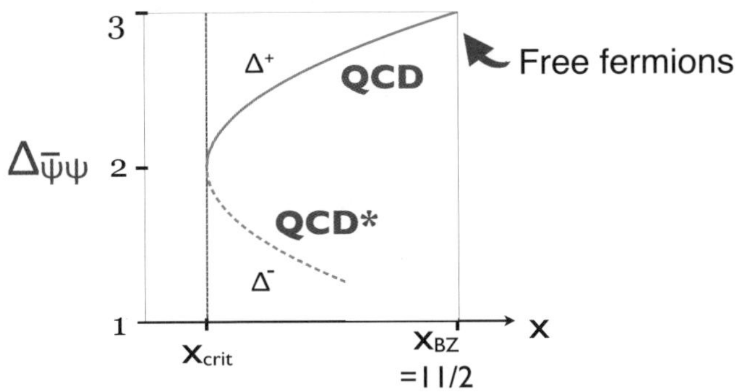

Fig. 2. A possible picture for the QCD chiral phase transition in N_f/N_c. The lines denotes the dependence of the dimension of the chiral condensate $\bar{\psi}\psi$ at the fixed point. The solid line is the IR fixed point, and the dashed line is the UV fixed point.

However, this picture implicitly assumes that the set of relevant operators in QCD* consist of just kinetic terms for the gauge fields and fermions, as is the case at weak coupling. At strong coupling other operators could be relevant as well, so it is natural to consider the possibility that a chirally symmetric four-fermion operator is relevant in QCD*

$$\mathcal{L}_{\mathrm{QCD}^*} = \mathcal{L}_{\mathrm{QCD}} - c(\bar{\psi}\gamma^\mu t^a \psi)^2 \tag{25}$$

and that the unstable fixed point exists at some value $\{g_*, c_*\}$ in the two-dimensional space of couplings. We then expect that the beta function for c contains linear, quadratic and constant terms,

$$\beta(c) = \gamma_1 c - \gamma_2 N_c c^2 + \gamma_0 g^2, \tag{26}$$

where the linear $\gamma_1 c$ term is due to the anomalous dimension of the four-fermi operator, the quadratic c^2 term is due to the one-loop graph involving two four-fermi vertices, and the constant g^2 term is due to, e.g., one-gluon exchange graph. This is essentially the picture advocated by Gies and Jaeckel.[16] The constants γ_0, γ_1, γ_2 depend on $x = N_f/N_c$; and x_{crit} is where $\beta(c)$ has a double zero.

We do not know where in x the fixed point QCD* exists. A particularly interesting possibility is that QCD* exists at weak coupling, say near $x = 11/2$. As described subsequently, we find many theories similar to QCD* in the perturbative regime, but none of them possess the full chiral symmetry of QCD, and hence cannot be QCD*. The possible "phase diagram" is illustrated in Fig. 2, the line corresponding to QCD* does not continue to the vicinity of $x = 11/2$.

In the rest of this section, we shall be looking for perturbative UV fixed points that flow to the BZ fixed point.

3.1. *Model A*

As the first step, we try to find a fixed point when *one* operator, $\bar{\psi}\psi$, has dimension different from its dimension at the BZ fixed point. From AdS/CFT experience, we expect that operator dimensions at the two fixed points satisfy

$$\Delta_+ + \Delta_- = d = 4\,. \tag{27}$$

At the BZ fixed point $\Delta_+ \approx 3$, therefore at the QCD* fixed point, $\Delta_- \approx 1$, which means that this operator is almost a free scalar. This suggests that we look for QCD* in the following Lagrangian, which will be called model A,

$$\mathcal{L}_{\text{model A}} = \mathcal{L}_{\text{QCD}} + \frac{1}{2}(\partial_\mu\phi)^2 - \frac{y}{\sqrt{2}}\bar{\psi}\psi\phi - \frac{\lambda}{24}\phi^4\,. \tag{28}$$

It is convenient to define the rescaled couplings,

$$a_s = \frac{g^2 N_c}{(4\pi)^2}\,, \qquad a_y = \frac{y^2 N_c N_f}{(4\pi)^2}\,, \qquad \hat{\lambda} = \frac{\lambda N_c N_f}{(4\pi)^2}\,. \tag{29}$$

These constants will be found to be $O(N_c^0)$ at the fixed point. In this regime, the beta functions are

$$\beta_{a_s} = -\frac{2}{3}\left[(11-2x)a_s^2 + (34-13x)a_s^3\right], \tag{30}$$

$$\beta_{a_y} = -6a_s a_y + 2a_y^2\,, \tag{31}$$

$$\beta_{\hat{\lambda}} = -12a_y^2 + 4a_y\hat{\lambda}\,. \tag{32}$$

The fixed point for the gauge coupling, a_{s*}, is the same as at the BZ fixed point to leading order in $1/N_c$. The fixed point for the Yukawa and four-scalar couplings are

$$a_{y*} = 3a_{s*}\,, \qquad \hat{\lambda} = 3a_{y*} = 9a_{s*}\,. \tag{33}$$

Thus, model A has a perturbative fixed point.

3.2. *Model C*

Model A preserves only a small subset of the SU(N_f)×SU(N_f) chiral symmetry of QCD. The simplest Lagrangian which preserves chiral symmetry is

$$\mathcal{L} = \mathcal{L}_{\rm QCD} - y(\bar{\psi} t^A \psi \phi^A + i\bar{\psi} t^A \gamma^5 \psi \pi^A) + {\rm Tr}\, \partial^\mu \Phi^\dagger \partial_\mu \Phi - \lambda_1 ({\rm Tr}\, \Phi^\dagger \Phi)^2 - \lambda_2 {\rm Tr}\, (\Phi^+ \Phi)^2 \,, \quad (34)$$

where $\Phi = (\phi^A + i\pi^A)t^A$, $A = 0, \ldots N_f^2$ are flavor Gell-Mann matrices, normalized so that ${\rm Tr}(t^A t^B) = \frac{1}{2}\delta^{AB}$. We need to find the fixed point of this theory. This fixed point should have only one IR unstable direction corresponding to the mass term for Φ.

The RG equations for g and y can be read out from Refs. [17–19]. We need the two-loop beta function for the gauge coupling (as the one-loop contribution has a small coefficient near $x = 11/2$), but for the Yukawa and scalar couplings one-loop beta function suffices. Moreover, the two-loop beta function for g and the one-loop beta function for y does not contain scalar self-couplings, thus one can first solve for the fixed point for g and y, and then look for the fixed points of scalar couplings. In this model we define a_y as

$$a_y = \frac{y^2 N_c}{(4\pi)^2} \,. \quad (35)$$

The beta functions are

$$\beta_{a_s} = -\frac{2}{3}(11 - 2x)a_s^2 - \frac{2}{3}(34 - 13x)a_s^3 - 2x^2 a_s^2 a_y \,, \quad (36)$$

$$\beta_{a_y} = -6a_s a_y + 2(1 + x)a_y^2 \,. \quad (37)$$

We work around $x = 11/2$. At fixed a_s, there is a zero of β_y, but there is no fixed point of both beta functions. Therefore model C does not have a perturbative fixed point.

4. Conclusion

There have recently been several lattice studies seeking to find the boundaries of the conformal window in QCD [20,21] and in other QCD-like theories.[22–28] Interest in the phase transition between conformal and nonconformal theories is motivated in part by the invocation of approximate conformal symmetry in numerous theories for physics beyond the Standard Model.

In this paper we have investigated the nature of such a phase transition, and we suggest that there is a wide class of theories where it is due to the merger and annihilation of fixed points. Several explicit examples of this phenomenon were given, and we speculate that this mechanism is also responsible for the chiral phase transition in QCD in the large N_c, large N_f regime, at some critical value for N_f/N_c. We show that this mechanism leads to the BKT scaling behavior of the chiral condensate at the phase transition, and also implies the existence of the conformal theory QCD* which annihilates with QCD at the lower end of the conformal window. We tried, unsuccessfully, to construct QCD* in the perturbative regime, and argued that it should exist in the nonperturbative regime. It would be interesting to search for evidence for QCD* on the lattice.

References

1. K. G. Wilson and J. B. Kogut, Phys. Rept. **12**, 75 (1974).
2. J. Kosterlitz, J. Phys. C **7**, 1046 (1974).
3. N. Seiberg, Nucl. Phys. B **435**, 129 (1995).
4. K. A. Intriligator and N. Seiberg, Nucl. Phys. Proc. Suppl. **45BC**, 1 (1996).
5. T. Banks and A. Zaks, Nucl. Phys. B **196**, 189 (1982).
6. J. Maldacena, Adv. Theor. Math. Phys. **2**, 231 (1998).
7. S. S. Gubser, I. R. Klebanov and A. M. Polyakov, Phys. Lett. B **428**, 105 (1998).
8. E. Witten, Adv. Theor. Math. Phys. **2**, 253 (1998).
9. P. Breitenlohner and D. Z. Freedman, Ann. Phys. **144**, 249 (1982).
10. V. A. Miransky, Nuovo Cim. A **90**, 149 (1985).
11. B. Holdom, Phys. Lett. B **213**, 365 (1988).
12. B. Holdom, Phys. Rev. Lett. **62**, 997 (1989).
13. T. Appelquist, J. Terning and L. C. R. Wijewardhana, Phys. Rev. Lett. **77**, 1214 (1996).
14. T. Appelquist, A. Ratnaweera, J. Terning and L. C. R. Wijewardhana, Phys. Rev. D **58**, 105017 (1998).
15. Y. Nishida and D. T. Son, Phys. Rev. D **76**, 086004 (2007).
16. H. Gies and J. Jaeckel, Eur. Phys. J. C **46**, 433 (2006).
17. M. E. Machacek and M. T. Vaughn, Nucl. Phys. B **222**, 83 (1983).
18. M. E. Machacek and M. T. Vaughn, Nucl. Phys. B **236**, 221 (1984).
19. M. E. Machacek and M. T. Vaughn, Nucl. Phys. B **249**, 70 (1985).
20. T. Appelquist, G. T. Fleming and E. T. Neil, Phys. Rev. Lett. **100**, 171607 (2008).
21. T. Appelquist, G. T. Fleming and E. T. Neil, Phys. Rev. D **79**, 076010 (2009).
22. F. Sannino and K. Tuominen, Phys. Rev. D **71**, 051901 (2005).
23. D. D. Dietrich and F. Sannino, Phys. Rev. D **75**, 085018 (2007).
24. S. Catterall and F. Sannino, Phys. Rev. D **76**, 034504 (2007).
25. S. Catterall, J. Giedt, F. Sannino and J. Schneible, JHEP **11**, 009 (2008).
26. Y. Shamir, B. Svetitsky and T. DeGrand, Phys. Rev. D **78**, 031502 (2008).
27. T. DeGrand, Y. Shamir and B. Svetitsky, Phys. Rev. D **79**, 034501 (2009).
28. A. J. Hietanen, J. Rantaharju, K. Rummukainen and K. Tuominen, JHEP **05**, 025 (2009).

Probing Universality in AdS/CFT

A. Ritz

Department of Physics and Astronomy, University of Victoria
Victoria BC, V8P 5C2, Canada

We discuss some universal relations that hold for conformal field theories which admit a dual classical gravitational description in anti-de Sitter space. These relations therefore apply in a suitable large N limit, and imply that various hydrodynamic transport coefficients for conserved currents and thermodynamic state functions are entirely fixed by central charges.

1. Introduction

From the Wilsonian viewpoint,[1] the primary building blocks of the space of quantum field theories are those special points which exhibit conformal invariance. Indeed, the quantum field theories which play a role in nature are all understood to be either conformal field theories (CFT) themselves, or relevant deformations of a CFT. Among other things, such field theories describe the interactions of all known elementary particles, and the universal scaling behaviour near critical points in statistical mechanics.

In attempting to study CFTs at the quantitative level, there are several avenues available beyond $d = 1 + 1$ where symmetry alone is sufficiently powerful to constrain the allowed space of theories. One approach to arrive at an interacting CFT is by taking a critical limit of a statistical-mechanical model (or, more generally, by following the renormalization group flow to the endpoint).[1] Another is to start with a Lagrangian formulation of a classically conformal theory, and use extra symmetry (such as supersymmetry) to argue that the quantum theory must be conformally invariant as well.[2] A prime example of such a CFT is the $\mathcal{N} = 4$ supersymmetric Yang-Mills theory in 3+1 dimensions. However, the last decade has exhibited another remarkable approach to arrive at a CFT, namely through a higher-dimensional theory of (quantum) gravity in Anti-de Sitter (AdS) space, and the AdS/CFT correspondence.[3–5]

In the limit where classical gravity is a reliable approximation, which involves taking both the number of CFT degrees of freedom and their interactions to be parametrically large, this latter approach is particularly tractable and provides a new class of tools. It is therefore natural to ask which class of CFTs, and thus which universality classes, can be described in this tractable classical gravity regime. In exploring this question, particularly in the condensed matter context, it is important to distinguish two notions of universality:

- *Static universality*: This reflects the universal scaling behaviour of thermodynamic functions, such as the specific heat, as one approaches the critical point. Static universality classes are generally determined by the spatial dimension and the symmetry of the order parameter distinguishing the phases. A classic example is the fact that the liquid-gas critical point and the ferromagnetic critical point are both in the same static universality class, that of the $d = 3$ Ising CFT.
- *Dynamic universality*: This goes beyond static universality and describes classes of systems in which dynamical transport coefficients also exhibit universal scaling laws.[6] Two fixed points in the same static universality class may exhibit quite distinct dynamical behaviour. Indeed, the liquid-gas and ferromagnetic critical points are in different dynamical universality classes.[6] The additional characteristic which plays a crucial role in defining these classes is the presence or otherwise of conserved currents, which provide additional hydrodynamic degrees of freedom.

Given this dichotomy, it is natural to ask whether there are special fixed points in which these two notions of universality are the same, so that all static and dynamical behaviour is described near the fixed point by a Lorentzian CFT, formulated in spacetime (rather than just space). Physical examples are indeed provided by quantum critical phenomena,[7,8] where the phase transition occurs at zero temperature, driven by quantum fluctuations (see Fig. 1). Such CFTs are relativistic, although their speed of "light" v need not equal 3×10^8m/s. The CFT then dictates static and dynamical quantities in any nearby regime dominated by a single dynamical scale, i.e. temperature, which breaks conformal invariance.

It has become clear that, for these reasons, the quantum critical regime is a primary target to be modeled using the AdS/CFT correspondence. Understanding CFTs in the language of classical gravity has already proved useful in studies of quantum critical transport,[9] and this motivates further

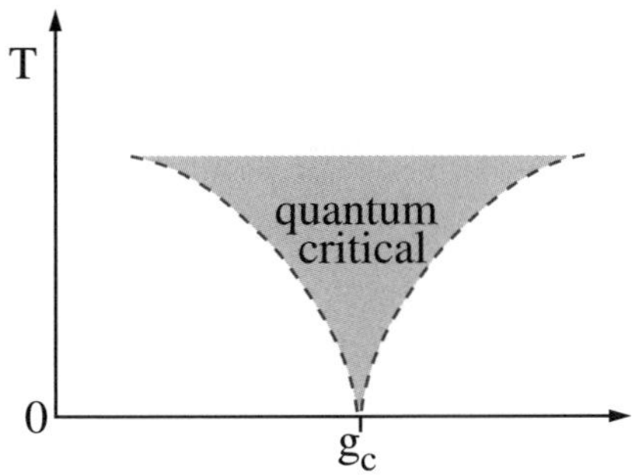

Fig. 1. A schematic illustration of the phase diagram near a quantum critical point at $T = 0$ and $g = g_c$, where g is some coupling or control parameter for the system. The shaded region is controlled by the CFT associated with the critical point.

exploration of the relation of the AdS/CFT correspondence to quantum criticality. In this contribution, we will discuss a number of constraints that naturally apply to any CFT dual to classical gravity in AdS (see [10–12] for more details). In broad terms, these constraints relate the thermodynamic and hydrodynamic peroperties of the CFT at finite temperature to each other, and furthermore to the central charges which determine the zero-temperature correlators of conserved currents.

Recall that at very short distances, the effects of temperature are irrelevant, and the natural physical questions involve the leading short-distance singularities of the correlation functions. On the other hand, at long distances the effects of temperature become important, and the natural questions are related to thermodynamics and hydrodynamic transport phenomena. In CFTs, however, short and long distances are related by a scaling symmetry, and one may anticipate a universal relation between the long-distance transport coefficients and the parameters which describe the short-distance singularities. Unfortunately, in general this expectation seems to be quantitatively true only in 1+1 dimensions. Nonetheless, this expectation does indeed hold true in CFTs dual to classical gravity in any dimension.

At zero temperature, the (Euclidean) vacuum correlation functions of the energy-momentum tensor $T_{\mu\nu}$ and a $U(1)$ conserved current J_μ in a CFT are fixed to be

$$\langle J_\mu(x) J_\nu(0)\rangle = \frac{k}{x^{2(d-1)}}\frac{1}{\omega_{d-1}^2} I_{\mu\nu}\,, \tag{1}$$

$$\langle T_{\mu\nu}(x) T_{\alpha\beta}(0)\rangle = \frac{c}{x^{2d}}\frac{1}{\omega_{d-1}^2}\left(I_{\mu\alpha}I_{\nu\beta} + I_{\mu\beta}I_{\nu\alpha} - \frac{2}{d}\delta_{\mu\nu}\delta_{\alpha\beta}\right), \tag{2}$$

where $I_{\mu\nu} \equiv \delta_{\mu\nu} - 2x_\mu x_\nu/x^2$, and k and c are central charges, which are dimensionless constants. [We use units in which $\hbar$=v=1 where v is the speed of "light" in the CFT.] The factors of $\omega_{d-1} \equiv 2\pi^{d/2}/\Gamma(d/2)$ are inserted for notational convenience. At non-zero temperature T, the equilibrium state is characterized by pressure P, as well as by the charge susceptibility $\chi = \langle Q^2\rangle/(VT)$, where Q is the conserved charge associated with the current J_μ, and we take the thermodynamic limit $V\to\infty$. The susceptibility can be evaluated by introducing a small chemical potential μ, so that $\chi(T) = \partial\rho/\partial\mu|_{\mu=0}$, where $\rho(T,\mu) = \langle Q\rangle/V$ is the charge density. In a CFT, temperature remains the only scale which dictates,

$$P(T) = c' T^d\,, \qquad \chi(T) = k' T^{d-2}\,, \tag{3}$$

where c' and k' are dimensionless constants. Physically, c and c' provide a measure of the total number of degrees of freedom in the system, while k and k' measure the number of charged degrees of freedom. In two dimensions, c is uniquely related to c' [13,14], while k is uniquely related to k':

$$c' = \frac{\pi}{6}\,c\,, \qquad k' = \frac{1}{2\pi}\,k\,, \tag{4}$$

which means that thermodynamics is uniquely fixed by the central charges. The reason is that in two dimensional CFTs, the vacuum state is related to the thermal state by a symmetry transformation. In $d > 2$, the conformal symmetry group is not large enough to enforce a relation similar to Eq. (4), and therefore thermodynamics is not determined by the central charges. However, there does exist a large class of CFTs in $d > 2$, whose pressure is determined by the central charge c, and whose susceptibility is determined by the central charge k, resembling the two-dimensional case.[10,11] The crucial property of these models is that they admit a dual description in terms of classical gravity on a (d+1) dimensional anti-de Sitter space (AdS). Namely, we have the following relations [10,11]:

$$\frac{c'}{c} = \frac{1}{4\pi^{d/2}}\left(\frac{4\pi}{d}\right)^d \frac{\Gamma(d/2)^3}{\Gamma(d)}\,\frac{(d{-}1)}{d(d{+}1)}\,, \qquad \frac{k'}{k} = \frac{1}{2\pi^{d/2}}\left(\frac{4\pi}{d}\right)^{d-2}\frac{\Gamma(d/2)^3}{\Gamma(d)}\,. \tag{5}$$

In $d = 2$ they reproduce the universal relations (4).

The CFTs which admit a dual gravitational description have many more universal properties beyond the above relation between thermodynamics and the central charges. A surprising feature of these CFTs (and of their relevant deformations) is that momentum transport in these models is completely determined by thermodynamics. In particular, their viscosity is given by $\eta = s/4\pi$ in any dimension,[15,16] where $s = (\partial P/\partial T)$ is the entropy density. This is surprising because transport coefficients are typically determined by the mean-free path even in CFTs,[17] and are not fixed by thermodynamics. Moreover, charge transport in these models is also completely determined by thermodynamics. In particular, the dc electrical conductivity σ obeys a similar relation, so we have

$$\frac{\eta}{s} = \frac{1}{4\pi}\,, \qquad \frac{\sigma}{\chi} = \frac{1}{4\pi T}\frac{d}{d-2}\,. \tag{6}$$

Even though the ratio σ/χ was derived for integer $d \geqslant 3$, it can be analytically continued to any real positive $d > 2$. The singularity at $d = 2$ is precisely what one expects in 1+1 dimensional CFTs.

In what follows, we describe the derivation of these relations between central charges and thermodynamics, and between thermodynamics and hydrodynamics. In the final section, we will discuss some comparisons to known systems, and deviations from these universal relations beyond the classical gravity limit.

2. Thermodynamics from central charges

We will focus on quantum field theories which admit a dual description in terms of classical gravity in Anti-de Sitter (AdS) space within the AdS/CFT correspondence.[3–5] For such field theories, a large-volume thermal state in a d-dimensional CFT is described by a $(d+1)$-dimensional black hole in AdS. The black hole solution follows from the Einstein-Maxwell action,

$$S = \frac{1}{16\pi G}\int \mathrm{d}^{d+1}x\,\sqrt{-g}\,\left[R + \frac{d(d-1)}{L^2}\right] - \frac{1}{4g_{d+1}^2}\int \mathrm{d}^{d+1}x\,\sqrt{-g}\,F^2\,, \tag{7}$$

where L^2 sets the value of the cosmological constant, and g_{d+1}^2 is the $(d+1)$-dimensional gauge coupling constant, which has dimension of $(\text{length})^{d-3}$. An equilibrium state at finite temperature and density is described by the Reissner-Nordstrom black hole in AdS. The thermodynamics of these black holes has been studied extensively, see for example [18]. The metric in the thermodynamic limit is given by

$$ds^2 = \frac{r^2}{L^2}\left(-V(r)dt^2 + d\mathbf{x}^2\right) + \frac{L^2}{r^2}\frac{dr^2}{V(r)}, \tag{8}$$

where $V(r) = 1 - m/r^d + m_q^2/r^{2d-2}$, and the boundary is at $r\to\infty$. The parameter m determines the mass of the black hole, and m_q determines its charge. The background gauge field is,

$$A_t = \mu - \frac{g_{d+1}^2 L^{d-1}\rho}{(d-2)r^{d-2}}, \tag{9}$$

where the charge density ρ is defined by the variation of the action with respect to the boundary value of the gauge field $A_t^{(b)} = A_t(r\to\infty)$, and is given by $\rho = \mu(d-2)r_0^{d-2}/(g_{d+1}^2 L^{d-1})$ in terms of the chemical potential μ. This relation is fixed by requiring that A_t vanishes at the horizon $r = r_0$.

We will actually work in the limit $\mu \to 0$, so that temperature is the dominant scale breaking conformal invariance, given by $T = r_0 d/(4\pi L^2)$. The charge density then vanishes, but the results above allow us to compute the susceptibility.

Central charges: To find the central charge c in dimension $d \geqslant 3$, we consider the vacuum geometry ($T = 0$) and a convenient momentum space representation of the correlator (2),

$$G_{\mu\nu,\alpha\beta}(k) = \left(P_{\mu\alpha}P_{\nu\beta} + P_{\mu\beta}P_{\nu\alpha} - \frac{2}{d-1}P_{\mu\nu}P_{\alpha\beta}\right) G(k^2), \tag{10}$$

where $P_{\mu\nu} = \delta_{\mu\nu} - k_\mu k_\nu/k^2$. The central charge is related to $G(k^2)$ by

$$\frac{c}{(x^2)^d} = \frac{d+1}{d-1}\omega_{d-1}^2 \int \frac{d^dk}{(2\pi)^d} e^{ikx} G(k^2). \tag{11}$$

Choosing k along x_i with $i \neq 1$ or 2, one has $G_{12,12}(k) = G(k^2)$, and therefore it suffices to evaluate $G_{12,12}$ to find the central charge. According to the AdS/CFT prescription, $G_{12,12}$ is given by the second variation of the gravitational action with respect to the boundary value of the metric perturbation h_{12}. The h_{12} perturbation decouples from all other perturbations, and obeys the equation of motion coming from the action of a massless scalar in the AdS_{d+1} background. The two-point correlation function for the massless scalar can be evaluated using the standard AdS/CFT prescription, and one finds for the central charge,

$$c = \frac{d+1}{d-1}\frac{L^{d-1}}{4\pi G_N^{d+1}}\frac{\Gamma(d+1)\pi^{d/2}}{\Gamma(d/2)^3}. \tag{12}$$

The value of the central charge k can be found in a similar manner using the Maxwell action for the gauge field dual to the current J^μ. However, we will skip the details here, as it was obtained previously in the work of Freedman et al. [19]:

$$k = \frac{L^{d-3}}{g_{d+1}^2}\frac{\Gamma(d)(d-2)}{2\pi^{d/2}\Gamma(d/2)}\omega_{d-1}^2. \tag{13}$$

Thermodynamics: The entropy is proportional to the $d-1$ dimensional area of the horizon, $S = A_{d-1}/4G_N$ where G_N is the $d+1$-dimensional Newton's constant. Dividing by the (infinite) $(d-1)$–volume V, and working with $T \gg \mu$ as described above, one finds for the entropy density,

$$s = \frac{1}{4G_N}\left(\frac{4\pi L}{d}\right)^{d-1} T^{d-1}. \tag{14}$$

To find the susceptibility, we need the relation between ρ and μ to linear order in μ. Expressing r_0 in terms of temperature when $\mu \to 0$, $T = r_0 d/(4\pi L^2)$, we find $\rho(T,\mu) = \chi(T)\,\mu$, where the susceptibility is

$$\chi = \frac{(d-2)L^{d-3}}{g_{d+1}^2}\left(\frac{4\pi}{d}\right)^{d-2} T^{d-2}. \tag{15}$$

Comparing the results above for s and χ to those for c and k, we deduce the universal relations for c'/c and k'/k given in Eq. (5).

3. Hydrodynamics from central charges

Momentum transport: The well-known universal result for the shear viscosity to entropy ratio, $\eta/s = 1/(4\pi)$ can now be translated one step further to a universal relation of η to the central charge c. Given,[20]

$$\eta = -\lim_{\omega \to 0} \mathrm{Im}\,\frac{1}{\omega}\,\langle T_{xy} T_{xy}\rangle(\omega, \mathbf{q} = \mathbf{0})_{\mathrm{ret}}\,, \tag{16}$$

one obtains

$$\eta = \frac{s}{4\pi} = \left[\frac{1}{16\pi^{d/2+1}}\frac{d-1}{d+1}\left(\frac{4\pi}{d}\right)^{d}\frac{\Gamma(d/2)^3}{\Gamma(d)}\right] c T^{d-1}. \tag{17}$$

This exhibits the universal relation between hydrodynamics, thermodynamics, and the vacuum in the shear channel associated with the conserved energy momentum tensor.

Charge transport: An analogous procedure allows us to compute the electrical conductivity of CFTs with a dual gravity description. To define conductivity, we imagine gauging a global $U(1)$ symmetry of the theory with a small coupling e. To leading order in e, the effects of the gauge field can be ignored, and the electromagnetic response can be determined from the original theory. This essentially amounts to sending $J_\mu \to eJ_\mu$, and a factor of e^2 will appear in both the conductivity and the susceptibility.

The dc conductivity is determined from the real-time retarded correlation function of U(1) currents in thermal equilibrium,

$$\sigma = e^2 \lim_{\omega\to 0} \mathrm{Im}\, \frac{1}{\omega} \langle J_x J_x \rangle(\omega, \mathbf{q}{=}0)_{\mathrm{ret}} \,. \tag{18}$$

To evaluate the retarded correlator, we use the standard AdS/CFT recipe of [20,21], and consider Maxwell fields propagating on the $(d{+}1)$ dimensional background (8) at $m_q{=}0$. Skipping the details,[11] the Kubo formula (18) then gives the conductivity,

$$\sigma = \frac{e^2}{g_{d+1}^2} \left(\frac{L}{z_0}\right)^{d-3} . \tag{19}$$

On the other hand, using the susceptibility (15), we arrive at the simple result (6) for the conductivity to susceptibility ratio. For systems in which charge transport proceeds by diffusion, conductivity is related to the diffusion constant D by the Einstein relation $\sigma = \chi D$. Therefore, our result can be interpreted as a remarkably simple diffusion constant in d spacetime dimensions,

$$D = \frac{\sigma}{\chi} = \frac{1}{4\pi T} \frac{d}{d{-}2} \,. \tag{20}$$

which agrees with the known results in $d{=}4$ [20], and $d{=}3,6$ [22].The electrical conductivity takes a particularly simple form in 2+1 dimensional CFTs where it is *frequency-independent* $\sigma(\omega) = e^2/g_4^2$ [9], and shows no crossover regime at $\hbar\omega \sim k_{\mathrm{B}}T$.

Putting the pieces together, we have

$$\sigma = \frac{\chi}{4\pi T} \frac{d}{d-2} = \left[\frac{1}{8\pi^{d/2+1}} \frac{d}{d-2} \left(\frac{4\pi}{d}\right)^{d-2} \frac{\Gamma(d/2)^3}{\Gamma(d)} \right] kT^{d-3} , \tag{21}$$

again exhibiting the relation between hydrodynamics, thermodynamics, and the vacuum, this time in relation to the conserved U(1) current.

4. Discussion

In this concluding section, we will comment on some corollaries of these results, some comparisons with known systems, and deviations from universality away from the classical gravity (i.e. large N) limit.

Transport coefficient ratios: The general relations, $\eta \sim c\,T^{d-1}$, and $\sigma \sim k\,T^{d-3}$ also imply that the ratio of viscosity to conductivity is proportional to the ratio of the central charges,

$$\frac{\eta\, e^2}{\sigma T^2} = \frac{8\pi^2(d-1)(d-2)}{d^3(d+1)}\,\frac{c}{k} \propto \frac{g_{d+1}^2}{G_N}\,. \tag{22}$$

If c and k indeed provide a suitable measure of the number of degrees of freedom in the system, it is natural to conjecture a bound of the form $\sigma T^2 \leqslant \lambda_d\, \eta\, e^2$, with some order one constant λ_d, given that k is only sensitive to charged degrees of freedom.[a] Another suggestive viewpoint is that the ratio in (22) is proportional to the ratio of bulk couplings, g_{d+1}^2/G_N, and from this dual gravitational point of view, such an inequality between c and k is related to a version of the "weak gravity conjecture" [24] in AdS space. However, one should keep in mind that the definition of σ (or η) involves an arbitrary choice of normalization for the corresponding current and therefore any bound on conductivity will more naturally involve a quantity which is independent of the normalization, such as σ/χ.[11]

Nonetheless, it is interesting to note that the ratio c/k also determines when the U(1) symmetry is unstable to condensation forming a superfluid (or superconducting) phase. In $d = 3$, the instability occurs when[25]

$$q^2 \geqslant \left(3 + 2\Delta(\Delta - 3)\right)\frac{k}{c} \tag{23}$$

with q (Δ) the charge (dimension) of the operator breaking U(1).

Free-field comparisons: It is natural to ask if there are "conventional" CFTs which obey the AdS/CFT relations for c'/c and k'/k. This cannot happen for free (or weakly-interacting) theories when d is odd, because c' is proportional to $\zeta(d)$, while c contains no such factors. In even dimensions, this is no longer the case and in $d = 4$, for a free theory with n_v vector, n_f fermionic and n_s scalar degrees of freedom, we find that

$$\left(\frac{c}{c'}\right)^{\rm free}_{4d} = \frac{60}{\pi^2}\,\frac{n_s + \frac{3}{2}n_f + 12 n_v}{n_s + \frac{7}{8}n_f + 2n_v}\,, \tag{24}$$

so that $3/8 \leqslant \frac{c/c'|_{\rm free}}{c/c'|_{\rm AdS}} \leqslant 9/4$. Note that the ratio is 3/4 for the free limit of $\mathcal{N} = 4$ SYM consistent with the known difference of the entropy (and thus c') at strong and weak coupling. Importantly, we observe that free theories

[a] There are, however, known counter-examples to the decrease of c and k along renormalization group trajectories in supersymmetric theories [23], but in these examples k corresponds to the R-current, and therefore $k \propto c$ by supersymmetry.

with $2n_s + n_f = 8n_v$ do satisfy the AdS relation. Thus we conclude that, at least in $d = 4$, the AdS relation c'/c is necessary but not sufficient for a given CFT to possess a gravity dual.

In $d = 3$, as noted above we do not expect free theories to exhibit the AdS relation, but we can consider an interacting example, namely the 3-dimensional $O(N)$ model at large N with fields ϕ^α, $\alpha = 1, .., N$ subject to a constraint $\phi^\alpha \phi^\alpha = 1$. In this system, the ratio of c'/c was computed in the large-N limit by Sachdev,[26] with the result,

$$\left(\frac{c'}{c}\right)_{O(N)} = \frac{8\zeta(3)}{15\pi} \approx 0.2041, \tag{25}$$

which differs by only a few percent from the holographic answer, $c'/c = \pi^3/162 \approx 0.1914$ [10]. With regard to the proposal of Klebanov and Polyakov [27] – that the large-N dual is a higher-spin gauge theory in AdS – this result amounts to a prediction for the bulk spin-two sector and implies a (small) quantitative difference with pure Einstein gravity. A similar comparison is possible for the ratio k'/k, where a natural variant of the vector current studied above lies in the adjoint, $J_\mu^{\alpha\beta} = (\phi^\alpha \partial_\mu \phi^\beta - \phi^\beta \partial_\mu \phi^\alpha)$. Computation of the central charge at large N leads to $k = 2$ [28], while the charge susceptibility was obtained in [29] leading to,

$$\left(\frac{k'}{k}\right)_{O(N)} = \frac{\sqrt{5}}{2\pi} \ln\left(\frac{\sqrt{5}+1}{2}\right) \approx 0.1713\,, \tag{26}$$

which differs by about 24% from the result $k'/k = \pi/24 \approx 0.1309$ for models with gravitational duals in AdS. On the other hand, the conductivity in the $O(N)$ model is large, σ/χ is $\mathcal{O}(N)$ [8], reflecting the fact that the model becomes weakly coupled at large N. Therefore, the comparison with the $O(N)$ model in $d = 3$ provides an example of a situation where two systems have very similar static thermodynamic properties, but vastly different transport properties.

Deviations from universality: We have argued that in a large class of CFTs in $d > 2$, there are universal relations between the thermodynamic and transport properties, and the central charges which dictate the short distance behaviour of current-current correlators. One way of defining this class of theories is that they possess dual descriptions within AdS at the level of classical gravity and Maxwell electrodynamics.

It is natural to ask about the regime of validity, or alternatively the constraints on the CFTs which may enter such universality classes. Indeed, the analysis we have performed using the AdS/CFT duality required the validity of a classical gravity approximation, and thus some kind of large-N limit. On the gravity side of the duality, universality follows from the uniqueness of the lowest dimension operator which determines the dynamics of the metric and/or the gauge field dual to the current in question, i.e. the uniqueness of the Einstein-Hilbert and Maxwell actions respectively. From this point of view, once we move to finite N, it appears that a large number of higher derivative corrections will also be required, thus limiting the possibility for universal behaviour.

Examples of higher curvature corrections to the Einstein-Hilbert action have been shown to correct the universal relation $\eta/s = 1/(4\pi)$ [30]. Here we will discuss an analogous correction to σ, arising from a bulk higher-derivative coupling to the Weyl tensor,[12]

$$\mathcal{L} = \frac{1}{4g_{d+1}^2}\left(F^{\mu\nu}F_{\mu\nu} - 4\gamma C^{\mu\nu\rho\sigma}F_{\mu\nu}F_{\rho\sigma} + \cdots\right), \tag{27}$$

This setting provides a very clean test of the universality relation (21) at higher order, because the geometry dual to the zero temperature state, namely pure AdS, is Weyl flat. It follows immediately that the 2-point function $\langle J_\mu(x)J_\nu(0)\rangle$ at $T = 0$, and thus the central charge k, are uncorrected by turning on the perturbation γ,

$$k(\gamma) = k(\gamma = 0)\,. \tag{28}$$

However, the Weyl tensor does not vanish for the AdS black hole and a straightforward calculation shows that [12]

$$\sigma = \frac{\pi LT}{g_5^2}\left(1 + \frac{8\gamma}{L^2}\right), \qquad D = \frac{\sigma}{\chi} = \frac{1}{2\pi T}\left(1 + \frac{16\gamma}{L^2} + \cdots\right), \tag{29}$$

and the universal relation is violated beyond the classical gravity limit. This result is not unexpected, but it is interesting that causality [30] provides limits on γ ensuring that deviations cannot be large.[12]

It would be interesting to see if there are other symmetries which might preserve these universality relations beyond the classical gravity limit, i.e. beyond leading order in N. It would also be interesting to explore other regimes, analogous to quantum critical points in which these techniques may be useful. Recall that the primary link was the presence of just one dimensionful scale breaking conformal invariance. One can easily imagine scenarios where this scale is something other than temperature. For example, it would be interesting to consider cold systems dominated instead by the chemical potential, which should have a dual description in terms of extremal black holes in AdS.

Acknowledgments: My best wishes to Misha Shifman on the occasion of his 60th birthday, and I'd like to thank him for very fruitful and enjoyable collaboration on several subjects over the past decade, and to thank the organizers for the invitation to this stimulating meeting. My thanks also to Pavel Kovtun and John Ward for enjoyable collaboration on the work described here, which was supported in part by NSERC, Canada.

References

1. K. G. Wilson and J. B. Kogut, Phys. Rept. **12**, 75 (1974).
2. See *e.g.* R. G. Leigh and M. J. Strassler, Nucl. Phys. B **447**, 95 (1995).
3. J. M. Maldacena, Adv. Theor. Math. Phys. **2**, 231 (1998).
4. S. S. Gubser et al., Phys. Lett. B **428**, 105 (1998).
5. E. Witten, Adv. Theor. Math. Phys. **2**, 253 (1998).
6. P. C. Hohenberg and B. I. Halperin, Rev. Mod. Phys. **49**, 435 (1977).
7. For a review, see S. Sachdev, Nature Physics **4**, 173 (2008).
8. S. Sachdev, *Quantum phase transitions,* Cambridge, 1999.
9. C. P. Herzog et al., Phys. Rev. D **75**, 085020 (2007).
10. P. Kovtun and A. Ritz, Phys. Rev. Lett. **100**, 171606 (2008).
11. P. Kovtun and A. Ritz, Phys. Rev. D **78**, 066009 (2008).
12. A. Ritz and J. Ward, Phys. Rev. D **79**, 066003 (2009).
13. H. W. J. Bloete et al., Phys. Rev. Lett. **56**, 742 (1986).
14. I. Affleck, Phys. Rev. Lett. **56**, 746 (1986).
15. P. Kovtun et al., Phys. Rev. Lett. **94**, 111601 (2005).
16. A. Buchel, Phys. Lett. B **609**, 392 (2005).
17. K. Damle and S. Sachdev, Phys. Rev. B **56**, 8714 (1997).
18. A. Chamblin et al., Phys. Rev. D **60**, 064018 (1999).
19. D. Z. Freedman et al., Nucl. Phys. B **546**, 96 (1999).
20. G. Policastro et al., JHEP **0209**, 043 (2002).
21. D. T. Son and A. O. Starinets, JHEP **0209**, 042 (2002).
22. C. P. Herzog, JHEP **0212**, 026 (2002).
23. D. Anselmi et al., Phys. Rev. D **57**, 7570 (1998).
24. N. Arkani-Hamed et al., JHEP **0706**, 060 (2007).
25. F. Denef and S. A. Hartnoll, Phys. Rev. D **79**, 126008 (2009).
26. S. Sachdev, Phys. Lett. B **309**, 285 (1993).
27. I. R. Klebanov and A. M. Polyakov, Phys. Lett. B **550**, 213 (2002).
28. A. Petkou, Annals Phys. **249**, 180 (1996).
29. A.V. Chubukov et al., Phys. Rev. B **49**, 11919 (1994).
30. A. Buchel et al., Nucl. Phys. B **707**, 56 (2005); M. Brigante et al., Phys. Rev. D **77**, 126006 (2008); Y. Kats and P. Petrov, JHEP **0901**, 044 (2009).

Holography and the Speed of Sound

Paul M. Hohler and Mikhail A. Stephanov

Department of Physics, University of Illinois, Chicago, IL 60607-7059, USA

We study the approach of the speed of sound towards the value $c_s^2 = 1/3$ at high temperatures in a general class of strongly interacting theories. This class includes theories holographically dual to a theory of gravity coupled to a single scalar field, representing the operator of the scale anomaly. We show that quite universally the conformal value $c_s^2 = 1/3$ is approached *from below* and give a generic argument for why this has to be so.

1. Introduction

The discovery of the holographic correspondence between string and gauge theories [1–3] has led to a growing number of applications ranging from hadronic physics to the physics of condensed matter. Among most fruitful is the application of the gauge/string correspondence to modeling properties of quark-gluon plasma of QCD in the regime relevant to heavy-ion collision experiments (see, e.g., Ref. [4] for a guide and references).

Although the precise form of the theory holographically dual to QCD is not known, many features of QCD can be implemented on the string theory side of the correspondence. Predictably, implementing such features comes at the expense of analytical control. Alternatively, one can pursue a bottom-up approach, by constructing an effective theory, or a model, limiting the description to a *minimal* set of operators relevant for the phenomena under consideration. In this paper, we consider a general class of such models, describing the physics of scaling violation in QCD. In Refs. [5, 6, 7] such theories have been considered with the aim to model, or mimic, the equation of state of QCD known numerically from the lattice calculations. This has been achieved by tuning the potential $V(\phi)$ of the scalar field in the model.

The main goal of this paper is to establish properties of the equation of state $p(\epsilon)$ which are *universal* with respect to the choice of the potential $V(\phi)$. An obvious universal property is that the conformal symmetry is gradually restored, and thus $p(\epsilon) \to \epsilon/3$, as the temperature $T \to \infty$. This is natural since the temperature becomes the only relevant scale, once it exceeds significantly the scale associated with conformality violation. It is a welcomed feature of the models since the same phenomenon occurs in QCD.

Here we shall show that the *deviation* of the speed of sound $c_s^2 = dp/d\epsilon$ from the conformal limit $c_s^2 = 1/3$ is universally *negative* in such holographic models, at least to the leading order in inverse temperature.

At the outset, we must point out that $c_s^2 = 1/3$ is by no means a universal upper bound on the sound velocity. A counterexample is given by the speed of sound in QCD at large isospin chemical potential,[8] where c_s can approach the speed of light in a certain limit. Nevertheless, a number of string theory examples[9–11] of holographically dual theories do indeed consistently show $c_s^2 \leq 1/3$.

In QCD, Monte Carlo lattice calculations [12] show that this inequality is fulfilled. In the regime $T \gg \Lambda_{\text{QCD}}$, this inequality follows from asymptotic freedom: $\beta(\alpha) < 0$. For a pure N_c-color Yang-Mills theory at high T, for example, [13]:

$$c_s^2 \simeq 1/3 + 5N_c/(36\pi)\,\beta(\alpha) < 1/3\,. \tag{1}$$

How general is the inequality $c_s^2 < 1/3$ and what are the prerequisites for it to hold? For example, does it hold in regimes where the theory is close to being conformal, but is still strongly interacting, as is apparently the case for QCD in the range of $T \sim (2-3)\,T_c$?

To investigate this question, we consider a class of theories, or models, which possess a holographically dual description. The minimal set of operators which we need to consider consists of the stress-energy tensor $T^{\mu\nu}$ and a scalar operator $\mathcal{O}$ necessary to provide a nonzero right-hand side of the scale (trace) anomaly equation. As an example, in QCD, the role of such an operator is played by $\text{Tr}F^2$: $\theta = \beta(\alpha)/(8\pi\alpha^2)\,\text{Tr}F^2$.

At this point it is helpful to express the energy density ϵ and pressure p in terms of the heat function (enthalpy) $w = \epsilon + p$ and the trace of the stress-energy tensor $\theta = \epsilon - 3p$:

$$\epsilon = (3w + \theta)/4\,; \qquad p = (w - \theta)/4\,. \tag{2}$$

Then

$$c_s^2 = \frac{dp}{d\epsilon} = \frac{1 - d\theta/dw}{3 + d\theta/dw}\,. \tag{3}$$

In conformal theories $\theta \equiv 0$ and $c_s^2 = 1/3$. The inequality $c_s^2 < 1/3$ is equivalent to $d\theta/dw > 0$. This inequality is somewhat reminiscent, but not equivalent, to the inequality $d(pT^{-4})/dT = \theta/T^5 > 0$ conjectured in Ref. [14] and the weaker constraint $\int dT\,\theta/T^5 > 0$ proposed in Refs. [15,16].

Since the enthalpy w appears to play here the role of a more natural thermal parameter than T, it is helpful to keep in mind that w is a monotonous function of T: $dw/dT = c_v + s > 0$.

2. The theory

The class of the 5-dimensional holographic theories which we consider contains a scalar field ϕ, holographically dual to the (scaling violating) operator $\mathcal{O}$, coupled to the metric $g_{\mu\nu}$ (dual to $T^{\mu\nu}$). The fields are governed by the action

$$S_5 = \frac{1}{2\kappa^2}\left(\int_M d^5x\,\sqrt{-g}\left(R - V(\phi) - \frac{1}{2}(\partial\phi)^2\right) - 2\int_{\partial M} d^4x\,\sqrt{-\gamma}\,K\right), \tag{4}$$

where R is the Ricci scalar, g is the determinant of the metric, γ is the determinant of the induced metric on the UV boundary ∂M, K is the extrinsic curvature on ∂M, and κ^2 is the 5D Einstein gravitational constant. The value of κ^2 is inversely proportional to the number of the degrees of freedom in the dual 4-dimensional theory, e.g., N_c^2 in a gauge theory with large number of colors N_c. The smallness of κ^2 (i.e., the largeness of the number of colors) controls the semiclassical approximation which we use. The last term in the action is the Gibbons-Hawking term, which removes the boundary terms arising upon integration by parts of the terms in R linear in second derivatives of the metric.[17] This boundary term does not affect classical equations of motion, but is essential for evaluating variations of the action with respect to the boundary values of the fields.

We assume that the potential $V(\phi)$ has an extremum at $\phi = 0$. The value of $V(0) = 2\Lambda = -12$, where Λ is the cosmological constant necessary to achieve proper asymptotics of the metric near the boundary.

It should be noted that we are going to treat the holographic theory with a method technically different from the one used in either Ref. [5, 6] or Ref. [7]. Our approach is complementary to the existing ones and has its own advantages, which might go beyond the specific application which we consider in this paper.

According to the holographic correspondence, the correlation functions of the dual 4D theory are equal to the variations of the 5D action under the changes of the boundary conditions on the 5D fields. We shall first determine the extremum of the action, and then consider first-order variations, which we can relate to energy density and pressure. The most general metric (up to general coordinate transformations) possessing 3D Euclidean isometry is given by:

$$ds^2 = \frac{1}{z^2}\left(-f(z)dt^2 + d\vec{x}^2\right) + e^{2B(z)}\frac{dz^2}{z^2 f(z)}\,. \tag{5}$$

The functions $B(z)$ and $f(z)$ depend on the holographic coordinate z and will be determined by extremizing the action. The boundary ∂M is located at $z = \varepsilon$ with ε acting as an UV regulator. In relations involving only physical quantities ε can be taken to 0.

Substituting the metric of Eq. (5) into the equations of motion we find:

$$\dot{B} = -\tfrac{1}{6}\,\dot{\phi}^2\,, \tag{6}$$

$$\ddot{f} = \left(4 + \dot{B}\right)\dot{f}\,, \tag{7}$$

$$-6\dot{f} + f\left(24 - \dot{\phi}^2\right) + 2e^{2B}\,V(\phi) = 0\,, \tag{8}$$

$$\ddot{\phi}f + \dot{\phi}\left(\dot{f} - f(4 + \dot{B})\right) - e^{2B}\,V'(\phi) = 0\,, \tag{9}$$

where a dot denotes a $\log z$ derivative, e.g., $\dot{\phi} = z\,d\phi/dz$, while $V' = dV/d\phi$. The holographic correspondence provides the boundary conditions at the UV boundary $z = \varepsilon$. The boundary condition on ϕ,

$$\phi(\varepsilon) = c\,\varepsilon^{\Delta_-}\,, \tag{10}$$

corresponds to introducing a term $c\,\mathcal{O}$ into the action of the 4D theory dual to the theory in Eq. (4), where c is the source of the scalar operator $\mathcal{O}$, dual to the field ϕ. The dimensions of the source and the operator are given by

$$[c] = \Delta_-; \quad [\mathcal{O}] = \Delta_+; \quad \Delta_+ + \Delta_- = 4\,. \tag{11}$$

The metric must approach the Lorentz invariant form at the UV boundary, hence,

$$f(\varepsilon) = 1\,. \tag{12}$$

Equation (7) can be integrated once to give,

$$\dot{f} = -wz^4 e^B. \tag{13}$$

The integration constant w must be positive if the metric is to possess a horizon $f(z_H) = 0$ at some value of z_H.

There are no more independent boundary conditions. The boundary value of B is determined by Eq. (8) which is algebraic in B. The role of the second boundary condition for Eq. (9) is played by the requirement that ϕ is finite at the horizon, $z = z_H$, which is a regular singular point of the second order differential equation (9).

Thus we find a two-parameter family of solutions. These parameters are c from Eq. (10) and w from Eq. (13). Different members of this family are related by rescaling $z \to z/\lambda$, $w \to \lambda^4 w$, $c \to \lambda^{\Delta_-} c$, which represents scale invariance inherent in the action given by Eq. (4).

Near the $z = \varepsilon \to 0$ boundary, $\phi \to 0$, $B \to 0$, $\dot{B} \to 0$ and $\dot{f} \to 0$. Equation (9) for ϕ can be linearized and the asymptotic behavior of ϕ near the boundary can be determined easily:

$$\phi(z) \to (c - d\,\varepsilon^{\Delta_+ - \Delta_-})\, z^{\Delta_-} + d\, z^{\Delta_+} + \ldots, \tag{14}$$

where the curvature of the potential $V''(0) \equiv m^2$ determines the indices $\Delta_\pm = 2 \pm \sqrt{4 + m^2}$. The coefficient of the first term is related to c by Eq. (10). The coefficient d of the second linearly independent solution should be determined by the finiteness condition at the horizon.

By calculating the derivative of the 5D action with respect to c and matching it, by holographic correspondence, to the expectation value $\langle \mathcal{O} \rangle$, one finds [18]

$$\langle \mathcal{O} \rangle = -\partial S_5/\partial c = -d\,(\Delta_+ - \Delta_-)\,. \tag{15}$$

3. Equation of state

Rather than following existing approaches to calculating the speed of sound, based on the relation [5,6] $c_s^2 = d\log T/d\log s$, or looking at the poles of the two-point function for the sound channel,[19,20] we shall directly calculate *one*-point functions $\langle T^{tt} \rangle$ and $\langle T^{xx} \rangle$ and use the relation $c_s^2 = dp/d\epsilon$. As we shall see, this quickly yields some closed-form results, not evident in the alternative approaches.

In order to calculate the one-point function, such as $\langle T^{tt} \rangle$, we observe that the generalization of the 4D theory to a curved 4D background with metric $h_{\mu\nu}$ corresponds, holographically, to imposing the following boundary condition on the bulk 5D metric:

$$g_{\mu\nu}(\varepsilon) = h_{\mu\nu}\, \varepsilon^{-2}\,. \tag{16}$$

Holographic correspondence then gives us

$$\langle T^{\mu\nu} \rangle = 2\,\delta S_5/\delta h_{\mu\nu}\,. \tag{17}$$

Variation of the boundary condition causes variations of the metric in the bulk, as it follows the equations of motion. But since we are varying around the extremum of the action, the only contribution to the first order variation comes from the boundary. This contribution consists of the variation of the boundary terms which appear during integration by parts while deriving equations of motion for, e.g., g_{tt}, as well as the variation of the Gibbons-Hawking term. Altogether this gives

$$\langle T^{tt}\rangle = -\frac{6}{\varepsilon^4}e^{-B(\varepsilon)}; \qquad \langle T^{xx}\rangle = w - \langle T^{tt}\rangle. \tag{18}$$

Both of these quantities are divergent as $\varepsilon \to 0$. This represents the familiar vacuum energy divergence in the quantum field theory. A simple vacuum subtraction

$$\begin{aligned} \epsilon &= \langle T^{tt}\rangle - \langle T^{tt}\rangle_{T=0}\,, \\ p &= \langle T^{xx}\rangle - \langle T^{xx}\rangle_{T=0}\,, \end{aligned} \tag{19}$$

takes care of this, and we find for the energy density and the pressure, after solving Eq. (8) for B at $z = \varepsilon \to 0$ with ϕ from Eq. (14):

$$\begin{aligned} \epsilon &= (\,3w - c\,d\,\Delta_-\,(\Delta_+ - \Delta_-)\,)/4\,, \\ p &= (\,w + c\,d\,\Delta_-\,(\Delta_+ - \Delta_-)\,)/4\,. \end{aligned} \tag{20}$$

We can now conclude that the integration constant introduced in Eq. (13) is the enthalpy, $w = \epsilon + p$, of the corresponding 4D field theory. We also observe that the scale anomaly

$$\theta = \epsilon - 3p = -c\,d\,\Delta_-\,(\Delta_+ - \Delta_-) = \Delta_-\,c\,\langle\mathcal{O}\rangle, \tag{21}$$

is related to the expectation value of the operator $\mathcal{O}$ (using Eq. (15)). This relation is easy to derive directly on the field theory side by observing that the violation of the scale invariance comes from the dimensionful parameter c. Naturally, it is proportional to the dimension of c, Δ_-.

From Eq. (20), the speed of sound can be expressed now in terms of the derivative $\partial d/\partial w$ at fixed c and dimensions $\Delta_\pm$:

$$c_s^2 = \frac{dp}{d\epsilon} = \frac{1 + c\,\Delta_-\,(\Delta_+ - \Delta_-)\,(\partial d/\partial w)}{3 - \,c\,\Delta_-\,(\Delta_+ - \Delta_-)\,(\partial d/\partial w)}\,. \tag{22}$$

It should be emphasized that this equation is an *exact* expression valid for *all* temperatures.

4. High temperature

The high-temperature limit is equivalent to the small c limit, since c is the only other dimensionful parameter in the theory, and high temperature expansion is controlled by the dimensionless parameter c/T^{Δ_-}, or $c\,w^{-\Delta_-/4}$.

In this limit,

$$c_s^2 = 1/3 + (4/9)\, c\, \Delta_- \,(\Delta_+ - \Delta_-)\,(\partial d/\partial w) + \dots \,. \tag{23}$$

We shall now show that, even though generally the dependence of d on w can be only found numerically, in the high-temperature (i.e., high w) limit an analytic expression can be found for arbitrary potential $V(\phi)$.

One can begin by observing that at large w the function f varies very rapidly according to Eq. (13). This means one can neglect variations of the function B between the boundary $z = \varepsilon$ and the horizon $z = z_H$, since z_H becomes small (as $w^{-1/4}$). Since on the boundary $B = 0$ (up to terms of order $\varepsilon^{2\Delta_-}$, negligible here, according to Eq. (8)), we find from Eq. (13):

$$f(z) = 1 - w\, z^4/4\,. \tag{24}$$

Another consequence is that ϕ, which is small at $z = \varepsilon$, remains small up to z_H ($\phi \sim c z_H^{\Delta_-} \sim c/T^{\Delta_-} \ll 1$), and the linearized approximation to Eq. (9) is valid not only near the boundary, but all the way to the horizon. With $B = 0$ and f from Eq. (24) we obtain

$$\left(1 - \frac{1}{4}\, w z^4\right)\phi'' - \left(\frac{3}{z} + \frac{w z^4}{4}\right)\phi' - \frac{m^2}{z^2}\,\phi = 0\,. \tag{25}$$

Equation (25) can be solved analytically

$$\begin{aligned}\phi(z) = {}& c\, z^{\Delta_-}\; {}_2F_1\left(\Delta_-/4,\, \Delta_-/4,\, \Delta_-/2,\, w z^4/\,4\right)\\ &+ d\, z^{\Delta_+}\; {}_2F_1\left(\Delta_+/4,\, \Delta_+/4,\, \Delta_+/2,\, w z^4/4\right),\end{aligned} \tag{26}$$

where the coefficients follow the notations of Eq. (14) (up to terms $\mathcal{O}(\varepsilon^{\Delta_+ - \Delta_-})$, here negligible). Both linearly independent solutions are logarithmically divergent at the horizon $z = z_H$, where $w z_H^4/4 = 1$. The condition $|\phi(z_H)| < \infty$ requires us to select the linear combination in which these divergences cancel. This fixes d in terms of c:

$$d = -c\, w^{(\Delta_+ - \Delta_-)/4}\, D(\Delta_-)\,, \tag{27}$$

where the function $D(\Delta_-) = 1/D(\Delta_+)$ is given by

$$D(\Delta_-) = \frac{\pi\, 2^{\Delta_-}}{2 - \Delta_-}\, \cot\left(\pi \Delta_-/4\right) \frac{\Gamma(\Delta_-/2)^2}{\Gamma(\Delta_-/4)^4}\,. \tag{28}$$

Substituting Eq. (27) into Eq. (22) we find our main result

$$c_s^2 = \frac{1}{3} - \frac{1}{9}\, c^2\, \Delta_- (\Delta_+ - \Delta_-)^2 w^{-\Delta_-/2} D(\Delta_-) + \dots \,. \tag{29}$$

It is clear that the correction term is *negative* for all values of $0 < \Delta_- < 2$ (i.e., $2 < \Delta_+ < 4$). As expected, the correction vanishes with the power of w dictated by the dimension of the operator $\mathcal{O}$.

5. Discussion

When this work was completed, the authors learned about a similar result,[21] obtained using the relation $c_s^2 = d\log T/d\log s$. In Ref. [21] the speed of sound is expressed in terms of the value ϕ_H of the scalar field at the horizon (since both s and T are calculated at the horizon): $c_s^2 = 1/3 - C\phi_H^2 + \ldots$. The coefficient C is then given by an integral from the boundary to the horizon of the square of a hypergeometric function, which is manifestly positive. We have verified that, upon evaluating the integral, the result of Ref. [21] coincides with Eq. (29). For completeness of comparison we evaluate ϕ_H explicitly by using Eq. (26) and applying relation Eq. (27):

$$\phi_H = c\, w^{\Delta/4-1}\, 2^{-\Delta/2}(2\Delta - 4)\, \frac{\Gamma(\Delta/4)^2}{\Gamma(\Delta/2)}\,. \tag{30}$$

where $\Delta = \Delta_+$. Using Eq. (29) we then find

$$c_s^2 = \frac{1}{3} - \frac{1}{18\pi}\,(4-\Delta)\,(4-2\Delta)\,\tan\left(\pi\Delta/4\right)\,\phi_H^2 + \ldots\,. \tag{31}$$

As we have pointed out already, and as Eq. (30) demonstrates explicitly, in the high temperature (large w) limit the value of ϕ remains small everywhere between the boundary and the horizon. This is the origin of the universality we find. The speed of sound near the high T limit depends only on the behavior of the scalar potential near the origin, i.e., specifically, on $V''(0) = m^2 = \Delta(\Delta - 4)$.

6. Generic argument

One can interpret and further generalize our results in the following way. In the high-temperature limit, Eq. (21) becomes $\theta = \Delta_-\, c^2\, \chi_{\mathcal{O}} + \ldots$. The susceptibility $\chi_{\mathcal{O}} \equiv \partial\langle\mathcal{O}\rangle/\partial c$ can be related to the effective potential for $\langle\mathcal{O}\rangle$, defined as the Legendre transform of the generating functional $W(c)$ (in holography $W = -S_5$), i.e., $\Gamma(\langle\mathcal{O}\rangle) \equiv W(c) + c\langle\mathcal{O}\rangle$, as $\chi_{\mathcal{O}} = 1/\Gamma''(0)$. Stability implies $\Gamma''(0) > 0$, and thus $\chi_{\mathcal{O}} > 0$. Consequently $\theta > 0$, as conjectured in [14]. Holographic models confirm these expectations according to Eqs. (15) and (27).

We also find that $d\theta/dw > 0$, which requires $d\chi_{\mathcal{O}}/dw > 0$. That means that the curvature $\Gamma''(0) = \chi_{\mathcal{O}}^{-1}$ *decreases* with temperature. This behavior is unusual, if one recalls that in a weakly coupled $\lambda\phi^4$ scalar theory, the leading perturbative temperature correction to the curvature ($\sim \lambda T^2$) increases with temperature (e.g., the restoration of a broken symmetry is a well-known manifestation of this). The opposite behavior of the curvature of the effective potential $\Gamma(\langle\mathcal{O}\rangle)$ can be understood by counting dimensions: $[\chi_{\mathcal{O}}^{-1}] = \Delta_- - \Delta_+ < 0$, which means $\chi_{\mathcal{O}}^{-1} \sim T^{\Delta_- - \Delta_+}$ and decreases with T.

In conclusion, we have shown that in a quite general class of gravity dual theories with a single scalar operator representing the scale anomaly the speed of sound always approaches the conformal value $c_s^2 = 1/3$ from *below*.

Acknowledgments

We are grateful to the organizers of *Shifmania: Crossing the Boundaries: Gauge Dynamics at Strong Coupling* for the opportunity to present this work, and wish Mikhail Shifman happy birthday and many more years of successful research.

We would like to thank Aleksey Cherman, Thomas Cohen, and Abhinav Nellore for discussions and for sharing the results of Ref. [21] prior to publication. The hospitality of the Institute of the Nuclear Theory at the University of Washington and partial support of the Department of Energy through the program "Strings and Things" during the early stages of this work is gratefully acknowledged. This work is supported by the DOE grant No. DE-FG0201ER41195.

References

1. J. M. Maldacena, Adv. Theor. Math. Phys. **2**, 231 (1998) [Int. J. Theor. Phys. **38**, 1113 (1999)] [arXiv:hep-th/9711200].
2. S. S. Gubser, I. R. Klebanov and A. M. Polyakov, Phys. Lett. B **428**, 105 (1998) [arXiv:hep-th/9802109].
3. E. Witten, Adv. Theor. Math. Phys. **2**, 253 (1998) [arXiv:hep-th/9802150].
4. S. S. Gubser and A. Karch, arXiv:0901.0935 [hep-th].
5. S. S. Gubser and A. Nellore, Phys. Rev. D **78**, 086007 (2008) [arXiv:0804.0434 [hep-th]].
6. S. S. Gubser, A. Nellore, S. S. Pufu and F. D. Rocha, Phys. Rev. Lett. **101**, 131601 (2008) [arXiv:0804.1950 [hep-th]].
7. U. Gursoy, E. Kiritsis, L. Mazzanti and F. Nitti, arXiv:0812.0792 [hep-th].
8. D. T. Son and M. A. Stephanov, Phys. Atom. Nucl. **64**, 834 (2001) [Yad. Fiz. **64**, 899 (2001)] [arXiv:hep-ph/0011365].
9. A. Buchel and J. T. Liu, JHEP **0311**, 031 (2003) [arXiv:hep-th/0305064].
10. P. Benincasa, A. Buchel and A. O. Starinets, Nucl. Phys. B **733**, 160 (2006) [arXiv:hep-th/0507026].
11. A. Buchel, Nucl. Phys. B **708**, 451 (2005) [arXiv:hep-th/0406200].
12. F. Karsch, PoS C **POD07**, 026 (2007) [arXiv:0711.0656 [hep-lat]].
13. J. I. Kapusta, "Finite-Temperature Field Theory," Cambridge University Press, 1993.

14. J. D. Bjorken, Phys. Rev. D **27**, 140 (1983).
15. T. Appelquist, A. G. Cohen and M. Schmaltz, Phys. Rev. D **60**, 045003 (1999) [arXiv:hep-th/9901109].
16. T. Appelquist, A. G. Cohen, M. Schmaltz and R. Shrock, Phys. Lett. B **459**, 235 (1999) [arXiv:hep-th/9904172].
17. G. W. Gibbons and S. W. Hawking, Phys. Rev. D **15**, 2752 (1977).
18. I. R. Klebanov and E. Witten, Nucl. Phys. B **556**, 89 (1999) [arXiv:hep-th/9905104].
19. P. K. Kovtun and A. O. Starinets, Phys. Rev. D **72**, 086009 (2005) [arXiv:hep-th/0506184].
20. G. Policastro, D. T. Son and A. O. Starinets, JHEP **0212**, 054 (2002) [arXiv:hep-th/0210220].
21. A. Cherman, T. D. Cohen and A. Nelore, arXiv:0905.0903 [hep-th].

Chiral Symmetry Breaking in a Soft-Wall Model of AdS/QCD

Joseph I. Kapusta and Thomas M. Kelley

School of Physics and Astronomy, University of Minnesota,
Minneapolis, MN 55455, USA
E-mail: kapusta@physics.umn.edu
kelley@physics.umn.edu

Tony Gherghetta

School of Physics, University of Melbourne,
Victoria 3010, Australia
E-mail: tgher@unimelb.edu.au

We incorporate chiral symmetry breaking in a soft-wall version of the AdS/QCD model by using a modified dilaton profile and a quartic term in the bulk scalar potential. This allows one to separate the dependence on spontaneous and explicit chiral symmetry breaking. The resulting mass spectra in the scalar, vector and axial-vector sectors compares favorably with the respective QCD resonances.

Keywords: Gauge/gravity duality, chiral symmetry, meson properties.

1. Introduction

The AdS/CFT correspondence [1–4] provides a remarkable mathematical tool with which to understand strongly-coupled gauge theories. Calculations performed on the gravity side can be reinterpreted as due to nonperturbative effects on the field theory side. The theory of the strong interactions, quantum chromodynamics (QCD), is a strongly-coupled gauge theory at low energies, and therefore a natural candidate with which to apply the gauge/gravity correspondence. This had led to a bottom-up approach, commonly known as AdS/QCD,[5,6] which relates QCD to a five-dimensional (5D) gravity theory. This simple model captures the essential features of the low-lying meson spectrum.

To incorporate more realistic features of the excited states, the AdS/QCD model can be modified to include a dilaton with a quadratic profile.[7] While the linear radial spectrum is indeed produced in this soft-wall version of the AdS/QCD model, the form of chiral symmetry breaking is not QCD-like. In particular, the bulk scalar field, dual to the quark bilinear operator $\bar{q}q$, whose vacuum expectation value is responsible for spontaneous chiral symmetry breaking, does not allow the spontaneous and explicit breaking to be independent. Moreover, chiral symmetry is restored for the highly excited states, a phenomenological feature that is not supported in the QCD mass spectrum.[8]

We [9] modify the existing soft-wall version of the AdS/QCD model in order to incorporate these two phenomenological features of QCD. This is done by adding a quartic term to the bulk scalar potential and changing the dilaton profile. By assuming that the bulk scalar field contains the desired limiting behavior for non-restoration of chiral symmetry,[4,8] we derive a new dilaton background profile. The extra parameter introduced by the quartic term decouples the quark mass from the chiral condensate, thereby allowing for spontaneous and explicit chiral symmetry breaking to occur independently.

2. The Model

We consider a modified version of the soft-wall AdS/QCD model first introduced in Ref. [7] and investigated further in Refs. [10–16]. The background geometry is assumed to be 5D AdS space with the metric

$$ds^2 = g_{MN}dx^M dx^N = a^2(z)\left(\eta_{\mu\nu}dx^\mu dx^\nu + dz^2\right) , \qquad (1)$$

where $a(z) = L/z$ is the warp factor, L is the AdS curvature radius and the Minkowski metric $\eta_{\mu\nu} = \mathrm{diag}(-1,+1,+1,+1)$. The conformal coordinate z has a range $0 \le z < \infty$. To obtain linear trajectories, Ref. [7] also introduced a background dilaton field $\phi(z) = \lambda z^2$ where λ sets the mass scale for the meson spectrum.

To describe chiral symmetry breaking in the meson sector the 5D action contains $\mathrm{SU}(2)_L\times \mathrm{SU}(2)_R$ gauge fields and a bifundamental scalar field X. As suggested by Ref. [7], we add a quartic term in the potential $V(X)$ of our 5D action,

$$S_5 = -\int d^5x\sqrt{-g}\,e^{-\phi(z)}\mathrm{Tr}\left[|DX|^2 + m_X^2|X|^2 - \kappa|X|^4 + \frac{1}{4g_5^2}(F_L^2 + F_R^2)\right], \qquad (2)$$

where $m_X^2 = -3/L^2$, κ is a constant, $g_5^2 = 12\pi^2/N_c$, and the covariant derivative is $D^M X = \partial^M X - iA_L^M X + iXA_R^M$. To describe the vector and axial-vector mesons we simply transform to the vector (V) and axial-vector (A) fields where $V^M = \frac{1}{2}(A_L^M + A_R^M)$ and $A^M = \frac{1}{2}(A_L^M - A_R^M)$.

The scalar field X, which is dual to the operator $\bar{q}q$, is assumed to obtain a z-dependent vacuum expectation value (VEV),

$$\langle X \rangle \equiv \frac{v(z)}{2} \begin{pmatrix} 1 & 0 \\ 0 & 1 \end{pmatrix}, \tag{3}$$

which breaks the chiral symmetry $\mathrm{SU}(2)_L \times \mathrm{SU}(2)_R to \mathrm{SU}(2)_V$. Assuming (3) we obtain a nonlinear equation for $v(z)$,

$$\partial_z(a^3 e^{-\phi}\partial_z v(z)) - a^5 e^{-\phi}(m_X^2 v(z) - \frac{\kappa}{2} v^3(z)) = 0. \tag{4}$$

When $\kappa = 0$, the solution of (4) which leads to a finite action in the limit $z \to \infty$ is given by [7,14]

$$v(z) = m_q \frac{z}{L} U\left(\frac{1}{2}, 0, \lambda \frac{z^2}{L^2}\right), \tag{5}$$

where $U(a, b, y)$ is the Tricomi confluent hypergeometric function.

As expected from the AdS/CFT dictionary established in Refs. [3,4], as $z \to 0$ the VEV should take the form

$$v(z) = \frac{z}{L}\left(m_q \zeta + \frac{\Sigma}{\zeta} z^2 + \cdots\right). \tag{6}$$

Here m_q is the quark mass, $\Sigma \equiv \langle \bar{q}q \rangle$ is the chiral condensate, and ζ is the normalization parameter introduced in Ref. [13]. For fixed values of m_q and Σ, the introduction of ζ still satisfies the Gell-Mann-Oakes-Renner relation, $m_\pi^2 f_\pi^2 = 2m_q\Sigma$. In actuality the normalization ζ is not a free parameter but is determined by QCD, as shown in Ref. [13], to be $\zeta = \sqrt{3}/g_5$. Expanding the solution (5) in the small z limit leads to $\alpha \propto m_q$ and $\beta \propto \Sigma \propto m_q$. Thus in the limit $m_q \to 0$ the model eliminates explicit and spontaneous chiral symmetry breaking in contradiction with QCD. It will be seen that the introduction of a quartic term in the potential $V(X)$ avoids the dependence of the chiral condensate on the quark mass encountered in Refs. [7,12,14].

The solution (5) has an asymptotic limit $v(z) \to$ constant for large values of z. This asymptotic behavior suggests chiral symmetry restoration in the mass spectrum, a phenomenon not supported in QCD (although speculation continues on whether such a restoration indeed exists[17,18]). As noted in Ref. [8] the highly excited mesons exhibit seemingly parallel slopes signifying that chiral symmetry is not restored. In order to incorporate this behavior $v(z)$ must behave linearly as z becomes large, $v(z \to \infty) \sim z$ causing the mass-squared difference between vector and axial-vector resonances to approach a constant as $z \to \infty$. By including a quartic term and requiring $v(z)$ to have this linear asymptotic behavior we aim to incorporate these QCD-like characteristics into the soft-wall model.

Instead of solving for $v(z)$ directly, we assume the VEV asymptotically behaves as expected, namely Eq. (6) and

$$v(z \to \infty) = \frac{\gamma}{L} z\,, \tag{7}$$

and then solve for the dilaton $\phi(z)$ using (4), which becomes

$$\phi'(z) = \frac{1}{a^3 v'(z)} \left[\partial_z (a^3 v'(z)) - a^5 (m_X^2 v(z) - \frac{\kappa}{2} v^3(z)) \right], \tag{8}$$

where the prime denotes the derivative with respect to z. Given the required behavior (6) and (7) we can uniquely determine the dilaton profile up to a constant. With this procedure the two sources of chiral symmetry breaking decouple while simultaneously allowing for linear trajectories in the meson spectrum.

A particularly simple parametrized form for $v(z)$ that satisfies (6) and (7) is

$$v(z) = \frac{z}{L}(A + B \tanh C z^2), \tag{9}$$

The parameters γ, A, B and C can be expressed in terms of the input parameters $m_q, \Sigma, \lambda, \kappa$ as

$$\gamma = \sqrt{\frac{4\lambda}{\kappa}}, \tag{10}$$

$$A = \frac{\sqrt{3} m_q}{g_5}, \tag{11}$$

$$B = \gamma - A, \tag{12}$$

$$C = \frac{g_5 \Sigma}{\sqrt{3} B}. \tag{13}$$

The parameter λ is determined by the average slope of the radial trajectories of the scalar, vector, and axial-vector mesons for radial quantum numbers $n \geq 3$. Its value was determined to be $\lambda = 0.1831\ \mathrm{GeV}^2$, as explained in the next section. The quark mass, quark condensate, pion decay constant, and pion mass are all related by the Gell-Mann-Oakes-Renner relation, $f_\pi^2 m_\pi^2 = 2 m_q \Sigma$. This relation holds in this model as a natural consequence of chiral symmetry.[5] We use the measured values of $f_\pi = 92.4$ MeV and $m_\pi = 139.6$ MeV, and adjust the quark mass to reproduce the input value of f_π from a solution to the axial-vector field equation in Section 4 for a given value of κ. The parameter κ essentially controls the mass splitting between the vector and axial-vector mesons. It is determined to be $\kappa = 15$ by a best fit to the radial spectra of the axial-vector mesons. This results in $m_q = 9.73$ MeV and therefore $\Sigma = (204.5 \text{ MeV})^3$. The inferred value of the quark mass is consistent with an average of the up and down quark masses as summarized in the Review of Particle Physics [19] as appropriate at the hadronic energy scale.

Other parameterizations of $v(z)$ which lead to qualitatively similar behavior as that required in (6) and (7) can also be used. Other forms for $v(z)/z$ include $(a_1 + a_2 z^2)/(1 + a_3 z^2)$ [Pade] and $c_1 - c_2 \exp(-c_3 z^2)$ [Gaussian]. These forms were all studied but the best results were found to be obtained using the form (9). These parameterizations are shown in Fig. 1 using the above parameters. The resulting dilaton profile $\phi(z)$ is shown in Fig. 2.

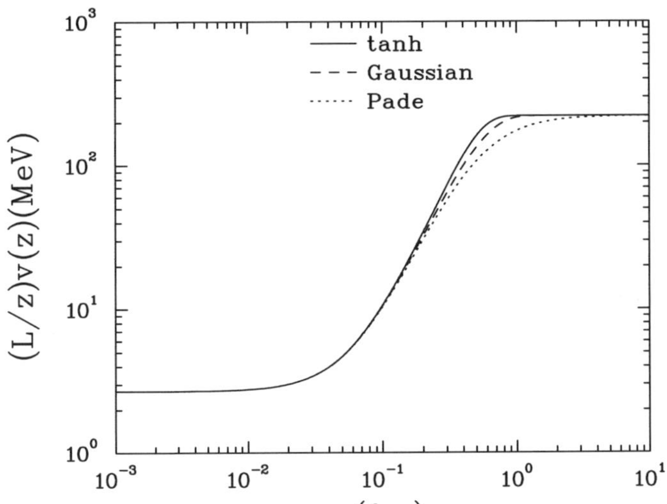

Fig. 1. *A plot of $v(z)/z$ for various parameterizations fitted to the mass spectra. The best fit to the mass spectra is obtained with the tanh form (9).*

3. Meson Mass Spectra

The soft wall model can be used to fit the observed scalar, vector, and axial-vector meson mass spectra.[19–22] All but one of the included states are

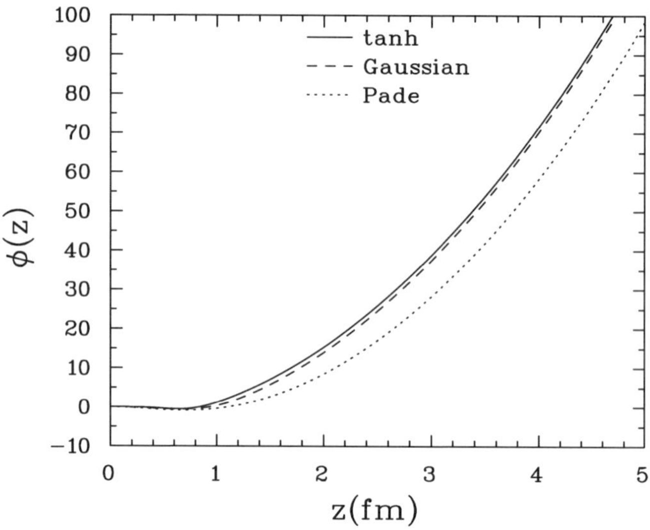

Fig. 2. *The dilaton profile $\phi(z)$ resulting from the various parameterizations of $v(z)$. The best fit to the meson spectra occurs with the tanh parameterization (9). For $z \lesssim 1$ the behavior deviates from the quadratic asymptotic form.*

listed in the Review of Particle Physics (RPP).[19] For an explanation of the masses used see Ref. [9]. A straight line is fitted to the m^2 versus n plot with $n \geq 3$ for all three mesons, assuming the same slope 4λ but different intercepts, namely $m_n^2 = 4\lambda n + m_0^2$. The results are: $\lambda = 0.1831 \pm 0.0059$ GeV^2, $m_{V,0}^2 = 0.0806 \pm 0.0104$ GeV^2, $m_{A,0}^2 = 1.5023 \pm 0.0366$ GeV^2, and $m_{S,0}^2 = -0.6634 \pm 0.0038$ GeV^2. See Fig. 3. We use this value of λ in our model calculations.

3.1. *Scalars*

Introducing a quartic term in the Lagrangian causes the scalar excitations to couple with their own VEV, giving a modified equation of motion unlike those in Refs. [6,14]. The equation of motion can be brought into a Schrödinger-like form

$$-\partial_z^2 s_n(z) + \left(\frac{1}{4}\omega_s'^2 - \frac{1}{2}\omega_s'' - \frac{3}{2}\frac{L^2}{z^2}\kappa v^2(z) - \frac{3}{z^2}\right) s_n(z) = m_{S_n}^2 s_n(z). \quad (14)$$

where $\omega_s = \phi(z) + 3\log z$. To solve this equation we implement a shooting method in which (14) is solved for various values of m_{S_n}. The eigenvalues are then those mass values that produce a solution for $S_n(z)$ that is

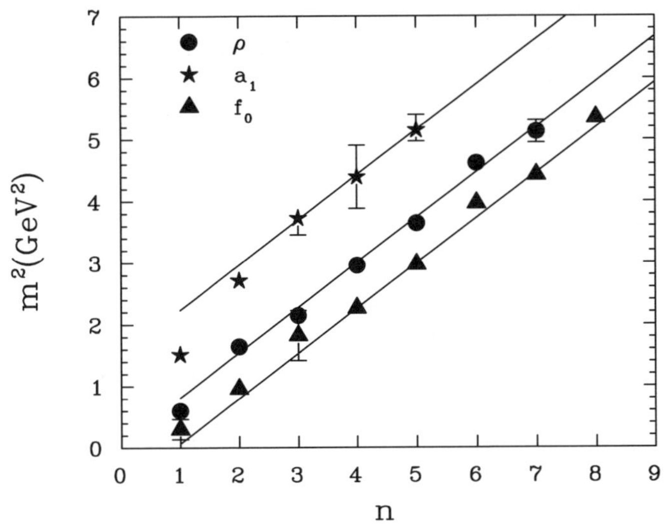

Fig. 3. *A straight-line fit to the measured scalar, vector and axial-vector mass spectra for $n \geq 3$ used to determine the dilaton mass parameter λ.*

bounded as $z \to \infty$. Applying the shooting method to (14) with the boundary conditions $\lim_{z_0 \to 0} s_n(z_0) = 0$, $\partial_z s_n(z \to \infty) = 0$ produces a scalar mass spectrum which has the correct slope but too large magnitude. This could well be a failure of this specific model. However, it has been argued[23] that the σ meson is not truly a particle but only an enhancement in the π-π cross section, and that the phase shift does not reach 90°. Removing the $f_0(550)$ would shift all the higher masses to the left by one unit of n, resulting in a better fit to the model. See Fig. 4. An obvious extension of this work would be to include strange quarks and glueballs and to determine the mixing among the resulting scalar states.

3.2. *Vectors*

From the action (2) the equation of motion of the vector field can be written in the Schrödinger form

$$-\partial_z^2 v_n + \left(\frac{1}{4}\omega'^2 - \frac{1}{2}\omega''\right) v_n = m_{V_n}^2 v_n \qquad (15)$$

where $\omega = \phi(z) + \log z$. Using the dilaton form $\phi = \lambda z^2$, the eigenvalues of (15) at large n can be solved analytically with the boundary conditions $\lim_{z_0 \to 0} v_n(z_0) = 0$, $\partial_z v_n(z \to \infty) = 0$ and agree with those found in

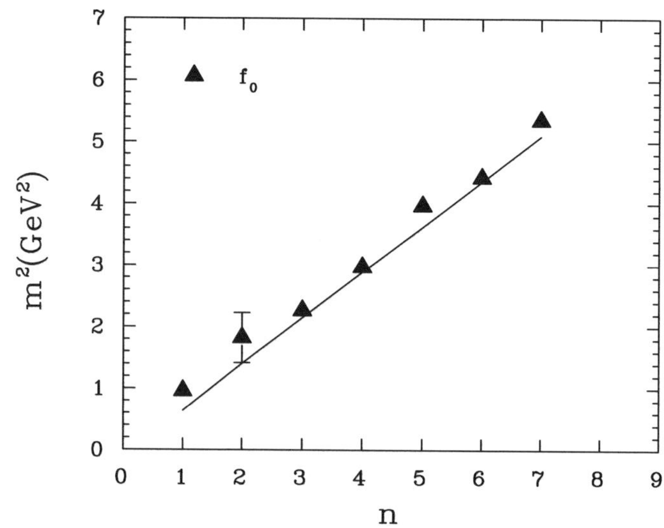

Fig. 4. *Comparison of the predicted scalar mass eigenvalues using the tanh form (9) of $v(z)$ (solid) with the QCD f_0 scalar mass spectrum with the σ meson removed [19].*

Ref. [7]. However, since the dilaton specified in (8) is modified for $z \lesssim 1$ there is a change in the slope of the mass spectrum around $n = 2$ which matches the behavior of the experimental data. The numerical vector mass spectrum is compared to the experimental data in Fig. 5. While the prediction for the $\rho(775)$ mass is low, the rest of the vector meson masses are in reasonable agreement with experiment. Most likely the agreement with the $\rho(775)$ could be improved upon by using a parameterization of $v(z)$ which rises more rapidly to its asymptotic value at large z. Nevertheless, the purpose of this paper is to incorporate QCD-like chiral symmetry breaking in soft-wall AdS/QCD models, not just to fit data.

3.3. *Axial-vectors*

Unlike the vector field, the axial-vector couples to the scalar field VEV, producing a z-dependent mass term in its equation of motion.

$$-\partial_z^2 a_n + \left(\frac{1}{4}\omega'^2 - \frac{1}{2}\omega'' + g_5^2 \frac{L^2}{z^2} v^2(z)\right) a_n = m_{A_n}^2 a_n. \tag{16}$$

The expression (16) for the axial-vector field matches that of the vector field except for the additional term, $g_5^2 v^2(z) L^2/z^2$. Using the boundary conditions $\lim_{z_0 \to 0} a_n(z_0) = 0$, $\partial_z a_n(z \to \infty) = 0$, the axial-vector meson spectrum is obtained for the fixed values of λ, m_q, Σ, and κ.

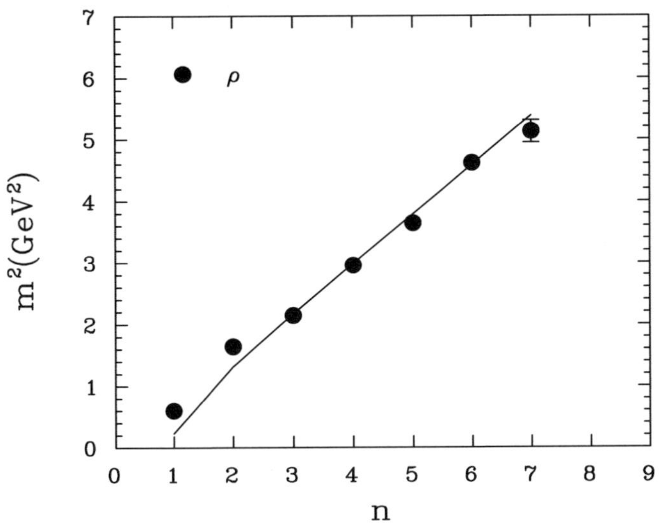

Fig. 5. *Comparison of the predicted vector mass eigenvalues using the tanh form (9) of $v(z)$ (solid) with the QCD ρ mass spectrum [19].*

The limiting behavior of $v(z)$ as $z \to \infty$ leads to a constant shift between the vector and axial-vector spectra at high mass values. Comparing the equations of motion (15) and (16) for these fields one finds the asymptotic behavior

$$\Delta m^2 \equiv \left(m_{A_n}^2 - m_{V_n}^2\right)_{n\to\infty} = g_5^2 \frac{L^2}{z^2} v^2(z \to \infty) = \frac{4g_5^2\lambda}{\kappa}. \tag{17}$$

Together with the slope λ, this determines the numerical value to be $\kappa \sim 30$, although the best visual global fit to all the data suggests $\kappa = 15$, which is the value used here. This is probably due to the small number of radial excitations to which we are fitting. The results of our analysis are plotted in Fig. 6. The $a_1(1260)$ resonance is predicted to within 5% and there is good agreement with the higher resonances of a_1.

Note that from (17), $\Delta m^2 > 0$ implies that $\kappa > 0$, which means that the potential in (2) is unbounded from below. To address the stability of the gravity-dilaton background requires a complete fluctuation analysis generalizing the work in Ref. [22]. Even though this is beyond the scope of the present work it does suggest that higher-order terms will be needed for stability.

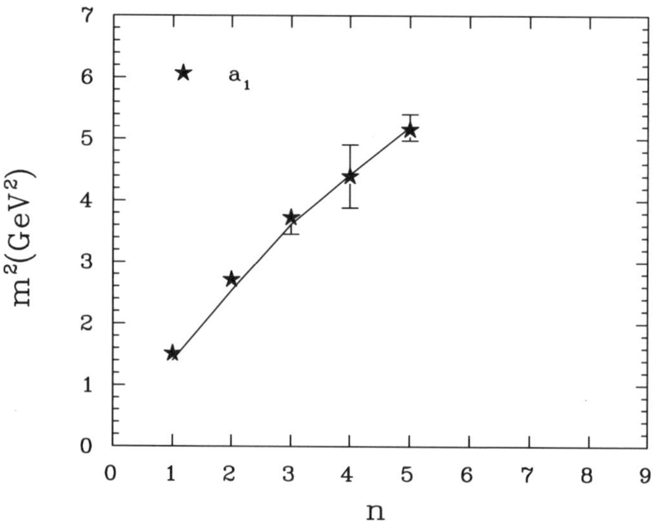

Fig. 6. *Comparison of the numerical results for the axial-vector mass eigenvalues using the tanh form (9) of $v(z)$ (solid) with the QCD a_1 mass spectrum [19].*

4. Pion Coupling

To confirm that our model is consistent with AdS/QCD model predictions, we calculate the pion decay constant using the formula [5]

$$f_\pi^2 = -\frac{1}{g_5^2} \lim_{\epsilon \to 0} \frac{\partial_z A_0(0,z)}{z}\bigg|_{z=\epsilon} . \tag{18}$$

Here $A_0(q,z)$ is the axial-vector bulk-to-boundary propagator with boundary conditions $A_0(0,\epsilon) = 1$ and $\partial_z A_0(0, z \to \infty) = 0$. The above formula returns the measured value of the pion decay constant if $m_q = 9.75$ MeV, $\kappa = 15$, and $\lambda = 0.1831$ GeV^2.

Once we calculate the $g_{\rho_n \pi\pi}$, the space-like pion form factor can easily be determined from a sum over vector meson poles. However the sum converges slowly and numerically it is much better to use the expression in terms of the vector and axial-vector bulk-to-boundary propagators as in Ref. [12]

$$F_\pi(q^2) = \int dz\, e^{-\phi(z)} \frac{V_0(q,z)}{f_\pi^2} \left(\frac{1}{g_5^2 z} (\partial_z \varphi(z))^2 + \frac{v^2(z)}{z^3} (\pi(z) - \varphi(z))^2 \right), \tag{19}$$

where $V_0(q,z)$ is the vector bulk-to-boundary propagator. The results of our $F_\pi(q^2)$ calculation are plotted in Fig. 7, and shows a slight improvement in matching the experimental values compared to that obtained in Ref. [12].

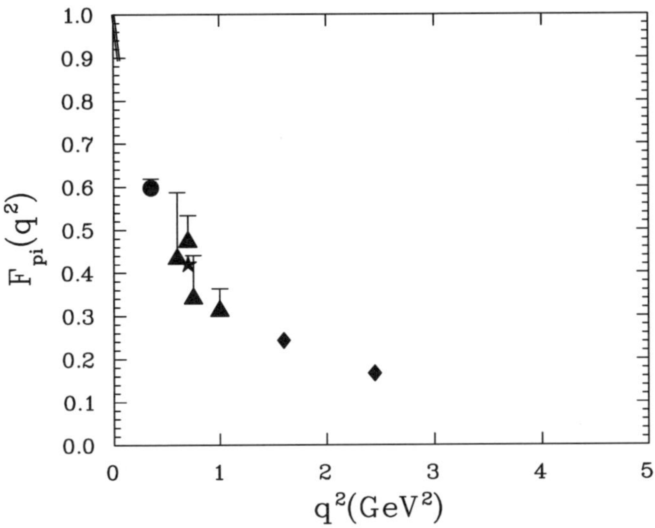

Fig. 7. *The line shows the predicted space-like behavior of the pion form factor $F_\pi(q^2)$ which is compared to the experimental data obtained from [12]. The triangles are data from DESY, reanalyzed by [24]. The diamonds are data from Jefferson Lab [25]. The circles [26] as well as the star [27] are also data obtained from DESY.*

5. Conclusion

We have shown how to incorporate chiral symmetry breaking into a soft-wall version of the AdS/QCD model with independent sources for spontaneous and explicit breaking. This is achieved by introducing a quartic term in the potential for the bulk scalar field dual to the quark bilinear operator $\bar{q}q$. This enables us to obtain reasonable agreement within 10% of the QCD meson mass spectra for scalar, vector and axial-vector fields, although the prediction for the lowest lying ρ is not as good.

Further progress must recognize the limitations of this type of phenomenological model. Genuine stringy behavior is most likely required to fully describe the characteristics of QCD. On the theoretical side it would be interesting to further understand the soft-wall model from the top-down including finding a dynamical solution of the features exhibited in our model along the lines considered in Ref. [15]. In addition the stability of the scalar potential will most likely require higher-order terms that can only be studied from the top-down. It is interesting that the simple 5D model contains QCD-like features and suggests that a further understanding of QCD can be obtained from the gauge/gravity correspondence.

Acknowledgments

This work was supported by the Research Corporation for Science Advancement, the Australian Research Council, the US Department of Energy under grant DE-FG02-87ER40328, and by the School of Physics and Astronomy at the University of Minnesota.

References

1. J. M. Maldacena, Adv. Theor. Math. Phys. **2**, 231 (1998) [Int. J. Theor. Phys. **38**, 1113 (1999)].
2. S. S. Gubser, I. R. Klebanov and A. M. Polyakov, Phys. Lett. B **428**, 105 (1998).
3. E. Witten, Adv. Theor. Math. Phys. **2**, 253 (1998).
4. I. R. Klebanov and E. Witten, Nucl. Phys. B **556**, 89 (1999).
5. J. Erlich, E. Katz, D. T. Son and M. A. Stephanov, Phys. Rev. Lett. **95**, 261602 (2005).
6. L. Da Rold and A. Pomarol, Nucl. Phys. B **721**, 79 (2005).
7. A. Karch, E. Katz, D. T. Son and M. A. Stephanov, Phys. Rev. D **74**, 015005 (2006).
8. M. Shifman and A. Vainshtein, Phys. Rev. D **77**, 034002 (2008).
9. T. Gherghetta, J. I. Kapusta and T. M. Kelley, Phys. Rev. D **79**, 076003 (2009).
10. N. Evans and A. Tedder, Phys. Lett. B **642**, 546 (2006).
11. H. R. Grigoryan and A. V. Radyushkin, Phys. Rev. D **76**, 095007 (2007).
12. H. J. Kwee and R. F. Lebed, Phys. Rev. D **77**, 115007 (2008).
13. A. Cherman, T. D. Cohen and E. S. Werbos, Phys. Rev. C **79**, 045203 (2009).
14. P. Colangelo, F. De Fazio, F. Giannuzzi, F. Jugeau and S. Nicotri, Phys. Rev. D **78**, 055009 (2008).
15. B. Batell and T. Gherghetta, Phys. Rev. D **78**, 026002 (2008).
16. M. Huang, Q. S. Yan and Y. Yang, "Confront Holographic QCD with Regge Trajectories" arXiv:0710.0988
17. T. D. Cohen and L. Y. Glozman, Mod. Phys. Lett. A **21**, 1939 (2006).
18. R. F. Wagenbrunn and L. Y. Glozman, Phys. Lett. B **643**, 98 (2006).
19. C. Amsler, *et al.*, [Particle Data Group], Phys. Lett. B **667**, 1 (2008).
20. D. V. Bugg, Phys. Rept. **397**, 257 (2004).
21. A. Bertin, *et al.* [OBELIX Collaboration], Phys. Lett. B **414**, 220 (1997).
22. P. Breitenlohner and D. Z. Freedman, Phys. Lett. B **115**, 197 (1982).
23. W. Lin and B. D. Serot, Phys. Lett. B **233**, 23 (1989).
24. V. Tadevosyan, *et al.*, [Jefferson Lab F_π Collaboration], Phys. Rev. C **75**, 055205 (2007).
25. T. Horn, *et al.*, [Jefferson Lab F_π-2 Collaboration], Phys. Rev. Lett. **97**, 192001 (2006).
26. H. Ackermann, *et al.*, Nucl. Phys. B **137**, 294 (1978).
27. P. Brauel, *et al.*, Phys. Lett. B **69**, 253 (1977).

Chiral Symmetry and Holographic Models of the Baryon: Testing Holographic Models of the Baryon with Model-Independent Relations

Thomas D. Cohen*

Maryland Center for Fundamental Physics and Department of Physics, University of Maryland College Park MD 20742, USA

**E-mail: cohen@physics.umd.edu*

A model-independent relation for the long-distance behavior of baryon form factors in the large N_c and chiral limits is introduced and used to probe the consistency of various models of the baryon including recently proposed holographic models. This relation is satisfied by the Skyrme model and all other 4D semiclassical chiral soliton models. The "bottom-up" Pomarol-Wulzer holographic model which treats baryons as 5D Skyrmions also satisfies the relation. However, the "top down" holographic model treating baryons as instantons in the Sakai-Sugimoto model fails to satisfy the relation. This failure can ultimately be traced to the imposition of a scale separation in the model between the curvature scale and the KK scale; such a scale separation has no counterpart in QCD.

Keywords: baryon models; holography

1. Introduction

It is a great honor to present a talk at a conference in honor of Misha Shifman. Misha is a truly first-rate theorist who has made numerous seminal contributions to particle physics. He also approaches physics—and life in general—with a zest that is always marvelous to behold. Thus before going on, let me wish Misha a very happy 60th birthday.

The backdrop to this work described in this talk is what is often described as "AdS/QCD" [2]. This is an attempt motivated on the AdS/CFT [1] correspondence to describe QCD using higher dimensional theories. However, while the AdS/CFT correspondence is on solid theoretical grounds, AdS/QCD is not. *A priori* there is no reason to believe that physics of QCD—even in a highly idealized limit such as the large N_c

limit—can be captured in such a framework. Still, it is not implausible that *some* aspects of QCD can; one might hope to make sensible holographic models capture much of the guts of QCD. There have been numerous attempts over the past several years to do just this; virtually all of these claim some measure of phenomenological success. While this is encouraging there is a nagging concern: How can one tell whether this apparent success is real, reflecting an accurate modeling of the physics, or whether it merely reflects good data fitting without much physics? One way to gain at least a measure of confidence in such models is to make sure at the very least that they correctly encode the part of the physics which they are designed to get right. Since these models are motivated by gauge/gravity duality which only holds in the large N_c limit, presumably the models ought to correctly encode large N_c QCD. Moreover, the models are explicitly designed to have chiral symmetry and thus should reproduce the chiral aspects of QCD. The question, then, is do these models correctly encode the physics of the large N_c and chiral limits of QCD? A useful place to look is in holographic models of baryons.[3–8]

Before proceeding it is useful to recall that the large N_c and chiral limits of QCD need not be uniform. Numerous examples are known for baryon properties in which the two limits do not commute.[9] One fundamental reason for this is that in the chiral limit the sole massless excitation is the pion; at large N_c , however, there is another class of zero energy excitation of the nucleon—namely the Δ. Since the overall structure of the holographic models incorporate the large N_c limit from the outset, it is reasonable to suppose that these models implicitly take the large N_c limit prior to the chiral limit. It is well known that analogous properties in the Skyrme model do indeed give the correct results for QCD in the combined large N_c and chiral limits (with the large N_c limit taken first).[9]

Recently,[11] it was suggested that the long-distance behavior of Fourier transformed electromagnetic form factors given by

$$\tilde{G}_E^{I=0,1}(r) = \int \frac{d^3q}{(2\pi)^3}\, e^{i\vec{q}\cdot\vec{r}} G_E^{I=0,1}(q)$$
$$\tilde{G}_M^{I=0,1}(r) = \frac{-i}{3}\int \frac{d^3q}{(2\pi)^3}\, e^{i\vec{q}\cdot\vec{r}}\; \vec{q}\cdot\vec{r}\; G_M^{I=0,1}(q)$$

(where $G(q)$ the standard experimentally accessible form factors and I refers to isospin) are a useful probe of the large N_c and chiral dynamics for baryons. A particularly useful combination is the ratio $(\tilde{G}_E^{I=0}(r)\tilde{G}_E^{I=1}(r))/(\tilde{G}_M^{I=0}(r)\tilde{G}_M^{I=1}(r))$, which in the large N_c and chiral limits (with the large N_c limit taken first) can be shown to have the property

$$\lim_{r\to\infty} \frac{\tilde{G}_E^{I=0}(r)\tilde{G}_E^{I=1}(r)}{\tilde{G}_M^{I=0}(r)\tilde{G}_M^{I=1}(r)} = \frac{18}{r^2} \,. \tag{1}$$

Note that this relation is a property of large N_c QCD itself; it is completely free of any model parameters.

The relation in Eq. (1) has a number of virtues aside from model independence. Firstly, as a configuration space quantity, it is useful in the exact chiral limit. The analogous information in momentum space is most easily found by studying how various derivatives diverge at $q^2 = 0$ as a function of m_π^{-1}. However, to date, the holographic baryon models have typically only been studied in the chiral limit, making the momentum space strategy difficult. Secondly, the $I = 1$ part comes from the anomalous coupling of the baryon to three pions and thus the ratio probes the ability of the models to correctly encode the anomaly.

It is a straightforward exercise to verify that the Skyrme model and all other chiral soliton models in three plus one dimensions satisfy this relation.[11] The reason is fairly simple: in all chiral soliton models the long distance behavior is dominated by the pion, and the strength of the pion field at large r is completely fixed by g_A. Thus the $I = 0$ form factors which are dominated by two pions at long distance scale as g_A^2, while the $I = I$ form factors which are dominated by three pions at long distance scale as g_A^3. In the ratio in Eq. (1) the factors of g_A cancel. Similarly the isovector electric and isoscalar magnetic form factors are proportional to the moment of inertia which arises in the semiclassical quantization. Again in the ratio this cancels. Thus at long distance all details of the model cancel as does any sensitivity to model parameters; by dimensional analysis the ratio must be proportional to r^{-2}. The value of the proportionality constant is fixed by the structure of the semiclassical quantization; an explicit calculation yields 18.

2. Holographic Baryon Models

How do models of baryons based on holography fare with regard to the ratio in Eq. (1)? *A priori* one would expect the relation to hold. After all, the relation is model independent and ultimately only depends on getting the large N_c and chiral physics right. The holographic models by construction should have the large N_c dynamics right and are specifically constructed to have chiral symmetry and have it spontaneously broken. However, it is important to check that the relation is satisfied as advertised. Note that the way chiral physics is implemented in these models is quite different than in, say, the Skyrme model where the effects are obvious. Thus, in order to gain insight into these models, it is useful to see how the various pieces of the model fit together to satisfy the relation. Moreover, given the relative complexity of these models, there is the concern that the pieces might not really fit together as they should. Thus, it is useful to check the consistency of these models using Eq. (1) as a probe.

First let me discuss the Pomarol-Wulzer holographic baryon model.[3] The model is "bottom up". That is, it is entirely phenomenological in origin. While it is motivated by gauge/gravity duality and associates 5-dimension gauge fields with 4-dimensional currents in the spirit of the AdS/CFT dictionary, there is no attempt to derive the model from any fundamental stringy construction. For the sake of brevity I will only briefly describe some critical aspects of the model here and refer you to the original literature for more detail. The model uses a 4-dimensional AdS geometry with a hard wall cutoff in the holographic coordinate to simulate the physics of confinement; the dynamical fields are two 5-dimensional gauge fields which are associated with the 4-dimensional left-handed and right-handed chiral currents. A Chern-Simons term is included to incorporate the physics of the anomaly. Baryons are described as slowly rotating solitons in the spirit of the Skyrme model. The Chern-Simons term plays the role of the Skyrme term and stabilizes the soliton. The principal difference with the usual Skyrme model is that these 5-dimensional Skyrmions, even after a hedgehog symmetry is imposed, involve the numerical solution of 2-dimensional PDEs as opposed to the ODE need in the 4-dimensional Skyrme model. However, as in the case of the 4-dimensional Skyrme model, the asymptotic form of the fields at large distance can be found analytically and can be related to g_A. From this the asymptotic form of the position-space form factors can be extracted in terms of g_A and the moment of inertia characterizing the slow rotations. Combining these together in the ratio, one finds that Eq. (1) is, in fact, satisfied. Whatever other virtues and vices the model has, at the very least it appears to get right the leading large N_c and chiral dynamics for baryons—including the anomalous effects.

The situation is quite different for recent treatments of baryons[4–8] in the Sakai-Sugimoto model.[12] The model at first glance appears to be on more firm theoretical ground than the Pomarol-Wulzer holographic baryon model. It is a "top-down" model from a sensible higher dimensional construction based on N_c D4 branes with $N_f = 2$ probe D8 branes. The underlying theory can be reduced to a tractable theory when λ, the 't Hooft coupling, is large; the model is based on the leading terms of a systematic λ^{-1} expansion. The model has two widely different length scales—the string scale and the inverse of the KK mass scale (which plays the role of the confinement scale). The 't Hooft coupling in this model can be interpreted as the of square of the ratio of these scales.[7] Thus the large λ limit corresponds to a large separation of scales.

This scale separation has been exploited in descriptions of baryons. The key point is that the intrinsic size of the baryon scales as λ^{-1}. In the large λ limit this becomes small enough so that non-linearities in the equations of motion become negligible, yielding an analytically tractable form: the baryons become instantons in the 5-dimensional space. Now, this is potentially problematic: an expansion in λ^{-1} was used to obtain the the Sakai-Sugimoto model in the first place [12] but the smallness of the instanton implies that higher derivative terms neglected in the derivation of the model are not parametrically suppressed. One can, quite legitimately, argue that as a *model* this need not be a cause for concern—after all the 4-dimensional Skyrme model has no systematic power counting in derivatives either. Still, this issue suggests that there is a profound need to check the self-consistency of the model. Equation (1) is a useful way to do precisely this.

Using the formalism in ref. [7] It is a straightforward matter to construct the relevant ratio:

$$\lim_{r\to\infty} \frac{\tilde{G}_E^{I=0}\tilde{G}_E^{I=1}}{\tilde{G}_M^{I=0}\tilde{G}_M^{I=1}} = \frac{\lambda\sqrt{40/3}}{\pi\rho_1^2 r^2}\,, \tag{2}$$

where λ is the 't Hooft coupling and ρ_1 is a numerical constant, independent of λ obtained from the mass of the low-lying vector meson. It is noteworthy that this is model *dependent*: the value explicitly depends on λ which is a parameter of the model. For arbitrary λ, it fails to satisfy Eq. (1); this implies that the structure of the model does not impose the correct large N_c and chiral behavior. The model is not generally consistent with the underlying QCD dynamics it is supposed to be describing. Moreover, one cannot evade this problem by arguing that one can simply use Eqs. (1) and (2) to fit the value of λ to be

$$\lambda = \sqrt{\frac{243}{10}}\,\pi\rho_1^2\,; \tag{3}$$

this is not theoretically consistent. Formally λ is treated as being parametrically large in the Sakai-Sugimoto model, while the right-hand side of Eq. (3) is simply a numerical constant of order unity in the expansion.

Thus, the treatments of the baryon as instantons in the Sakai-Sugimoto model appear to be fundamentally diseased. It is not hard to understand the origin of this disease at the hadronic level. If one looks at the calculations of the pion decay constant and the axial coupling in the model one finds

$$\frac{g_A}{f_\pi} = \frac{12\pi}{\sqrt{5}} \sqrt{\frac{N_c}{\lambda}} \frac{1}{M_{KK}} \,. \tag{4}$$

By formally working at large λ—with the large λ limit implicitly taken before the large N_c limit —one sees that the coupling to pions is formally suppressed. The effects of virtual pions are thus lost—thereby severing the connection with the model-independent predictions of large N_c chiral perturbation theory. Thus, it seems that the instanton treatment of baryons *is* implicitly taking the large λ limit first in the expansion.

Is this a minor curiosity of little significance or does it reflect a fundamental flaw in the Sakai-Sugimoto model as a model for QCD? I would argue that it is the latter. Note the difficulty here is caused by an expansion about the large λ limit. If one does not impose this limit the theoretical connection of the model to the underlying stringy construction is lost. However, this limit has no analog in QCD. Recall that λ is given by the ratio of the string scale and the KK scale. Thus the model explicitly is constructed in terms of *two* distinct mass scales, whereas QCD in the chiral limit has only one. This means that the model is fundamentally different from QCD unless the string scale completely decouples from all physical quantities. That is, the Sakai-Sugimoto model can only map on to QCD if the $\lambda \to \infty$ limit has the feature that *all* physical quantities in large N_c QCD neither diverge nor vanish. However, this is not the case. For example, $f_\pi \sim \lambda^{1/2}$. Thus, the failure of the model-independent result of Eq. (1) is, in effect, a statement that QCD and the Sakai-Sugimoto model are not in the same universality class.

Acknowledgments

The author thanks the US DOE for financial support. The work discussed in this talk was largely done in collaboration with Marina Nielsen and Aleksey Cherman. Their contributions to this work are gratefully acknowledged.

References

1. J. M. Maldacena, Adv. Theor. Math. Phys. **2**, 231 (1998); S. S. Gubser, I. R. Klebanov and A. M. Polyakov, Phys. Lett. B **428**, 105 (1998); E. Witten, Adv. Theor. Math. Phys. **2**, 253 (1998).
2. J. Erlich, E. Katz, D. T. Son and M. A. Stephanov, Phys. Rev. Lett. **95**, 261602 (2005); L. Da Rold and A. Pomarol, Nucl. Phys. B **721**, 79 (2005); G. F. de Teramond and S. J. Brodsky, Phys. Rev. Lett. **94**, 201601 (2005).
3. A. Pomarol and A. Wulzer, Nucl. Phys. B **809**, 347 (2009).
4. H. Hata, T. Sakai, S. Sugimoto and S. Yamato, "Baryons from instantons in holographic QCD," arXiv:hep-th/0701280; D. K. Hong, M. Rho, H. U. Yee and P. Yi, Phys. Rev. D **76**, 061901 (2007).
5. K. Nawa, H. Suganuma and T. Kojo, Phys. Rev. D **75**, 086003 (2007); H. Hata, M. Murata and S. Yamato, Phys. Rev. D **78**, 086006 (2008).
6. D. K. Hong, M. Rho, H. U. Yee and P. Yi, JHEP **0709**, 063 (2007); D. K. Hong, M. Rho, H. U. Yee and P. Yi, Phys. Rev. D **77**, 014030 (2008).
7. K. Hashimoto, T. Sakai and S. Sugimoto, Prog. Theor. Phys. **120**, 1093 (2008).
8. K. Y. Kim and I. Zahed, JHEP **0809**, 007 (2008).
9. T. D. Cohen and W. Broniowski, Phys. Lett. B **292**, 5 (1992); T. D. Cohen, Rev. Mod. Phys. **68**, 599 (1996).
10. T. D. Cohen, Phys. Lett. B **359**, 23 (1995).
11. A. Cherman, T. D. Cohen and M. Nielsen, Phys. Rev. Lett. **103**, 022001 (2009).
12. T. Sakai and S. Sugimoto, Prog. Theor. Phys. **113**, 843 (2005); T. Sakai and S. Sugimoto, Prog. Theor. Phys. **114**, 1083 (2005).

SECTION 7.

QCD, COLD AND HOT

QCD and Supersymmetry

Adi Armoni

Department of Physics, Swansea University
Singleton Park, Swansea, SA2 8PP, UK

We review the connection between QCD and supersymmetric theories. We focus on the non-perturbative large-N (planar) correspondence between one-flavor QCD and pure supersymmetric Yang-Mills theory ($\mathcal{N} = 1$ SYM). We explain how non-perturbative quantities in QCD, such as the quark condensate, can be evaluated by using the corresponding non-perturbative results in supersymmetric gauge theories. The review consists of three parts. The first part is devoted to a review of pure $\mathcal{N} = 1$ SYM. In the second part we introduce "orientifold planar equivalence". The third part is devoted to the implications of planar equivalence for QCD.

1. Preface - Misha and I

This article is about a subject that is dear to Misha: the connection between QCD and supersymmetry. Misha has contributed enormously to both areas from the early days of QCD and supersymmetric gauge theories. I had the pleasure of being involved in a collaboration with Misha and Gabriele Veneziano which led to the subject of these lectures: how to use the non-perturbative results that Misha and others established in SUSY Yang-Mills for real QCD.

I met Misha Shifman in January 2003 when he started his sabbatical at the theory division of CERN. I remember clearly the day that I met Misha first. It was in the office of the secretaries. I introduced myself to Misha by asking: "Are you professor Shifman?" "Yes I am," he replied and asked who I am. I told him my name and Misha said in a loud voice "You are Israeli, right?" I confirmed and suggested to discuss physics together. We've decided to meet the next morning in Misha's office. I knew that Misha is interested in non-perturbative planar equivalence between non-supersymmetric orbifold field theories and supersymmetric theories. We both thought that such an equivalence has little chance to be valid. I told Misha about a different option which is now called "orientifold field theories". I was amazed by the great intuition and deep insights of Misha. Soon, he became convinced that the refined version is valid and explained to me how we can use the enormous knowledge that was accumulated about susy theories in more than 20 years for non-supersymmetric theories. We started to think how to prove the relation we have just found. Later in the afternoon we have met Gabriele Veneziano and told him with great excitement about our finding. Together, based on a brilliant idea of Gabriele, we were able to sketch a proof of planar equivalence.

Three weeks after our first meeting, we posted our first paper and a new friendship began. Happy $60^{\rm th}$ birthday Misha!

2. Introduction

Extracting results in the non-perturbative regime of QCD is an important and challenging task. The only reliable tool that we have at hand is the lattice. Although at the conceptual level the lattice can provide a 'solution' of QCD, in practice lattice results are still limited.

The situation in supersymmetric gauge is much better: we can calculate exact non-perturbative quantities, such as the gluino condensate. The underlying reason is holomorphicity. Holomorphic quantities are protected against deformations and as a result it is possible to calculate them analytically.

In this short review we will describe a relatively new tool called "orientifold planar equivalence", which enables us to copy exact non-perturbative results from supersymmetric gauge theories to non-supersymmetric theories. Moreover, since QCD belongs to the class of non-supersymmetric theories under consideration, we will be able to estimate non-perturbative quantities in QCD. The prime example is a calculation of the chiral condensate.

Let us introduce the basic idea. Consider $\mathcal{N} = 1$ SYM with a gauge group $SU(N)$. The theory consists of a gluon and a Majorana fermion in the adjoint representation. The statement about "orientifold planar equivalence" is that in the large-N limit and in a well defined common sector, $\mathcal{N} = 1$ SYM becomes equivalent to a non-supersymmetric $SU(N)$ theory with a Dirac fermion in either the symmetric or the antisymmetric representation of $SU(N)$. This relation is the main subject of these talks.

The statement, while interesting within field theory per-se, is useful for QCD: for $SU(3)$ the antisymmetric representation is equivalent to the anti-fundamental representation and therefore we can use planar equivalence to estimate (within $1/N$ error) non-perturbative quantities in QCD.

In this review we will discuss in detail the above relation and some of its generalizations and applications. The review is based on three talks delivered in the 2009 Schladming school for theoretical physics.

In the first part we will review $\mathcal{N} = 1$ SYM, from the first non-perturbative results: the Witten index, the Veneziano-Yankielowicz effective action until the most recent results about domain-walls dynamics in this theory.

The second part, "orientifold planar equivalence" is dedicated to the large-N relation between $\mathcal{N} = 1$ SYM and "orientifold field theories" (non-supersymmetric theories with fermions in two-index symmetric or antisymmetric representations).

In the third part we will use the knowledge from the first two parts to estimate non-perturbative quantities in QCD.

3. What do we know about $\mathcal{N} = 1$ SYM ?

3.1. *The $\mathcal{N} = 1$ Super Yang-Mills Lagrangian and the $U_R(1)$ symmetry*

$\mathcal{N} = 1$ Super Yang-Mills is a theory which consists of a gluon and a gluino. The gluino is the "superpartner" of the gluon. It is a four-component Majorana (real) fermion transforming in the adjoint representation of the gauge group. We will focus on $SU(N)$. The Lagrangian of $\mathcal{N} = 1$ SYM is

$$\mathcal{L} = -\frac{1}{4g^2} F^a_{\mu\nu} F^a_{\mu\nu} + \frac{i}{2g^2} \bar{\lambda}^a D_\mu \gamma^\mu \lambda^a \,. \tag{1}$$

The Lagrangian looks exactly as the Lagrangian of QCD with one quark flavor, except that in the present case the quark field (denoted by λ) transforms in the *adjoint* representation. The theory is expected to confine and to have a mass-gap of degenerate color-singlets ("glueballs" and "glueballinos").

The SYM theory admits, at the classical level, a $U(1)$ axial symmetry (similar to the $U_A(1)$ in QCD), often called "R-symmetry", of the form (using the Weyl notation)

$$\lambda^\alpha \to e^{i\beta} \lambda^\alpha \tag{2}$$

The associated Noether current is $J_\mu = \frac{1}{g^2} \bar{\lambda} \gamma_\mu \gamma_5 \lambda$. The $U_R(1)$ symmetry is broken by the triangle anomaly (see figure 1 below)

$$\partial^\mu J_\mu = \frac{N}{16\pi^2} F^a_{\mu\nu} \tilde{F}^a_{\mu\nu} \,. \tag{3}$$

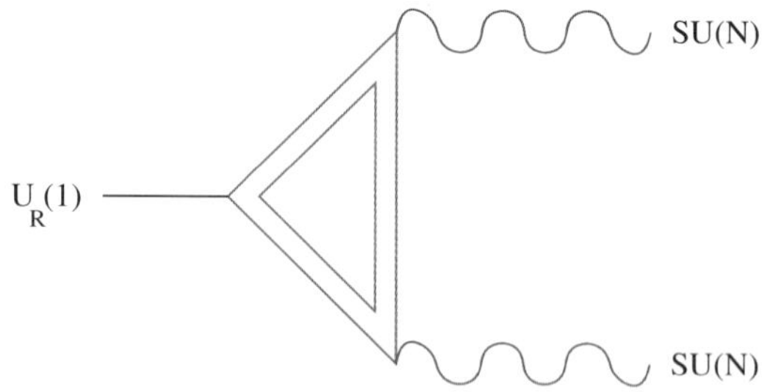

Fig. 1. The triangle diagram responsible for the $U_R(1)$ anomaly

Notice that the anomaly is proportional to N, since the gluino transforms in the adjoint representation and $\operatorname{tr} T^a T^b = N\delta^{ab}$ (in contrast to QCD where the quarks are in the fundamental and the axial anomaly is *not* proportional to N, since $\operatorname{tr} t^a t^b = \frac{1}{2}\delta^{ab}$).

Thus,

$$U_R(1) \to Z_{2N} \tag{4}$$

Namely, a residual transformation of the form

$$\lambda \to e^{i\pi \frac{k}{N}} \lambda \tag{5}$$

leaves the partition function invariant.

Witten conjectured[1] in 1982 that the Z_{2N} symmetry is spontaneously broken down to Z_2. That indeed happens and the result is a theory with N vacua, where the gluino condensate serves as the order parameter of the breaking, namely

$$\langle \lambda^\alpha \lambda^\alpha \rangle_k = \text{const.}\, \Lambda^3 e^{i2\pi \frac{k}{N}} \tag{6}$$

where $k = 0, ..., N-1$ labels the various vacua.

How do we know that a gluino condensate is formed and that $Z_{2N} \to Z_2$? An evidence for the existence of a gluino condensate follows from the Veneziano-Yankielowicz effective action.

3.2. *The Veneziano-Yankielowicz effective action and the gluino condensate*

In 1982, from anomaly considerations, Veneziano and Yankielowicz wrote an effective action for $\mathcal{N} = 1$ SYM.[2] The fundamental (super)-field that appears in their Lagrangian is $S = \lambda\lambda$. The Lagrangian for S is

$$I = N(S \log S/\Lambda^3 - S)\,. \tag{7}$$

The equation of motion for S is

$$N \log S/\Lambda^3 = c\,. \tag{8}$$

Hence

$$(S/\Lambda^3)^N = c'\,, \tag{9}$$

or

$$\langle S \rangle = \langle \lambda\lambda \rangle = c'' \Lambda^3 e^{i2\pi \frac{k}{N}}\,. \tag{10}$$

The Veneziano-Yankielowicz Lagrangian does not fix the value of the gluino condensate, it only demonstrates its non-vanishing. In order to fix the value of the gluino condensate a calculation is needed.

There are various methods of calculating the gluino condensate. In particular the method of Shifman and Vainshtein [3] and the method of Davies et. al.[4] All the methods rely on holomorphy. I like the last method where the authors compactified the theory on a small circle to make the theory weakly coupled. They calculated the gluino condensate by showing that $\langle \lambda\lambda \rangle$ is saturated solely by monopoles. Then they showed that the value does not depend on the radius of the circle, due to holomorphicity. So, they could de-compactify the theory to obtain

$$\langle \lambda\lambda \rangle_k = -6N\Lambda^3 e^{i2\pi \frac{k}{N}}\,. \tag{11}$$

The gluino condensate is analogous to the quark condensate $\langle \bar{q}q \rangle$ in QCD. The calculation of such a quantity is a great achievement !

By analyzing the response of the coupling to the variation of the UV-cutoff in the expression for the gluino condensate, NSVZ obtained an important result: the *exact* beta function [5]

$$\beta(g) = -\frac{g^3}{16\pi^2} \frac{3N}{1 - \frac{g^2 N}{8\pi^2}} \tag{12}$$

The NSVZ beta function is exact to all orders in perturbation theory. The first two coefficients are universal (namely scheme independent) and they match the explicit one and two loop calculations.

The generalized NSVZ beta function for super-QCD played a central role in fixing the conformal window in SQCD and in the derivation of Seiberg duality .[6]

3.3. *Domain Walls*

When a discrete symmetry, such as Z_{2N} is spontaneously broken, there exists domain walls which interpolate between the various vacua of the theory, see figure 2.

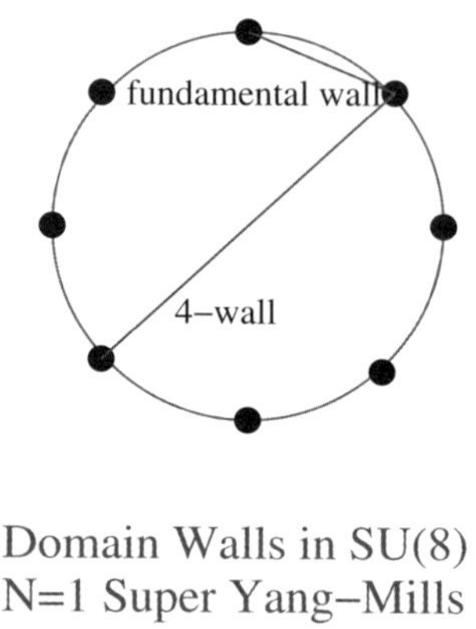

Fig. 2. The vacuum structure and domain walls in $\mathcal{N}=1$ SYM.

The domain walls are (2+1) dimensional objects, localized at, say $x_3 = z$. The "fundamental" wall interpolates between neighbouring vacua, whereas a k-wall "skips" k vacua, as depicted in the figure.

We can think about the k-wall as a bound state of k elementary walls. The tension of a k-wall is given by the difference between the values of the gluino condensate in the given vacua [7]

$$T_k = \frac{N}{8\pi^2} \left| \langle\lambda\lambda\rangle_{l+k} - \langle\lambda\lambda\rangle_l \right| = \frac{3N^2}{2\pi^2} \Lambda^3 \sin\left(\pi \frac{k}{N}\right) \tag{13}$$

Note that when k is kept fixed and $N \to \infty$, $T_k \sim k T_1$ and also $T_1 \sim N\Lambda^3$. It means that in the large-N limit the k-wall becomes a collection of k non-interacting fundamental domain walls. The tension of each fundamental domain wall is proportional to N. This is surprising since solitons in a theory with adjoint matter should carry a tension $\sim N^2$. In fact, every quantity (at least in perturbation theory, or semiclassically) should depend on N^2. This observation led to a bold conjecture by Witten. In ref. [8] Witten argued that the $\mathcal{N}=1$ SYM domain walls are *QCD D-branes*. What does it mean ?

Usually we think about (a large-N) gauge theory as a string theory of closed and open strings. The closed strings are identified with flavor singlets - the glueballs of the theory, whereas the open strings are identified with mesons.

But now we know that in string theory there exists extended objects which are called "branes". Thus, if $\mathcal{N} = 1$ SYM is described by (or dual to a) string theory, where are the D-branes in field theory ? Witten argue that these are the domain walls.

In string theory D-branes carry the following characters:

i. They are non-perturbative objects. Their tension is $T \sim \frac{1}{g_{string}}$.
ii. An open string can end on a D-brane, see figure 3.
iii. D-branes interact by an exchange of closed strings.

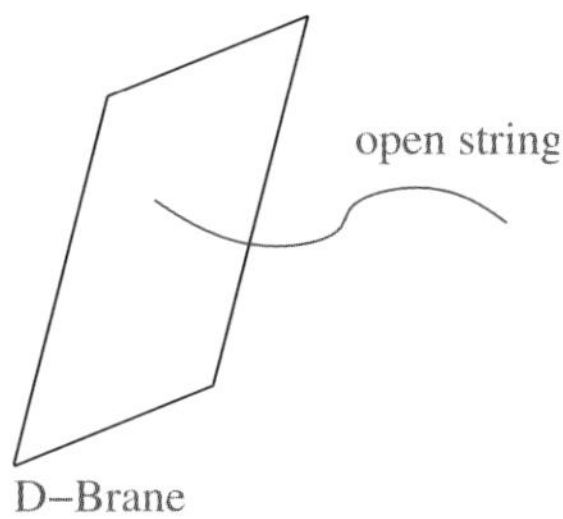

Fig. 3. An open string can end on a D-brane. Similarly, the QCD-string can end on a domain wall.

Witten noticed that the tension of the fundamental domain wall is $T_1 \sim N$ (at large-N). By using the relation between field theory and string theory $g_{string} \sim \frac{1}{N}$, we observe that the domain wall tension matches the expectation from a D-brane.

Moreover, Witten argued that the QCD-string (flux tube) can end on the domain wall, in the same way that the open-string ends on a D-brane.

Finally, together with Shifman [9] we showed how domain walls interact via an exchange of glueballs. In fact, at large-N there is a cancellation between the attraction due to an exchange of even-parity glueballs and the repulsion due to an exchange of odd-parity glueballs. This is similar to the cancellation of the interaction between parallel D-branes (dilaton and graviton against RR-fields).

3.4. *The Acharya-Vafa theory*

It is well known in string theory, that when Dp-branes coincide there appears a $(p + 1)$ dimensional field theory "on the branes". If domain walls are the "QCD D-branes", does something similar happen on the domain wall ?

It was argued by Acharya and Vafa that this is indeed the case.[10] The field theory on the k wall was argued to be a $U(k)$ gauge theory which contains a level N Chern-Simons term

$$\mathcal{L}_{\text{Acharya-Vafa}} = \frac{1}{2g^2}\text{tr}\Big(-\frac{1}{2}F^2_{mn} - (D_i\phi)^2 + N\epsilon_{ijk}(A^i\partial^j A^k + \frac{1}{3}A^iA^jA^k) + \text{fermions}\Big) \tag{14}$$

By using the Acharya-Vafa theory, it is possible to calculate the force between domain walls.[11,12] As in standard D-branes physics, the vev of the scalar, $x \equiv \langle\phi\rangle$, parametrizes the distance between domain walls. A Coleman-Weinberg effective potential for ϕ is interpreted as the potential between a pair of parallel domain walls.

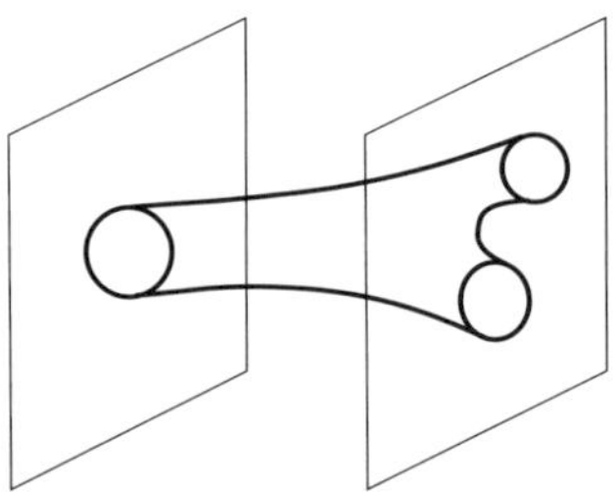

Fig. 4. The string theory diagram that gives rise to a potential between two parallel domain walls.

It turns out that the one-loop effective potential vanishes and we had to perform a two-loops calculation. The string diagram which is responsible for the potential is depicted in figure 4. The result is

$$V(x) \sim \frac{1}{N}\frac{x^2}{1+x^2}\,. \tag{15}$$

3.5. $\mathcal{N}=1$ *SYM summary*

Let us summarize what We learnt quite a lot about $\mathcal{N}=1$ SYM

i. The theory posses N degenerate vacua.
ii. Gluino condensation is formed $\langle\lambda\lambda\rangle_k = -6N\Lambda^3 e^{i2\pi\frac{k}{N}}$.
iii. $\beta_{\text{NSVZ}}(g) = -\frac{g^3}{16\pi^2}\frac{3N}{1-\frac{g^2N}{8\pi^2}}$.
iv. There exists domain walls whose tension is
$T_k = \frac{N}{8\pi^2}\left|\langle\lambda\lambda\rangle_{l+k} - \langle\lambda\lambda\rangle_l\right| = \frac{3N^2}{2\pi^2}\Lambda^3\sin(\pi\frac{k}{N})$.
v. Domain walls are "QCD D-branes".

In the next section we will learn one more thing about $\mathcal{N}=1$ SYM: We can use some of the above results for QCD !

4. What is orientifold planar equivalence ?

In this part we review the relation between large-N $\mathcal{N}=1$ SYM and the "orientifold field theory". The motivation is to approximate QCD by a supersymmetric theory. We also review the basic ideas behind the $1/N$ expansion.

4.1. *Motivation*

Statement:

In the limit of infinite number of colors, $N \to \infty$, and in a well-defined common sector, $SU(N)$ gauge theory + a Dirac fermion in the antisymmetric representation becomes equivalent to $SU(N)$ gauge theory + a Majorana fermion in the adjoint representation.

The first theory, often called "orientifold field theory" (due to a relation with string theory) contains an antisymmetric Dirac fermion $\Psi_{[ij]}$ ($\Psi_{ij} = -\Psi_{ji}$). It is a non-supersymmetric theory. The other theory is $\mathcal{N}=1$ SYM.

So, the claim is that in the limit $N \to \infty$, a supersymmetric theory and a non-supersymmetric theory become equivalent ! *Why is it useful for QCD ?*

For $SU(3)$ the antisymmetric representation is identical to the anti-fundamental representation

$$q^i = \frac{1}{2}\epsilon^{ijk}\Psi_{[jk]} \tag{16}$$

So, we can start with one-flavor QCD, think about the quark as an antisymmetric fermion, generalize to an $SU(N)$ theory with an antisymmetric fermion and finally take the $N \to \infty$ limit.

By using the above procedure, and the relation with $\mathcal{N}=1$ SYM we will be able to approximate (one-flavor) QCD by SUSY YM and to copy results from the latter to QCD !

4.2. *What is the large-N limit ?*

In 1973 't Hooft suggested that a generalization of QCD from $SU(3)$ to $SU(N)$ may result in a simpler theory in the large N limit.[13] More precisely, 't Hooft showed that when taking the limit $N \to \infty$, while keeping fixed the combination $\lambda \equiv g^2 N$ (now called the 't Hooft coupling), the theory is controlled by *planar diagrams*.

Let us focus on $U(N)$ pure Yang-Mills theory, namely on

$$\mathcal{L} = -\frac{1}{4g^2} F^a_{\mu\nu} F^a_{\mu\nu} \tag{17}$$

't Hooft showed that Feynman diagrams are classified according to their topology. For simplicity let us focus on 'vacuum diagram':

1. Planar diagram, admitting the topology of a sphere are $\mathcal{O}(N^2)$, see figure 5.

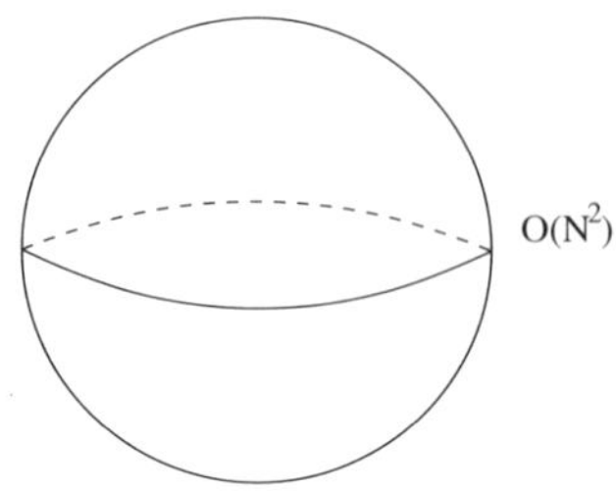

Fig. 5. Planar vacuum diagrams can be drawn on a sphere.

2. Non-planar diagrams carry a weight of $\mathcal{O}(N^{(2-2h)})$, where h is the number of holes in the diagram. For an example, see figure 6.

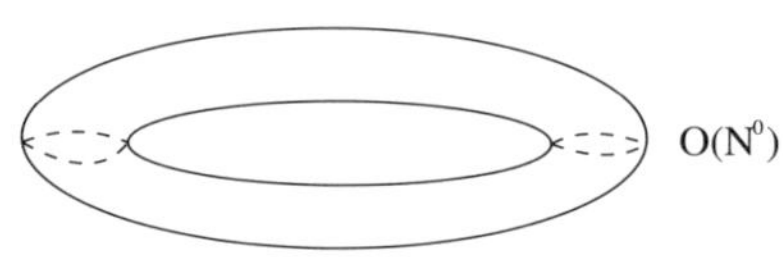

Fig. 6. Non-planar vacuum diagrams are drawn on a Riemann surface.

Thus, in the limit $N \to \infty$ only vacuum diagrams with sphere topology dominate.*For this reason the large-N limit is simpler: it selects only a subset of Feynman diagrams.*

In order to understand 't Hooft's idea, let us introduce the 't Hooft notation: The gluon propagator is represented by two lines with arrows that point in opposite directions, see figure 7. These lines represent the flow of

Fig. 7. The gluon propagator.

color in the Feynman diagram. We should think about the adjoint representation as the tensor product of the fundamental and the anti-fundamental representations. The three and four gluons vertices are depicted in figure 8.

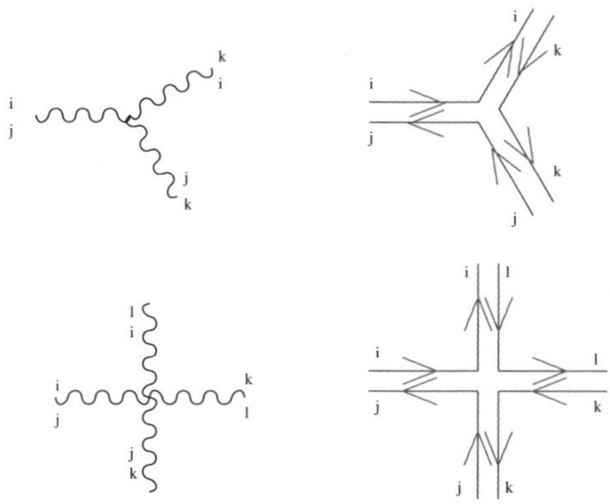

Fig. 8. Feynman rules for pure Yang-Mills theory. The ghosts are ignored.

Let us consider two examples. The first is a one-loop vacuum diagram, see figure 9. Second, a three-loop vacuum diagram which contains both

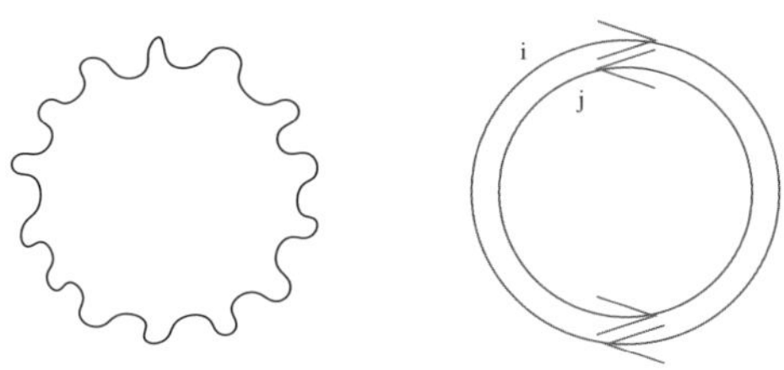

Fig. 9. A one-loop vacuum diagram

planar and non-planar contributions, see figure 10.

$\mathcal{N}=1$ SYM contains in addition to the gluon a gluino. Since the gluino transform in the adjoint representation, its Feynman rules look exactly as the Feynman rules of the gluon. What about the antisymmetric fermion of the "orientifold theory" ? Since $\Psi_{[ij]}$ carries two fundamental indices, it is denoted by two lines with arrows that point in the same direction, since the two-index antisymmetric representation is obtained by the tensor product of the fundamental representation with itself.

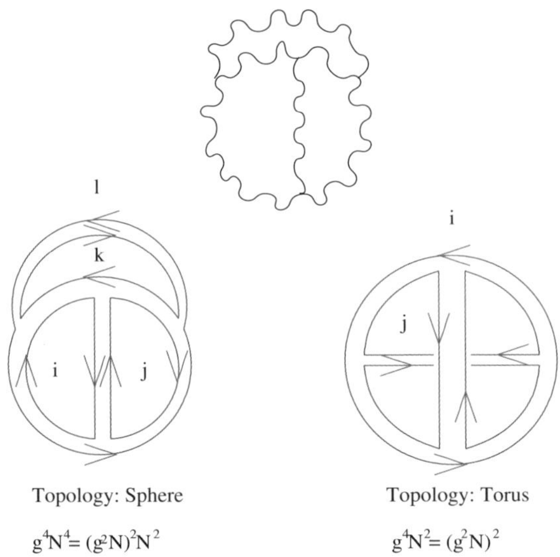

Fig. 10. A three-loop vacuum diagram.

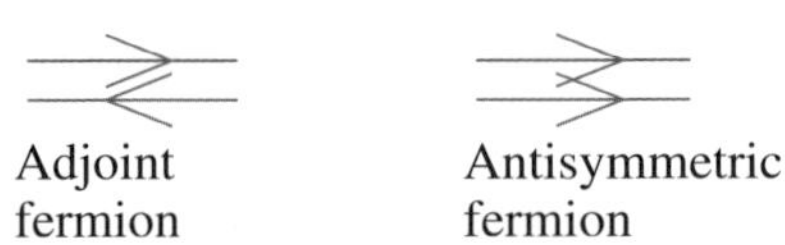

Fig. 11. The propagator of fermions in the adjoint and in the antisymmetric representations.

4.3. *Perturbative planar equivalence*

It was shown that all *planar diagrams* of $\mathcal{N} = 1$ SYM and the "orientifold" theory are equivalent to each other.[14] Why ? in order to map a graph in $\mathcal{N} = 1$ SYM to a graph in the "orientifold theory", simply reverse the orientation of one of the lines along each fermionic loop. An example is depicted in figure 12.

The general rule is as follows: draw the planar diagram in the 't Hooft notation on a sphere. Fermionic loops will divide the sphere to two regions: "interiors" and "exteriors", see figure 13. In order to relate the two theories, simply reverse the arrows on all the lines in the "interiors" (or the "exteriors").

So there is a one-to-one correspondence between the planar graphs of the two theories. Hence the theories are equivalent at the planar level.

The full non-perturbative equivalence is more difficult to prove.[14–16]

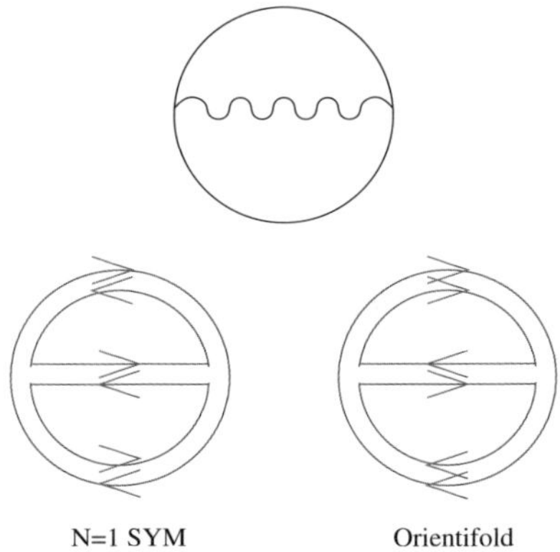

Fig. 12. An example of a two-loop vacuum diagram in $\mathcal{N}=1$ SYM and in the "orientifold field theory".

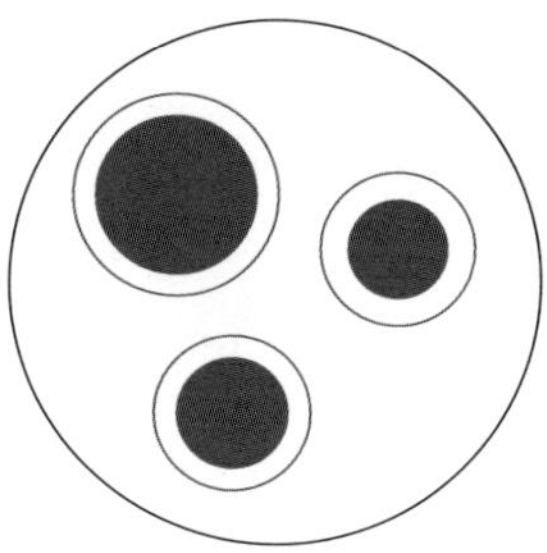

Fig. 13. Fermion lines divide the sphere to "interior" and "exterior" regions.

The idea is to show the equivalence of the partition functions of the two theories at the planar level. The partition function of $\mathcal{N}=1$ SYM is

$$\mathcal{Z}_0 = \int \mathcal{D}A \, \exp\left(iS[A]\right) \det\left(i \, \partial\!\!\!/ + A\!\!\!/^a T^a_{\mathrm{adj}}\right) \tag{18}$$

Next, we will make use of the fact that

$$T^a_{\mathrm{adj}} \sim T^a_{\square \times \overline{\square}} = T^a_{\square} \otimes 1 + 1 \otimes T^a_{\overline{\square}}, \tag{19}$$

to write the partition function as follows

$$\mathcal{Z}_0 = \int \mathcal{D}A \, \mathcal{D}B \, \delta(A-B) \, \exp -\tfrac{1}{2}(S[A]+S[B]) \\ \times \det\left(i \, \partial\!\!\!/ + A\!\!\!/^a (T^a_{\square} \otimes 1) + B\!\!\!/^a (1 \otimes T^a_{\overline{\square}})\right) \tag{20}$$

The determinant is gauge invariant and hence it can be written formally by using Wilson loops in the fundamental representation

$$\mathcal{Z}_0 = \int \mathcal{D}A\,\mathcal{D}B\,\delta(A - B)\,\exp - \tfrac{1}{2}(S[A] + S[B]) \sum \prod \mathcal{W}(A)\mathcal{W}(B). \quad (21)$$

Or

$$\mathcal{Z}_0 = \sum \langle \prod \mathcal{W}(\mathcal{A})\mathcal{W}(\mathcal{B}) \rangle \quad (22)$$

The partition function of the "orientifold field theory" is similar, but with reversed orientation for one of the Wilson loops, namely $\mathcal{W}(\mathcal{B}) \to \mathcal{W}^\star(\mathcal{B})$. Thus the partition function of the orientifold theory takes the form

$$\mathcal{Z}_0 = \sum \langle \prod \mathcal{W}(\mathcal{A})\mathcal{W}^\star(\mathcal{B}) \rangle \quad (23)$$

At large-N gauge invariant amplitudes factorise. For example

$$\langle \mathcal{W}(A)\mathcal{W}(B) \rangle = \langle \mathcal{W}(\mathcal{A}) \rangle \langle \mathcal{W}(\mathcal{B}) \rangle \quad (24)$$

As a result

$$\langle \mathcal{W}(A)\mathcal{W}(B) \rangle = \langle \mathcal{W}(A)\mathcal{W}^\star(B) \rangle \quad (25)$$

since $\langle \mathcal{W} \rangle = \langle \mathcal{W}^\star \rangle$ (We assume here that C-parity is not broken). More generally, we proved that at large-N $\langle \prod \mathcal{W}_{\mathrm{adj.}} \rangle_{\mathrm{conn.}} = \langle \prod \mathcal{W}_{\mathrm{anti-symm.}} \rangle_{\mathrm{conn.}}$ So, the partition functions of the two theories coincide.

An alternative proof was given by Ünsal and Yaffe. They have shown that: *"Non-perturbative planar equivalence between $\mathcal{N} = 1$ SYM and the orientifold theory holds if and only if charge conjugation is not broken"*

4.4. *Brief description of the implications*

- Since the partition function of the two theories coincide it follows that the $\mathcal{O}(N^2)$ contribution to the vacuum energy of the non-supersymmetric theory vanishes.
- We can repeat the proof in the presence of sources to conclude that the value of the quark condensate is identical to the value of the gluino condensate.
- The bosonic color-singlet spectra is the same (however, there are no fermionic color singlets in the orientifold theory).
- Bosonic Green functions should also coincide.
- The centre of the gauge group of the orientifold theory is enhanced in the common sector from $\mathcal{Z}_2$ to $\mathcal{Z}_N$.
- A degeneracy between bosonic and fermionic open QCD-strings and an approximate meson-baryon degeneracy.

4.5. *The relation with string theory*

The AdS/CFT (gravity/gauge theory) correspondence suggests a classical description of the large-N gauge theory. This is a modern version of the old idea that large-N theories should be controlled by a classical master field.

The supersymmetric theory is dual to the type IIB superstring. Interestingly, the "orientifold theory" is dual to a non-supersymmetric string theory called type 0'B.[19,20]

The type 0'B closed string spectrum is the bosonic truncation of type IIB spectrum. This is in agreement with the fact that the color-singlet spectrum of the "orientifold gauge theory" is the bosonic truncation of the $\mathcal{N} = 1$ spectrum.

5. Supersymmetry relics in QCD

In the first two lectures we learnt about the non-perturbative properties of $\mathcal{N} = 1$ SYM and that a certain "orientifold gauge theory" becomes equivalent at large-N to $\mathcal{N} = 1$ SYM. For $SU(3)$ the "orientifold" theory becomes ordinary one-flavor QCD, so we can use the connection with $\mathcal{N} = 1$ SYM to estimate quantities in QCD.

5.1. *SUSY Relics in the QCD Spectrum*

In supersymmetric theories the color-singlets spectrum is bose/fermi degenerate. Moreover, massive color-singlet bosons are also parity even/odd degenerate.

In the large-N orientifold theory there are only bosonic color-singlets, since it is impossible to form a "light" fermionic color-singlet from antisymmetric constituents.

We thus predict that in one flavor QCD the color singlets are approximately even/odd parity degenerate.[21] Later, together with Imeroni,[22] we argued by using some insight from type 0' string theory that the ratio between the η' and the σ masses is given by the ratio of the axial and conformal anomalies

$$M_{\eta'}/M_\sigma \sim (N-2)/N \tag{26}$$

Hence for one-flavor QCD $M_{\eta'}/M_\sigma \sim 1/3$!

The prediction that the η' is much lighter than the σ is rather surprising. In real QCD the η' is heavier than the σ (although the σ is broad. This is not the case in one-flavor QCD). In order to check this prediction, a lattice simulation of one-flavor QCD was carried out by the Munster collaboration.[23] Their result for the masses of η' and σ, extrapolated to the chiral limit, is

$$M_{\eta'}/M_\sigma = 0.410(32)(25)\,, \tag{27}$$

In a reasonable agreement with our estimate.

5.2. *The QCD quark condensate from SUSY*

The QCD quark condensate $\langle \bar{q}q \rangle$ is the order parameter for chiral symmetry breaking. It is a non-perturbative quantity and hence very difficult to compute. The only reliable method of calculating it is the lattice. Since we know the value of the gluino condensate, we can use it to estimate the value of the quark condensate.[24]

Let us define a renormalization group invariant (RGI) quantity

$$(g^2)^{\gamma/\beta}\bar{\Psi}_{[ij]}\Psi^{[ij]}\,. \tag{28}$$

The above quantity coincides with the gluino condensate when $N \to \infty$. In order to estimate the (large !) $1/N$ corrections, notice that for the orientifold $SU(2)$ theory the quark condensate vanishes, since for $SU(2)$ the antisymmetric and the singlet are the same representation. We thus argue that

$$(g^2)^{\gamma/\beta}\bar{\Psi}_{[ij]}\Psi^{[ij]} = -6(N-2)\Lambda^3_{\overline{\rm MS}}\mathcal{K}(1/N)\,, \tag{29}$$

where $\mathcal{K}(1/N) = 1 + O(1/N)$. In order to compare our estimate with lattice results, which are often quoted at a scale 2 GeV, we evaluated the quark condensate at that scale.

Our result (including 30% due to $1/N$ corrections) is

$$\langle \bar{q}q \rangle_{2\,\rm GeV} = -(270 \pm 30\,\rm MeV)^3\,. \tag{30}$$

This number has to be compared with the results of [25]

$$\langle \bar{q}q \rangle_{2\,\rm GeV} = -(269(9)\,\rm MeV)^3\,. \tag{31}$$

The agreement is more than satisfactory.

5.3. *A lattice check of planar equivalence*

It fairly easy to check some predictions of planar equivalence by using a lattice simulation of the *quenched theory*[26]

$$\langle \bar{\psi}\psi \rangle = \int DA_\mu (\exp iS_{\rm YM})\,\mathrm{tr}\,\frac{1}{i\gamma_\mu D_\mu - m}\,. \tag{32}$$

Planar equivalence implies

$$\lim_{N\to\infty}\left[\frac{\langle\lambda\lambda\rangle_{\text{Adj}}}{N^2}-\frac{\langle\bar\psi\psi\rangle_{\text{Sym}}+\langle\bar\psi\psi\rangle_{\text{Asym}}}{2N^2}\right]=0\,. \tag{33}$$

Let us parametrise the condensates as follows

$$\frac{\langle\bar\psi\psi\rangle_{\text{Sym/Asym}}}{N^2}=f\left(\frac{1}{N^2},m\right)\pm\frac{1}{N}g\left(\frac{1}{N^2},m\right) \tag{34}$$

$$\frac{\langle\lambda\lambda\rangle_{\text{Adj}}}{N^2}=\tilde f\left(\frac{1}{N^2},m\right)-\frac{1}{2N^2}\langle\bar\psi\psi\rangle_{\text{free}} \tag{35}$$

We expect

$$f\left(\frac{1}{N^2},m\right)-\tilde f\left(\frac{1}{N^2},m\right)=\frac{c_1(m)}{N^2}+\frac{c_2(m)}{N^4}\,. \tag{36}$$

The lattice simulation was conducted for $SU(2)$, $SU(3)$, $SU(4)$, $SU(6)$ and $SU(8)$, for various values of masses. The results are presented in figure 14. As N increases the difference $f-\tilde f$ goes to zero.

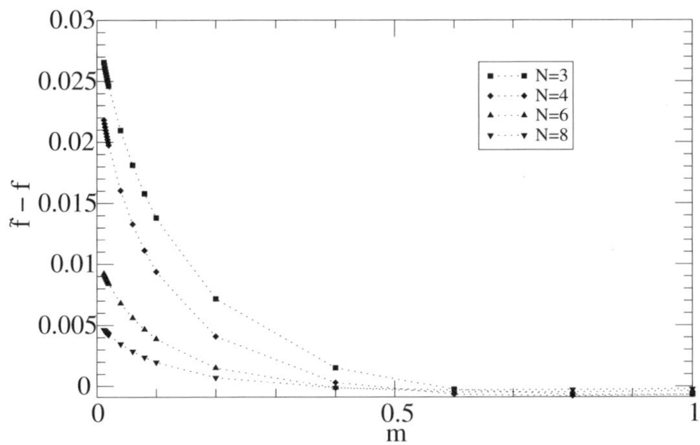

Fig. 14. The difference between the values of the quark condensate in a theory with a fermion in the reducible two-index representation and a theory with an adjoint fermion as a function of the mass for various values of N. The difference goes to zero as N increases.

5.4. *Three-flavors QCD*

It is possible to generalize[27] the analysis to three-flavors QCD. The idea is to consider an $SU(N)$ theory with one-flavor in the antisymmetric representation and two additional flavors in the fundamental representation.

This model reduces to three flavors QCD for $SU(3)$, becomes SUSY in the bosonic sector in the large N limit, but differs from the previous model by the $1/N$ corrections. The result for the quark condensate as a function of the 't Hooft coupling $\lambda \equiv \frac{g^2 N}{2\pi}$, at 2 GeV is

$$\langle \bar{q}q \rangle_{2\,\mathrm{GeV}} = -\frac{12}{\pi^2} \lambda^{-\frac{44}{27}} e^{-\frac{1}{\lambda}} . \tag{37}$$

Assuming (taken from the PDG) $\lambda(2\,\mathrm{GeV}) = 0.148 \pm 0.010$, we obtain

$$\langle \bar{q}q \rangle_{2\,\mathrm{GeV}} \quad = \quad -(317 \pm 30 \pm 36\ \mathrm{MeV})^3 . \tag{38}$$

The two errors are $\pm 30\%$ due to $1/N$ corrections and the experimental uncertainty of λ at 2 GeV.

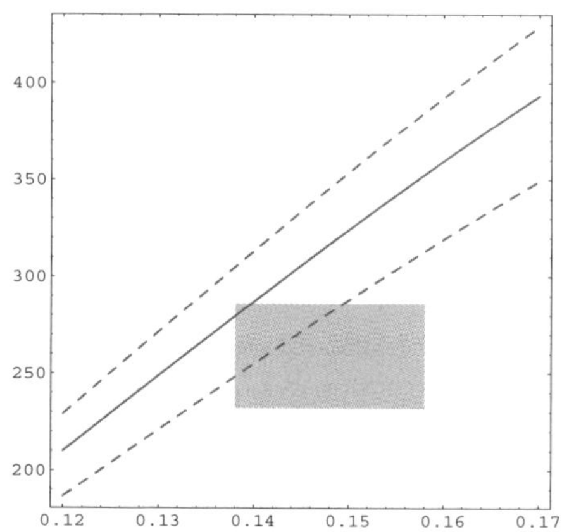

Fig. 15. The quark condensate in QCD with three massless quark, as a function of the 't Hooft coupling.

The plot (15) depicts the quark condensate in three flavor QCD, expressed as $-(y\,\mathrm{MeV})^3$, as a function of the 't Hooft coupling at 2 GeV. The blue area represents the lattice data.[28] The solid line is the prediction of the quark condensate. The two dashed lines represent the $1/N$ (30%) error.

5.5. *NSVZ beta function for QCD*

Another prediction of planar equivalence is an NSVZ beta function for one-flavor QCD

$$\beta(g) = -\frac{g^3}{16\pi^2} \frac{3N}{1 - \frac{g^2 N}{8\pi^2}} . \tag{39}$$

Of course the above expression is only an approximation and it suffers from $1/N$ corrections.

As we shall see, the NSVZ beta function will be used to support the existence of a Seiberg duality in a non-supersymmetric theory.

5.6. *Supersymmetry between Mesons and Baryons*

It is well known that both mesons and baryons admit a linear Regge trajectory of the form

$$J = \alpha_0 + \alpha' M^2 \,, \tag{40}$$

with an almost identical Regge slope $\alpha' \sim 0.89\,\mathrm{GeV}^{-2}$. The coincidence of the Regge slopes is explained by the diquark picture. The meson consists of a quark anti-quark connected by a string and similarly the baryon consists of a quark diquark connected by a string.

Let us propose a different explanation.[29] In our model the meson is a standard quark anti-quark connected by a string,

$$M = \bar{Q}^j(x_0) \left(P \exp i \int_{x_0}^{x_1} \vec{A}(y) d\vec{y} \right)_j^i Q_i(x_1) \,. \tag{41}$$

The baryon is made of two quarks which are connected to a smeared antisymmetric antiquark $\bar{\Psi}^{[kl]}$ by a string,

$$B_\alpha = \int d\vec{z}\, Q_j^C(x_0) \Big(P \exp i \int_{x_0}^{z} \vec{A}(y) d\vec{y} \Big)_k^j \vec{\gamma}_{\alpha\beta} \bar{\Psi}_\beta^{[kl]}(z) \Big(P \exp i \int_z^{x_1} \vec{A}(y) d\vec{y} \Big)_l^i Q_i(x_1) \,. \tag{42}$$

By replacing the antisymmetric fermion by a gluino, we map the baryon of the orientifold theory to a fermion in $\mathcal{N} = 1$ SYM. The latter fermion in SYM is a superpartner of the meson.

Thus, by using planar equivalence, we can argue that up to $1/N$, heavy mesons of the form $\bar{Q}Q$ and baryons of the form QQq are mass degenerate. The degeneracy improves as we neglect the end-points quarks. We thus predict the same Regge slope for asymptotically heavy mesons and baryons.

5.7. *SUSY relics in pure Yang-Mills ?*

We argue that one-flavor QCD can be approximated by a supersymmetric theory. In the old 't Hooft large-N expansion fundamental flavor is quenched and hence the one flavor QCD becomes at large-N pure Yang-Mills. So, perhaps there are remnants of supersymmetry in pure Yang-Mills ...

In supersymmetric theories

$$\langle \mathrm{tr}\, F^2 + i \mathrm{tr}\, F\tilde{F} \rangle = M_{\mathrm{UV}}^4 \exp(-\tau/3) \tag{43}$$

where τ is the holomorphic coupling. We can estimate $F\tilde{F}$ from either Witten-Veneziano or lattice simulations. The value is

$$\frac{1}{8\pi^2} F^a \tilde{F}^a = 1.3 \times 10^{-2}\, \mathrm{GeV}^4 \,. \tag{44}$$

The gluon condensate in pure Yang-Mills was estimated[30]

$$\frac{1}{8\pi^2} F^a F^a = 1.2 \times 10^{-2}\, \mathrm{GeV}^4 \,. \tag{45}$$

5.8. *More applications ...*

I haven't discussed many other applications of planar equivalence. In the past years planar equivalence was discussed in relation with

- Centre symmetry ($\mathcal{Z}_2$ or $\mathcal{Z}_N$?) and confinement.[31–33]
- Non-supersymmetric type 0' string theory.[34]
- Baryons.[35]
- Confinement/de-confinement transition.[36]
- Eguchi-Kawai reductions.[37]
- The Hagedorn transition.[38]
- Non-supersymmetric Seiberg duality.[39]
- M-theory membranes.[40]

6. Summary

I hope that I have convinced you that planar equivalence is a new and an exciting tool for studying the non-perturbative regime of QCD.

I would like to thank all my collaborators on this subject: *Sasha Gorsky, Emiliano Imeroni, Dan Israel, Biagio Lucini, Gregory Moraitis, Asad Naqvi, Vasilis Niarchos, Agostino Patella, Claudio Pica, Misha Shifman, Graham Shore, Mithat Ünsal, Gabriele Veneziano.*

And also to thank the organizers of the Schladming school and especially to Leonid Glozman for the kind invitation to deliver talks and for a fascinating meeting.

References

1. E. Witten, "Constraints On Supersymmetry Breaking," Nucl. Phys. B **202**, 253 (1982).

2. G. Veneziano and S. Yankielowicz, "An Effective Lagrangian For The Pure N=1 Supersymmetric Yang-Mills Theory," Phys. Lett. B **113**, 231 (1982).
3. M. A. Shifman and A. I. Vainshtein, "On Gluino Condensation in Supersymmetric Gauge Theories. SU(N) and O(N) Groups," Nucl. Phys. B **296**, 445 (1988) [Sov. Phys. JETP **66**, 1100 (1987)].
4. N. M. Davies, T. J. Hollowood, V. V. Khoze and M. P. Mattis, "Gluino condensate and magnetic monopoles in supersymmetric gluodynamics," Nucl. Phys. B **559**, 123 (1999) [arXiv:hep-th/9905015].
5. V. A. Novikov, M. A. Shifman, A. I. Vainshtein and V. I. Zakharov, "Exact Gell-Mann-Low Function Of Supersymmetric Yang-Mills Theories From Instanton Calculus," Nucl. Phys. B **229**, 381 (1983).
6. N. Seiberg, "Electric - magnetic duality in supersymmetric nonAbelian gauge theories," Nucl. Phys. B **435**, 129 (1995) [arXiv:hep-th/9411149].
7. G. R. Dvali and M. A. Shifman, "Domain walls in strongly coupled theories," Phys. Lett. B **396**, 64 (1997) [Erratum-ibid. B **407**, 452 (1997)] [arXiv:hep-th/9612128].
8. E. Witten, "Branes and the dynamics of QCD," Nucl. Phys. B **507**, 658 (1997) [arXiv:hep-th/9706109].
9. A. Armoni and M. Shifman, "The cosmological constant and domain walls in orientifold field theories and N = 1 gluodynamics," Nucl. Phys. B **670**, 148 (2003) [arXiv:hep-th/0303109].
10. B. S. Acharya and C. Vafa, "On domain walls of $\mathcal{N}$=1 supersymmetric Yang-Mills in four dimensions," arXiv:hep-th/0103011.
11. A. Armoni and T. J. Hollowood, "Sitting on the domain walls of N = 1 super Yang-Mills," JHEP **0507**, 043 (2005) [arXiv:hep-th/0505213].
12. A. Armoni and T. J. Hollowood, "Interactions of domain walls of SUSY Yang-Mills as D-branes," JHEP **0602**, 072 (2006) [arXiv:hep-th/0601150].
13. G. 't Hooft, "A Planar Diagram Theory for Strong Interactions," Nucl. Phys. B **72**, 461 (1974).
14. A. Armoni, M. Shifman and G. Veneziano, "Exact results in non-supersymmetric large N orientifold field theories," Nucl. Phys. B **667**, 170 (2003) [arXiv:hep-th/0302163].
15. A. Armoni, M. Shifman and G. Veneziano, "Refining the proof of planar equivalence," Phys. Rev. D **71**, 045015 (2005) [arXiv:hep-th/0412203].
16. M. Ünsal and L. G. Yaffe, "(In)validity of large N orientifold equivalence," Phys. Rev. D **74**, 105019 (2006) [arXiv:hep-th/0608180].
17. T. DeGrand and R. Hoffmann, "QCD with one compact spatial dimension," JHEP **0702**, 022 (2007) [arXiv:hep-lat/0612012].
18. B. Lucini, A. Patella and C. Pica, "Baryon currents in QCD with compact dimensions," Phys. Rev. D **75**, 121701 (2007) [arXiv:hep-th/0702167].
19. A. Sagnotti, "Some properties of open string theories," arXiv:hep-th/9509080.
20. A. Sagnotti, "Surprises in open-string perturbation theory," Nucl. Phys. Proc. Suppl. **56B**, 332 (1997) [arXiv:hep-th/9702093].
21. A. Armoni, M. Shifman and G. Veneziano, "SUSY relics in one-flavor QCD from a new 1/N expansion," Phys. Rev. Lett. **91**, 191601 (2003) [arXiv:hep-

th/0307097].
22. A. Armoni and E. Imeroni, "Predictions for orientifold field theories from type 0' string theory," Phys. Lett. B **631**, 192 (2005) [arXiv:hep-th/0508107].
23. F. Farchioni, I. Montvay, G. Munster, E. E. Scholz, T. Sudmann and J. Wuilloud, "Hadron masses in QCD with one quark flavour," Eur. Phys. J. C **52**, 305 (2007) [arXiv:0706.1131 [hep-lat]].
24. A. Armoni, M. Shifman and G. Veneziano, "QCD quark condensate from SUSY and the orientifold large-N expansion," Phys. Lett. B **579**, 384 (2004) [arXiv:hep-th/0309013].
25. T. DeGrand, R. Hoffmann, S. Schaefer and Z. Liu, "Quark condensate in one-flavor QCD," Phys. Rev. D **74**, 054501 (2006) [arXiv:hep-th/0605147].
26. A. Armoni, B. Lucini, A. Patella and C. Pica, "Lattice Study of Planar Equivalence: The Quark Condensate," Phys. Rev. D **78**, 045019 (2008) [arXiv:0804.4501 [hep-th]].
27. A. Armoni, G. Shore and G. Veneziano, "Quark condensate in massless QCD from planar equivalence," Nucl. Phys. B **740**, 23 (2006) [arXiv:hep-ph/0511143].
28. C. McNeile, "An estimate of the chiral condensate from unquenched lattice QCD," Phys. Lett. B **619**, 124 (2005) [arXiv:hep-lat/0504006].
29. A. Armoni and A. Patella, "Degeneracy Between the Regge Slope of Mesons and Baryons from Supersymmetry," arXiv:0901.4508 [hep-th].
30. V. A. Novikov, M. A. Shifman, A. I. Vainshtein and V. I. Zakharov, "Are All Hadrons Alike? Technical Appendices," Nucl. Phys. B **191**, 301 (1981).
31. F. Sannino, "Higher representations: Confinement and large N," Phys. Rev. D **72**, 125006 (2005) [arXiv:hep-th/0507251].
32. A. Armoni, M. Shifman and M. Ünsal, "Planar Limit of Orientifold Field Theories and Emergent Center Symmetry," Phys. Rev. D **77**, 045012 (2008) [arXiv:0712.0672 [hep-th]].
33. L. Del Debbio and A. Patella, "Center symmetry and the orientifold planar equivalence," JHEP **0903**, 071 (2009) [arXiv:0812.3617 [hep-th]].
34. P. Di Vecchia, A. Liccardo, R. Marotta and F. Pezzella, "On the gauge / gravity correspondence and the open/closed string duality," Int. J. Mod. Phys. A **20**, 4699 (2005) [arXiv:hep-th/0503156].
35. S. Bolognesi, "Skyrmions in Orientifold and Adjoint QCD," arXiv:0901.3796 [hep-th].
36. T. J. Hollowood and A. Naqvi, "Phase transitions of orientifold gauge theories at large N in finite volume," JHEP **0704**, 087 (2007) [arXiv:hep-th/0609203].
37. P. Kovtun, M. Ünsal and L. G. Yaffe, "Necessary and sufficient conditions for non-perturbative equivalences of large N(c) orbifold gauge theories," JHEP **0507**, 008 (2005) [arXiv:hep-th/0411177].
38. T. D. Cohen, "QCD and the Hagedorn spectrum," arXiv:0901.0494 [hep-th].
39. A. Armoni, D. Israel, G. Moraitis and V. Niarchos, "Non-Supersymmetric Seiberg Duality, Orientifold QCD and Non-Critical Strings," Phys. Rev. D **77**, 105009 (2008) [arXiv:0801.0762 [hep-th]].
40. A. Armoni and A. Naqvi, "A Non-Supersymmetric Large-N 3D CFT And Its Gravity Dual," JHEP **0809**, 119 (2008) [arXiv:0806.4068 [hep-th]].

Confining Gauge Fields

Frieder Lenz*
Institute for Theoretical Physics III, University of Erlangen-Nürnberg
91058 Erlangen, Germany
**E-mail: flenz@theorie3.physik.uni-erlangen.de*

By superposition of regular gauge instantons or merons, ensembles of gauge fields are constructed which describe the confining phase of SU(2) Yang-Mills theory. Various properties of the Wilson loops, the gluon condensate and the topological susceptibility are found to be in qualitative agreement with phenomenology or results of lattice calculations. Limitations in the application to the glueball spectrum and small size Wilson loops are discussed.

Introduction

Despite convincing evidence for confinement in Yang-Mills theories from lattice gauge calculations and despite a host of analytical studies and conjectures, the intricacies of non-abelian gauge fields have prevented an unambiguous identification of the mechanism of confinement. In particular, the gauge dependence of many of the mechanisms proposed constitutes a severe obstacle in reaching this goal. I will choose a different path towards understanding confinement. I will describe construction and analysis of ensembles of gauge fields which exhibit confinement and account for other facets of the dynamics of SU(2) Yang Mills theory. The studies to be presented are based on the ideas that instantons [1] or merons,[2] solutions of the classical Yang-Mills equations, play a prominent role also in the quantum theory.[3,4] They are related to investigations which led to the instanton gas and liquid models [5,6] which successfully describe a variety of phenomena of strong interaction physics.[7,8] These models however miss confinement, the trademark of the dynamics of Yang-Mills fields. I will discuss an alternative construction of ensembles of gauge-fields obtained by superposition of "regular gauge" instantons or of merons.[9] Although irrelevant for single instanton properties, the choice of either the regular or the singular gauge in the superposition of instantons leads to two different phases, the gas or liquid phase of singular gauge instantons with a well defined low density limit and the strongly correlated confining phase of regular gauge instantons (or merons).

Gauge fields of instantons and merons

After appropriate choice of the coordinate system in color space and after regularization of the singularity, the SU(2) gauge field for a meron or an (regular gauge) instanton in Lorenz gauge with the center at the origin, is given by

$$a_\mu(x) = \xi \, \frac{\eta_{a\mu\nu} x_\nu}{x^2 + \rho^2} \frac{\sigma^a}{2} \,, \tag{1}$$

with the Pauli matrices σ^a, the 't Hooft tensor $\eta_{a\mu\nu}$,[7] with $\xi = 1$ denoting merons and $\xi = 2$ instantons. The size parameter ρ controls the short distance behavior of instanton and meron fields. For instantons, and for merons with vanishing ρ, $a_\mu(x)$ is a solution of the Euclidean classical field equations.[1,2]

Important for the following are the infrared properties of the constituents. The gauge fields of a single meron and of a single regular gauge instanton behave asymptotically as $1/x$ while the asymptotics of the field strengths,

$$F^a_{\mu\nu}[A] = \partial_\mu A^a_\nu - \partial_\nu A^a_\mu + \epsilon^{abc} A^b_\mu A^c_\nu \,,$$

is different for merons and instantons

$$x \to \infty \,, \quad F_{\xi=1} \to x^{-2} \,, \quad F_{\xi=2} \to x^{-4} \,, \tag{2}$$

due to the cancellation of the abelian and non-abelian contributions to leading order for instantons. Therefore the action density

$$s(x) = \frac{1}{2} \operatorname{tr} F_{\mu\nu} F^{\mu\nu} \tag{3}$$

behaves asymptotically as x^{-4} and x^{-8} for merons and instantons respectively and gives rise to an infinite action for merons. The topological charge density is defined by

$$\tilde{s}(x) = \frac{1}{2} \operatorname{tr} F_{\mu\nu} \tilde{F}^{\mu\nu} \,, \tag{4}$$

and yields a finite value for the topological charge $\nu = \xi/2$. The fast asymptotic decrease of the field strength of an instanton is the basis for the alternate representation of instantons in "singular gauge"

$$a^{\text{sing}}_\mu = \frac{2\rho^2}{x^2} \frac{\bar{\eta}_{a\mu\nu} x_\nu}{x^2 + \rho^2} \frac{\sigma^a}{2} \,, \tag{5}$$

where the gauge field decays asymptotically as x^{-3} and $\bar{\eta}$ denotes the anti-self-dual 't Hooft tensor.

Gauge fields from superposition of merons or instantons

The SU(2) gauge fields to be considered are superpositions of instanton or meron fields

$$A_\mu(x) = \sum_{i=1}^{N_P} h(i) a_\mu(x - z(i)) h^{-1}(i). \tag{6}$$

The dynamical variables of these fields are the positions of the centers $z(i)$ of the instantons or merons and their color orientations

$$h(i) = h_0(i) + \mathrm{i}\mathbf{h}(i) \cdot \boldsymbol{\sigma}\,, \qquad h_0^2(i) + \mathbf{h}^2(i) = 1\,. \tag{7}$$

In general, the action of gauge fields generated by superposition of either merons or instantons is logarithmically divergent in the infrared. By a judicious choice of the color orientation, finite values of the total action can be obtained. This is achieved, e.g., by the following superposition of 4 instanton or merons

$$A(x) = \sum_{i=1}^{3} R_i^\pi a(x - z(i)) + a(x - z(4))\,,$$

with the color rotations R_i^π by π around the three color directions i. With this choice

$$x \to \infty : \quad A \sim \frac{1}{x^2}\,, \quad F[a] \sim \frac{1}{x^3}\,, \quad s(x) \sim \frac{1}{x^6}\,, \tag{8}$$

the system is "neutral" and the action is finite. The constituents, instantons or merons, are "confined". Removal of a single constituent gives rise to an infinite action. Thus ensembles of instantons and merons exhibit similar infrared properties and in both cases strong correlations between the constituents are required to guarantee a finite action.

The ensembles of gauge fields (cf. Eq.(6)) to be constructed in the following are defined by the partition function

$$Z = \int \mathrm{d}z_i \mathrm{d}h_i \mathrm{e}^{-\frac{1}{g^2} S[A(z_i,h_i)]}\,, \tag{9}$$

with the effective action S identified with the standard (Euclidean) action (cf. Eq. (3))

$$S = \int_V \mathrm{d}^4 x s(x)\,. \tag{10}$$

This definition of the partition function guarantees that for finite g^2 a non-zero weight is assigned to neutral configurations only, i.e. to fields with the asymptotic behavior (8). After construction of ensembles of field configurations using the Metropolis algorithm, vacuum expectation values of observables $\mathcal{O}$ can be computed

$$\langle \mathcal{O} \rangle = \frac{1}{Z} \int \mathrm{d}z_i \mathrm{d}h_i \mathrm{e}^{-\frac{1}{g^2} S[A(z_i,h_i)]} \mathcal{O}[A(z_i,h_i)] . \tag{11}$$

In the numerical determination of these ensembles, the location $z(i)$ is restricted to a hypercube

$$-1 \le z_\mu(i) \le 1 , \tag{12}$$

an equal number of instanton and anti-instantons (merons and antimerons) is used, and in most of the applications the values of size and coupling constant are chosen as

$$\rho = 0.16 , \quad g^2 = 32 . \tag{13}$$

Characteristics of instanton and meron ensembles

Action densities

In this section I will characterize qualitative properties of gauge fields in the ensembles defined by (9). In Fig. 1 is shown a typical landscape of the

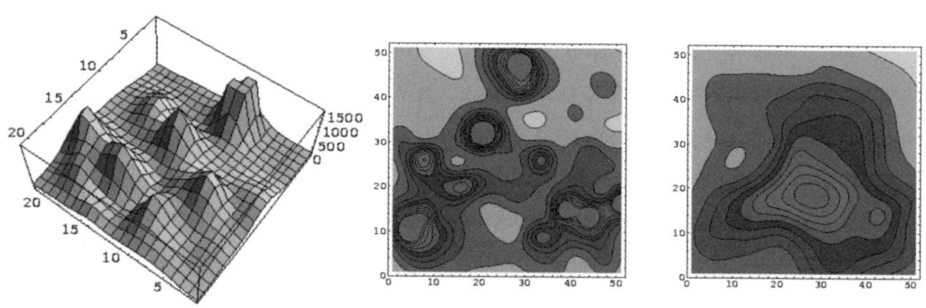

Fig. 1. Action density along a plane for a gauge field of 500 merons (left) and lines of equal action density ($g^2 = 32$, middle) and ($g^2 = \infty$ right).

action density displaying peaks of single merons on top of a background. Background and single meron contributions to the total action are of the same order of magnitude. Unlike in the superposition of singular instantons, the appearance of single meron or instanton structures is not automatic, it rather reflects the dynamics defined by the partition function (9). This becomes evident in the comparison (Fig. 1) with the action density of a meron field of the corresponding stochastic ensemble ($g^2 = \infty$). The values of the action density of the two ensembles

$$\begin{aligned} g^2 &= 32\,, \quad \langle s \rangle = 1500\,, \quad 200 < s(x) < 11000\,, \\ g^2 &= \infty\,, \quad \langle s \rangle = 110000\,, \quad 15000 < s(x) < 330000\,, \end{aligned} \tag{14}$$

differ by almost two orders of magnitude. As a result of of the huge background generated by the stochastic superposition of the constituents no single meron contribution ($s_{\max} = 18000$) can be identified. The maximum close to the center of Fig. 1 is a result of the stochastic superposition as can be verified in an analytical calculation. The reduction of the action from the strong coupling value is a first quantitative signature of the importance of correlations in the ($g^2 \neq \infty$) ensembles. Another measure of the strength of the correlations between constituents [10] is provided by the response of the action to changes of the color orientation or of the position of a single meron or instanton. As shown in Fig. 2, under variations of the color orientation of a single constituent the total action is increased by up to 60% and by up to 20% under variations of the position. In a weakly interacting system of 50 instantons (instanton gas) one would expect a 2% decrease of the action when removing a single instanton and even smaller changes under variations of the color orientation.

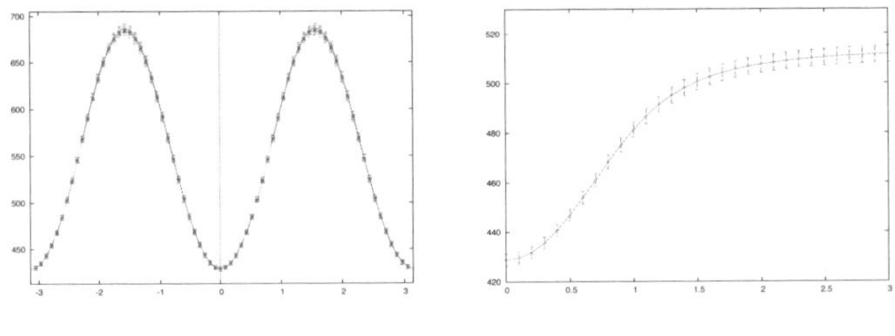

Fig. 2. Response of the action of 50 instantons to changes of the color orientation along 3 independent geodesics on the manifold S^3 of the color rotations (7) (left) of the instanton closest to the center and to changes of its position (right).

Confinement

In this section I will present evidence that the phase of highly correlated constituents is confining. In Yang Mills theories, the area law of Wilson loops is the signature of confinement. The Wilson loop is defined by

$$W = \frac{1}{2}\,\mathrm{tr}\Big\{P\exp ig \oint_{\mathcal{C}} A_\mu(x)\mathrm{d}x^\mu\Big\}\,, \tag{15}$$

with the integral ordered along the closed path $\mathcal{C}$. The following results refer to rectangular paths with aspect ratio 2:1. For sufficiently large loops the results for the logarithm of the expectation value of the Wilson loop can be parametrized as a sum of constant, perimeter ($\mathcal{P}$) and area ($\mathcal{A}$) terms,

$$\ln\langle W\rangle = \omega + \tau\,\mathcal{P} - \sigma\mathcal{A}\,, \tag{16}$$

with σ denoting the string tension. The parameters of the fit for ensembles of 500 merons ($\rho = 0.16$) are

$$\omega = -2.1,\ \tau = 2.3,\ \sigma = 22.9 \tag{17}$$

for the stochastic ensemble ($g^2 = \infty$). For ensembles with $g^2 = 32$ the following values

$$\omega = -0.7,\ \tau = 1.1,\ \sigma = 11.8 \tag{18}$$

are obtained if the color orientations are dynamical variables while the positions are randomly chosen but fixed, and

$$\omega = -0.7,\ \tau = 1.1,\ \sigma = 11.5 \tag{19}$$

for the ensemble with dynamical color orientations and dynamical positions. These 3 ensembles give rise to an area law with a value of the string tension differing by a factor of 2 between stochastic and dynamically correlated ensembles. The large difference in the dimensionless ratios formed by action density and string tension,

$$\frac{g^2\langle s\rangle}{\sigma^2} \approx 500,\ \ (g^2 = \infty)\,,\qquad \frac{g^2\langle s\rangle}{\sigma^2} \approx 11\ \ (g^2 = 32)\,, \tag{20}$$

reflects the different dynamics of the ensembles. As in lattice gauge theories, confinement in the strong coupling limit is a result of disorder of the system generated by the unconstrained fluctuations of the gauge fields. Suppression of the fluctuations if $g^2 \neq \infty$ affects more strongly the local action density than the non local-Wilson loop and in turn leads to the strong decrease in the ratio $g^2\langle s\rangle/\sigma$. Only minor changes of Wilson loops are obtained if, in addition to the color orientations, also the meron positions are treated as

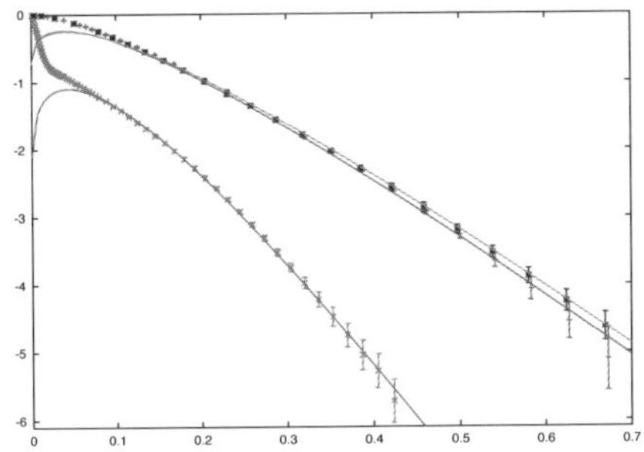

Fig. 3. Logarithm of a Wilson loop as a function of its area for the three meron ensembles of Eqs. (17) (green $\times$), (18) (pink $-$), and (19) (blue $*$).

dynamical variables. This result is in accordance with the above findings concerning the response to changes in the color orientation and position (Fig. 2) of single constituents. It suggests that the confining phase is a close relative of the nematic phase of liquid crystals with strong correlations in internal and comparatively weak correlations in position space.[11]

Quantitative results

Scaling and confinement

Quantitative predictions for observables evaluated in the ensembles defined above require definition of a scale. As in lattice gauge calculations the scale is defined by the phenomenological value $\sigma = 4.4\,\text{fm}^{-2}$ of the string tension. Rescaling the different contributions to the Wilson loop (16),

$$\ln\langle W\rangle = \omega + \tau\,\sqrt{\lambda}\,\mathcal{P} - \sigma\,\lambda\,\mathcal{A}\,, \tag{21}$$

and fitting the scaling parameter λ separately for each ensemble gives rise to a universal curve as shown in Fig. 4. The values of the universal parameters and of the unit of length are for the instanton ensembles

$$\omega = -0.52, \quad \tau = 1.18, \quad \sigma = 19.0, \quad 1\,\text{u.l.} = 2.08\sqrt{\lambda}\,\text{fm}\,. \tag{22}$$

Similar results have been obtained for the meron ensembles.

As the Wilson loops, also other observables exhibit approximate independence of the number of degrees of freedom. Table 1 summarizes for ensembles with N_I instantons the results for the vacuum expectation values of

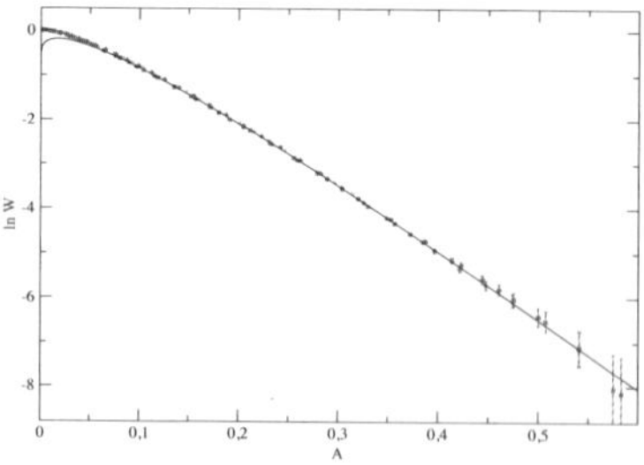

Fig. 4. Logarithm of a Wilson loop as a function of its area for instanton ensembles $N_I = 50$ (star), 100 (diamond), 200 (square), 500 (circle). The area has been rescaled with λ given in Table 1. Also shown is the curve corresponding to the parametrization (21) with the values of the parameters given in (22).

the action density s and of the topological susceptibility (also in physical units),

$$\chi = \left(\frac{1}{8\pi^2}\right)^2 \int d^4x \langle \tilde{s}(x)\tilde{s}(0)\rangle \,, \tag{23}$$

which is a measure of the strength of the fluctuations of the topological charge density (4). In comparison to the values obtained from sum rules [12,13] and from lattice gauge calculations [14]

$$85\,\mathrm{fm}^{-4} \leq \langle s \rangle \leq 260\,\mathrm{fm}^{-4} \,, \tag{24}$$

the expectation value of the action density i.e. of the gluon condensate is of the correct order of magnitude. It receives about equal contributions from the background field generated by the superposition of the constituents and from the peaks of the fields of single constituents as suggested by Fig. 1. Also the value of the topological susceptibility is in reasonable agreement with the lattice result [14] $\chi^{1/4} \sim 215\,\mathrm{MeV}$. The topological susceptibility is dominated by the topological charge density of single constituents. Keeping only the single constituent contributions in the evaluation of the integral (23)

$$\chi_I^{1/4} = 0.83\, n_I^{1/4} \,, \quad \chi_M^{1/4} = 0.52 n_M^{1/4} \,,$$

reproduces within 10 % the results of the numerical evaluation (Table 1). Due to the dominance of single instanton properties similar values of this observable in the liquid phase [7] of singular gauge and in the confining phase of regular gauge instantons are obtained.

Table 1. Properties of instanton ensembles:

N_I	$\langle s \rangle$	λ	n_I	ρ	$\langle s \rangle$	$\chi^{1/4}$
-	-	-	[fm^{-4}]	[fm^{1}]	[fm^{-4}]	[MeV]
500	5430	1.0	1.68	0.33	291	162
200	2490	0.66	1.54	0.27	307	164
100	1350	0.48	1.45	0.23	314	180
50	651	0.32	1.64	0.19	340	190

Table 1 shows that ensembles with increasing numbers of constituents (N_I) and located in a hypercube of fixed volume (12) describe, after rescaling, systems of essentially the same density n_I of instantons or merons. The increase in number is accounted for by the increase in volume of the hypercube. The value of the string tension determines the density of constituents. Variations in the constituent size in the interval $0.06 \le \rho \le 0.25$ have been considered and found to leave the scaling properties essentially unperturbed. Origin of the scaling properties is the scale independence of the asymptotics of regular gauge instantons or merons (Eq. (1)). In ensembles of singular gauge instantons on the other hand, ρ controls the strength of the asymptotics of these fields (Eq. (5)) and therefore the overlap and in turn the strength of the interaction of these constituents.[16]

Wilson loops and glueball masses

In addition to the area law, also more detailed properties of Wilson loops have been found to agree with those of lattice gauge calculations. I mention the Wilson loop distributions shown in Fig. 5 which, as similar lattice results,[17] can be interpreted as distributions of a diffusion process on S^3, the group manifold of SU(2), given by

$$p(\cos\vartheta, t) = \frac{2}{\pi}\,\theta(t)\sum_{n=1}^{\infty} n \sin n\vartheta\, \mathrm{e}^{-(n^2-1)t}\,, \tag{25}$$

with the diffusion time t determined by the expectation value of the Wilson loop. Diffusion on S^3 entails "Casimir scaling" of the Wilson loop expectation values in higher representations which has been observed in lattice gauge calculations.[18] Also in agreement with lattice gauge calculations is the breakdown of Casimir scaling observed for sufficiently large loops for ensembles of gauge fields similar to the ones discussed here.[19]
In a study of correlation functions of circular loops of equal radius r which

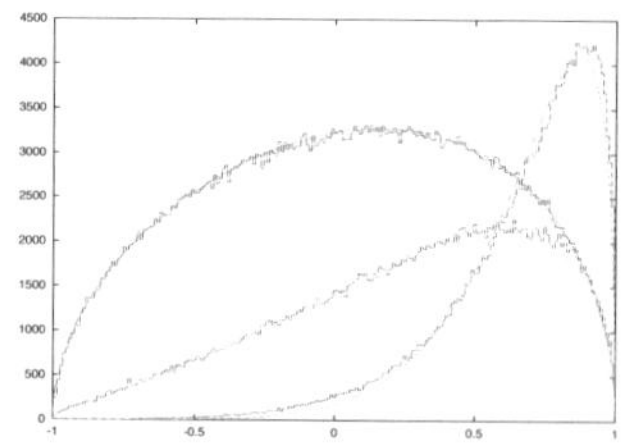

Fig. 5. Distribution of the values ($\cos\vartheta$) of Wilson loops for an ensemble of configurations containing 500 merons with sizes $0.48 \times 0.24, 0.72 \times 0.36, 1. \times 0.5$ in comparison with the distributions (25).

are parallel to each other and orthogonal to the direction of separation (t)

$$C_r(t) = \langle W_r(\mathbf{n}, \mathbf{x}_0, t) W_r(\mathbf{n}, \mathbf{x}_0, 0) \rangle \,, \tag{26}$$

indications for formation of flux tubes in connection with the area law

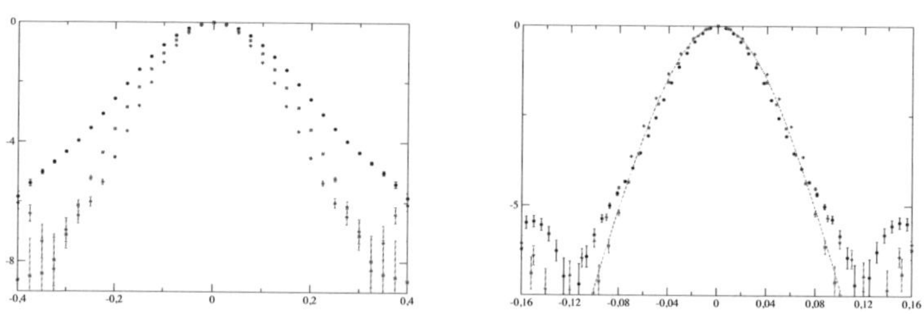

Fig. 6. Left: Logarithm of correlation functions of circular Wilson loops for an ensemble of 500 instantons of size $\rho = 0.08$ as a function of separation in time t for 3 values of the Wilson loop radius $r = 0.25$ (black circles), $r = 0.32$ (red squares), $r = 0.40$ (blue diamonds) . Right: The same as a function of the variable rt.

have been found. If gauge strings are important degrees of freedom the application of the Wilson loop operator to the vacuum should generate such a gauge string with energy $E \approx 2\pi\sigma r$, and the relevant variable for describing the correlation function should be rt. Indeed the three correlation functions are described approximately by a universal curve in terms of this variable as shown in Fig. 6. A rough estimate of the string tension based on this calculation yields $\sigma = (3.7 \pm 1.3)\,\text{fm}^{-2}$.

As the last topic I discuss the glueball spectrum calculated via (Euclidean) correlation functions

$$\langle \mathcal{O}(x)\mathcal{O}(0)\rangle \sim \sum_n \langle \Omega|\mathcal{O}|n\rangle \mathrm{e}^{-E_n x}\langle n|\mathcal{O}|\Omega\rangle \,.$$

of appropriately chosen operators carrying the quantum numbers of the

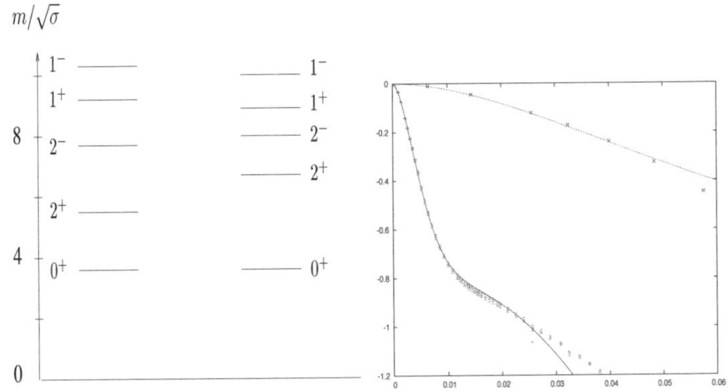

Fig. 7. Left: Spectra of glueball states. Lattice gauge theory [20] (left part) Instanton ensembles after rescaling (right part). Right: Enlarged portion of Fig. 3 for small size Wilson loops. Upper curve obtained for the ensemble with dynamical meron positions and color orientations, lower curve for the stochastic ensemble ($g^2 = \infty$).

glueballs. For sufficiently large separations, the correlation function is dominated by the state of lowest energy which can be excited by the operator $\mathcal{O}$. For determination of the masses local and non-local operators have been used and various ensembles differing in the size parameter and the number of constituents have been studied. The final result is shown in Fig. 7. Qualitative agreement with respect to the ordering of the level has been reached. The scale however is not compatible with the scale set by the string tension. The values of the masses have been multiplied with a factor of 1.8 (1.5 for meron ensembles) to reproduce the 0^+ glueball mass obtained in lattice gauge calculations. Determined by the requirement of the infrared finiteness of the action, the degrees of freedom do not adequately account for the dynamics on the scale of the size of the glueballs (0.2 -0.5 fm).[21] Missing strength in the ultraviolet also shows up in the Wilson loops of small size as is shown in the right part of Fig. 7. While the stochastic ensemble with its unconstrained fluctuations generates a Coulomb-like behavior at small distances, the ensembles of dynamical merons or instantons fail in the description of Wilson loops in the perturbative regime. A remedy for the shortcoming in the calculation of the glueball spectrum could be the inclusion of perturbative contributions in addition to the non-perturbative correlation functions, as applied for ensembles of singular gauge instantons.[22]

Conclusion

The central achievement of these studies is the construction of ensembles of gauge fields which exhibit confinement. These fields are obtained by superposition of regular gauge instantons or merons. It is remarkable that superposition of the gauge equivalent regular and singular instantons gives rise to two distinctively different phases, the gas or liquid phase generated by singular and the confining phase by regular gauge instantons respectively. The confining phase owes its existence to the non-trivial requirement of finite action in the presence of the slow asymptotic $1/x$ decay of the regular gauge instanton or meron fields. Unlike the disordered strong coupling limit, the confining phase is an ordered phase similar to the nematic phase of liquid crystals. The infrared behavior of the constituents is not only the source of confinement it's scale independence also induces the scaling properties of observables under changes in the number of constituents. Besides various properties of Wilson loops, such as the area law, Casimir scaling or flux tube formation, also the action density (gluon condensate) and the topological susceptibility are reasonably well reproduced. These latter quantities are determined by the relative strength of the gauge fields close to the center of the constituents and of the background gauge field generated by the superposition of the constituents. If observables such as the topological susceptibility are dominated by single instanton properties similar results in the liquid and in the confining phase can be expected. Limits in the applicability of meron and instanton ensembles are encountered if short distance properties are important as is the case for the glueball spectrum and for small size Wilson loops.

References

1. A. A. Belavin, A. M. Polyakov, A. S. Schwartz and Yu. S. Tyupkin, Phys. Lett. B **59**, 85 (1975).
2. V. de Alfaro, S. Fubini and G. Furlan, Phys. Lett. B **65**, 163 (1976).
3. C. G. Callan, R. F. Dashen and D. J. Gross, Phys. Rev. D **17**, 2717 (1978).
4. C. G. Callan, R. F. Dashen and D. J. Gross, Phys. Lett. B **66**, 375 (1977).
5. E. V. Shuryak, Nucl. Phys. B **203**, 93 (1982).

6. D. Diakonov and V. Y. Petrov, Nucl. Phys. B **245**, 259 (1984).
7. T. Schäfer and E. V. Shuryak, Rev. Mod. Phys. **70**, 323 (1998) [arXiv:hep-ph/**9610451**].
8. D. Diakonov, Prog. Part. Nucl. Phys. **51**, 173 (2003) [arXiv:hep-ph/**0212026**].
9. F. Lenz, J. W. Negele and M. Thies, Phys. Rev. D **69**, 074009 (2004) [arXiv:hep-th/**0306105**]; Ann. Phys. **323**, 1536 (2008) [arXiv:hep-ph/**0708.1687**].
10. F. Lenz and C. Szasz, to be published.
11. F. Lenz, "Topological Concepts in Gauge Theory," in "Topology and Geometry in Physics," Lecture Notes in Physics, (Springer Berlin, 2005) [arXiv:hep-th/**0403286**].
12. M. A.Shifman, A. I. Vainshtein and V. I. Zakharov, Nucl. Phys. B **147**, 385, 448 (1979).
13. S. Narison, Phys. Lett. B **387**, 162 (1996) [arXiv:hep-ph/**9512348**].
14. M. Campostrini, A. Di Giacomo and G. Paffuti, Z. Phys. C **22**, 143 (1984).
15. B. Lucini and M. Teper, JHEP **0106**, 050 (2001) [arXiv:hep-lat/**0103027**].
16. D. I. Diakonov, V. Y. Petrov and P. V. Pobylitsa, Phys. Lett. B **206**, 372 (1989).
17. A. M. Brzoska, F. Lenz, J. W. Negele and M. Thies, Phys. Rev. D **71**, 034008 (2005) [arXiv:hep-th/**0412003**].
18. G. S. Bali, Phys. Rev. D **62**, 114503 (2000) [arXiv:hep-lat/**0006022**].
19. Ch. Szasz and M. Wagner, Phys. Rev. D **78**, 036006 (2008) [arXiv:hep-ph/**0806.1977**].
20. M. J. Teper, "Glueball masses and other physical properties of SU(N) gauge theories in D = 3+1: A review of lattice results for theorists," arXiv:hep-th/**9812187**.
21. P. de Forcrand and K.-F. Liu, Phys. Rev. Lett. **69**, 245 (1992) [arXiv:hep-lat/**9211054**].
22. T. Schäfer and E. V. Shuryak, Phys. Rev. Lett. **75**, 1707 (1995) [arXiv:hep-lat/**9410372**].

Quark Counting Rules: Old and New Approaches

A. Radyushkin

Old Dominion University, Jefferson Lab and JINR, Dubna

I discuss the subject of powerlike asymptotic behavior of hadronic form factors in pre-QCD analyses of soft (Feynman/Drell-Yan) and hard (West) mechanisms, and also recent derivation of $1/Q^2$ asymptotics of meson form factors in AdS/QCD. At the end, I briefly comment on "light-front holography" ansatz.

1. Hadronic form factors

Introduction. Experimental evidence that (exclusive) form factors of hadrons consisting of n_q quarks behave like $(1/Q^2)^{n_q-1}$ for large Q^2, provokes expectations that there is a fundamental and/or easily visible reason for such a phenomenon, scale invariance being the most natural suspect.[1] Indeed, hard rescattering in a theory with spinor constituents and dimensionless coupling constant for their interaction with an intermediary boson field provides a specific dynamical mechanism[2] that produces the $(1/Q^2)^{n_q-1}$ behavior. In this approach, $n_q - 1$ is just the number of hard exchanges. Another property apparently correlated with the number of quarks in the hadron is the $\sim (1-x)^{2n_q-3}$ behavior of the (inclusive) quark distributions functions in the $x \to 1$ region. This observation suggests to look for connection between these exclusive and inclusive observables. Below in this section we discuss scenarios which display two versions of exclusive-inclusive correlation. In subsequent sections, we discuss derivation of the $1/Q^2$ behavior for meson form factors in AdS/QCD.

Soft mechanism. Powerlike behavior of hadronic form factors due to Feynman mechanism[3] can be derived from the Drell-Yan formula[4]

$$F(Q^2) = \int_0^1 dx \int d^2\mathbf{k}_\perp \, \Psi^*(x, \mathbf{k}_\perp + (1-x)\mathbf{q}_\perp)\Psi(x, \mathbf{k}_\perp) \; , \qquad (1)$$

which represents form factor in terms of the light-front wave function $\Psi(x, \mathbf{k}_\perp)$ and light-front variables x and $\mathbf{k}_\perp$. When the wave function

$\Psi(x, \mathbf{k}_\perp)$ rapidly (say, exponentially) decreases for $k_\perp \gtrsim \Lambda$, it is natural to consider the region where both $\Psi(x, \mathbf{k}_\perp)$ and $\Psi^*_M(x, \mathbf{k}_\perp + \bar{x}\mathbf{q}_\perp)$ are maximal: *i*) $|\mathbf{k}_\perp \lesssim \Lambda$ is small and *ii*) $\bar{x} \equiv 1 - x$ is close to 0, so that $|\bar{x}\mathbf{q}_\perp| \lesssim \Lambda$. If $|\Psi(x, \mathbf{k}_\perp \lesssim \Lambda)|^2 \sim (1-x)^{2n_q-3}$ then

$$F(Q^2) \sim \int_0^{\Lambda/Q} \bar{x}^{2n-3}\, d\bar{x} \sim (1/Q^2)^{n_q-1} \; . \tag{2}$$

The parton distribution functions in this formalism are given by the integral of $|\Psi(x, \mathbf{k}_\perp)|^2$ over $\mathbf{k}_\perp$. The latter is dominated by $k_\perp \lesssim \Lambda$, hence $f(x) \sim (1-x)^{2n_q-3}$. Thus, changing the shape of $f(x)$, one would change the result for form factor. In other words, there is a *causal relation* between the $x \to 1$ shape of the distribution function $f(x)$ and the $Q^2 \to 1$ behavior of the form factor $F(Q^2)$: form of $f(x)$ determines $F(Q^2)$.

Hard mechanism. For the Feynman/DY mechanism it was important that the fraction $\bar{x} \equiv 1 - x$ vanishes in the $Q^2 \to 0$ limit. Consider now the regions in DY formula (1), in which the fraction $\bar{x}$ is finite, while the transverse momentum argument of one of the wave functions is small, e.g., the region $|\mathbf{k}_\perp| \ll \bar{x}|\mathbf{q}_\perp|$, where $\Psi(x, \mathbf{k}_\perp)$ is maximal. Then

$$F(Q^2) \sim \int_0^1 |\Psi^*(x, \bar{x}\mathbf{q}_\perp)\, \varphi(x)|\, dx \; , \tag{3}$$

where

$$\varphi(x) = \int \Psi(x, \mathbf{k}_\perp)\, d^2\mathbf{k}_\perp \tag{4}$$

is the relevant distribution amplitude. In this scenario, the form factor repeats large-$\mathbf{k}_\perp$ behavior of the hadron wave function, e.g., if $\Psi(x, \mathbf{k}_\perp) \sim (1/\mathbf{k}_\perp^2)^n$, then $F(Q^2) \sim (1/Q^2)^n$. This mechanism was proposed by G.B. West,[5] who used, in fact, a covariant Bethe-Salpeter (BS) formalism rather than light-front variables, writing the form factor as[5]

$$F(Q^2) \sim \int f(p)\, f(p+q)\, d^4p \; , \tag{5}$$

where $f(p)$ is treated as a function of the active parton virtuality $t \equiv p^2$ and spectator mass M^2. Assuming that $f(t, M^2) \sim t^{-n} g(M^2)$ for large t, West concludes that $F(Q^2) \sim (1/Q^2)^n$: form factor repeats the large-Q^2 behavior of the BS wave function $f(p+q)$. For the deep inelastic structure function, West obtains[5]

$$\nu W_2(x) \sim \int_{t_{\min}}^{t_{\max} \sim -2\nu} f^2(t, M^2)\, dt \sim (t_{\min})^{-2n+1} \; , \tag{6}$$

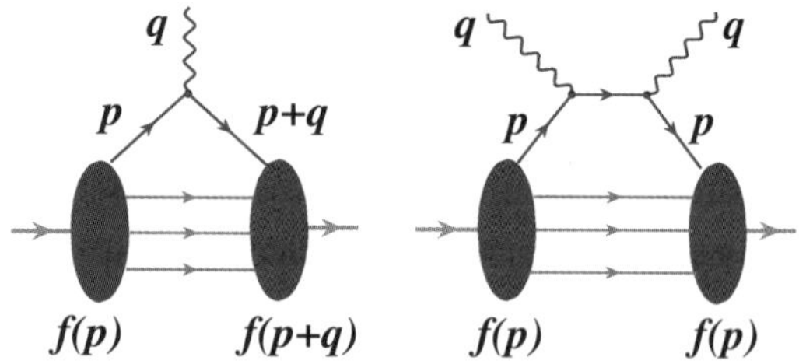

Fig. 1. Form factors and structure functions in West's model.

where $t_{\min} = \left(\frac{-x}{1-x}\right)\left[M^2 - (1-x)M_N^2\right]$, M_N being the nucleon mass. As a result, $\nu W_2(x) \sim (1-x)^{2n-1}$.

DY vs West's model. If $n = n_q - 1$, the power-law predictions of the two models formally coincide. However, these results were obtained from completely different assumptions. In DY picture, the active parton is "on-shell" both before and after the collision: both $|\mathbf{k}_\perp|$ and $|\mathbf{k}_\perp + \bar{x}\mathbf{q}_\perp|$ are of order Λ, and form factor $F(Q^2)$ reflects the *size of phase space* region in which $1 - x \sim \Lambda/Q$. On the other hand, in West's model, the active parton is highly virtual either in initial or final state, and $F(Q^2)$ reflects the t-dependence of WF for large virtualities $t = p^2$. Still, though the two mechanisms are completely different, the connection $(1/Q^2)^n \Leftrightarrow (1-x)^{2n-1}$ ("Drell-Yan-West relation") holds in both models![a] It should be also emphasized that in West's model, $(1/Q^2)^n$ and $(1-x)^{2n-1}$ have the same cause (large-t behavior of $f(p)$), but they are not "causing" each other.

West's hard mechanism & pQCD. In DY model, n is not necessarily integer. Integer values of n naturally appear in West's hard scenario, where they are related to the number of hard propagators. In particular, hard exchange in a theory with a dimensionless coupling constant gives $n = n_q - 1$ [2], which is a consequence of scale invariance.[1] In quantum chromodynamics, each hard gluon exchange is accompanied by effective coupling constant α_s, i.e., $F_{n_q}(Q^2) \sim (\alpha_s/Q^2)^{n_q-1}$. According to explicit calculation,[6,7] the asymptotic prediction for the pion form factor in pQCD is $F_\pi(Q^2) \to (2\alpha_s/\pi)\, s_0/Q^2$, where $s_0 = 4\pi^2 f_\pi^2 \approx 0.7\,\text{GeV}^2 \sim m_\rho^2$. Compared to the VMD expectation $F_\pi(Q^2) \sim m_\rho^2/Q^2$, pQCD prediction is suppressed by $2\alpha_s/\pi$ factor. It is well known that the factor $\alpha_s/\pi \sim 0.1$ is penalty for an extra loop, which suggests that the hard one-gluon-exchange contribu-

[a] This is apparently why the two models are confused up to the point that Eq. (1) is often referred to as "Drell-Yan-West formula", which is absolutely incorrect because its crucial feature is incorporation of light-front variables that West did not use.

tion is an $\mathcal{O}(\alpha_s)$ correction to some $\mathcal{O}(\alpha_s^0)$ term. The only candidate is the Feynman/DY soft contribution, which should be calculated in a nonperturbative way. In particular, in holographic AdS/QCD models considered in Refs. [8, 9] one has $F_\pi(Q^2) \to s_0/Q^2$, without a suppression factor.

2. Vector meson form factors in AdS/QCD

Models based on AdS/CFT correspondence are often claimed to provide nonperturbative explanation of quark counting rules for form factors that is based on conformal invariance and short-distance behavior of normalizable modes $\Phi(\zeta)$ playing the role of wave functions of initial and final hadrons. Namely, in the model of Polchinski and Strassler[10,11] (that involves on the AdS side scalar fields only) one has[12]

$$F(Q^2) = \int_0^{1/\Lambda} \Phi_{P'}(z) J(Q,z) \Phi_P(z)\, dz/z^3 \,, \tag{7}$$

where $J(Q,z) = zQK_1(zQ) \equiv \mathcal{K}_1(zQ)$ is nonnormalizable mode describing the probing EM current, and normalizable modes for mesons are given by $\Phi(z) = Cz^2 J_{L+1}(\beta_{L,k} z\Lambda)$, with K_1 and J_{L+1} being standard Bessel functions. For large Q, one may approximate $\mathcal{K}_1(zQ) \sim e^{-zQ}$, and it is clear that only small $z \lesssim 1/Q$ contribute. As a result, $F_{L=0}(Q^2) \to 1/Q^4$ for the ground state. But this is not the $1/Q^2$ power that one is longing to get! To bring the result of this AdS/CFT-based model in agreement with pQCD expectations, Brodsky and de Teramond proposed[12] to modify the basic principle of AdS/CFT correspondence, requiring that the dimension of the operator on the AdS side should be equal to the twist of the corresponding current in the 4-dimensional theory rather than to its dimension. In our papers with H.R. Grigoryan[8,9,13–15] we demonstrated that in more realistic AdS/QCD models of Refs. [16, 17] it is possible to get $F_{L=0}(Q^2) \to 1/Q^2$ for (leading) meson form factors without challenging the Maldacena[18] correspondence principle.

Hard-wall model is formulated in 5-dimensional space $\{x^\mu, z\} \equiv X^M$ having AdS_5 metric $ds^2 = \left(\eta_{\mu\nu} dx^\mu dx^\nu - dz^2\right)/z^2$ with a hard wall: $0 \le z \le z_0 = 1/\Lambda$. The basic object is the 5-dimensional (5D) vector gauge field $A_M(X)$ $(M = \mu, z)$ which produces 4D field $A_\mu(x) = A_\mu(x, z=0)$. at the UV boundary of AdS space. The 5D gauge action for the vector field is given by

$$S_{\text{AdS}} = -\frac{1}{4g_5^2} \int d^4x\, dz\, \sqrt{g}\, \text{Tr}\left(F_{MN}F^{MN}\right) \,, \tag{8}$$

where F_{MN} is the field-strength tensor. The coupling constant $g_5^2 = 6\pi^2/N_c$ is small in large-N_c limit. The free field satisfies $\Box_5 A(X) = 0$ or

$$\Box_4 A(x,z) + z\partial_z \left(\frac{1}{z}\, \partial_z A(x,z) \right) = 0 \; . \tag{9}$$

In 4D momentum representation this gives

$$z\partial_z \left(\frac{1}{z}\, \partial_z \tilde{A}(p,z) \right) + p^2 \tilde{A}(p,z) = 0 \; . \tag{10}$$

According to AdS/QCD correspondence

$$\tilde{A}_\mu(p,z) = \tilde{A}_\mu(p)\, V(p,z)/V(p,0) \equiv \tilde{A}_\mu(p)\, \mathcal{V}(p,z) \; , \tag{11}$$

where the bulk-to-boundary propagator $V(p,z)$ satisfies Eq.(10). The gauge-invariant boundary condition (b.c.) $F_{\mu z}(x, z_0) = 0$ on the infrared (IR) wall results in Neumann b.c. $\partial_z V(p, z_0) = 0$, with solution

$$V(p,z) = Pz\,[Y_0(Pz_0)J_1(Pz) - J_0(Pz_0)Y_1(Pz)] \; . \tag{12}$$

Using Kneser-Sommerfeld formula[19] gives bound state expansion

$$\mathcal{V}(p,z) = -\sum_{n=1}^{\infty} \frac{g_5 f_n}{p^2 - M_n^2}\, \psi_n(z) \tag{13}$$

with masses $M_n = \gamma_{0,n}/z_0$ determined by zeros $J_0(\gamma_{0,n}) = 0$ of Bessel functions, while the "coupling constants" f_n are given by

$$f_n = \frac{\sqrt{2}M_n}{g_5 z_0 J_1(\gamma_{0,n})} \; . \tag{14}$$

They are accompanied by "ψ" wave functions

$$\psi_n(z) = \frac{\sqrt{2}}{z_0 J_1(\gamma_{0,n})}\, z J_1(M_n z) \tag{15}$$

coinciding with nonnormalizable modes of Polchinski-Strassler model.[10,11] These "ψ" wave functions (w.f.) obey equation of motion (10) with $p^2 = M_n^2$, satisfy $\psi_n(0) = 0$ at UV boundary, and $\partial_z \psi_n(z_0) = 0$ at IR boundary. They are normalized according to

$$\int_0^{z_0} |\psi_n(z)|^2\, \frac{dz}{z} = 1 \; . \tag{16}$$

However, they do not look like bound state w.f. in quantum mechanics, see Fig.2, left. To this end, it makes sense to introduce "ϕ" wave functions

$$\phi_n(z) \equiv \frac{1}{M_n z}\, \partial_z \psi_n(z) = \frac{\sqrt{2}}{z_0 J_1(\gamma_{0,n})}\, J_0(M_n z) \; . \tag{17}$$

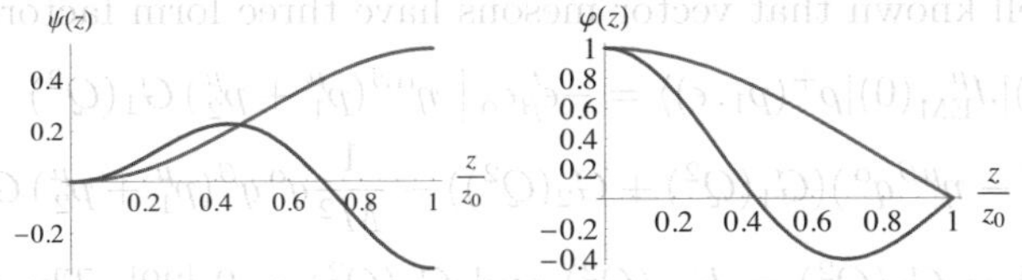

Fig. 2. ψ and ϕ wave functions for vector mesons.

According to Sturm-Liouville equation (10), they are reciprocal to "ψ" w.f.:

$$\psi_n(z) = -z\,\partial_z \phi_n(z)/M_n\,. \tag{18}$$

The ϕ w.f. give couplings $g_5 f_n/M_n$ as their values at the origin, they satisfy Dirichlet b. c. $\phi_n(z_0) = 0$ at confinement radius, and are normalized by

$$\int_0^{z_0} |\phi_n(z)|^2 z\,dz = 1\,. \tag{19}$$

The "ϕ" w.f. (see Fig.2,right) are thus analogous to bound state wave functions in quantum mechanics. The difference between the two types of AdS/QCD wave functions can be easily understood: ψ w.f. correspond to vector-potential A_M, while ϕ w.f. correspond to field strength tensor F_{MN}.

Three-point function should be introduced to study form factors. It has a "Mercedes-Benz" form

$$W(p_1, p_2, q) = \int_0^{z_0} \mathcal{V}(p_1, z)\mathcal{V}(p_2, z)\,\mathcal{V}(q, z)\,\frac{dz}{z}\,. \tag{20}$$

For spacelike q (with $q^2 = -Q^2$) we have $\mathcal{V}(iQ, z) \equiv \mathcal{J}(Q, z)$ The form factors for diagonal $n \to n$ transitions may be written

$$F_{nn}(Q^2) = \int_0^{z_0} \mathcal{J}(Q, z)\,|\psi_n(z)|^2\,\frac{dz}{z} \tag{21}$$

either in terms of ψ functions or in terms of ϕ functions[13]

$$F_{nn}(Q^2) = \frac{1}{1 + Q^2/2M_n^2}\int_0^{z_0} \mathcal{J}(Q, z)\,|\phi_n(z)|^2\,z\,dz\,. \tag{22}$$

The overlap integral here is a direct analogue of form factors in quantum mechanics, so we define

$$\mathcal{F}_{nn}(Q^2) \equiv \int_0^{z_0} \mathcal{J}(Q, z)\,|\phi_n(z)|^2\,z\,dz\,. \tag{23}$$

The hard-wall model calculation gives

$$\begin{aligned}&\langle \rho^+(p_2, \epsilon')|J^\mu_{\rm EM}(0)|\rho^+(p_1, \epsilon)\rangle \\ &= -\epsilon'_\beta \epsilon_\alpha \left[\eta_{\alpha\beta}(p_1 + p_2)_\mu + 2(\eta_{\alpha\mu} q_\beta - \eta_{\beta\mu} q_\alpha)\right] F_{nn}(Q^2)\,.\end{aligned} \tag{24}$$

But it is well known that vector mesons have three form factors:

$$\langle\rho^+(p_2,\epsilon')|J^\mu_{\rm EM}(0)|\rho^+(p_1,\epsilon)\rangle = -\epsilon'_\beta\epsilon_\alpha\big[\,\eta^{\alpha\beta}(p_1^\mu+p_2^\mu)\,G_1(Q^2) \tag{25}$$
$$+\,(\eta^{\mu\alpha}q^\beta-\eta^{\mu\beta}q^\alpha)(G_1(Q^2)+G_2(Q^2)) - \frac{1}{M^2}q^\alpha q^\beta(p_1^\mu+p_2^\mu)\,G_3(Q^2)\,\big]\,,$$

i.e., $G_1(Q^2)=G_2(Q^2)=F_{nn}(Q^2)$ and $G_3(Q^2)=0$ [20]. The form factor (23) is projected by taking the "+++" component of 3-point correlator,

$$\mathcal{F}(Q^2) = G_1(Q^2)+\frac{Q^2}{2M^2}\,G_2(Q^2)-\left(\frac{Q^2}{2M^2}\right)^2 G_3(Q^2)\,. \tag{26}$$

For ρ-meson, this combination coincides with the IMF "LL" transition having $\sim\alpha_s/Q^2$ behavior in pQCD.[21] Taking the hard-wall model prediction (23) and using that $z\sim 1/Q$ dominate in the large-Q limit because $\mathcal{J}(Q,z)\to zQK_1(Qz)\sim e^{-Qz}$, we may substitute $\phi(z)$ by $\phi(0)$. Thus,

$$\mathcal{F}(Q^2)\to\frac{|\phi(0)|^2}{Q^2}\int_0^\infty d\chi\,\chi^2\,K_1(\chi) = 2\,\frac{|\phi(0)|^2}{Q^2}\,, \tag{27}$$

and we get the same power of $1/Q^2$ as in pQCD, but without α_s/π factor.

Soft-wall model[17] corresponds to z^2 barrier, and bulk-to-boundary propagator $\mathcal{V}(p,z)$ can be written $(a=-p^2/4\kappa^2)$ as[14]

$$\mathcal{V}(p,z) = a\int_0^1 dx\,x^{a-1}\,\exp\left[-\frac{x}{1-x}\,\kappa^2z^2\right]\,. \tag{28}$$

The propagator poles are located at $p^2=4(n+1)\kappa^2\equiv M_n^2$ [17]:

$$\mathcal{V}(p,z) = \kappa^2z^2\sum_{n=0}^\infty\frac{L_n^1(\kappa^2z^2)}{a+n+1} = \sum_{n=0}^\infty\frac{g_5f_n}{M_n^2-p^2}\,\psi_n(z)\,. \tag{29}$$

Just like in the hard-wall case, we deal with ψ wave functions and coupling constants g_5f_n given by their derivatives at the origin

$$g_5f_n = \frac{1}{z}\,e^{-\kappa^2z^2}\,\partial_z\psi_n(z)\Big|_{z=\epsilon\to 0} = \sqrt{8(n+1)}\kappa^2\,. \tag{30}$$

Again, we can introduce the (Sturm-Liouville-) conjugate ϕ wave functions:

$$\phi_n(z) = \frac{1}{M_nz}e^{-\kappa^2z^2}\partial_z\psi_n(z) = \frac{2}{M_n}e^{-\kappa^2z^2}L_n^0(\kappa^2z^2)\,. \tag{31}$$

Taking the diagonal form factor for the lowest state

$$\mathcal{F}_{00}(Q^2) = 2\int_0^\infty e^{-\kappa^2z^2}\mathcal{J}(Q,z)\,z\,dz \tag{32}$$

and using representation (28) for $\mathcal{J}(Q,z)$ gives $\mathcal{F}_{00}(Q^2)=1/(1+Q^2/M_0^2)$, i.e., exact vector meson dominance. Large-Q^2 behavior of $\mathcal{F}$ form factor is given by the same expression (27) as in hard-wall model, the only difference being in the value of w.f. at the origin. As a result, we have

$$\mathcal{F}_\rho^{\rm H}(Q^2) \to 2.56\, m_\rho^2/Q^2 \quad , \quad \mathcal{F}_\rho^{\rm S}(Q^2) \to m_\rho^2/Q^2 \; . \tag{33}$$

3. Pion Form Factors in AdS/QCD

The full action of hard-wall model[16] is given by

$$S_{\rm AdS}^B = {\rm Tr} \int d^4x \int_0^{z_0} dz \left[\frac{1}{z^3} (D^M X)^\dagger (D_M X) + \frac{3}{z^5} X^\dagger X \right. \\ \left. - \frac{1}{8 g_5^2 z} (B_{(L)}^{MN} B_{(L)MN} + B_{(R)}^{MN} B_{(R)MN}) \right] , \tag{34}$$

where $DX = \partial X - iB_{(L)}X + iXB_{(R)}$, $B_{(L,R)} = V \pm A$ and $X(x,z) = v(z)U(x,z)/2$ involves the chiral field: $U(x,z) = \exp\left[i\sigma^a \pi^a(x,z)\right]$, with the pion field $\pi^a(x,z)$. The chiral symmetry is broken by the term $v(z) = (m_q z + \sigma z^3)$, with $m_q \sim$ quark mass and σ playing the role of quark condensate. The longitudinal component of the axial field $A^a_{\parallel M}(x,z) = \partial_M \psi^a(x,z)$ gives another pion field $\psi^a(x,z)$. The model satisfies Gell-Mann–Oakes–Renner relation $m_\pi^2 \sim m_q$. In the chiral limit $m_q = 0$, it is possible to get the analytic result[8,22] for $\Psi(z) \equiv \psi(z) - \pi(z)$

$$\Psi(z) = z\,\Gamma\left(2/3\right) \left(\frac{\alpha}{2}\right)^{1/3} \left[I_{-1/3}\left(\alpha z^3\right) - I_{1/3}\left(\alpha z^3\right) \frac{I_{2/3}\left(\alpha z_0^3\right)}{I_{-2/3}\left(\alpha z_0^3\right)} \right] , \tag{35}$$

where $\alpha = g_5 \sigma/3$. $\Psi(z)$ satisfies $\Psi(0) = 1$, Neumann b.c. $\Psi'(z_0) = 0$ and

$$f_\pi^2 = -\frac{1}{g_5^2} \left(\frac{1}{z} \partial_z \Psi(z) \right)_{z=\epsilon \to 0}$$

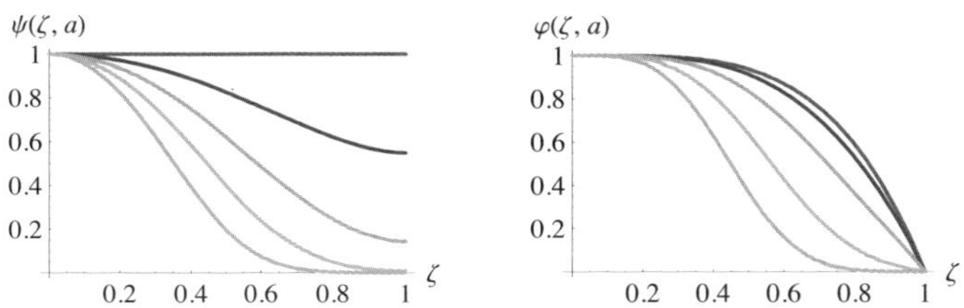

Fig. 3. Pion wave functions $\Psi(z) \to \psi(\zeta, a)$ and $\Phi(z) \to \phi(\zeta, a)$ as functions of $\zeta \equiv z/z_0$ and $a \equiv \alpha z_0^3$ for $a = 0$, $a = 1$, $a = 2.26$, $a = 5$ and $a = 10$.

The conjugate wave function is given by

$$\Phi(z) = -\frac{1}{g_5^2 f_\pi^2} \left(\frac{1}{z}\, \partial_z \Psi(z) \right) = -\frac{2}{s_0} \left(\frac{1}{z}\, \partial_z \Psi(z) \right) , \tag{36}$$

where $s_0 = 4\pi^2 f_\pi^2 \approx 0.67\,\mathrm{GeV}^2$ is the usual characteristic scale for pion. The function $\Phi(z)$ satisfies $\Phi(0) = 1$ and Dirichlet b.c. $\Phi(z_0) = 0$.

Pion EM form factor written in terms of $\Psi(z)$ looks like

$$F_\pi(Q^2) = \frac{1}{g_5^2 f_\pi^2} \int_0^{z_0} \mathcal{J}(Q,z) \left[\left(\frac{\partial_z \Psi}{z} \right)^2 + \frac{g_5^2 v^2}{z^4} \Psi^2(z) \right] z\, dz \; . \tag{37}$$

To analyze form factor at large Q^2, we write it in terms of $\Psi(z)$ and $\Phi(z)$:

$$F_\pi(Q^2) = \int_0^{z_0} \mathcal{J}(Q,z) \left[g_5^2 f_\pi^2 \Phi^2(z) + \frac{9\alpha^2}{g_5^2 f_\pi^2} z^2\, \Psi^2(z) \right] z\, dz \; . \tag{38}$$

For large Q, only $z \sim 1/Q$ part of $\Phi^2(z)$ term works, which gives

$$F_\pi(Q^2) \to \frac{2\, g_5^2 f_\pi^2 \Phi^2(0)}{Q^2} = \frac{4\pi^2 f_\pi^2}{Q^2} \equiv \frac{s_0}{Q^2} \; . \tag{39}$$

The curve we obtained from the AdS/QCD model (see Ref. [8]) goes above existing experimental data that give $Q^2 F_\pi(Q^2) \approx 0.4\mathrm{GeV}^2$, which means that the pion in this model is too small.

We remind that pQCD result[6,7] has $2\alpha_s/\pi$ factor

$$F_\pi^{\mathrm{pQCD}}(Q^2) \to \frac{2\alpha_s}{\pi} \cdot \frac{s_0}{Q^2} \sim 0.2\, F_\pi^{\mathrm{AdS/QCD}}(Q^2) \tag{40}$$

due to one-gluon exchange.

Anomalous amplitude of the $\gamma^*\gamma^*\pi^0$ transition is defined by

$$\int \langle \pi, p | T \left\{ J^\mu_{\mathrm{EM}}(x)\, J^\nu_{\mathrm{EM}}(0) \right\} |0\rangle e^{-iq_1 x} d^4x$$
$$= \epsilon^{\mu\nu\alpha\beta} q_{1\,\alpha} q_{2\,\beta}\, \frac{N_c}{12\pi^2 f_\pi}\, K_{\gamma^*\gamma^*\pi^0}\left(Q_1^2, Q_2^2\right) \; , \tag{41}$$

where $p = q_1 + q_2$ and $q_{1,2}^2 = -Q_{1,2}^2$. Its value for real photons is fixed in QCD by axial anomaly: $K_{\gamma^*\gamma^*\pi^0}(0,0) = 1$. To consider this form factor, the AdS/QCD model should be extended. We need isoscalar fields, which is achieved by gauging $U(2)_L \otimes U(2)_R$ and introducing the field $\mathcal{B}_\mu = t^a B^a_\mu + \mathbf{1}\, \frac{\hat{B}_\mu}{2}$, and we also need the Chern-Simons term

$$S^{(3)}_{\mathrm{CS}}[\mathcal{B}] = \frac{N_c}{24\pi^2} \epsilon^{\mu\nu\rho\sigma} \mathrm{Tr} \int d^4x\, dz\, (\partial_z \mathcal{B}_\mu) \left[\mathcal{F}_{\nu\rho} \mathcal{B}_\sigma + \mathcal{B}_\nu \mathcal{F}_{\rho\sigma} \right] . \tag{42}$$

The anomalous form factor conforming to QCD anomaly is given by

$$K(Q_1^2, Q_2^2) = \Psi(z_0) \mathcal{J}(Q_1, z_0) \mathcal{J}(Q_2, z_0)$$
$$- \int_0^{z_0} \mathcal{J}(Q_1, z) \mathcal{J}(Q_2, z)\, \partial_z \Psi(z)\, dz \; . \tag{43}$$

For large Q_1 and/or Q_2 we may write

$$K(Q_1^2, Q_2^2) \simeq \frac{s_0}{2} \int_0^{z_0} \mathcal{J}(Q_1, z)\mathcal{J}(Q_2, z)\, \Phi(z)\, z\, dz \, . \tag{44}$$

If one of the photons is real, we have

$$K(0, Q^2) \to \frac{\Phi(0) s_0}{2Q^2} \int_0^{\infty} d\chi\, \chi^2\, K_1(\chi) = \frac{s_0}{Q^2} \, . \tag{45}$$

For comparison, in pQCD $\gamma^* \gamma \pi^0$ form factor is given by

$$K^{\rm pQCD}(0, Q^2) = \frac{s_0}{3Q^2} \int_0^1 \frac{\varphi_\pi(x)}{x}\, dx \equiv \frac{s_0}{3Q^2}\, I^\varphi \, .$$

The pQCD result agrees with AdS/QCD model if $I^\varphi = 3$, e.g., for $\varphi_\pi(x) = 6x(1-x)$ (asymptotic DA). Our model[9] is very close to Brodsky-Lepage interpolation $K^{\rm BL}(0, Q^2) = 1/(1 + Q^2/s_0)$ which goes above CLEO data. However, next-to-leading pQCD correction is negative which allows to fit CLEO data if one takes distribution amplitudes with $I^\varphi \approx 3$.

In case of large and equal photon virtualities, the AdS/QCD result is

$$K(Q^2, Q^2) \to \frac{\Phi(0) s_0}{Q^2} \int_0^{\infty} d\chi\, \chi^3\, [K_1(\chi)]^2 = \frac{s_0}{3Q^2} \, . \tag{46}$$

Note that pQCD result in this kinematics does not depend on pion DA

$$K^{\rm pQCD}(Q^2, Q^2) = \frac{s_0}{3} \int_0^1 \frac{\varphi_\pi(x)\, dx}{xQ^2 + (1-x)Q^2} = \frac{s_0}{3Q^2} \tag{47}$$

and *coincides* with AdS/QCD model!

For non-equal large photon virtualities, we write $Q_1^2 = (1+\omega)Q^2$ and $Q_2^2 = (1-\omega)Q^2$. The leading-order pQCD then gives

$$K^{\rm pQCD}(Q_1^2, Q_2^2) = \frac{s_0}{3Q^2} \int_0^1 \frac{\varphi_\pi(x)\, dx}{1 + \omega(2x-1)} \equiv \frac{s_0}{3Q^2}\, I^\varphi(\omega) \, , \tag{48}$$

while the AdS/QCD model result reads

$$\begin{aligned} K(Q_1^2, Q_2^2) &\to \frac{\Phi(0) s_0}{2Q^2} \sqrt{1-\omega^2} \int_0^{\infty} d\chi\, \chi^3\, K_1(\chi\sqrt{1+\omega}) K_1(\chi\sqrt{1-\omega}) \\ &= \left(\frac{s_0}{3Q^2}\right) \left\{ \frac{3}{4\omega^3} \left[2\omega - (1-\omega^2) \ln\left(\frac{1-\omega}{1+\omega}\right) \right] \right\} . \end{aligned} \tag{49}$$

Note, that the term enclosed in curly brackets *coincides* with pQCD $I^\varphi(\omega)$ for $\varphi(x) = 6x(1-x)$. Indeed, using representation

$$\chi K_1(\chi) = \int_0^{\infty} e^{-\chi^2/4u - u}\, du \, , \tag{50}$$

and integrating over χ we get

$$K(Q_1^2, Q_2^2) \to \frac{s_0}{Q^2} \int_0^\infty \int_0^\infty \frac{u_1 u_2 \, e^{-u_1-u_2} du_1 du_2}{u_2(1+\omega) + u_1(1-\omega)} \, . \tag{51}$$

Changing $u_2 = x\lambda$, $u_1 = (1-x)\lambda$ and integrating over λ gives

$$K(Q_1^2, Q_2^2) \to \frac{s_0}{3Q^2} \int_0^1 \frac{6\,x(1-x)\,dx}{1+\omega(2x-1)} \, . \tag{52}$$

Comment on "Light-Front Holography". The AdS/CFT form factor expression (7) has structure similar to that of DY light-front formula (1), especially when the latter is written in terms of the impact parameter space w.f. $\widetilde{\Psi}(x, \mathbf{b}_\perp)$. Brodsky and de Teramond[12] noticed that, identifying z with $|\mathbf{b}_\perp|\sqrt{x(1-x)}$ and taking a special form of the light-front w.f.

$$\widetilde{\Psi}(x, \mathbf{b}_\perp) = \frac{1}{\sqrt{2\pi}} \frac{\Phi(|\mathbf{b}_\perp|\sqrt{x(1-x)})}{\mathbf{b}_\perp^2 \sqrt{x(1-x)}} \, , \tag{53}$$

one can convert the 3D DY formula (1) into the 1D AdS/CFT integral (7). This observation is the basis of the "Light-Front Holography" approach.[24] However, it is easy to check that if one would calculate the meson couplings f_n (14), (30) from the light-front w.f. fixed by this ansatz, the results would have an extra $\sqrt{6}\pi/8$ factor (see Eqs.(88),(89) of Ref. [23]) compared to exact AdS/QCD results (14), (30). Furthermore, this ansatz gives $8\sqrt{x(1-x)}/\pi$ for meson distribution amplitude, while we demonstrated above that AdS/QCD results for $\gamma^*\gamma^* \to \pi^0$ form factor correspond to asymptotic $6x(1-x)$ distribution amplitude. In general, the light-front holography ansatz[12] is not consistent with AdS/QCD for any observable that depends linearly on the w.f. (rather than bilinearly as in DY formula).

4. Summary

Summarizing, we established that meson form factors in AdS/QCD are given by formulas similar to those in quantum mechanics. For large Q, there is only one mechanism $z \sim 1/Q$. For vector mesons, the leading (LL) IMF form factor $\mathcal{F}(Q^2)$ indeed behaves like $1/Q^2$ for large Q^2. In soft-wall model, $\mathcal{F}(Q^2)$ demonstrates exact ρ-dominance. For pion, large-Q^2 asymptotics is s_0/Q^2 vs. pQCD result $(2\alpha_s/\pi)s_0/Q^2$. We included the anomalous amplitude into the AdS/QCD analysis, extending it to $U(2)_L \otimes U(2)_R$ and adding the Chern-Simons term. Fixing normalization by conforming to QCD anomaly, we observed that large-Q^2 behavior coincides then with pQCD calculations for asymptotic pion DA, the result contradicting the claim of "light-front holography" approach that meson distribution amplitude is given by $8\sqrt{x(1-x)}/\pi$. In conclusion, AdS/QCD is an instructive model for what may happen with form factors in real-world QCD.

Acknowledgments

I am very grateful to Organizers for invitation to Workshop honoring 60th anniversary of M. Shifman and their hospitality. Happy birthday, Misha!

I thank H.R. Grigoryan for collaboration on the studies of form factors in AdS/QCD. This paper is authored by Jefferson Science Associates, LLC under U.S. DOE Contract No. DE-AC05-06OR23177.

References

1. V. A. Matveev, R. M. Muradian and A. N. Tavkhelidze, Lett. Nuovo Cimento **7**, 719 (1973).
2. S. J. Brodsky and G. R. Farrar, Phys. Rev. Lett. **31**, 1153 (1973).
3. R. P. Feynman, "Photon-Hadron Interactions," Reading, 282p (1972).
4. S. D. Drell and T. M. Yan, Phys. Rev. Lett. **24**, 181 (1970).
5. G. B. West, Phys. Rev. Lett. **24**, 1206 (1970).
6. A. V. Radyushkin, JINR-P2-10717 (1977); English translation: arXiv:hep-ph/0410276.
7. G. P. Lepage and S. J. Brodsky, Phys. Lett. B **87**, 359 (1979).
8. H. R. Grigoryan and A. V. Radyushkin, Phys. Rev. D **76**, 115007 (2007).
9. H. R. Grigoryan and A. V. Radyushkin, Phys. Rev. D **78**, 115008 (2008).
10. J. Polchinski and M. J. Strassler, Phys. Rev. Lett. **88**, 031601 (2002).
11. J. Polchinski and M. J. Strassler, JHEP **0305**, 012 (2003).
12. S. J. Brodsky and G. F. de Teramond, Phys. Rev. Lett. **96**, 201601 (2006).
13. H. R. Grigoryan and A. V. Radyushkin, Phys. Lett. B **650**, 421 (2007).
14. H. R. Grigoryan and A. V. Radyushkin, Phys. Rev. D **76**, 095007 (2007).
15. H. R. Grigoryan and A. V. Radyushkin, Phys. Rev. D **77**, 115024 (2008).
16. J. Erlich, E. Katz, D. T. Son and M. A. Stephanov, Phys. Rev. Lett. **95**, 261602 (2005).
17. A. Karch, E. Katz, D. T. Son and M. A. Stephanov, Phys. Rev. D **74**, 015005 (2006).
18. J. M. Maldacena, Adv. Theor. Math. Phys. **2**, 231 (1998); Int. J. Theor. Phys. **38**, 1113 (1999).
19. J. C. C. A. Kneser, Math. Ann. **LXIII**, 447 (1907); A. J. W. Sommerfeld, Jahresber. Deutsch. Mat. Ver., **XXI**, 309 (1912).
20. D. T. Son and M. A. Stephanov, Phys. Rev. D **69**, 065020 (2004).
21. S. J. Brodsky and J. R. Hiller, Phys. Rev. D **46**, 2141 (1992).
22. L. Da Rold and A. Pomarol, Nucl. Phys. B **721**, 79 (2005); JHEP **0601**, 157 (2006).
23. S. J. Brodsky and G. F. de Teramond, Phys. Rev. D **77**, 056007 (2008).
24. G. F. de Teramond and S. J. Brodsky, Phys. Rev. Lett. **102**, 081601 (2009).

Hadronic Form Factors: Combining QCD Calculations with Analyticity

A. Khodjamirian*

Theoretische Physik 1, Fachbereich Physik, Universität Siegen, D-57068 Siegen, Germany
**E-mail: khodjam@hep.physik.uni-siegen.de*

I discuss recent applications of QCD light-cone sum rules to various form factors of pseudoscalar mesons. In this approach both soft and hard contributions to the form factors are taken into account. Combining QCD calculation with the analyticity of the form factors, one enlarges the region of accessible momentum transfers.

1. Introduction

At the beginning of this talk let me quote Misha Shifman:
"Unlike some models whose relation to Nature is still a big question mark, Quantum Chromodynamics will stay with us forever."[1]

In the strong coupling domain, the quark-gluon gauge dynamics of QCD manifests itself in a form of hadrons. A comprehensive analytic description of hadrons and their interactions in QCD remains an unsolved problem that may also stay with us forever. Today we have at our disposal only approximate methods and effective theories of hadrons, with an impressive progress achieved in the numerical simulation of QCD on the lattice.

An approximate analytical calculation of hadronic observables has been made possible with the advent of QCD sum rules.[2] Not only the original SVZ method is still being extensively used, but also its "offspring", the light-cone sum rules (LCSR).[3] In what follows, I overview applications of LCSR to hadronic form factors. I will also discuss a possibility to enlarge the region of accessible momentum transfers by employing the analyticity of the form factors.

2. Hadronic form factors

The simplest hadronic form factors parameterize electroweak transitions between two ground-state pseudoscalar mesons. A well-known example is the pion form factor generated by the quark electromagnetic current:

$$\langle \pi(p+q)|j_\mu^{em}|\pi(p)\rangle = (2p+q)_\mu F_\pi(Q^2)\,, \tag{1}$$

where $j_\mu^{em} = \frac{2}{3}\bar{u}(x)\gamma_\mu u(x) - \frac{1}{3}\bar{d}(x)\gamma_\mu d(x)$ and $Q^2 = -q^2$. This form factor remains a popular test ground of QCD models, and is being used nowadays to probe the AdS/QCD approach (see, e.g., [4,5]).

An important constraint on any model of the pion form factor is the $Q^2 \to \infty$ asymptotics [6,7] :

$$F_\pi(Q^2)|_{Q^2\to\infty} = \frac{8\pi\alpha_s f_\pi^2}{9Q^2}\left|\int_0^1 du \frac{\varphi_\pi(u,\mu)}{1-u}\right|^2, \tag{2}$$

determined by the perturbative gluon exchange between the quark and antiquark constituents of the pion. In this factorization formula, the long-distance dynamics below the separation scale $\mu \sim \sqrt{Q^2}$ is encoded in the the pion distribution amplitude (DA) $\varphi_\pi(u,\mu)$. The latter is defined as the twist-2 part of the quark-antiquark vacuum-pion matrix element expanded near the light-cone:

$$\langle \pi(p)|\bar{u}(x)[x,0]\gamma_\mu\gamma_5 d(0)|0\rangle_{x^2=0} = -ip_\mu f_\pi \int_0^1 du\, e^{iup\cdot x}\varphi_\pi(u,\mu)\,, \tag{3}$$

where $[x,0]$ is the gauge factor and $\mu \sim 1/\sqrt{|x^2|}$. The pion DA is a universal object, in a sense that it also enters factorization formulae for other pion form factors. The knowledge of $\varphi_\pi(u,\mu)$ is however not sufficient for a form factor calculation at finite Q^2. A major question remains to be addressed: how large are the soft contributions to the form factor (1) which start from $O(1/Q^4)$.

Switching from the electromagnetic to the weak flavour-changing current $\bar{u}\gamma_\mu b$, one gets the $B \to \pi$ transition form factor:

$$\langle \pi(p)|\bar{u}\gamma_\mu b|B(p+q)\rangle = (2p+q)_\mu f^+_{B\pi}(q^2) + \ldots\,, \tag{4}$$

where ellipses indicate the presence of the second form factor, due to non-conservation of the current. The $c \to d, s$ weak transitions generate similar $D \to \pi, K$ form factors. The role of large scale Q^2 is taken now by the heavy quark mass m_Q ($Q = b, c$). In these heavy-light form factors there is again an interplay of hard and soft quark-gluon interactions. One can write an asymptotic factorization formula, similar to (2), containing $\varphi_\pi(u,\mu)$ together with the heavy-meson DA. Again, such an ansatz is phenomenologically incomplete, because the soft contributions have to be included. They play an even more important role in heavy-to-light form factors, since they are not power suppressed (in $1/m_Q$) with respect to the hard factorizable part.

Heavy-light form factors are not only interesting hadronic objects. Their knowledge is of primary importance for flavour physics. For example the $B \to \pi$ form factor is needed to extract the quark mixing parameter $|V_{ub}|$ from data on semileptonic decay $B \to \pi l \nu_l$, whereas $D \to \pi$ and $D \to K$ form factors are used to determine $|V_{cd}|$ and, respectively, $|V_{cs}|$. The exploration of current and future experimental data on exclusive B and D decays demands accurate theoretical predictions for these form factors.

3. QCD light-cone sum rules

The method of LCSR was described in detail in many papers, (see, e.g. [8–12] and the review [13]), hence I only give a very short outline. The central object is the correlation function of two quark currents between the vacuum and on-shell pion state. To give a familiar example, the calculation of the $B \to \pi$ form factors starts from the following expression:

$$\int d^4x e^{iqx} \langle \pi(p)|T\{\overline{u}(x)\gamma_\mu b(x), \overline{b}(0) i m_b \gamma_5 d(0)\}|0\rangle = F(q^2,(p+q)^2) p_\mu + \dots , \quad (5)$$

where only the invariant amplitude multiplying p_μ is relevant for the form factor $f^+_{B\pi}(q^2)$. At $q^2, (p+q)^2 \ll m_b^2$, the b-quark propagating in the correlation function has a large virtuality and the product of b-quark fields is expanded near the light-cone $x^2 \sim 0$, leading to OPE for the correlation function, schematically:

$$F_{OPE}(q^2, (p+q)^2) = \sum_{\Gamma_a} \int d^4x e^{iqx} C_{(\Gamma_a)}(x^2, m_b) \langle \pi(p)|\bar{u}(x)\Gamma_a d(0)|0\rangle$$
$$+ \sum_{\Gamma_b} \int d^4x e^{iqx} \int_0^1 dv \, \tilde{C}^{\alpha\beta}_{(\Gamma_b)}(x^2, v, m_b) \langle \pi(p)|\bar{u}(x) G_{\alpha\beta}(vx) \Gamma_b d(0)|0\rangle , \quad (6)$$

where gauge factors are not shown, $\Gamma_{a,b}$ are various combinations of Dirac matrices. In the above, the short-distance coefficients $C_{(\Gamma_a)}$ and $\tilde{C}_{(\Gamma_b)}$ stemming from the b-quark propagator are calculated perturbatively, whereas the vacuum-pion matrix elements are expressed via pion light-cone DA's. The term with $\Gamma_a = \gamma_\mu\gamma_5$ yields the twist-2 DA defined in (3) and, in addition, the twist-4 DA's. The terms with $\Gamma_a = i\gamma_5, \sigma_{\mu\nu}\gamma_5$ yield twist-3 DA's. Furthermore, the soft gluon emitted from the b-quark propagator enters quark-antiquark-gluon operators, shown in the second line in (6). These operators are decomposed in pion three-particle DA's of twist-3 and 4. The dominant twist-2 and twist-3 contributions are calculated to $O(\alpha_s)$, obeying collinear factorization.[9–11] Subdominant twist-4 and three-particle contributions are power suppressed, justifying the truncated twist/Fock-state expansion. It is a future task to assess also the twist-5 and 6 terms of this expansion.

Matching OPE and hadronic dispersion relation for the correlation function at $|(p+q)^2| \sim m_b\chi$, where χ is an intermediate scale $\gg \Lambda_{QCD}$, one uses quark-hadron duality for the hadronic states above B meson, obtaining:

$$F_{OPE}(q^2,(p+q)^2) = \frac{2m_B^2 f_B f_{B\pi}^+(q^2)}{m_B^2 - (p+q)^2} + \int\limits_{s_0^B}^{\infty} ds \frac{[\mathrm{Im}F(s,q^2)]_{OPE}}{s-(p+q)^2}\,. \tag{7}$$

Substituting in (7) the amplitude F_{OPE} calculated from (6), one obtains an approximate analytical answer for the form factor in terms of pion DA's convoluted with calculable coefficients. Inputs in this relation include the b quark mass (in the $\overline{MS}$ scheme), α_s, and the set of low-twist universal pion DA's. The decay constant f_B is determined from two-point QCD (SVZ) sum rule. For other details, see, e.g. [11] .

Importantly, in the resulting LCSR for the form factor $f_{B\pi}^+$, the leading twist-2 and 3 terms start from α_s^0, with NLO corrections being $O(\alpha_s)$ suppressed. Thus, in LCSR both soft (nonfactorizable) and hard (factorizable) contributions are taken into account and the soft one dominates. Importantly, higher twist contributions are suppressed by (Λ_{QCD}/m_b) and/or (Λ_{QCD}/χ). Since the calculation is done at finite m_b, a transition from b to c quark in (5) is straightforward, yielding LCSR for $D \to \pi, K$ form factors.[8,12,14,15]

The universality of the method goes even further. Forming a vacuum→pion correlation function of the currents j_μ^{em} and $\bar{u}\gamma_\rho\gamma_5 d$, one obtains LCSR for the pion form factor $F_\pi(Q^2)$ [16–18], valid at $Q^2 \gg \Lambda_{QCD}^2$. This sum rule predicts a substantial soft contribution at intermediate Q^2 and, simultaneously, reproduces the QCD asymptotics (2). Furthermore, employing vacuum→ baryon correlation functions with a baryonic interpolating current and DA's, one is able to calculate various baryon form factors, e.g., the nucleon form factors.[19]

The method of LCSR has certain limitations. First of all, there is no "direct access" to the hadronic form factor. The analytical expression in a form of factorized OPE is obtained for the correlation function, whereas the form factor enters the pole term of the dispersion relation. One isolates it from the contributions of higher states to the dispersion relation, estimated using quark-hadron duality. This approximation introduces a sort of "systematic" uncertainty of the method. To keep it under control, one fits the threshold parameter s_0^B in (7) by calculating the B-meson mass from the same LCSR. A typical accuracy of the form factors, calculated by this method is estimated at the level of $\sim \pm 15\%$, by varying all input parameters and scales within their adopted interval. One of the main uncertainties is the shape of the twist-2 DA $\varphi_\pi(u)$ expressed via Gegenbauer moments.

The region of momentum transfer q^2 accessible for LCSR is restricted: $f^+_{B\pi}(q^2)$ is calculated at $q^2 \ll (m_B - m_\pi)^2$, practically at $q^2 < 12 - 14$ GeV^2, $f^+_{D\pi}(q^2)$ at $q^2 \simeq 0$, and $F_\pi(Q^2)$ at $Q^2 \geq 1$ GeV^2. It is therefore an important task to access other regions of q^2 where more data are available. This is possible, provided one uses the analyticity of the form factors.

4. Employing the analyticity

Hadronic form factors are analytic functions of the momentum-transfer variable q^2. A typical dispersion relation which follows from the analyticity of the $B \to \pi$ form factor:

$$f^+_{B\pi}(q^2) = \frac{m_{B^*} f_{B^*} g_{B^*B\pi}}{2(m^2_{B^*} - q^2)} + \frac{1}{\pi} \int\limits_{(m_B+m_\pi)^2}^{\infty} ds \frac{\mathrm{Im} f^+_{B\pi}(s)}{s - q^2}, \tag{8}$$

takes into account the singularities located on the positive real q^2 axis, the lowest one being the ground-state B^* pole (with the $B^*B\pi$ coupling defined as in [8]). Starting from the threshold at $q^2 = (m_B + m_\pi)^2$, there are branch points and poles, generated by hadronic continuum states and excited resonances with B^* quantum numbers. Note that, due to the QCD asymptotics $f^+_{B\pi}(q^2 \to \infty) \sim 1/q^2$ (similar to (2)), there are no subtractions in (8). Importantly, this dispersion relation is valid at any q^2. Hence, a practical way to enlarge the accessible q^2-region is to match the LCSR result to (8) at $q^2 \ll m_B^2$ and analytically continue the dispersion relation. This however can only be done if a model/ansatz is introduced for the integral over the spectral density of higher states in (8), e.g., an effective pole.[20]

A less model-dependent approach employs conformal mapping (for earlier uses see [21]). One maps the complex q^2-plane where (8) is valid, onto the unit circle $|z| < 1$ in the plane of the new variable: $z(q^2, t_0) = \frac{\sqrt{t_+ - q^2} - \sqrt{t_+ - t_0}}{\sqrt{t_+ - q^2} + \sqrt{t_+ - t_0}}$, where $t_+ = (m_B + m_\pi)^2$, and $t_0 < t_+$ is a parameter. There are many applications[22] of this approach to $B \to \pi$ and other form factors, combined with perturbative QCD bounds obtained from the unitarity for the 2-point correlation function. In fact, the bounds are usually not restrictive, and, since the maximal values of $|z|$ corresponding to kinematical boundaries of $B \to \pi$ or $D \to \pi, K$ transitions are rather small, a simple Taylor series near $z = 0$ suffices to parameterize the form factor. The last version of this series parameterization [23] advocates a truncated power expansion:

$$f^+_{B\pi}(q^2) = \frac{1}{1 - q^2/m^2_{B^*}} \sum_{k=0}^{k_{max}} a_k \Big(z(q^2, t_0) \Big)^k , \quad (9)$$

(with certain constraints on a_k), where the B^* -pole near the threshold is isolated. Three or four parameters are sufficient for a reasonably accurate parameterization. Employing (9), it is possible to go beyond the region where a LCSR calculation is valid. For that one has to fit the coefficients a_k to the LCSR result for the form factor in the "trusted "region of q^2. After that, transforming $q^2 \to z$, one continues (9) over z beyond the initial region and finally switch the variable back to q^2.

5. Recent results for heavy-light form factors

5.1. $B \to \pi$ form factor and $|V_{ub}|$

Let me first quote our update of the $B \to \pi$ form factor calculated from LCSR [11] with the result: $f^+_{B\pi}(0) = 0.26^{+0.04}_{-0.03}$, where a recent very accurate determination of the $\overline{MS}$ b-quark mass [24] was used. In this calculation no attempt yet was done to use the analytical continuation. On the contrary, in order to diminish the theoretical error for the form factor at $q^2 = 0$, the calculated form factor shape at $0 < q^2 < 12$ GeV^2 (the estimated region of validity of LCSR) was fitted to the q^2 -distribution in $B \to \pi l \nu_l$, measured by BABAR collaboration.[25] This fit allowed to tighten the constraints on the Gegenbauer moments of the twist-2 pion DA, yielding :

$$\varphi_\pi(u, \mu) = 6u(1-u)\Big(1 + a_2(\mu) C_2^{3/2}(2u-1) + a_4(\mu) C_4^{3/2}(2u-1)\Big), \quad (10)$$

with $a_2(1\text{GeV}) = 0.16 \pm 0.01$, $a_4(1\text{GeV}) = 0.04 \pm 0.01$ (neglecting $a_{6,...}$). Note that these intervals are quite narrow and are within broader "world averages" (see e.g., [26]). Deviation of the DA from its asymptotic form is thus rather mild. Finally, in [DKMMO] the form factor at $q^2 = 0$ was used to extract $|V_{ub}|$ from the data on $B \to \pi l \bar{\nu}$. This and other recent $|V_{ub}|$ determinations from $B \to \pi l \nu_l$ are summarized in the table1.

Importantly, both LCSR and lattice QCD determinations are in a good agreement with $|V_{ub}| = (3.5^{+0.15}_{-0.14}) \times 10^{-3}$ inferred from the CKM unitarity triangle fits [30]. Let me also mention a LCSR calculation [31] of $B \to K$ and $B_s \to K$ form factors (needed e.g., for models of rare exclusive B decays).

Table 1.

[Ref.]	$f^+_{B\pi}(q^2)$ calculation	$f^+_{B\pi}(q^2)$ input	$\|V_{ub}\| \times 10^3$
[27]	lattice	-	3.78±0.25±0.52
[28]	lattice	-	3.55±0.25±0.50
[10]	LCSR	-	$3.5 \pm 0.4 \pm 0.1$
[29]	-	lattice $\oplus$ LCSR	$3.47 \pm 0.29 \pm 0.03$
[11]	LCSR	-	$3.5 \pm 0.4 \pm 0.2 \pm 0.1$
[23]	-	lattice$\oplus$ LCSR	3.54 ± 0.24

5.2. $D \to \pi, K$ form factors and $|V_{cd}|$, $|V_{cs}|$

Following the same calculational scheme as for $B \to \pi$ [12] and $B \to K$ [31] transitions, using the same set of pion and kaon DA's and $\overline{MS}$ c-quark mass from [24], we calculated [12] the $D \to \pi$ and $D \to K$ form factors. In the latter, the $SU(3)_{fl}$ violation is taken into account in $O(m_s)$. Predictions for these form factors from lattice QCD and LCSR are presented in this table:

Method	[Ref.]	$f^+_{D\pi}(0)$	$f^+_{DK}(0)$
Lattice QCD	[33]	$0.57 \pm 0.06 \pm 0.02$	$0.66 \pm 0.04 \pm 0.01$
	[34]	$0.64 \pm 0.03 \pm 0.06$	$0.73 \pm 0.03 \pm 0.07$
	[35]	$0.74 \pm 0.06 \pm 0.04$	$0.78 \pm 0.05 \pm 0.04$
LCSR	[14]	0.65 ± 0.11	$0.78^{+0.2}_{-0.15}$
	[15]	0.63 ± 0.11	0.75 ± 0.12
	[12]	$0.67^{+0.10}_{-0.07}$	$0.75^{+0.11}_{-0.08}$

Our calculation was then used to determine $|V_{cd}|$ and $|V_{cs}|$ with an improved accuracy from the recent CLEO data [32] on semileptonic D decays. (see [12] for more details and numerical results).

In addition, the series parameterization similar to (9) was used to enlarge the accessible q^2 region. The form factors were calculated at $q^2 < 0$ (still within the region of validity of LCSR), and fitted to the z-parameterization. The result was then analytically continued, to cover the whole kinematical region $0 \leq q^2 \leq (m_D - m_{\pi(K)})^2$ of $D \to \pi(K) l\nu_l$ decay. The predicted shape of $f^+_{D\pi}(q^2)$ normalized at $q^2 = 0$ is compared with the one measured by CLEO [32] in Fig. 1.

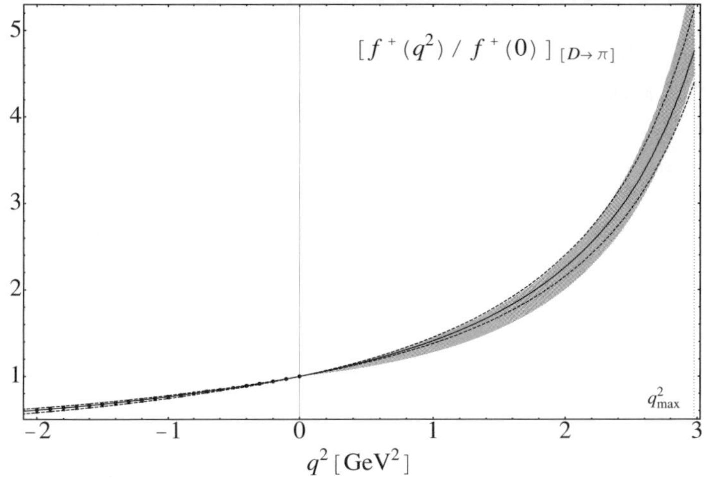

Fig. 1. The shape of the $D \to \pi$ form factor obtained from LCSR [12] at $q^2 \leq 0$, fitted to series parameterization (solid line and dashed lines indicating uncertainties), and compared with the shape measured by CLEO [32] at $0 \leq q^2 \leq (m_D - m_\pi)^2$ (shaded region) .

6. Pion form factor

As already mentioned, the LCSR approach allows one to calculate any hadronic form factor, provided the momentum transfer is sufficiently below the hadronic threshold in the channel of the transition current. It is therefore interesting to return to the pion e.m. form factor $F_\pi(Q^2)$ which is quite sensitive to the pion twist-2 DA and recalculate it from LCSR.[17,18] In this sum rule the twist-2 term with $O(\alpha_s)$ corrections and twist-4,6 terms are taken into account. With the pion DA given by (10) and taking the remaining input from [18], I recalculated the pion e.m. form factor at $Q^2 = 1.0-5.0$ GeV^2. The result is presented in Fig. 2, for the central values of the input. Conservatively, a $\pm 15\%$ uncertainty still has to be added. The agreement with the recent accurate data at $Q^2 \leq 2.45$ GeV^2, obtained at Jefferson Lab [36] is encouraging.

Furthermore, it is easy to write down an analog of series parameterization (9) for $F_\pi(Q^2)$, which is simply a polynomial in z in this case (the lowest ρ-meson pole is far above the threshold) and continue the form factor to small Q^2. The result shown in Fig. 2 is consistent with the direct measurement of the pion form factor in this region [37] and correctly reproduces $lim_{Q^2\to 0}F_\pi(Q^2) = 1$ (within the accuracy of the sum rule).

I conclude that the pion DA (10) used in calculating the heavy-light form factors successfully passes the pion form factor test. A more detailed

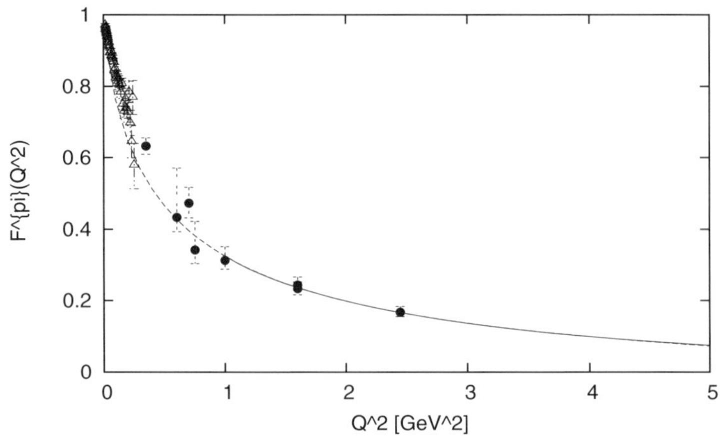

Fig. 2. The pion e.m. form factor calculated from LCSR [17,18] at $Q^2 = 1 - 5$ GeV^2 (solid), fitted to series parameterization and analytically continued to smaller Q^2 (dashed), compared with JLAB data [36] (filled points) and NA7 data [37] (open triangles).

study will be published elsewhere.

7. Photon-pion transition form factor
(a comment added after the talk)

The pion twist-2 DA also determines the QCD asymptotics [7,38] of the $\gamma^*\gamma^* \to \pi^0$ amplitude:

$$F^{\gamma^*\pi}(Q^2,k^2) = \frac{\sqrt{2}f_\pi}{3}\int\limits_0^1 du \frac{\varphi_\pi(u,\mu)}{Q^2(1-u)+|k^2|u} \oplus O(\alpha_s)\,, \tag{11}$$

where both photon virtualities, $Q^2 = -q^2$ and $|k^2| \neq Q^2$, are sufficiently large. The $\gamma^*\gamma^* \to \pi^0$ amplitude is however only measured [44–46] when one of the photons is almost real ($k^2 \simeq 0$), in which case it reduces to the photon-pion transition form factor $F^{\gamma\pi}(Q^2) = F^{\gamma^*\pi}(Q^2,0)$. Note that at $k^2 = 0$ the factorization formula (11) is incomplete, because the real photon has a long-distance, hadronlike component. Hence, substituting a certain model of $\varphi_\pi(u,\mu)$ in (11) and comparing the result at $k^2 = 0$ with experimental data on $F^{\gamma\pi}(Q^2)$, is not yet a conclusive test of QCD. Calculations which take into account long-distance component of the photon can be found in the literature, (see, e.g. [39]).

A method to calculate $F^{\gamma\pi}(Q^2)$, combining LCSR with the analytical continuation was suggested in [40]. The $\gamma^*\gamma^* \to \pi^0$ amplitude was calculated using light-cone OPE, with both $Q^2 = -q^2$ and $|k^2|$ large, adding also

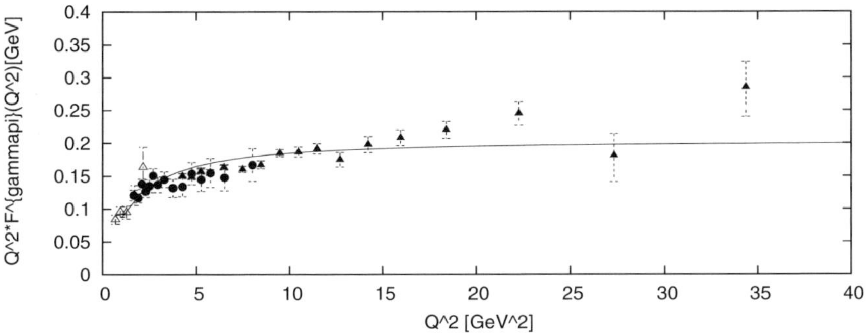

Fig. 3. Photon-pion form factor $Q^2F^{\gamma\pi}(Q^2)$ calculated from a combination of LCSR and dispersion relation [40], including $O(\alpha_s)$ corrections [41], compared with the CELLO [44] (open triangles), CLEO [45] (full points) and BABAR [46] data (full triangles).

the twist-4 term to (11). (A calculation of the next, twist-6 term remains a future task.) Equating this amplitude to the dispersion relation in k^2:

$$F_{OPE}^{\gamma^*\pi}(Q^2,k^2) = \frac{\sqrt{2}f_\rho F_{\rho\pi}(Q^2)}{m_\rho^2-k^2} + \frac{1}{\pi}\int_{s_0}^{\infty}\frac{\mathrm{Im}F^{\gamma^*\pi}(Q^2,s)}{s-k^2}, \qquad (12)$$

where ρ and ω contributions are combined in one resonance term with $m_\rho \simeq m_\omega$, one calculates the $\rho \to \pi$ form factor using the technique of LCSR and employs quark-hadron duality for the integral above the threshold s_0. After that a smooth transition to $k^2 \to 0$ in the dispersion relation is possible, yielding the form factor $F^{\gamma\pi}(Q^2)$. The $O(\alpha_s)$ corrections in this approach were calculated in [41], and some recent updates can be found in [42,43].

Intrigued by the new BABAR data,[46] I returned to the calculation of $F^{\gamma\pi}(Q^2)$ from (12), using the twist-2 pion DA (10), the remaining input from [40], and adding the $O(\alpha_s)$ corrections from [41]. The result is shown in Fig. 3 (for the central values of the input). At large Q^2, as expected, $Q^2F^{\gamma\pi}(Q^2)$ tends to a constant.[a] In spite of a reasonable agreement with the old CELLO and CLEO data at low Q^2 and with the new BABAR data up to $q^2 \leq 15$ GeV^2, this calculation (as well as the one in [43]) does not reproduce the tendency of the increasing $Q^2F^{\gamma\pi}(Q^2)$ at larger Q^2 visible in the data, although the experimental errors in this region are rather large. An additional measurement, e.g., by Belle collaboration is important to confirm these very interesting data.

[a] Note that the $O(\alpha_s \ln^2 Q^2)$ terms in the NLO part of the form factor [41] and the effect of their resummation still have to be assessed.

Acknowledgments

I am very grateful to FTPI for inviting me to take part in the celebration of Misha Shifman's extraordinary achievements in particle theory. Due to his important and enlightening contributions, deep connections between QCD and hadron phenomenology were found and are being extensively used.
I acknowledge a useful discussion with Vladimir Braun. This work is supported by the Deutsche Forschungsgemeinschaft under the contract No. KH205/1-2.

References

1. M. Shifman, *Persistent challenges of quantum chromodynamics*, Lilienfeld Prize Lecture, Int. J. Mod. Phys. A **21**, 5695 (2006) [arXiv:hep-th/0606015].
2. M. A. Shifman, A. I. Vainshtein and V. I. Zakharov, Nucl. Phys. B **147**, 385, 448 (1979).
3. I. I. Balitsky, V. M. Braun and A.V. Kolesnichenko, Nucl. Phys. B **312**, 509 (1989); V. M. Braun and I. E. Filyanov, Z. Phys. C **44**, 157 (1989); V. L. Chernyak and I. R. Zhitnitsky, Nucl. Phys. B **345**, 137 (1990).
4. A. Radyushkin, "Quark Counting Rules: Old and New Approaches," arXiv:0907.4585 [hep-ph], talk at this workshop.
5. J. Erlich, "How Well Does AdS/QCD Describe QCD?," arXiv:0908.0312 [hep-ph], talk at this workshop.
6. V. L. Chernyak and A. R. Zhitnitsky, JETP Lett. **25**, 510 (1977); A. V. Efremov and A. V. Radyushkin, Phys. Lett. B **94**, 245 (1980).
7. G. P. Lepage and S. J. Brodsky, Phys. Lett. B **87**, 359 (1979); Phys. Rev. D **22**, 2157 (1980).
8. V. M. Belyaev, V. M. Braun, A. Khodjamirian and R. Rückl, Phys. Rev. D **51**, 6177 (1995).
9. A. Khodjamirian, R. Rückl, S. Weinzierl and O. I. Yakovlev, Phys. Lett. B **410**, 275 (1997); E. Bagan, P. Ball and V. M. Braun, Phys. Lett. B **417**, 154 (1998).
10. P. Ball and R. Zwicky, Phys. Rev. D **71**, 014015 (2005).
11. G. Duplancic, A. Khodjamirian, T. Mannel, B. Melic and N. Offen, JHEP **0804**, 014 (2008).
12. A. Khodjamirian, C. Klein, T. Mannel, N. Offen, "Semileptonic charm decays $D \to \pi l \nu_l$ and $D \to K l \nu_l$ from QCD Light-Cone Sum Rules," arXiv:0907.2842 [hep-ph].
13. P. Colangelo and A. Khodjamirian, In Shifman, M. (ed.): At the frontier of particle physics/Handbook of QCD, vol. 3, 1495-1576 [arXiv:hep-ph/0010175].
14. A. Khodjamirian, R. Ruckl, S. Weinzierl, C. W. Winhart and O. I. Yakovlev, Phys. Rev. D **62**, 114002 (2000).
15. P. Ball, Phys. Lett. B **641**, 50 (2006).
16. V. M. Braun and I. E. Halperin, Phys. Lett. B **328**, 457 (1994).
17. V. M. Braun, A. Khodjamirian and M. Maul, Phys. Rev. D **61**, 073004 (2000).

18. J. Bijnens and A. Khodjamirian, Eur. Phys. J. C **26**, 67 (2002).
19. V. M. Braun, "Nucleons on the light-cone: Theory and phenomenology of baryon distribution amplitudes," arXiv:hep-ph/0608231, in Minneapolis 2006, Continuous advances in QCD*, pp 42-57;
V. M. Braun, A. Lenz and M. Wittmann, Phys. Rev. D **73**, 094019 (2006).
20. D. Becirevic and A. B. Kaidalov, Phys. Lett. B **478**, 417 (2000).
21. N. N. Meiman, Sov. Phys. JETP **17**, 830 (1963); B. V. Geshkenbein and B. L. Ioffe, Sov. Phys. JETP **17**, 820 (1963); S. Okubo, Phys. Rev. D **4**, 725 (1971); C. Bourrely, B. Machet and E. de Rafael, Nucl. Phys. B **189**, 157 (1981).
22. C. G. Boyd, B. Grinstein and R. F. Lebed, Phys. Rev. Lett. **74**, 4603 (1995); L. Lellouch, Nucl. Phys. B **479**, 353 (1996); M. C. Arnesen, B. Grinstein, I. Z. Rothstein and I. W. Stewart, Phys. Rev. Lett. **95** (2005) 071802; T. Becher and R. J. Hill, Phys. Lett. B **633**, 61 (2006).
23. C. Bourrely, I. Caprini and L. Lellouch, Phys. Rev. D **79**, 013008 (2009).
24. J. H. Kuhn, M. Steinhauser and C. Sturm, Nucl. Phys. B **778**, 192 (2007); R. Boughezal, M. Czakon and T. Schutzmeier, Phys. Rev. D **74**, 074006 (2006).
25. B. Aubert *et al.* [BABAR Collaboration], Phys. Rev. Lett. **98**, 091801 (2007).
26. P. Ball, V. M. Braun and A. Lenz, JHEP **0708**, 090 (2007).
27. M. Okamoto, PoS **LAT2005**, 013 (2006) [arXiv:hep-lat/0510113].
28. E. Dalgic *et al.* , Phys. Rev. D **73**, 074502 (2006) [Erratum-ibid. D **75**, 119906 (2007)].
29. J. M. Flynn and J. Nieves, Phys. Rev. D **76**, 031302 (2007).
30. J. Charles *et al.* [CKMfitter Group], Eur. Phys. J. C **41**, 1 (2005), Moriond 2009 update at http://ckmfitter.in2p3.fr .
31. G. Duplancic and B. Melic, Phys. Rev. D **78**, 054015 (2008).
32. D. Besson *et al.* [CLEO Collaboration], Phys. Rev. D **80**, 032005 (2009).
33. A. Abada *et al.* Nucl. Phys. B **619**, 565 (2001).
34. C. Aubin *et al.* [Fermilab Lattice Collaboration, MILC Collaboration and HPQCD Collaboration], Phys. Rev. Lett. **94**, 011601 (2005).
35. A. Al-Haydari *et al.* [QCDSF Collaboration], "Semileptonic form factors $D \to \pi, K$ and $B \to \pi, K$ from a fine lattice," arXiv:0903.1664 [hep-lat].
36. G. M. Huber *et al.* [Jefferson Lab Collaboration], Phys. Rev. C **78**, 045203 (2008).
37. S. R. Amendolia *et al.* [NA7 Collaboration], Nucl. Phys. B **277**, 168 (1986).
38. S. J. Brodsky and G. P. Lepage, Phys. Rev. D **24**, 1808 (1981).
39. A. V. Radyushkin and R. T. Ruskov, Nucl. Phys. B **481**, 625 (1996).
40. A. Khodjamirian, Eur. Phys. J. C **6**, 477 (1999).
41. A. Schmedding and O. I. Yakovlev, Phys. Rev. D **62**, 116002 (2000).
42. S. S. Agaev, Phys. Rev. D **72**, 114010 (2005) [E. ibid. **73**, 059902 (2006)].
43. S. V. Mikhailov and N. G. Stefanis, Nucl. Phys. B **821**, 291 (2009).
44. H. J. Behrend *et al.* [CELLO Collaboration], Z. Phys. C **49**, 401 (1991).
45. J. Gronberg *et al.* [CLEO Collaboration], Phys. Rev. D **57** (1998) 33.
46. B. Aubert *et al.* [The BABAR Collaboration], Phys. Rev. D **80**, 052002 (2009).

High Energy Hadronic Scattering – the Quantum Evolution

Alex Kovner*

Physics Department, University of Connectictut,
Storrs, CT 06269
**E-mail: kovner@phys.uconn.edu*

I review the modern approach to quantum evolution of hadronic observables with energy. Recent developments relating the (eikonal) evolution equation to the QCD Reggeon field theory and the approach to including the Pomeron loops is discussed.

Keywords: High energy scattering, Quantum evolution.

1. The eikonal scattering

The aim of this talk is to describe how to derive the evolution equations governing the evolution of hadronic observables, such as forward S-matrix element, with energy.

Consider an hadronic projectile with the wave function $|P\rangle$ impinging on a target with the wave function $|T\rangle$. We are working in the projectile light cone gauge. In this gauge (and frame where the projectile moves with close to velocity of light) the wave function of the projectile is simplest thought of as being built of gluons. At very high energy the projectile is Lorentz contracted to an infinitely thin disc, and thus the only important degree of freedom is the distribution of charge density $j^a(X)$ in the transverse plain. The target on the other hand is conveniently described as some distribution of chromoelectric fields which from the point of view of the fast moving projectile gluons, are simply external fields $\alpha^a(x^-, X)$ in which the gluons scatter.

The scattering of a projectile gluon in such external field is given by the eikonal formula

$$|X, a\rangle \to S^{ab}(X)|X, b\rangle \tag{1}$$

with

$$S^{ab}(X) = \mathcal{P} \exp\left\{ig \int_{x^-} T^a \alpha^a(x^-, X)\right\} \tag{2}$$

where X is the coordinate of the projectile gluon in the transverse (impact parameter) plane. The x^- dependence of the field α is not important and I will drop it in the following.

This means that the second quantized S-matrix operator in the eikonal approximation can be written in terms of the color charge density operator as

$$\hat{S} = \exp\left\{i \int d^2X \hat{j}^a(X) \hat{\alpha}^a(X)\right\} . \tag{3}$$

Remembering that the fields α are distributed with some probability density $W[\alpha]$, while the color charge have their own probability density function $W[j]$ we can write for the forward scattering S-matrix element in the form

$$\mathcal{S} = \langle T|\langle P|\hat{S}|P\rangle|T\rangle = \int D\alpha^a \; W^T[\alpha(X)] \; \Sigma^{PP}[\alpha(X)] \tag{4}$$

with

$$\Sigma^{PP}[\alpha] \equiv \langle P|\, \hat{S}\, |P\rangle = \int d\rho \; W^P[j] \;\; \exp\left\{i \int d^2X \, j^a(X) \; \alpha^a(X)\right\} . \tag{5}$$

2. The evolution

Now consider boosting the projectile wave function while keeping the target unchanged. This physically corresponds to increasing the energy of the collision. When the projectile wave function is boosted, the longitudinal momenta of all the gluons are scaled by the boost factor. Additionally new gluons with small longitudinal momenta are produced from the longitudinal fields which are associated with the color charges of the "valence" gluons. They are "pulled out" from the "bottomless pit" of the light cone zero mode. Denoting the separation line between the valence and soft gluons by Λ, pictorially under boost the projectile wave function evolves as in Fig. 1.

This physical mechanism of the evolution can be translated in perturbation theory into the evolution equation for the color charge density distribution $W[j]$. Such an evolution equation was derived some time ago [1] and is given below. One has to keep in mind however that the derivation of [1] assumes not only that the coupling constant is small, but also that the target on which the projectile is scattering is a perturbative, small object - its color fields are weak. Alternatively one can consider the projectile as

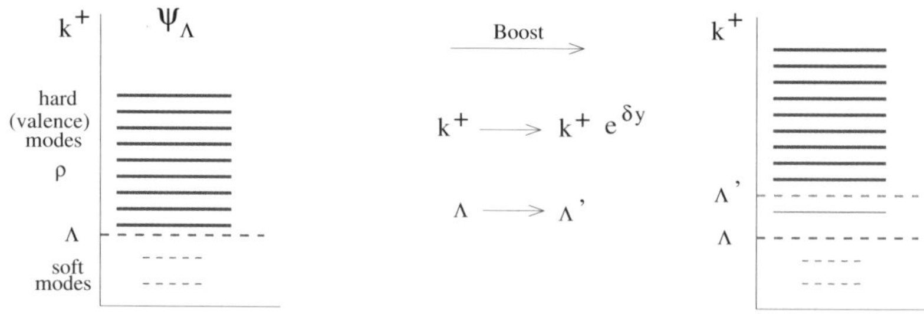

Fig. 1. Evolution of the hadronic wave function under boost.

a perturbative object, while allow the target fields to be arbitrarily large (but still not parametrically larger than $1/g$). Then the JIMWLK evolution equation when written for the target probability distribution takes the Schrödinger-like form

$$\frac{\partial}{\partial\eta} W[\alpha], = -\, H^{JIMWLK}\left[\frac{\delta}{\delta\alpha}, \alpha\right] W[\alpha]\,. \tag{6}$$

Here η is the rapidity (which is the natural parameter for the evolution) and the Hamiltonian is

$$H^{JIMWLK} = \frac{\alpha_s}{2\pi^2}\int d^2Z\, Q_i^a(Z)\, Q_i^a(Z) \tag{7}$$

with the hermitian amplitudes

$$Q_i^a(Z) = \int d^2X \frac{(X-Z)_i}{(X-Z)^2}[S^{ab}(Z) - S^{ab}(X)]J_R^b(X) \tag{8}$$

and the generators of the right $SU(N)$ rotations

$$J_R^a(X) = -\mathrm{tr}\left\{S(X)T^a \frac{\delta}{\delta S^\dagger(X)}\right\}\,. \tag{9}$$

3. The QCD Reggeon Field Theory

Formulated in this way, the problem of the QCD high energy evolution looks like a Quantum Field theory problem, where rapidity plays the role of "time" - or evolution parameter, while the space is two dimensional. Although this problem has not been approached in this fashion so far, one can certainly entertain the idea of the Hamiltonian Schrödinger approach. The road to solution would then consist of several steps.

A. Find the eigenfunctions of H^{JIMWLK}, $|\Psi^l\rangle$ coresponding to eigenvalues ω_l

$$H^{JIMWLK}\,|\Psi^l\rangle = \omega_l\,|\Psi^l\rangle\,. \tag{10}$$

The individual eigenfunctions evolve according to

$$|\Psi^l\rangle_\eta = e^{-\omega_l\,\eta}\,|\Psi^l\rangle\,. \tag{11}$$

B. Expand Σ^P and W^T in the complete basis

$$\Sigma^P = \sum_l \gamma_l^P\,|\Psi^l\rangle\,, \qquad W^T = \sum_l \gamma_l^{T\,*}\,\langle\Psi^l|\,. \tag{12}$$

C. Finally, calculate the scattering amplitude

$$\mathcal{S}(\eta) = \sum_l \gamma_l^P\,\gamma_l^{T\,*}\,e^{-\omega_l\,\eta}\,. \tag{13}$$

Unsurprisingly, but rather importantly we see that this procedure preserves unitarity explicitly. The JIMWLK Hamiltonian is clearly nonnegative definite, and thus has only nonnegative eigenvalues. Thus quite generally we see that at large energy the S matrix vanishes, which corresponds to unit scattering probability. The target therefore becomes black, unless the spectrum of H^{JIMWLK} contains a "grey" eigenstate with eigenvalue $\omega_l = 0$. It was argued in [2] that such a grey state does not exist in the spectrum.

The setup of a quantum field theory with the rapidity as the evolution parameter is of course an old one. It goes back to ideas of Gribov [3] and is sometimes called the Reggeon Field Theory. The JIMWLK Hamiltonian therefore should be considered as defining the QCD Reggeon Field Theory. One should be aware of the physics differences between Gribov's ideas and the theory defined by H^{JIMWLK}. The degrees of freedom in the former are scattering amplitudes of color invariant physical objects, and it thus is meant to describe directly the evolution of physical amplitudes. On the other hand when dealing with QCD one unavoidably deals with gluons as the building blocks of physical states. Thus the JIMWLK Hamiltonian defines a field theory whose degree of freedom is the color nonsinglet unitary matrix S which has the meaning of the eikonal amplitude of a (color nonsinglet) gluon parton. The scattering amplitude of a physical state is then calculated by choosing the color invariant initial probability density distribution $W[\alpha]$. Thus the RFT we are talking about here is the theory actively sought, for example, in the works of Bartels and collaborators [4] rather than the original Gribov's framework.

To convince ourselves that H^{JIMWLK} indeed defines the RFT, the easiest is to examine its small color charge limit. In this limit

$$Q \to Q_i^{\mathrm{BFKL}\,a}(Z) = \int d^2X\,\frac{(X-Z)_i}{(X-Z)^2}\,f^{abc}\left[\frac{\delta}{\delta j^b(X)} - \frac{\delta}{\delta j^b(Z)}\right] j^c(X) \tag{14}$$

and

$$H_{JIMWLK} \to H_{BFKL} = \frac{\alpha_s}{2\pi^2} \int d^2Z\, Q_i^{\mathrm{BFKL}\, a}(Z)\, Q_i^{\mathrm{BFKL}\, a}(Z)\,. \tag{15}$$

This now defines a (in principle) solvable theory. Since the Hamiltonian is homogeneous in j and $\delta/\delta j$ it preserves the "particle number" and thus can be solved by solving the problem separately in the sector with fixed number of gluons. This reduces to solution of the famous BKP hierarchy.[5] The eigenstates of H^{JIMWLK} have the following form

$$\begin{aligned}
|\Psi_1\rangle &= f_1^a(X) j^a(X)\,,\\
|\Psi_2\rangle &= f_2^{ab}(X,Y) j^a(X) j^b(Y)\,,\\
|\Psi_n\rangle &= f_n^{ab...c}(X,Y...Z) j^a(X) j^b(Y)...j^c(Z)\,,
\end{aligned} \tag{16}$$

where the wavefunctions f_n are solutions of the n - gluon BKP equation. The one gluon state is the Reggeized gluon, the two gluon state is the BFKL Pomeron and so on. Thus at leading order at small j the spectrum of the JIMWLK Hamiltonian contains all the BKP states - the QCD "Reggeons" with the BKP eigenvalues. Including higher order terms in j leads to appearance of the $2 \to n$ interaction vertices and the eigenstates accordingly will mix the states with different number of Reggeized gluons. For example the would be Pomeron will be changed to

$$|\Psi_2\rangle = |\text{BFKL POMERON}\rangle + |\text{4 REGGEIZED GLUON BKP}\rangle + \ldots\,. \tag{17}$$

One should note however that the utility of such perturbation theory is very questionable, as we discuss below. The relation between the JIMWLK Hamiltonian and RFT was discussed in detail in [7].

4. JIMWLK vs BFKL

The previous discussion raises a question. The weak charge limit of JIMWLK is the BFKL Hamiltonian. However there is a great difference between the two. As discussed above, the JIMWLK Hamiltonian has a nonnegative spectrum, and therefore the scattering probability is unitary and saturates. On the other hand BFKL is known to have negative eigenvalues, which leads to the violation of unitarity.[6] How does this happen? The answer is that the eigenfunctions of the BFKL Hamiltonian, being monomials in j are not normalizable, and thus the positivity of spectrum fails.

The situation is somewhat similar to the double well anharmonic potential

$$H = \frac{1}{2}\,p^2 - \frac{1}{2}\,x^2 + \frac{\lambda}{4!}\,x^4\,. \tag{18}$$

Suppose we want to follow the evolution of the wavepacket which initially is peaked around $x = 0$. We can expand the time dependent wave function in the basis of exact eigenfunctions of the double well Hamiltonian Ψ_n

$$\Psi(t) = \sum_n \alpha_n e^{-\epsilon_n t}\Psi_n\,. \tag{19}$$

In this expansion all the eigenvalues ϵ_n are positive. On the other hand, since for small times the wave packet stays in the vicinity of the unstable equilibrium point $x = 0$, it makes sense to expand the potential itself around $x = 0$ and then to expand the wave function in terms of the eigenfunctions Φ_n of the upside down harmonic oscillator $H^{BFKL} = \frac{1}{2}\,p^2 - \frac{1}{2}\,x^2$

$$\Psi(t) = \sum_n \beta_n e^{-\bar{\epsilon}_n t}\Phi_n\,. \tag{20}$$

In this expansion many $\bar{\epsilon}_n$ are negative and the eigenfunctions Φ_n are nonnormalizable. The eigenfunctions are in fact not decaying Gaussians, but rather oscillating Gaussians with imaginary frequency multiplied by polynomials. The higher the polynomial, the more negative is the corresponding eigenvalue and the spectrum is not bounded from below. One can ignore the wave functions with very large negative eigenvalues initially, since they have tiny overlaps with the initial wave packet, β_n. However if taken literally, the contributions of these states take over very quickly and violate unitarity. This is analogous to the contributions of high BKP states in the BFKL approach, since they come with arbitrarily large negative eigenvalues.[8]

This is very analogous to the BFKL approximation. Thus unitarizing BFKL is like looking for how does the potential turn over, and how does the evolution of the wave function change once it spreads to the minimum of the potential. It also suggests that the nonnormalizable zeroth order BFKL eigenfunctions have very little to do with the normalizable eigenfunctions of H^{JIMWLK}, and thus trying to perturbatively improve them is not a useful approach.

5. The Issue of Pomeron Loops

For all its good features, the JIMWLK equation is not complete. As mentioned earlier, it is derived under the assumption that the projectile has large color charge density while the target is a small perturbative object. In terms of the Reggeon terminology, it contains only the Pomeron merging diagrams, but no Pomeron splittings (Fig. 2 is the artists impression of the Pomeron trees).

Fig. 2. Red Pomeron Trees, by Natalija Krisciuniene

In other words, the JIMWLK evolution includes the effects of the non-linearities in the projectile wave function, but does not fully account for multiple scattering corrections. Interestingly these two types of corrections are dual to each other. In precise terms,[9] the following transformation transforms the situation with large projectile and small target into the situation of small projectile and large target:

$$S^{ab}(X) \to R^{ab}(X) \equiv [\exp\{T^c \frac{\delta}{\delta j^c(X)}\}]^{ab}. \tag{21}$$

We call this Dense-Dilute Duality (or DDD) transformation. Thus the evolution in the case of the low density projectile that scatters on the dense target is given by

$$H^{KLWMIJ} = \frac{\alpha_s}{2\pi^2} \int d^2 Z\, \bar{Q}_i^a(Z)\, \bar{Q}_i^a(Z) \tag{22}$$

with the Hermitian amplitudes

$$\bar{Q}_i^a(Z) = \int d^2 X \frac{(X-Z)_i}{(X-Z)^2} \left[R^{ab}(Z) - R^{ab}(X)\right] j^b(X). \tag{23}$$

Clearly, the RFT Hamiltonian should contain both the effects of the nonlinearity and of multiple scattering, and should thus be invariant under the DDD transformation. Recently we have derived the Hamiltonian which does indeed contain both effects.[10] For completeness I give below the expression for H^{RFT}

$$H_{RFT} = \frac{1}{8\pi^3}\int_{x,y,z,\bar{z}} \left[b^b_{Ri}(x)R^{\dagger ba}(x) - b^a_{Li}(x)\right]\left[\frac{\delta_{ij}}{(x-z)^2} - 2\,\frac{(x-z)_i\,(x-z)_j}{(x-z)^4}\right]$$
$$\times\left[\delta^{ac} + [S^\dagger_L(x)\,S_L(z)]^{ac}\right]\;K^{cd}_{\perp jk}(z,\bar{z})\left[\delta^{de} + [S^\dagger_R(\bar{z})\,S_R(y)]^{de}\right]$$
$$\times\left[\frac{\delta_{kl}}{(y-\bar{z})^2} - 2\,\frac{(y-\bar{z})_k\,(y-\bar{z})_l}{(y-\bar{z})^4}\right]\left[b^e_{Rl}(y) - R^{\dagger ef}(y)\,b^f_{Ll}(y)\right] \quad (24)$$

with

$$K^{ab}_{\perp ij}(x,y) = \frac{1}{2\,\pi^2}\int_z\left[\frac{\delta_{ik}}{(x-z)^2} - 2\,\frac{(x-z)_i\,(x-z)_k}{(x-z)^4}\right]\left[\frac{\delta_{kj}}{(z-y)^2}\right.$$
$$\left. -2\,\frac{(z-y)_k\,(z-y)_j}{(z-y)^4}\right]\left\{R^{ab}(z) + \left[S^\dagger_L(x)\,S_L(z)\,R(z)\,S^\dagger_R(z)\,S_R(y)\right]^{ab}\right\}, \quad (25)$$

and the classical Weizsacker-Williams fields "before" and "after" the scattering are given by

$$b^a_{L(R)i} = -\frac{1}{g}\,f^{abc}\,S^{\dagger bd}_{L(R)}\,\partial_i\,S^{dc}_{L(R)}\,. \quad (26)$$

This is a very recent result and it is not understood well yet. We know that it reproduces the known expressions in the appropriate limits (JIMWLK and KLWMIJ) and goes beyond both limits in that it includes the multiple scattering effects as well as the nonlinear effects in the wave function. Thus it includes the effects of Pomeron loops.

Fig. 3. Pomeron Loops by Natalija Krisciuniene.

It is now merely a question of solving it. Anybody? At any rate, once it is solved, the Pomeron loops will be summed (Fig. 3). And the life (just like the picture) will be so much brighter.

References

1. I. Balitsky, Nucl. Phys. B **463** 99 (1996), Phys. Rev. Lett. **81** 2024 (1998), Phys. Rev. D **60** 014020 (1999); Y. V. Kovchegov, Phys. Rev. D **61**, 074018 (2000) [arXiv:hep-ph/9905214]; J. Jalilian Marian, A. Kovner, A. Leonidov and H. Weigert, Nucl. Phys. B **504**, 415, 1997 [arXive:hep-ph/9701284], Phys. Rev. D **59**, 014014, 1999 [arXiv:hep-ph/9706377]; J. Jalilian Marian, A. Kovner and H. Weigert, Phys. Rev. D **59**, 014015, 1999 [arXiv: hep-ph/9709432]; A. Kovner and J. G. Milhano, Phys. Rev. D **61**, 014012, 2000 [arXiv: hep-ph/9904420]. A. Kovner, J. G. Milhano and H. Weigert, Phys. Rev. D **62**, 114005 (2000) [arXiv:hep-ph/0004014]; H. Weigert, Nucl. Phys. A **703**, 823 (2002) [arXiv:hep-ph/0004044]; E. Iancu, A. Leonidov and L. McLerran, Nucl. Phys. A **692**, 583 (2001), Phys. Lett. B **510**, 133 (2001); E. Ferreiro, E. Iancu, A. Leonidov and L. McLerran, Nucl. Phys. A **703**, 489 (2002).
2. A. Kovner and M. Lublinsky, Nucl. Phys. A **779**, 220 (2006).
3. V. N. Gribov, Sov. Phys. JETP **26**, 414 (1968) [Zh. Eksp. Teor. Fiz. **53**, 654 (1967)]; D. Amati, M. Le Bellac, G. Marchesini and M. Ciafaloni, Nucl. Phys. B **112**, 107 (1976); V. Alessandrini, D. Amati and M. Ciafaloni, Nucl. Phys. B **130**, 429 (1977); D. Amati, G. Marchesini, M. Ciafaloni and G. Parisi,

Nucl. Phys. B **114**, 483 (1976); M. Ciafaloni and G. Marchesini, Nucl. Phys. B **109**, 261 (1976); M. Baker and K. A. Ter-Martirosian, Phys. Rept. **28**, 1 (1976).
4. J. Bartels, Nucl. Phys. B **151**, 293 (1979); J. Bartels, Z. Phys. C **60**, 471 (1993); J. Bartels and M. Wusthoff, Z. Phys. C **66**, 157 (1995); J. Bartels and C. Ewerz, JHEP **9909**, 026 (1999) [arXiv:hep-ph/9908454].
5. J. Bartels, Nucl. Phys. B **175**, 365 (1980); J. Kwiecinski and M. Praszalowicz, Phys. Lett. B **94**, 413 (1980).
6. E. A. Kuraev, L. N. Lipatov, and F. S. Fadin, Sov. Phys. JETP **45**, 199 (1977); Ya. Ya. Balitsky and L. N. Lipatov, Sov. J. Nucl. Phys. **28**, 22 (1978).
7. A. Kovner and M. Lublinsky, JHEP **0702**, 058 (2007) [arXiv:hep-ph/0512316].
8. I thank Genya Levin for stressing this point to me.
9. A. Kovner and M. Lublinsky, Phys. Rev. D **71**, 085004 (2005), Phys. Rev. Lett. **94**, 181603 (2005).
10. T. Altinoluk, A. Kovner, M. Lublinsky and J. Peressutti, JHEP **0903**, 109 (2009).

Deconfinement, Monopoles and New Phenomena in Heavy Ion Collisions

Edward Shuryak

Department of Physics and Astronomy, University at Stony Brook, Stony Brook NY 11794 USA
**E-mail: shuryak@tonic.physics.sunysb.edu*

We discuss various manifestations of the "magnetic scenario" for the quark-gluon plasma viewed as a mixture of two plasmas, of electrically (quark and gluons) as well as magnetically charged quasiparticles. Near the deconfinement phase transition, $T \approx T_c$ very small density of free quarks should lead to negligible screening of electric field while magnetic screening remains strong. The consequence of this should be existence of a "corona" of the QGP, in a way similar to that of the Sun, in which electric fields influence propagation of perturbations and even form metastable flux tubes. The natural tool for its description is (dual) magnetohydrodynamics: among observable consequences is splitting of sound into two modes, with larger and smaller velocity. The latter can be zero, hinting for formation of pressure-stabilized flux tubes. Remarkably, recent experimental discoveries at RHIC show effects similar to expected for "corona structures". In dihadron correlation function with large-p_t trigger there are a "cone" and a "hard ridge", while the so called "soft ridge" is a similar structure seen without hard trigger. They seem to be remnants of flux tubes, which – contrary to naive expectations – seem to break less often in near-T_c matter than do confining strings in vacuum.

Keywords: confinement, monopoles, quark-gluon plasma, flux tubes

1. Introduction

Let me start my written version of the talk in the same way as the oral one: with comments on my interaction with Misha Shifman. It is more than 30 years since he start visiting Novosibirsk, working with Arkady on QCD sum rules, while I occasionally managed to visit ITEP. 1970's were formative years for all of us, and I learned a lot from those discussions. While we shared many interests over these years – to QCD correlators, heavy quark hadrons, instantons, now monopoles and confinement – we have so different styles that we never actually worked on exactly the same thing.

That produced some distance, a different perspective and often mutual criticism. And yet, it only strengthened our friendship: we are in agreement on so many other things in life. So, it is a delight to see Misha at another conference, of his beloved "Continuous Advances" series in particular, full of life and as usual having a joke or a couple of good (if skeptical) questions to any talk.

This talk fits into this long-time pattern perfectly. While in the years 2000-2005 I was preoccupied by RHIC data and then various strongly coupled plasmas, AdS/CFT and other issues not close to Misha, my talk at Continuous Advances 06 included a part about "magnetic scenario" for temperatures near T_c, on a plasma side. Presently both of us study different aspects of the "magnetic side" of QCD, trying to understand confinement mechanism: while Misha looks at models with enough supersymmetry to have rigid theoretical control, I looks at RHIC and lattice data plus some generic theoretical arguments to get the picture. Recent results and experimental discoveries look quite exciting: I hope that makes a good addition to this proceeding, a kind of our collective birthday present.

2. Dual electric/magnetic plasmas and transport

Let me thus start by reminding what I was speaking about at CA06. Among very few things we can agree on with Misha is a general "dual superconductor" view of confinement by 't Hooft and Mandelstam. If so, the confinement transition must be Bose-Einstein condensation (BEC) of certain magnetically charged objects. Furthermore, right above T_c there should be sufficient density of them in a "normal" (uncondensed) state.

Dirac's quantization condition demands the product of electric and magnetic couplings to be an integer: thus those two couplings *must run in the opposite directions* (as it happens e.g. in Seiberg-Witten $\mathcal{N}$=2 theory). So Liao and myself [2] proposed a view of CD plasma as of dual mixture, with two Coulomb-interacting subsystems, both at strong coupling whose magnitude is running with T in the *opposite* directions. (Chernodub and Zakharov [6] have independently proposed a similar picture, with large role of monopoles in the near-T_c region.) But how to see if this is true?

On the lattice one can identify monopoles by "the end of the Dirac string" in a certain gauge[a]: so one can look at monopole-antimonopole correlations in space. An example [9] for two temperatures is shown in

[a]Let me note that thus what is called a monopole actually include dyons, and that small size of the core and coarse lattices does not allowed for detail knowledge of its structure.

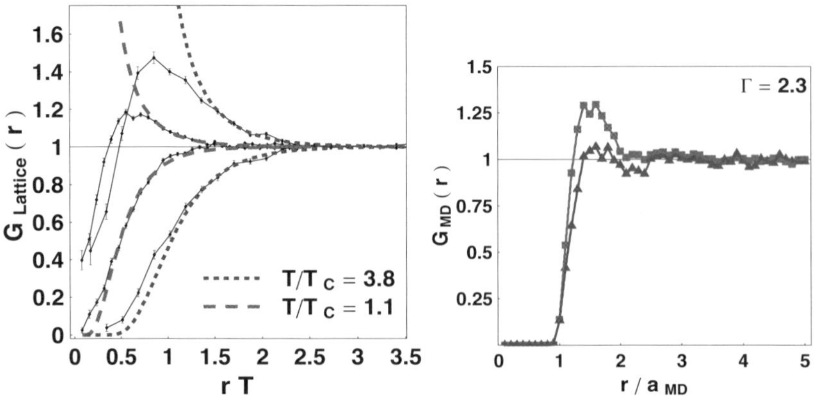

Fig. 1. (left) Monopole-antimonopole (the upper two curves) and monopole-monopole (the lower two curves) correlators at $1.1T_c$(red long dashed) and $3.8T_c$(blue dot dashed): points with error bars are lattice data [9], the dashed lines are our fits. (right) Monopole-antimonopole (boxes with red curve) and monopole-monopole (triangles with blue curve) correlators from MD simulations at $\Gamma = 2.3$.

Fig. 1(left). Note characteristic peaks which are found in strongly correlated liquids. Note also that the higher T has stronger correlation: this implies that magnetic Coulomb coupling *does* grow with T, contrary to asymptotic freedom for electric coupling. The right figure is our molecular dynamics (MD) simulations for plasmas containing monopoles: note that the red one (monopole-antimonopole) has the same shape and magnitude as the lattice data on the left. So, a picture two dual plasmas, electric and magnetic ones, does work.

What is the contribution of the monopoles to transport? Before we used classical molecular dynamics [2] while this time I report results by Ratti and myself.[3] We solved a beautiful quantum-mechanical problem of gluon-monopole scattering. Asymptotically, at $T \to \infty$, this process is subleading by one power of $log(T)$. Yet at T few times T_c, when the monopole density is still much smaller than that of quarks and gluons, we found that gluon-monopole scattering actually dominates the transport. This happens because of characteristic large backward scattering, absent in charge-charge case. Scattering rate and (η/s) are shown in Fig. 2(left), together with the ones obtained for the usual perturbative gluon-gluon scattering. There is a qualitative agreement between these results and the experimental value for η/s observed at RHIC (the box on the left), as well as with the famous AdS/CFT result $\eta/s = 1/4\pi$. LHC will create (in PbPb collisions) QGP

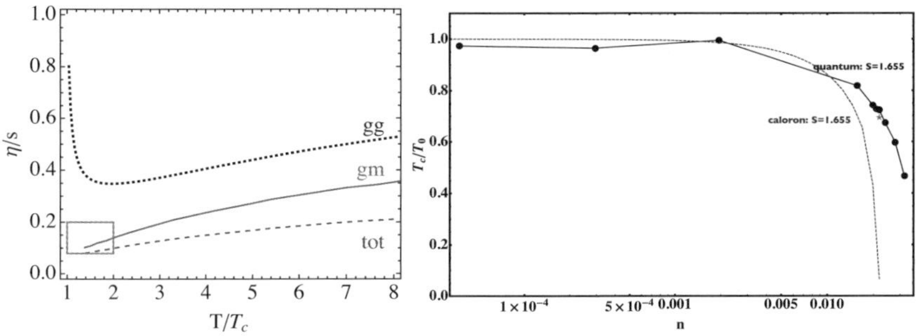

Fig. 2. (left) The contributions of the gluon-monopole and gluon-gluon scatterings into viscosity-to-entropy ratio. (right) Density dependence of the critical temperature for liquid ^{4}He: black points are our calculation while the red star is the experimental zero-T critical point.

with $T \sim 4T_c$: we thus have our prediction of how good a liquid QGP will be there.

3. Universal Feynman's criterion for Bose-Einstein condensation and monopoles at deconfinement

If one wanders whether confinement is indeed a BEC of monopoles, looking at their behavior on the lattice close to T_c is a good idea. But before doing it, one should know what to look at. In a paper with Marco Cristoforetti [4] we had revived an old idea by Feynman,[5] in which he introduced the notion of the *universal* critical value for the per-particle action in large supercurrent clusters. Note that in Matsubara formalism bosons should make periodic paths: but it may be not for each particle separately but for say k particles, called k-cluster. Feynman argued that the sum over cluster size k should diverge at the critical point, when action suppressing "jumps" from one particle location to the next is balanced by multiplicity of clusters. Feynman included kinetic energy only and thus approximated it by the straight line, so his expression for the amplitude is

$$y_F = \exp\left[-\frac{m^* T d^2}{2\hbar^2}\right] \tag{1}$$

Marco and I decided to work it out more, and include also potential energy. We had built the "moving string" model for the supercurrent, a line of particles in a crystal (a reasonably good model for liquid He^4) and calculated the action of all the string moving as a whole using realistic atomic potentials. Semiclassical calculation (another use of periodic instantons or calorons) or numerical path integration have given close actions. In

Fig. 2(right) we plot the ratio of the calculated critical temperature T_c/T_0, where T_0 is Einstein's critical temperature for free Bose gas. We found that for strongly interacting liquid He the critical action is $S_c = 1.655$, which is *exactly the same* value as one would get for noninteracting gas. So, Feynman was right about universality of his criterion: he just have not done a complete calculation to verify it for liquid He!

Now, with confidence about universal BEC criterion, we applied it for QCD monopoles. Since we know approximately their density, we can set upper limits on the monopole mass and magnetic Coulomb coupling at T_c. It turns out they must be as light as about 200 MeV or so.

Does it all agree with the lattice behavior of monopoles? As we were told by D'Elia, clusters with k=1,2,3 indeed have equal probability at T_c. The monopole mass is not yet accurately measured, but it indeed strongly decreases toward T_c about as low as predicted. In summary: monopoles do behave similarly to He atoms at their respective BEC transition temperatures.

4. QGP corona: theory

The word "corona" used here comes from the physics of the Sun. Let me briefly remind that it was started by Galileo Galilei, who in 1612 spent some time observing the motion of the black spots on the Sun and correctly concluded from motion of the spots that they must resign on a surface of a rotating sphere: he thus argued the spots were not shadows of some planets passing in front, as it was thought of before. In due time relation between the spots and solar magnetism was understood: modern telescopes allows one to see the fine structure of solar spots, resolving individual magnetic flux tubes. Better understanding of solar magnetism came with the advance of plasma physics in 1940's and development of MHD, which explained both the influence of diffuse magnetic field on plasma and formation and mechanical stability of the flux tubes.

Early stages of heavy ion collisions are believed to be described by the so called "glasma", a set of random color fields created by color charges of partons of the two colliding nuclei at the moment of the collision. However as two discs with charges move away from each other, those classical field are getting smaller and (in a still poorly understood process) rather quickly create the quark-gluon plasma, in which the occupation numbers are thermal, $O(1)$, with the classical fields disappearing.

Transition from field to plasma has been subject of multiple works, too many to be mentioned. One notable idea is existence of instabilities [11] such

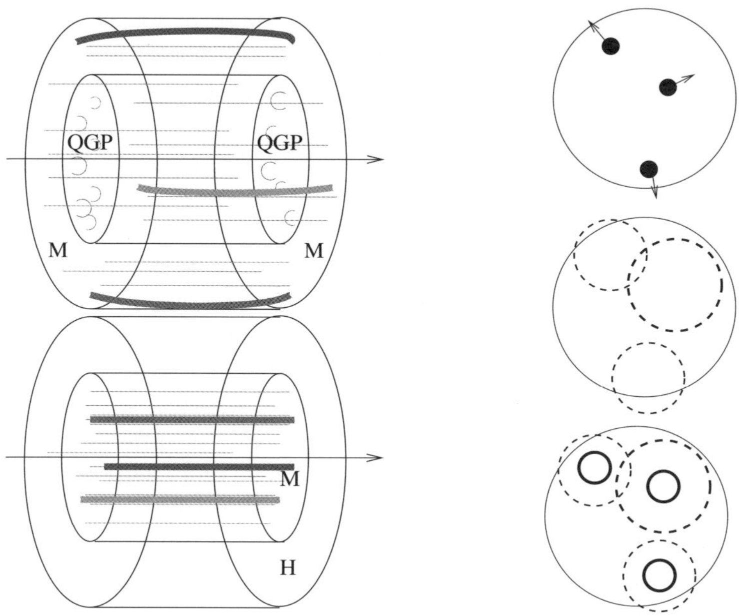

Fig. 3. (left) A snapshot of unscreened electric (dual-magnetic) field in the M (near-T_c) region of the fireball. Upper and lower ones correspond to full RHIC energy and the reduced energy (analogous to SPS). (right) A sketch of the transverse plane of the colliding system: the "spots" of extra density (a) are shown as black disks, to be moved by collective radial flow (arrows). Naive sound expansion (b) would produce large-size and small amplitude wave: yet the correct solution includes also brighter *secondary* wave (c) of smaller radius.

as the so called Weibel instability, producing filamentation of the longitudinal flow and transverse magnetic field $B_\perp$. Asakawa et al[12] argued that such chaotic magnetic field would remain in plasma, leading to "abnormal viscosity". Unfortunately those ideas, derived perturbatively and basically taken from electromagnetic plasma contexts, cannot possibly be valid for QGP at RHIC.

The crucial difference between the QED and QCD plasmas lies in the existence of magnetically charged quasiparticles – monopoles and dyons – leading to nonzero magnetic screening mass $M_M \sim g^2T$ first suggested by Polyakov long ago[13] and by now well confirmed by subsequent lattice studies. Thus hot QGP, unlike the electromagnetic plasmas, screen *both* the electric and the magnetic fields at some microscopic scales, although at a bit different ones. As at RHIC the central part of the produced fireball reaches relatively high temperature $T \sim 2T_c$, we expect both E, B fields to be effectively screened there, see the central cylindrical part of Fig. 3 (upper left) marked QGP. But in the near-T_c region there is magnetic plasma in the outer cylindrical part of Fig. 3 marked M (mixed or magnetic).It is very important to emphasize that although this region on the phase diagram is represented by a very narrow strip $|T - T_c| \ll T_c$, it corresponds to more than order of variation of the energy or entropy density, and the corresponding space-time volume in the expansion of the fireball is by no means small. Another snapshot of the geometry of the M region is shown in Fig. 3(lower left): it corresponds to lower collision energy for which the M-phase resign inside the fireball: obviously this case is very different, it is planned to be investigated in a specialized RHIC run in 2011.

Multiple lattice studies, e.g. ref. [8], have shown that at $T < 1.4T_c$ the relation between electric and magnetic screening masses get inverted, namely $M_M > M_E$. As $T \to T_c$ the electric screening mass strongly decreases, partly because of heavy quark and gluon quasiparticles and partly because of their suppression by small Polyakov loop expectation value $< L >$. So, near T_c $M_M(T \to T_c) \approx 3T_c, M_E(T \to T_c) \approx 0$. The most important consequence is that in the M-phase of the collision there is dense magnetic plasma, in which magnetic field is well screened while *electric* one remains *unscreened* for a significant time. This leads to the central idea: that QGP should have a "dual corona" in which *electric* (rather than magnetic) fields coexist with the plasma, affecting both the overall expansion and the propagation of perturbations. We will suggest to use "dual magnetohydrodynamics" (DMHD) for the description of diffuse electric fields in the M-phase, in particular study their effect on the velocity of propagation of small perturbations. We will further argue below that like solar corona, that of QGP should have metastable flux tubes, although microscopically thin ones which cannot be directly described by DMHD approximation.

Magnetohydrodynamics is a well known part of plasma physics, developed by Alfven, Fermi, Chandrasekhar and many others. It is an approximation which keeps only one (magnetic) field nonzero and half of Maxwell equations, while the other (electric) field is assumed to be totally screened. Ideal MHD approximation is the limit of *infinite* conductivity of plasma $\sigma \to \infty$, similar to *zero* viscosity approximation for ideal hydrodynamics. In MHD the coupling between the field and and matter is obtained by inclusion of the (magnetic) field contribution into the stress tensor of the medium.

Due to space limitations I will not put a complete set of resulting equations here, jumping directly to the main consequences. In a "(dual)magnetized" plasma small amplitude perturbations are split into two modes, known as Alfven waves, propagating with two different speeds

$$u_{\pm}^2 = \frac{1}{2}\Big(\frac{\vec{E}^2}{\epsilon} + c_s^2\Big) \pm \frac{1}{2}\left[\Big(\frac{\vec{E}^2}{\epsilon} + c_s^2\Big)^2 - \frac{4\vec{E}^2 cos^2\theta c_s^2}{\epsilon}\right]^{1/2} \tag{2}$$

where E, c_s, ϵ are the electric field, the speed of sound in "unmagnetized" medium without field and plasma energy density, θ is the angle between the field strength and the the direction of the wave propagation. Note a case in which the wave goes transverse to the field ($cos\theta = 0$) in which the lower mode has *zero* speed. In the case of jet quenching – when the jet direction is more or less up to experimentalist to pick – general shape of these waves can be complicated. However when the jet (or original charge fluctuation) propagates longitudinally, in the same direction as the field, the problem is axially symmetric and results in general in two cones. The angles of their propagation can be obtained from the previous expression, in which the l.h.s. is substituted by Mach relation $u \to cos(\theta)v$, v is the velocity of the jet, and solve it for the $cos\theta$ (see the paper for more details on solutions).

If instead the electric field to be constant in space there is a a spot, localized in transverse plane, inevitable there is nonzero $\partial\vec{E}_z/\partial r$, which is a part of $curl(\vec{E})$ and by the (dual) Maxwell equation it is proportional to (dual, or magnetic) current $\tilde{j}_\phi$. This tells us that a flux tube solution must have a "coil" with a current running around and cancelling the field outside the spot. Ideal DMHD has simple axially symmetric solution – the macroscopic flux tube, with the field pressure balancing that of the plasma. We do not discuss them in detail here because we think the flux tubes appeared in QGP corona are actually *microscopically thin*, and so should be considered in a way worked out by Liao and myself previously.[7,10] Let me emphasize it again: we don't speak hear about well known flux tubes at zero or low T but about flux tubes which exist at $T > T_c$, in the deconfined phase with magnetic constituents. As the current is not a supercurrent in this case, they are not stable forever but are metastable. We worked out quantum mechanics of the monopole-flux tube scattering and conditions for its mechanical stability. Perhaps surprising to many, those flux tubes are *tighter* (have larger tension) than the vacuum ones: this comes because uncondensed monopoles have thermal momenta and their scattering creates larger pressure on the tube. As we will discuss at the end of the talk, those tubes in the M-phase seem to have smaller breaking rate as well: so they are tighter in any sense.

The original hints for their existence came from lattice data on finite-T potentials between static quark and antiquark, studied on the lattice. In brief, the central observation is large difference found between the *free* energy $F(T,r)$ and the *potential* energy $V(T,r) = F(T,r) + TS(T,r)$ associated with quark pair at distance r. While F has no linear part (in r) above T_c, both V and S have it. The physical difference between the two is that F corresponds to adiabatically slow motion of the quarks, slow enough to produce maximal entropy possible and reach thermal equilibrium at any r. However when quarks are moving with certain time scale, only a fraction x of the maximal entropy can be produced (because of Landau-Zener argument on level crossing): thus the effective potential is $V_{eff}(T,r) = F(T,r) + (1-x)TS(T,r)$. For relatively rapid motion in which no entropy is produced, $x = 0$, and one returns to $V(T,r)$. The difference between the two potentials can be also explained in other way [7]: $F(T,r)$ is related to a stable flux tube with a "supercurrent coil" while $V(T,r)$ has just "normal" coil and thus is metastable .

5. QGP corona: experiment

Three different correlation phenomena have been discovered in heavy ion collisions at RHIC:

(i) The "cone" has been discovered in the 2-particle azimuthal correlations . The "away-side" peak at $\Delta\phi = \pi$ due to a partner jet, balancing hard trigger particle, disappears, substituted by new peaks at completely different angle.After discovery of those effects there was extensive studies of the 3-particle correlations, which confirmed that the observed structure is indeed cone-like, and not e.g. a reflected jet.

(ii) the hard ridge is also seen in 2-particle correlators, but plotted on the two-dimensional $\Delta\phi - \Delta\eta$ plane, the differences between the azimuthal angles and pseudorapidities of the two particles. The jet remnants make a peak near $\Delta\phi = 0, \Delta\eta = 0$, which was found to sit on top of the "ridge", with comparable width in $\Delta\phi$ but stretched in rapidity to nearly all 10 units (between beam rapidities at RHIC). For plots and various features one can consult the original talk by Putschke.[15]

(iii) the "soft ridge" is found by STAR collaboration [16,17] without a trigger, in the 2-particle correlations. For experimental details and phenomenological considerations seetalks at recent specialized workshop.[14]

We will return to these observations below, turning now to their suggested explanations:

(i) Stoecker et al, as well as Casalderrey, Teaney and myself [18] have proposed that the energy deposited by a quenched jet goes into two hydrodynamical excitation modes, the sound and the so called diffusion or wake modes. The sound from the propagating jet should thus create the famous Mach cone, in qualitative agreement with the conical structure observed. However surprising experimental fact is quite large amplitude of this signal, as well as large value of the cone angle which has not been reproduced by hydrodynamics (or AdS/CFT).The cone angle is in the range $\theta_{cone} = 1.2 - 1.4$ radians (not too far from $\pi/2 = 90^o$ or cylindrical waves!) which by Mach formula gives the speed of pertinent perturbation to be about

$$< v_{wave} >= cos(\theta_M) \approx 0.2 \tag{3}$$

while the expected speed of sound is .3 at its lowest point near T_c and about .5 in QGP.
(ii) One early model for "hard ridge" has been introduced in my paper.[19] It relates it with the forward-backward jets accompanying any hard scattering, providing extra particles ("hot spot") widely distributed in rapidity. This idea is then combined with the one suggested previously by Voloshin,[20] namely that extra particles deposited in the fireball would be moved transversely by the radial hydrodynamical flow, should produce a peak at certain azimuthal angle corresponding to the position of the hot spot, see Fig. 3(upper right). While particles of the ridge are separated by large rapidity gaps and cannot communicate during the expansion process, their azimuthal emission angles remain correlated with each other because they originate from the same "hot spot" in the transverse plane.
(iii) Similarly, transverse hydro boost of "hot spots" was used for the explanation of the "soft ridge" by McLerran and collaborators.[21,22] They have pointed out that the initial state color fluctuations in the colliding nuclei would create "spots" at some positions in the transverse plane The idea is shown in Fig. 3(upper right): the spots can be carried by hydro expansion and get visible at certain azimuthal angles. Further confirmation of hydro origin of ridges comes from the centrality dependence of the *angular width* of the ridge: the peak in azimuth *sharpens* for more central collisions.

So, at a very qualitative level the origin of all three phenomena seem to be explained: yet at more qualitative level a lot of puzzles appear. As an example, consider the simplest of them, the "soft ridge". As discussed in [21, 22], the initial stage (proper time $\tau \sim 1/Q_s \sim 0.2\, fm/c$ where $Q_s \sim 1\, GeV$ is the so called saturation scale at RHIC) can be discussed using classical Yang-Mills equations: thus color fluctuations naturally appear. However, the observed pions come from final freezeout time, separated from the initial "glasma" era by much longer time $\tau \sim 10\,fm$. This is certainly so, as the explanation heavily relies on radial hydro velocity and thus it has to wait till the hydro velocity is being created. As we will argue below, there are many reasons why one might have expected nearly complete disappearance of this signal during this time.

Common to all three cases is deposition of some additional energy (or entropy), on top of the "ambient matter". The number of correlated particles in all of them constitute a small ($\sim 10^{-3}$) fraction of the total multiplicity: thus they can only be seen in a high-statistics correlation analysis.

Similar to circles from a stone thrown into a pond, initial perturbation should become some expanding waves, and hydrodynamics predicts that basically nothing is left at the original location at later times. Even ignoring any dissipation and using ideal hydrodynamics one finds the radius of those waves at final freezeout to be given by the "sound horizon"

$$R_h = \int_0^{\tau_f} d\tau c_s(\tau)\,. \tag{4}$$

So, by the the freezeout proper time $\tau_f \sim 10 - 15\,fm/c$, this distance is large, $\sim 6fm$ or so, see Fig. 3(right middle). The amplitude of the sound wave is decreasing accordingly, and the width of ϕ distribution grows. Simple calculation show that in such case the width of the peak would be larger than observed, and the amplitude smaller. This is a common puzzle for all three phenomena mentioned: hydro fails to explain why all of them remain observable, as if nothing happened to them during rather long time of the hydro process. iThe second set of waves provided by DMHD, with smaller velocity of propagation, or better yet stabilized flux tubes is thus suggested as an explanation, see Fig. 3(bottom right) .

Let me end this brief wandering into experiment-generated puzzles by mentioning recent RHIC data which provided direct experimental indications for enhanced stability of the flux tube in matter relative to pp. Those are from PHOBOS collaboration (now with a detector no longer physically existing) which has large rapidity coverage of their silicon detector. They have studied clustering of secondaries in rapidity and found that the number of charged particles in a cluster observed in AuAu collisions is about twice that seen in correlation studies of the pp collisions, reaching the size of about 10 particles. The rapidity width of the cluster is also larger than for those in pp: they are not consistent with isotropically decaying resonances but apparently elongated longitudinally. We basically see fragments of the flux tubes, in pp and AuAu collisions: and apparently the tubes decay less frequently in the latter case, into larger pieces. I take it as the most direct indication to existence of the "magnetic plasma" from the data.

References

1. E. Shuryak, "The Fate of the Initial State Perturbations in Heavy Ion Collisions," arXiv:0903.3734 [nucl-th].
2. J. Liao and E. Shuryak, Phys. Rev. C **75**, 054907 (2007) [arXiv:hep-ph/0611131]; Phys. Rev. Lett. **101**, 162302 (2008) [arXiv:0804.0255 [hep-ph]].
3. C. Ratti and E. Shuryak, Phys. Rev. D **80**, 034004 (2009) [arXiv:0811.4174 [hep-ph]].
4. M. Cristoforetti and E. Shuryak, "Bose-Einstein Condensation of strongly interacting bosons: from liquid ^{4}He to QCD monopoles," arXiv:0906.2019 [hep-ph].
5. R. P. Feynman, Phys. Rev. 90, 1116 (1953); Phys. Rev. 91, 1291 (1953).
6. M. N. Chernodub and V. I. Zakharov, Phys. Rev. Lett. **98**, 082002 (2007) [arXiv:hep-ph/0611228].
7. J. Liao and E. Shuryak, "Electric $\bar{Q}Q$ potentials at $T \approx T_c$ and magnetic component of QCD plasma," arXiv:0804.4890 [hep-ph].
8. A. Nakamura, T. Saito and S. Sakai, Phys. Rev. D **69**, 014506 (2004) [arXiv:hep-lat/0311024].
9. A. D'Alessandro and M. D'Elia, Nucl. Phys. B **799**, 241 (2008) [arXiv:0711.1266 [hep-lat]].
10. J. Liao and E. Shuryak, Phys. Rev. C **77**, 064905 (2008) [arXiv:0706.4465 [hep-ph]].
11. S. Mrowczynski, Phys. Lett. B **314**, 118 (1993).
12. M. Asakawa, S. A. Bass and B. Muller, J. Phys. G **34**, S839 (2007) [arXiv:nucl-th/0702007].
13. A. M. Polyakov, Phys. Lett. B **72**, 477 (1978).
14. CATHIE workshop "Critical assessment of theory and experiments on correlations at RHIC," Feb.25-26, BNL. http://www.bnl.gov/cathie-riken .
15. J. Putschke, J. Phys. G **34**, S679 (2007) [arXiv:nucl-ex/0701074].
16. J. Adams *et al.* [STAR Collaboration], J. Phys. G **32**, L37 (2006) [arXiv:nucl-ex/0509030].
17. M. Daugherity [STAR Collaboration], J. Phys. G **35**, 104090 (2008) [arXiv:0806.2121 [nucl-ex]].
18. J. Casalderrey-Solana, E. V. Shuryak and D. Teaney, J. Phys. Conf. Ser. **27**, 22 (2005) [Nucl. Phys. A **774**, 577 (2006)] [arXiv:hep-ph/0411315]; L. M. Satarov, H. Stoecker and I. N. Mishustin, Phys. Lett. B **627**, 64 (2005) [arXiv:hep-ph/0505245].
19. E. V. Shuryak, Phys. Rev. C **76**, 047901 (2007) [arXiv:0706.3531 [nucl-th]].
20. S. A. Voloshin, Phys. Lett. B **632**, 490 (2006) [arXiv:nucl-th/0312065].
21. A. Dumitru, F. Gelis, L. McLerran and R. Venugopalan, Nucl. Phys. A **810**, 91 (2008) [arXiv:0804.3858 [hep-ph]].
22. S. Gavin, L. McLerran and G. Moschelli, Phys. Rev. C **79**, 051902 (2009) [arXiv:0806.4718 [nucl-th]].

Confinement–Deconfinement Phase Transition in Hot and Dense QCD at Large N

Ariel R. Zhitnitsky

Department of Physics and Astronomy,
University of British Columbia, Vancouver, Canada
E-mail: arz@phas.ubc.ca

We conjecture that the confinement–deconfinement phase transition in QCD at large number of colors N and $N_f \ll N$ at $T \neq 0$ and $\mu \neq 0$ is triggered by the drastic change in θ behavior. The conjecture is motivated by the holographic model of QCD where confinement–deconfinement phase transition indeed happens precisely at $T = T_c$ where θ dependence experiences a sudden change in behavior. The conjecture is also supported by quantum field theory arguments when the instanton calculations (which trigger the θ dependence) are under complete theoretical control for $T > T_c$, suddenly break down immediately below $T < T_c$ with sharp changes in the θ dependence. Finally, the conjecture is supported by a number of numerical lattice results. We employ this conjecture to study confinement–deconfinement phase transition of hot and dense QCD in large N limit by analyzing the θ dependence. We estimate the critical values for T_c and μ_c where the phase transition happens by approaching the critical values from the hot and/or dense regions where the instanton calculations are under complete theoretical control. We also describe some defects of various codimensions within a holographic model of QCD by focusing on their role around the phase transition point.

Keywords: phase transition, large N expansion, θ dependence, instantons.

1. Introduction

This talk is based on series of recent papers [1, 2, 3]. Understanding the phase diagram at nonzero external parameters T, μ is one of the most difficult problem in QCD. Obviously, this area is a prerogative of numerical lattice computations. However, some insights about the basic features of the phase diagram may be inferred by using some analytical approaches. In particular, some qualitative questions can be formulated and answered by considering a theory with large number of colors N. We conjecture that the confinement–deconfinement phase transition in QCD at large number of colors N and $N_f \ll N$ at $T \neq 0$ and $\mu \neq 0$ is triggered by the drastic change in θ behavior. This criteria is motivated by the observation that in holographic model of QCD the confinement–deconfinement phase transition happens precisely at the value of temperature $T = T_c$ where θ dependence experiences a sudden change in behavior.[1,4] Secondly, the proposal is supported by the numerical lattice results, see e.g. review article [5], which unambiguously suggest that the topological fluctuations are strongly suppressed in deconfined phase, and this suppression becomes more severe with increasing N. Finally, our new criteria is based on a physical picture which can be shortly summarized as follows.

For sufficiently high temperatures $T > T_c$ the instanton gas is dilute with density $\sim e^{-\gamma(T)N}$ which implies a strong suppression of the topological fluctuations at large N where $\gamma(T) > 0$, see below for details on structure of $\gamma(T)-$ function. The calculations in this region are under complete theoretical control and the vacuum energy has a nice analytic behavior $\sim \cos\theta e^{-\gamma(T)N}$ as function of θ. At the critical value of temperature, $T = T_c$ where $\gamma(T)$ changes the sign, the instanton expansion breaks down and one should naturally expect that at $T = T_c$ there should be a sharp transition in θ behavior as simple formula $\sim \cos\theta$ can only be valid when the instanton gas is dilute and semiclassical calculations are justified which is obviously not the case for $T < T_c$. Therefore, it is naturally to associate sharp changes in θ behavior with confinement–deconfinement transition, just as in the holographic model.[1]

The plan of the paper is as follows. I start in Section 2 by considering the phase transition in hot QCD.[1] In section 3 the same technique is applied to the dense QCD. I argue that the confinement–deconfinement phase transition happens at very large quark chemical potential $\mu_c \sim \sqrt{N}\Lambda_{QCD}$, where $\mu = \mu_B/N$ is already properly scaled quark chemical potential.[2] Finally, in section 4 some defects of various codimensions are described within a holographic model of QCD by focusing on their role around the phase transition point.[3]

2. Deconfinement Phase Transition in hot QCD at large N.

We start with analysis of nonzero temperature[1] when a sharp transition in θ dependence at the phase transition is indeed observed in the holographic model of QCD. From quantum field theory viewpoint such a transition can be understood as follows. Instanton calculations are under complete theoretical control in the region $T > T_c$ as the instanton density is parametrically suppressed at large N in deconfined region,[1]

$$V_{\text{inst}}(\theta) \sim e^{-\gamma N} \cos\theta, \qquad \gamma = \Big[\frac{11}{3} \ln\Big(\frac{\pi T}{\Lambda_{QCD}}\Big) - 1.86\Big]. \tag{1}$$

It is assumed that a higher order corrections may change the numerical coefficients in $\gamma(T)$, but they do not change the structure of eq. (1). The critical temperature is determined by condition $\gamma = 0$ where exponentially small expansion parameter $e^{-\gamma N}$ suddenly blows up and becomes exponentially large. Numerically, it happens at

$$\gamma = \Big[\frac{11}{3} \ln\Big(\frac{\pi T_c}{\Lambda_{QCD}}\Big) - 1.86\Big] = 0 \quad \Rightarrow \quad T_c(N=\infty) \simeq 0.53\Lambda_{QCD}, \tag{2}$$

where Λ_{QCD} is defined in the Pauli -Villars scheme. Our computations are carried out in the regime where the instanton density $\sim \exp(-\gamma N)$ is parametrically suppressed at any small but finite $\gamma(T) = \epsilon > 0$ when $N = \infty$. From eq. (1) one can obtain the following expression for instanton density in vicinity of $T > T_c$,

$$V_{\text{inst}}(\theta) \sim \cos\theta \cdot e^{-\alpha N(\frac{T-T_c}{T_c})}, \qquad 1 \gg \Big(\frac{T-T_c}{T_c}\Big) \gg 1/N. \tag{3}$$

where $\alpha = \frac{11}{3}$ and $T_c(N=\infty) \simeq 0.53\Lambda_{QCD}$ are estimated at one loop level. Such a behavior does imply that the dilute gas approximation is justified even in close vicinity of T_c as long as $\frac{T-T_c}{T_c} \gg \frac{1}{N}$. Therefore, the θ dependence, which is sensitive to the topological fluctuations is determined by (3) all the way down to the temperatures very close to the phase transition point from above, $T = T_c + O(1/N)$. The topological susceptibility is order of one for $T < T_c$ in confined phase while it vanishes $\sim e^{-\gamma N} \to 0$ for $T > T_c$ in deconfined phase. Nontopological quantum fluctuations on the other hand could be quite large in this region, but they do not effect the structure of eq. (3).

There are three basic reasons for a generic structure (1,2,3) to emerge:
a. The presence of the exponentially large "$T-$ independent" contribution (e.g. $e^{+1.86N}$ in eq. (1)). This term basically describes the entropy of the configuration. It is due to a number of contributions such as a number of embedding $SU(2)$ into $SU(N)$ etc;
b. The presence of the "$T-$ dependent" contribution to $V_{\text{inst}}(\theta)$ which comes from $\int n(\rho)d\rho$ integration. It is proportional to the standard factor,

$$\Big(\frac{\Lambda_{QCD}}{\pi T}\Big)^{\frac{11}{3}N} = \exp\Big[-\frac{11}{3}N \cdot \ln\Big(\frac{\pi T}{\Lambda_{QCD}}\Big)\Big]. \tag{4}$$

c. The fermion related contributions such as a chiral condensate, diquark condensate or non-vanishing mass term enter the instanton density as follows $\sim \langle \bar{\psi}\psi \rangle^{N_f} \sim e^{N\cdot\left(\kappa \ln |\langle \bar{\psi}\psi \rangle|\right)}$. For $\kappa \equiv \frac{N_f}{N} \to 0$ this term obviously leads to a sub leading effects $1/N$ in comparison with two main terms in the exponent (1) and will be neglected for the rest of this paper.

The crucial element in this analysis is that both leading contributions (items 1 and 2 above) have exponential e^N dependence, and therefore at $N \to \infty$ for $T > T_c$ the instanton gas is dilute with density $e^{-\gamma N}$, $\gamma > 0$ which ensures a nice $\cos\theta$ dependence (3), while for $T < T_c$ the expansion breaks down, and θ dependence must sharply change at $T < T_c$. We have identified such sharp changes with first order phase transition.

One more comment on this proposal. Our conjecture (that the confinement–deconfinement phase transition in QCD is triggered by the drastic change in θ at the same point $T = T_c$) implicitly implies that the configurations which are responsible for sharp θ changes must also drastically change their properties at $T = T_c$. For $T > T_c$ the objects which describe the θ behavior are the instantons, while at $T < T_c$ we believe, they are the instanton quarks, the quantum objects with fractional topological charges $\pm 1/N$. Therefore, we interpret this transition as dissociation of the instanton into instanton-quarks which become the dominant quasi-particles at $T < T_c$, see more comments on this interpretation in concluding section.

3. Phase Transition in dense QCD at large N.

In this section we estimate the value of μ_c where the instanton expansion breaks down and therefore, the θ dependence should experience a sharp change. According to our conjecture we should identify this position with the phase transition point. Similar arguments have been put forward previously[6] for numerical estimation of μ_c for small $N, N_f = 2, 3$. Our goal here is quite different: we want to understand an analytical dependence of $\mu_c(N, N_f)$ as a function of N, N_f at very large N and finite $N_f \ll N$ in order to compare with results of ref. [7] where the authors presented a very strong argument suggesting a very large $\mu_c \sim \sqrt{N}$ where the phase transition could happen.

We follow the same logic as before, and study the θ dependence in order to make a prediction about the phase transition point μ_c. In the regime $\mu > \mu_c$ the θ dependence is determined by the dilute instanton gas approximation. We expect that the expansion breaks down only in close vicinity of μ_c at large N as it happens in our previous analysis with phase transition at $T = T_c$. According to the conjecture this point will be identified with confinement–deconfinement phase transition point μ_c. In the present case of analyzing μ_c rather than T_c discussed previously we do not have any support from the lattice computations, nor from holographic models. Still, the basic governing principle remains the same. Therefore we identify the point where instanton expansion breaks down (and correspondingly a point where a simple $\cos\theta$ sharply changes to something else) with the point μ_c where the phase transition happens. In our estimates below we assume that the color superconducting phase is realized in deconfined phase for all N, see e.g. recent review [8].

As we shall see below, the instanton density in deconfined phase has the following generic behavior, $\sim \cos\theta \exp\left[-N\gamma(\mu)\right]$, where $\gamma(\mu) \sim const. + 0(1/N)$ in large N limit. Such a behavior implies that for any small (but finite) positive $\gamma > 0$ the instanton density is exponentially suppressed and our calculations are under complete theoretical control. In contrast: at arbitrary small and negative $\gamma < 0$ the instanton expansion obviously breaks down, theoretical control is lost as an exponential growth $\sim \exp\left(|\gamma| N\right)$ for the instanton density makes no sense. The θ behavior must drastically change at this point. Therefore, the value of μ_c is determined by the following condition,

$$\gamma(\mu = \mu_c) = 0 \qquad \Longrightarrow \qquad \mu_c = c\Lambda_{QCD}. \tag{5}$$

Our goal is to compute the coefficient c by approaching the critical point μ_c from deconfined side of phase boundary. Therefore, we will be interested in the instanton density in the dilute gas regime at $\mu > \mu_c$ where analytical instanton calculations are under control.

As is well known, the θ dependence goes away in full QCD in both phases: confined as well as deconfined in the presence of the massless chiral fermions. However we are interested in the magnitude of the instanton contribution $\sim V_{\rm inst}(\theta)$ in deconfined phase rather than in θ dependence of full QCD. Precisely this coefficient triggers the point where the instanton expansion suddenly blows up. The sharp changes in $V_{\rm inst}(\theta)$ we identify with complete reconstruction of the ground state, drastic changes of the relevant gluon configurations, and finally, with confinement–deconfinement phase transition. To avoid identical vanishing of $V_{\rm inst}(\theta)$ in the presence of massless fermions one can assume a non zero chiral condensate in deconfined phase or one can assume a non-vanishing masses $m_q \neq 0$ for the fermions, or non-vanishing diquark condensate $\langle\psi\psi\rangle \neq 0$ to avoid identical vanishing of $V_{\rm inst}(\theta)$ for dense matter at large μ. None of these assumptions effects any numerical estimates given below in the limit $N \to \infty$, $N_f \ll N$, as explained in the previous section.

The result of the computations can be represented as follows,

$$V_{\rm inst}(\theta) \sim e^{-\gamma N}\cos\theta, \qquad \gamma = \left[\frac{11}{6}\ln\left(\frac{N_f\bar{\mu}^2}{\Lambda^2_{QCD}}\right) - 1.1\right], \quad \mu^2 \equiv N\bar{\mu}^2, \quad (6)$$

where we introduced reduced chemical potential $\bar{\mu} \equiv \mu/\sqrt{N}$ and neglected all powers N^p in front of $e^{-\gamma N}$. The crucial difference in comparison with similar computation at nonzero temperature (1) is emerging of parameter $\bar{\mu}$ instead of the original quark chemical potential $\mu \equiv \sqrt{N}\bar{\mu}$. It implies that the critical chemical potential where γ changes the sign (and therefore where the phase transition is expected) is parametrically large $\mu_c \sim \sqrt{N}$ because $\bar{\mu}_c \sim 1$, see below for numerical estimates. The origin for this phenomenon can be understood from the following observation: the temperature -dependent factor in the instanton density before integrating over ρ is proportional to $\sim N$ while chemical potential enters this expression with factor $\sim N_f \ll N$, see e.g. the review paper [9] . Therefore, a very large chemical potential $\mu \sim \sqrt{N}\Lambda_{QCD}$ is required in order to achieve the same effect as temperature $T \sim \Lambda_{QCD}$. The physics of this phenomenon can be explained as follows: at $T \sim 1$ a large number of gluons $\sim N^2$ can get excited while at $\mu \sim 1$ only a relatively small number of quarks in fundamental representation $\sim N$ can get excited. Therefore, it requires a very large chemical potential $\mu^2 \sim N$ in order for fundamental quarks play the same role as gluons do at $T \sim 1$. As explained above, the critical chemical potential is determined by condition $\gamma = 0$ where exponentially small expansion parameter $e^{-\gamma N}$ at $\mu > \mu_c$ suddenly blows up at $\mu < \mu_c$. Numerically, it happens at

$$\gamma = \left[\frac{11}{6}\ln\left(\frac{N_f\bar{\mu}^2}{\Lambda^2_{QCD}}\right) - 1.1\right] = 0 \;\Rightarrow\; \mu_c(N\!=\!\infty) \simeq 1.4\cdot\Lambda_{QCD}\sqrt{\frac{N}{N_f}}, \quad (7)$$

where Λ_{QCD} is defined in the Pauli-Villars scheme. The topological susceptibility vanishes $\sim e^{-\gamma N} \to 0$ for $\mu > \mu_c$ while it must be drastically different for $\mu < \mu_c$ as θ dependence must experience some drastic changes in this region as the instanton expansion breaks down, and therefore simple $\cos\theta$ dependence must be replaced by something else. It is very likely that the standard Witten's arguments (valid for the confined phase) still hold in this region $\mu < \mu_c$ in which case the topological susceptibility is order of one.

The Λ_{QCD} in the Pauli-Villars scheme which enters our formula (7) is not well-known numerically. Therefore, for numerical estimates one can trade Λ_{QCD} in favor of $T_c(N = \infty)$ at $\mu = 0$ estimated above (2). Therefore, our final numerical estimate for $\mu_c(N = \infty)$ can be presented as follows,

$$\mu_c(N = \infty) \simeq 2.6 \cdot \sqrt{\frac{N}{N_f}} \cdot T_c(N = \infty, \mu = 0), \qquad N_f \ll N. \qquad (8)$$

If one uses the numerical value for $T_c(N = 3) \simeq 260$ MeV,[10,11] one arrives to $\mu_c(N = \infty) \simeq 690\sqrt{N/N_f}$ MeV which is our final numerical estimate for the critical chemical potential where deconfined phase transition is predicted for very large N. Few remarks are in order:

a. The most important result of the present studies is the observation that the confinement–deconfinement phase transition according to (7) happens at very large $\mu_c \sim \sqrt{N}$ if $N_f \ll N$. This is consistent with the results of [7] where parametrically large scale for $\mu_c \sim \sqrt{N}$ had been predicted. However, the technique of ref. [7] does not allow to answer the question whether the transition would be the first order or it would be a crossover. Within our framework at $N \gg 1$ and $N_f \ll N$ the entire phase transition line (which starts at $T = T_c \sim \Lambda_{QCD}$ at $\mu = 0$ and ends at $\mu = \mu_c \sim \sqrt{N}\Lambda_{QCD}$ at $T = 0$) is predicted to be the first order phase transition at large N and $N_f \ll N$. This is because the nature for the phase transition along the entire line is one and the same: it is drastic changes of θ dependence when the phase transition line is crossed.

b. Our computations are carried out in the regime where the instanton density $\sim \exp(-\gamma N)$ is parametrically suppressed at $N = \infty$. From eq. (6) one can obtain the following expression for instanton density in vicinity of $\mu > \mu_c$,

$$V_{\text{inst}}(\theta) \sim \cos\theta \cdot e^{-\alpha N\left(\frac{\mu-\mu_c}{\mu_c}\right)}, \qquad \frac{1}{N} \ll \left(\frac{\mu - \mu_c}{\mu_c}\right) \ll 1\,, \qquad (9)$$

where α is $11/3$ at one loop level, but the perturbative corrections could be large and they may considerably change this numerical coefficient. Such a behavior (9) does imply that the dilute gas approximation is justified even in close vicinity of μ_c as long as $\frac{\mu-\mu_c}{\mu_c} \gg \frac{1}{N}$. In this case the diluteness parameter remains small. We can not rule out, of course, the possibility that the perturbative corrections may change our numerical estimate for μ_c. However, we expect that a qualitative picture of the phase transition advocated in this paper remains unaffected as a result of these perturbative corrections in dilute gas regime.

c. Once μ_c is fixed one can compute the entire segment of the phase transition line $\mu_c(T)$ for relatively small T. Indeed, in the dilute gas regime at

$\mu > \mu_c$ the T dependence of the instanton density is determined by a simple insertion $\sim \exp[-2/3N\pi^2T^2\rho^2]$ in the expression for the density, see e.g. the review paper [9] . The result in the leading loop order $\mu_c(T)$ can be presented as follows,

$$\mu_c(T) = \mu_c(T=0)\Big[1 - \frac{N\pi^2T^2}{3N_f\mu_c^2(T=0)}\Big], \qquad \sqrt{N}T \ll \mu_c\,. \qquad (10)$$

4. QCD defects via holography

Our discussions in the previous sections were based on the instanton calculus in vicinity of the phase transition. The instanton in the holographic description can be identified with the D0 brane extended along x_4, see refs [1, 3, 4] for notations and introduction into the subject. The holographic picture gives a very natural way to represent the deconfinement phase transition as the Hawking-Page phase transition in which case the two metrics with the same asymptotics get interchanged. As our main goal in this talk is to understand the phase transition, it would be interesting to study other gauge configurations from the holographic viewpoint. In what follows we limit ourselves only with two examples. For a more complete list of different defects which can be studied by this technique, see the original paper [3].

D4 particle

There are several possible embeddings of D4 branes. One possibility is to wrap D4 brane around S^4 and extended along the Euclidean time τ, see [3] for the details. The key point here is that due to this wrapping around S^4 the "electric charge" N is induced on the D4 brane. The resulting 3 dimensional system is a static and topologically stable configuration. If we had quarks we could make a baryon by attaching quarks to open strings. In the pure YM case there are no flavor branes describing quarks to make a gauge invariant object. The most simple way to achieve this goal is to add N D0 branes yielding the D0-D4 open strings. Hence we get the D4 particle which is not sensitive to the θ-term. It is important that this configuration is well-defined below the critical temperature $T < T_c$ in contrast with instantons discussed in [1, 4] when D0 branes were sensitive to the θ term and were well defined above the critical temperature $T > T_c$. We can combine the " D4 particle" with "D4 anti-particle" to form a gauge invariant object, see Fig. 1. The mass of this object scales as $\sim N$ and is much heavier than the usual glueballs. In fact, one can construct gauge invariant objects with any even number of vertexes such that the total charge vanishes. It is a new family of glueballs with mass $\sim NV$, where V is the number of D4 and anti-D4 particles which form a desired configuration. It is amusing that such

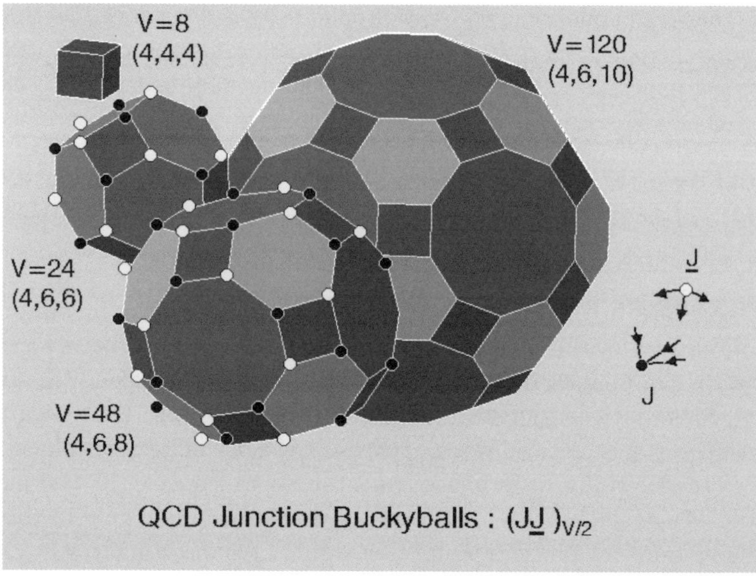

Fig. 1. Buckyballs with $N = 3$ and $V = 8, 24, 48, 120$ from [12]. The construction combines "D4 particle" with "D4 anti-particle" to form a gauge invariant object with zero baryon charge. The mass of this object scales as $\sim N$.

kind of structure in QCD had been previously discussed[12] motivated by the discovery of the carbonic Fullerenes C_{60} and C_{70} in 1985, which are nano-scale objects.[13] The QCD objects, similar to the carbonic Fullerenes with femto-meter scale were named Buckyballs. It has been also demonstrated that the "magic" numbers for Buckyballs are $V = 8, 24, 48, 120$ which correspond to the most symmetric, and likely, most stable configurations.[12] The properties of this configuration are not sensitive to θ. The possibility to discover such kind of configurations at RHIC was discussed in [12].

D0-D2.

In this subsection we want to address the following question: what happens to instantons (represented by D0 branes in holographic description) in the confined phase? Naively, one could think that the system becomes unstable at $T < T_c$, and therefore the instantons simply disappear from the system. However, one should speak about effectively zero action for formation of such kind of objects rather than about their instability. Therefore, numerous number of these objects can emerge in the system without any suppression. In fact, it has been argued in [1] that this is precisely what is happening when one crosses the phase transition line from above.

It turns out that there are configurations in holographic picture which exactly describe dissociation of the instanton into N different objects with fractional topological charge $1/N$. The construction is described in [3] and requires N additional D2 domain walls localized at some points along x_4 coordinate. In this construction an object with a fractional topological charge $1/N$ may emerge. Indeed, one can follow the construction of ref. [14] for SUSY case when N D2 branes located symmetrically split the instanton into N constituents stretched between pairs of domain walls. Each constituent has fractional instanton number $1/N$ as well as the fractional monopole number and has no reason to condense. In our system we have precisely appropriate D2 branes which are needed for this construction. These monopoles are instantons in the 3d gauge theory on the D2 worldvolume theory which involves the scalar corresponding to the position of D2 branes. As we speculate below these magnetic monopoles may play an important role in the region close to the phase transition $0 < |T - T_c|/T_c \leq 1/N$.

5. Conclusion. Speculations.

A general comment on this proposal can be formulated as follows. Our conjecture which relates two apparently unrelated phenomena (phase transition vs sharp changes in θ behavior) implicitly implies that topological configurations which are linked to θ must play a crucial role in the dynamics of the phase transition. For $T > T_c$ such configurations are well-known: they are dilute instantons with density $\sim e^{-\gamma N} \cos\theta$. We presented arguments in [1] (see also earlier references therein) suggesting that at $T < T_c$ the instantons do not disappear from the system, but rather dissociate into fractionally charged constituents, the so-called instanton quarks. In this sense the phase transition can be understood as a phase transition between molecular phase (deconfined) and plasma phase (confined) of these fractionally charged constituents. A similar conclusion on sharp changes in θ behavior at $T = T_c$ was also observed in ref. [15] where the authors studied the D2 branes in confined and deconfined phases at $T \neq 0$. The topological objects (sensitive to θ) were identified in ref. [15] as magnetic strings.

If the picture advocated in the present work about the nature of the transition turns out to be correct, it would strongly suggest that fractionally charged constituents (which carry the magnetic charges as discussed in [1]) may play a very important role in dynamics in deconfined phase in close vicinity of the transition $0 < (T - T_c) \leq 1/N$. For large N this region shrinks to a point, however for finite N it could be an extended region in temperatures. In this region the instantons are not formed yet, and our semiclassical analysis is not justified yet as eq. (3) suggests. However, the constituents in this region are already not in condensed form. Therefore they may become an important magnetic degrees of freedom which may contribute to the equation of state similar to analysis on wrapped monopoles in ref. [16].

To conclude this talk I want to make a few comments on two recent papers [17, 18] where the authors advocate the picture similar to the one presented here.

I start with [17]. In that work the authors consider a specifically deformed $SU(N)$ gluodynamics at $T \neq 0$. It has been shown that such a deformation supports a reliable analysis in the weak coupling regime in the confining phase. The results of the corresponding calculations imply that the relevant degrees of freedom in the confined phase are the self dual magnetic monopoles with action $\frac{8\pi^2}{g^2 N}$ and topological charges $Q = \pm 1/N$ which are precisely the features of the instanton quarks discussed above.

In contrast, the starting point of ref. [18] is semiclassical calculations in the background of calorons where the weak coupling regime can not be guaranteed. While the calculations are semiclassical in nature, and therefore, can not be trusted in the strong coupling regime, still, the corresponding analysis shows how well localized instantons with integer topological charges at $T > T_c$ may dissociate into the fractional constituents at $T < T_c$, and become the key players in the confining phase. This is precisely the picture we are advocating in the present work based on analysis of sharp θ changes at $N = \infty$. It is impressive how complicated semiclassical calculations carried out in [18] lead to the expression for the vacuum energy $E_{vac} \cos(\frac{\theta}{N})$ advocated in [19] using completely different technique.

Our technique does not allow us to make any dynamical calculations in this phase as all color degrees of freedom have been integrated out in obtaining the low energy effective lagrangian. In other words, we can not study the dynamics of fractionally charged constituents in contrast with papers [17, 18].[a] However, the fact that the constituents carry fractional topological charge $1/N$ can be recovered in our approach because the color-singlet η' field enters the effective lagrangian as $\cos(\frac{\theta-\varphi}{N})$ and serves as a

[a] In particular, we do not see a beautiful picture of a multi-component color Coulomb plasma with nearest-neighbor interactions in the Dynkin space advocated in [17, 18]. Still, we do see the color- singlet Coulomb interaction of the fractionally charged $\pm 1/N$ constituents due to η' at very large distances where color already confined.

perfect probe of the topological charges of the constituents. One should also emphasize that the procedure of the recovering of the fractional topological charge $1/N$ (which has been used here) is not based on the weak coupling expansion.

It was great honor and pleasure to participate in the Workshop honoring 60th anniversary of Misha Shifman. Happy birthday, Misha! This work was supported, in part, by the Natural Sciences and Engineering Research Council of Canada.

References

1. A. Parnachev and A. R. Zhitnitsky, Phys. Rev. D **78**, 125002 (2008) [arXiv:0806.1736 [hep-ph]].
2. A. R. Zhitnitsky, Nucl. Phys. A **813**, 279 (2008) [arXiv:0808.1447 [hep-ph]].
3. A. S. Gorsky, V. I. Zakharov and A. R. Zhitnitsky, Phys. Rev. D **79**, 106003 (2009) [arXiv:0902.1842 [hep-ph]].
4. O. Bergman and G. Lifschytz, JHEP **0704**, 043 (2007) [arXiv:hep-th/0612289].
5. E. Vicari and H. Panagopoulos, Phys. Rept. **470**, 93 (2009) [arXiv:0803.1593 [hep-th]].
6. D. Toublan and A. R. Zhitnitsky, Phys. Rev. D **73**, 034009 (2006) [arXiv:hep-ph/0503256].
7. L. McLerran and R. D. Pisarski, Nucl. Phys. A **796**, 83 (2007) [arXiv:0706.2191 [hep-ph]].
8. M. G. Alford, A. Schmitt, K. Rajagopal and T. Schafer, Rev. Mod. Phys. **80**, 1455 (2008) [arXiv:0709.4635 [hep-ph]].
9. T. Schafer and E. V. Shuryak, Rev. Mod. Phys. **70**, 323 (1998).
10. B. Lucini, M. Teper and U. Wenger, JHEP **0401**, 061 (2004) [arXiv:hep-lat/0307017].
11. B. Lucini, M. Teper and U. Wenger, JHEP **0502**, 033 (2005) [arXiv:hep-lat/0502003].
12. T. Csorgo, M. Gyulassy and D. Kharzeev, J. Phys. G **30**, L17 (2004) [arXiv:hep-ph/0112066].
13. H. W. Kroto, R. E. Smalley and R. F. Curl, Nature **318**, 165 (1985).
14. N. M. Davies, T. J. Hollowood, V. V. Khoze and M. P. Mattis, Nucl. Phys. B **559**, 123 (1999) [arXiv:hep-th/9905015].
15. A. Gorsky and V. Zakharov, Phys. Rev. D **77**, 045017 (2008) [arXiv:0707.1284 [hep-th]].
16. M. N. Chernodub and V. I. Zakharov, Phys. Rev. Lett. **98**, 082002 (2007) [arXiv:hep-ph/0611228].
17. M. Ünsal and L. G. Yaffe, Phys. Rev. D **78**, 065035 (2008) [arXiv:0803.0344 [hep-th]].
18. D. Diakonov and V. Petrov, Phys. Rev. D **76**, 056001 (2007) [arXiv:0704.3181 [hep-th]].
19. I. Halperin and A. R. Zhitnitsky, Phys. Rev. Lett. **81**, 4071 (1998) [arXiv:hep-ph/9803301]; Phys. Rev. D **58**, 054016 (1998) [hep-ph/9711398].

SECTION 8.

DARK MATTER AXIONS AND OTHER EXOTICS

Dark Matter Axions

P. Sikivie

Department of Physics, University of Florida,
Gainesville, FL 32611, USA
E-mail: sikivie@phys.ufl.edu

The hypothesis of an 'invisible' axion was made by Misha Shifman and others, approximately thirty years ago. It has turned out to be an unusually fruitful idea, crossing boundaries between particle physics, astrophysics and cosmology. An axion with mass of order 10^{-5} eV (with large uncertainties) is one of the leading candidates for the dark matter of the universe. It was found recently that dark matter axions thermalize and form a Bose-Einstein condensate (BEC). Because they form a BEC, axions differ from ordinary cold dark matter (CDM) in the non-linear regime of structure formation and upon entering the horizon. Axion BEC provides a mechanism for the production of net overall rotation in dark matter halos, and for the alignment of cosmic microwave anisotropy multipoles. Because there is evidence for these phenomena, unexplained with ordinary CDM, an argument can be made that the dark matter is axions.

Keywords: axion; dark matter; Bose-Einstein condensation

1. Introduction

It is a great pleasure and honor to be part of Misha Shifman's 60th birthday celebration. Among Misha's many outstanding contributions to particle physics is his well-known proposal, in collaboration with Arkady Vainshtein and Valentine Zakharov, that the axion may be very light and very weakly coupled.[1–3] Here is the abstract of their paper:

P- and T-invariance violation in quantum chromodynamics due to the so-called θ-term is discussed. It is shown that irrespectively of how the confinement works there emerge observable P- and T-odd effects. The proof is based on the assumption that QCD resolves the U(1) problem, i.e., the mass of the singlet pseudoscalar meson does not vanish in the chiral limit. We suggest a modification of the axion scheme which restores the natural P and T invariance of the theory and cannot be ruled out experimentally.

The θ-term mentioned by Shifman et al. is

$$\mathcal{L}_\theta = + \frac{\theta g^2}{32\pi^2} G^a_{\mu\nu} \tilde{G}^{a\mu\nu} \tag{1}$$

where $G^a_{\mu\nu}$ are the QCD field strengths, g is the QCD coupling constant and θ is a parameter. A θ-term is generally present in the action density of the Standard Model of elementary particles.[4] Its existence raises a puzzle, called the Strong CP Problem. As Shifman et al. explain in their paper, the physics of QCD necessarily depends on the value of θ, if none of the quark masses vanish, because otherwise QCD wouldn't solve the $U_A(1)$ problem (of explaining why the mass of the singlet pseudoscalar meson does not vanish in the chiral limit) and hence couldn't be the correct theory of strong interactions. This is an important point. If it were possible for QCD to be independent of θ, the Strong CP Problem wouldn't be so urgent.

One can show that QCD physics depends on the value of θ only through the combination $\bar{\theta} \equiv \theta - \arg\det m_q$ where m_q is the quark mass matrix. If $\bar{\theta} \neq 0$ the strong interactions violate P and CP. Such P and CP violation is incompatible with the experimental upper bound on the neutron electric dipole moment [5] unless $|\bar{\theta}| < 10^{-10}$. In the Standard Model, P and CP violation are introduced by letting the elements of the quark mass matrix m_q be arbitrary complex numbers.[6] In that case, $\bar{\theta}$ is of order one. The Strong CP Problem is the problem of explaining why $|\bar{\theta}| < 10^{-10}$ instead.

The Strong CP Problem is solved if the term (1) in the Standard Model action density is replaced by

$$\mathcal{L}_{\text{axion}} = -\frac{1}{2}\partial_\mu\varphi\partial^\mu\varphi + \frac{g^2}{32\pi^2}\frac{\varphi(x)}{f} G^a_{\mu\nu}\tilde{G}^{a\mu\nu} \tag{2}$$

where $\varphi(x)$ is a new scalar field, and f is a constant with dimension of energy. In the modified theory, $\bar{\theta} = \frac{\varphi(x)}{f} - \arg\det m_q$ depends on the expectation value of $\varphi(x)$. This field settles to a value that minimizes the effective potential. The Strong CP Problem is solved because the minimum of the QCD effective potential $V(\bar{\theta})$ occurs at $\bar{\theta} = 0$ [7]. The $\varphi G \cdot \tilde{G}$ interaction in Eq. (2) is not renormalizable. However, there is a recipe for constructing renormalizable theories whose low energy effective action density is of the form of Eq. (2): construct the theory in such a way that it has a $U_{PQ}(1)$ symmetry which is a global symmetry of the classical action density, is broken by the color anomaly, and is spontaneously broken. Such a symmetry is called Peccei-Quinn symmetry after its inventors.[8] Weinberg and Wilczek [9] pointed out that a theory with $U_{PQ}(1)$ symmetry has a light pseudo-scalar particle, called the axion. The axion field is $\varphi(x)$. f is of order the expectation value that breaks $U_{PQ}(1)$, and is called the "axion decay constant".

The axion mass is given in terms of f by [9]

$$m \simeq 6 \text{ eV} \frac{10^6 \text{ GeV}}{f} . \tag{3}$$

All axion couplings are inversely proportional to f. The axion coupling to two photons is:

$$\mathcal{L}_{a\gamma\gamma} = -g_\gamma \frac{\alpha}{\pi} \frac{\varphi(x)}{f} \vec{E} \cdot \vec{B} \quad , \tag{4}$$

where $\vec{E}$ and $\vec{B}$ are the electric and magnetic fields, α is the fine structure constant, and g_γ is a model-dependent coefficient of order one. It had first been thought that f is of order the electroweak scale, in which case the axion couplings have strength typical of neutrinos and the axion mass is relatively large, in the 10 keV to 10 MeV range. Such axions were quickly ruled out by particle physics (beam dumps and rare decays) and nuclear physics experiments. But Shifman and others showed [1–3] that f can be made arbitrarily large, and hence the axion can be made arbitrarily light and weakly coupled. Such an axion is unconstrained by the aforementioned experiments, and was dubbed 'invisible'. May an invisible axion really exist?

The axion has been searched for in many places, and has not been found.[10] Axion masses larger than about 50 keV are ruled out by the aforementioned particle and nuclear physics experiments. The next range of axion masses, in decreasing order, is ruled out by stellar evolution arguments. The longevity of red giants rules out 200 keV $> m >$ 0.5 eV [11,12] in case the axion has negligible coupling to the electron (such an axion is called 'hadronic'), and 200 keV $> m > 10^{-2}$ eV [13] in case the axion has a sizable coupling to electrons. The duration of the neutrino pulse from supernova 1987a rules out 2 eV $> m > 3 \cdot 10^{-3}$ eV [14]. Finally, there is a lower limit, $m \gtrsim 10^{-6}$ eV, from cosmology which is discussed in the next section. This leaves open an "axion window": $3 \cdot 10^{-3} > m \gtrsim 10^{-6}$ eV. The lower edge of this window (10^{-6} eV) is much softer than its upper edge.

2. Axion production in the early universe

There are two populations of axions produced in the early universe, which we may call 'hot' (or thermal) and cold. Hot axions are produced in thermal processes such as $q + g \rightarrow q + a$ where q is a quark and g a gluon, or $\pi + \pi \rightarrow \pi + a$ where π is a pion.[15–17] The number density of thermal axions today (time t_0) is

$$n_a^{\rm th}(t_0) \simeq \frac{7.5}{\rm cm^3}\left(\frac{106.75}{\mathcal{N}_{\rm D}}\right)^{\frac{1}{3}} \tag{5}$$

where $\mathcal{N}_{\rm D}$ is the effective number of thermal degrees of freedom at the time axions decouple from the thermal bath. The Standard Model has $\mathcal{N}_{\rm D} = 106.75$. Thermal axions are a form of hot dark matter, similar to neutrinos, in the context of large scale structure formation.

The cold axions are produced when the potential $V(\bar{\theta})$, and hence the axion mass, turns on near the QCD phase transition.[18] The critical time, defined by $m(t_1)t_1 = 1$, is $t_1 \simeq 2 \cdot 10^{-7}$ sec $(f/10^{12}{\rm GeV})^{\frac{1}{3}}$. Cold axions are the quanta of oscillation of the axion field that result from the turn on of the axion mass. The average number density of cold axions at time t_1 is

$$n_a(t_1) \simeq \frac{1}{2} m(t_1)\langle\varphi^2(t_1)\rangle \simeq \pi f^2 \frac{1}{t_1} \quad . \tag{6}$$

In Eq. (6), we used the fact that the axion field $\varphi(x)$ is approximately homogeneous on the horizon scale t_1, because wiggles in $\varphi(x)$ which entered the horizon long before t_1 have been red-shifted away.[19] We also used the fact that the initial departure of $\varphi(x)$ from the CP conserving minimum is of order f. The axions of Eq. (6) are non-relativistic. Assuming that the ratio of the axion number density to the entropy density is constant from time t_1 till today, one finds[18]

$$\Omega_a \simeq \frac{1}{2}\left(\frac{f}{10^{12}{\rm GeV}}\right)^{\frac{7}{6}}\left(\frac{0.7}{h}\right)^2 \tag{7}$$

for the ratio of the axion energy density to the critical density for closing the universe. h is the present Hubble rate in units of 100 km/s.Mpc. The requirement that axions do not overclose the universe implies the constraint $m \gtrsim 6 \cdot 10^{-6}$ eV. For a more detailed discussion of the production and properties of dark matter axions, the reader may wish to consult refs. [17, 20].

3. Dark matter caustics

It has been established from a variety of observational inputs that approximately 23% of the energy density of the universe is "cold dark matter" (CDM). The CDM particles must be non-baryonic, and cold. "Cold" means that their primordial velocity dispersion is small enough that it can be set equal to zero for all practical purposes when discussing the formation of large scale structure. The leading candidates for the CDM particles are axions, weakly interacting massive particles (WIMPs), e.g. the neutralino in supersymmetric extensions of the Standard Model, and sterile neutrinos with mass in the keV range.

A central problem in dark matter studies is the question how CDM is distributed in the halos of galaxies, and in particular in the halo of our Milky Way galaxy. Indeed, knowledge of this distribution is essential for understanding galactic dynamics and for predicting signals in direct and indirect searches for dark matter on Earth.

Galactic halos are thought to be collisionless fluids and must therefore be described in 6-dimensional phase space. A full description gives the phase space distribution $f(\vec{r}, \vec{v}; t)$ of the dark matter particles in the halo, i.e. their velocity ($\vec{v}$) distribution at every position $\vec{r}$. An important simplification occurs in the case of *cold* dark matter because CDM particles lie in phase space on a thin 3-dimensional hypersurface. This fact implies that the velocity distribution is everywhere discrete[21] and that there are surfaces in physical space, called caustics, where the density of dark matter is very large.

Galactic halos have two types of caustics, outer and inner. The outer caustics are simple fold (A_2) catastrophes located on nested topological spheres surrounding the galaxy. The catastrophe structure of the inner caustics depends on the angular momentum distribution of the infalling dark matter particles. If that angular momentum distribution is dominated by net overall rotation, implying $\vec{\nabla} \times \vec{v} \neq 0$, the inner caustics are a set of 'tricusp rings'. A tricusp ring is a closed tube whose cross section is a section of the *elliptic umbilic* (D_{-4}) catastrophe.[22–24] The rings are located in the plane of the galaxy. In the self-similar infall model,[25] generalized to include the effect of angular momentum,[26,27] the caustic ring radii a_n ($n = 1, 2, 3..$) are predicted to obey the law[22,27]

$$a_n \simeq \frac{40\text{kpc}}{n} \left(\frac{v_{\text{rot}}}{220\text{km/s}} \right) \left(\frac{j_{\max}}{0.18} \right) \quad , \tag{8}$$

where v_{rot} is the rotation velocity of the galaxy and $j_{\max}$ is a parameter characterizing the amount of angular momentum that the dark matter particles carry.

Observational evidence for caustic rings at the radii a_n predicted by Eq. (8) was found in the rotation curves of external galaxies,[28] the rotation curve of our own galaxy,[29] and an IRAS map of the Galactic plane in the direction of the nearest caustic ring ($n = 5$).[29] A summary of the evidence can be found in ref. [27]. The evidence implies that the distribution of $j_{\max}$

values over nearby spiral galaxies, including the Milky Way, is peaked at $j_{\rm max} \simeq 0.18$. It also implies that we on Earth are close to a cusp in the nearest caustic ring of dark matter. As a result, the dark matter velocity distribution on Earth is dominated by a single flow, of known velocity vector. That single flow, called the "Big Flow" has density of order 1 GeV/cc, which is two or three times larger than the commonly cited estimates of the *total* local dark matter density.

Finally, the evidence for caustic rings of dark matter halos implies that the dark matter particles fall in with net overall rotation, and hence that their velocity field has non-zero curl: $\vec{\nabla} \times \vec{v} \neq 0$. If their velocity field were irrotational ($\vec{\nabla} \times \vec{v} = 0$), the inner caustics would have a tent-like structure [24] which is quite distinct from that of the caustic rings for which evidence was found.

This raises a puzzle. Indeed if the dark matter is cold and collisionless, as is the case for weakly interacting massive particles (WIMPs) and was thought to be the case for axions, the velocity field remains irrotational at all times because it is the outcome of gravitational forces proportional to the gradient of the Newtonian potential.[24] (General relativistic effects allow the creation of rotational velocity fields but are subdominant because the velocities involved are much less than the speed of light.) Thus, if the dark matter is cold and collisionless, one expects the inner caustics of galactic halos to be the tent-like structures of the $\vec{\nabla} \times \vec{v} = 0$ case, instead of the rings for which evidence was found. This puzzle has bothered me for a number of years, but Qiaoli Yang and I may now have found a solution to it.[30] As explained below, cold dark matter axions form a Bose-Einstein condensate. As a result, their properties differ from those of ordinary CDM.

4. Bose-Einstein condensation of dark matter axions [a]

The number density of cold axions implied by Eq. (6) is

$$n(t) \sim \frac{4 \cdot 10^{47}}{\rm cm^3} \left(\frac{f}{10^{12}\ \rm GeV}\right)^{\frac{5}{3}} \left(\frac{a(t_1)}{a(t)}\right)^3 \tag{9}$$

where $a(t)$ is the cosmological scale factor. Because the axion momenta are of order $\frac{1}{t_1}$ at time t_1 and vary with time as $a(t)^{-1}$, the velocity dispersion of cold axions is

$$\delta v(t) \sim \frac{1}{m t_1} \frac{a(t_1)}{a(t)} \tag{10}$$

[a] All the material in this section is taken from ref. [30].

if each axion remains in whatever state it is in, i.e. if axion interactions are negligible. Let us refer to this case as the limit of decoupled cold axions. If decoupled, the average state occupation number of cold axions is

$$\mathcal{N} \sim n \, \frac{(2\pi)^3}{\frac{4\pi}{3}(m\delta v)^3} \sim 10^{61} \left(\frac{f}{10^{12}\ \mathrm{GeV}}\right)^{\frac{8}{3}} . \tag{11}$$

Clearly, the effective temperature of cold axions is much smaller than the critical temperature

$$T_{\mathrm{c}} = \left(\frac{\pi^2 n}{\zeta(3)}\right)^{\frac{1}{3}} \simeq 300\ \mathrm{GeV} \left(\frac{f}{10^{12}\ \mathrm{GeV}}\right)^{\frac{5}{9}} \frac{a(t_1)}{a(t)} \tag{12}$$

for BEC. Axion number violating processes, such as their decay to two photons, occur only on time scales vastly longer than the age of the universe. The only condition for axion BEC that is not clearly satisfied is thermal equilibrium.

Axions are in thermal equilibrium if their relaxation rate Γ is large compared to the Hubble expansion rate $H(t) = \frac{1}{2t}$. At low phase space densities, the relaxation rate is of order the particle interaction rate $\Gamma_s = n\sigma\delta v$ where σ is the scattering cross-section. Axions have self interactions described by the action density $\mathcal{L}_{\mathrm{self}} = +\frac{1}{4}\lambda\varphi^4$ where $\lambda \simeq 0.35(\frac{m}{f})^2$. The cross-section for $\varphi + \varphi \to \varphi + \varphi$ scattering due to axion self interaction is *in vacuum*

$$\sigma_0 = \frac{1}{64\pi}\frac{\lambda^2}{m^2} \simeq 1.5 \cdot 10^{-105}\mathrm{cm}^2 \left(\frac{m}{10^{-5}\ \mathrm{eV}}\right)^6 . \tag{13}$$

If one substitutes σ_0 for σ, Γ_s is found much smaller than the Hubble rate, by many orders of magnitude. However, in the cold axion fluid background, the scattering rate is enhanced by the average quantum state occupation number of both final state axions, $\sigma \sim \sigma_0\mathcal{N}^2$, because energy conservation forces the final state axions to be in highly occupied states if the initial axions are in highly occupied states. In that case, the relaxation rate is multiplied by *one* factor of $\mathcal{N}$ [31]

$$\Gamma \sim n \, \sigma_0 \, \delta v \, \mathcal{N} . \tag{14}$$

Combining Eqs. (9-11,13), one finds $\Gamma(t_1)/H(t_1) \sim \mathcal{O}(1)$, suggesting that cold axions thermalize at time t_1 through their self interactions, but only barely so.

It may seem surprising that the huge and tiny factors on the RHS of Eq. (14) cancel each other. In fact the cancellation is not an accident. Consider a generic axion-like particle (ALP) whose mass m and decay constant

f are unrelated to each other. Its self interaction coupling strength $\lambda \sim \frac{m^2}{f^2}$. Cold ALPs appear at a time $t_1 \sim \frac{1}{m}$ with number density $n(t_1) \sim f^2 m$, and velocity dispersion $\delta v(t_1) \sim 1$. Substituting these estimates in Eqs. (11), (13) and (14), one finds that the thermalization rate is of order the Hubble rate at t_1, for all f and m.

A critical aspect of axion BEC phenomenology is whether the BEC continues to thermalize after it has formed. Axion BEC means that (almost) all axions go to one state. However, only if the BEC continually rethermalizes does the axion state track the lowest energy state.

The particle kinetic equations that yield Eq. (14) are valid only when the energy dispersion $\frac{1}{2}m(\delta v)^2$ is larger than the thermalization rate.[31] After t_1 this condition is no longer satisfied. One enters then a regime where the relaxation rate due to self interactions is of order

$$\Gamma_\lambda \sim \lambda \, n \, m^{-2} \, . \tag{15}$$

$\Gamma_\lambda(t)/H(t)$ is of order one at time t_1 but decreases as $t \; a(t)^{-3}$ afterwards. Hence, self interactions are insufficient to cause axion BEC to rethermalize after t_1 even if they cause axion BEC at t_1. However gravitational interactions, which are long range, come in to play. The relaxation rate due to gravitational interactions is of order

$$\Gamma_{\rm g} \sim G \, n \, m^2 \, \ell^2 \tag{16}$$

where $\ell \sim (m\delta v)^{-1}$ is the correlation length. $\Gamma_{\rm g}(t)/H(t)$ is of order $4 \cdot 10^{-8}(f/10^{12}\text{ GeV})^{\frac{2}{3}}$ at time t_1 but grows as $ta^{-1}(t) \propto a(t)$. Thus gravitational interactions cause the axions to thermalize and form a BEC when the photon temperature is of order 100 eV $(f/10^{12}\text{ GeV})^{\frac{1}{2}}$.

The process of axion Bose-Einstein condensation is constrained by causality. Thus one expects overlapping condensate patches with typical size of order the horizon. As time goes on, say from t to $2t$, the axions in t-size condensate patches rethermalize into $2t$-size patches. The correlation length is then of order the horizon at all times, implying $\delta v \sim \frac{1}{mt}$ instead of Eq. (10), and $\Gamma_{\rm g}/H \propto t^3 a^{-3}(t)$ after the BEC has formed. Therefore gravitational interactions rethermalize the axion BEC on ever shorter time scales compared to the age of the universe.

The axion field may be expanded in modes labeled $\vec{\alpha}$:

$$\varphi(x) = \sum_{\vec{\alpha}} \; [a_{\vec{\alpha}} \; \Phi_{\vec{\alpha}}(x) \; + \; a^\dagger_{\vec{\alpha}} \; \Phi^\star_{\vec{\alpha}}] \tag{17}$$

where the $\Phi_{\vec{\alpha}}(x)$ are the positive frequency c-number solutions of the

Heisenberg equation of motion for the axion field

$$D^\mu D_\mu \varphi(x) = g^{\mu\nu}[\partial_\mu \partial_\nu - \Gamma^\lambda_{\mu\nu}\partial_\lambda]\varphi(x) = m^2\varphi(x) \ , \tag{18}$$

and the $a_{\vec{\alpha}}$ and $a^\dagger_{\vec{\alpha}}$ are creation and annihilation operators satisfying canonical commutation relations. We neglect the self-interaction term which would otherwise appear on the RHS of Eq. (18), because it is of order $\frac{\rho}{f^2}\varphi$, where ρ is the axion density, and hence smaller by the factor $\left(\frac{a(t_1)}{a(t)}\right)^3 \frac{t}{t_1}$ than the relevant terms (of order $\frac{m}{t}\varphi$) in that equation. BEC means that all cold axions, except for a small fraction, go to a single state which we label $\vec{\alpha} = 0$. The corresponding $\Phi_0(x)$ is the axion wavefunction. In the spatially flat, homogeneous and isotropic Robertson-Walker space-time,

$$\Phi_0 = \frac{A}{a(t)^{\frac{3}{2}}}\, e^{-imt} \tag{19}$$

where A is a constant. The state of the axion field is $|N >= (1/\sqrt{N!})\ (a^\dagger_0)^N|0>$ where $|0>$ is the empty state, defined by $a_{\vec{\alpha}}\ |0> = 0$ for all $\vec{\alpha}$, and N is the number of axions.

To compare axion BEC with CDM, let us divide the observations into three arenas: 1) the behaviour of density perturbations on the scale of the horizon, 2) their behaviour during the linear regime of evolution within the horizon, and 3) their behaviour during the non-linear regime. CDM provides a very successful description in arena 2. However, axion BEC and CDM are indistinguishable in arena 2 on all scales of observational interest.[30,32] In particular, the equation governing the evolution of axion BEC perturbations is

$$\partial_t^2\delta + 2H\partial_t\delta - \left(4\pi G\rho_0 - \frac{k^4}{4m^2a^4}\right)\delta = 0 \tag{20}$$

where k is co-moving wavevector. The last term in Eq. (20) is absent for CDM. Eq. (20) implies that the axion BEC has Jeans length

$$\begin{aligned} k_{\rm J}^{-1} &= (16\pi G\rho m^2)^{-\frac{1}{4}} \\ &= 1.02\cdot 10^{14}\ \text{cm}\left(\frac{10^{-5}\ \text{eV}}{m}\right)^{\frac{1}{2}}\left(\frac{10^{-29}\ \text{g/cm}^3}{\rho}\right)^{\frac{1}{4}}\ . \end{aligned} \tag{21}$$

However, the Jeans length is small compared to the smallest scales (~ 100 kpc) for which we have observations on the behavior of density perturbations in the linear regime.

In the non-linear regime of structure formation (arena 3) and in the absence of rethermalization, axion BEC and CDM again differ only on length scales smaller than de Broglie wavelength. This follows from the WKB approximation and has also been shown by numerical simulation.[33] Since the axion de Broglie wavelength (of order 10 meters in galactic halos) is negligibly small compared to all length scales of observational interest, we again find that axion BEC and CDM are indistinguishable when there is no rethermalization of the BEC.

However, it was found above that gravitational interactions do rethermalize the axion BEC continually so that the axion state tracks the lowest energy state. This is relevant to the angular momentum distribution of dark matter axions in galactic halos. The angular momentum of galaxies is caused by the gravitational torque of nearby galaxies early on when protogalaxies are still close to one another.[34] As was mentioned in Section III, CDM presents us with a puzzle. The velocity field of ordinary cold dark matter, such as WIMPs, remains irrotational whereas the evidence for caustic rings of dark matter implies that the dark matter falls in with net overall rotation. The puzzle is solved if the dark matter is an axion BEC which rethermalizes while tidal torque is applied to it. Indeed, the lowest energy state for given total angular momentum is one in which each particle carries an equal amount of angular momentum. In that case there is net overall rotation. $\vec{\nabla} \times \vec{v} \neq 0$ is accommodated in the BEC through the appearance of vortices. The phenomenon is observed in quantum liquids and well understood.[35]

Finally let's consider the behaviour of density perturbations as they enter the horizon (arena 1). Here too axion BEC differs from CDM. The CDM perturbations evolve linearly at all times. The axion BEC perturbations do not evolve linearly when they enter the horizon because the condensates which prevailed in neighboring horizon volumes rearrange themselves, through their gravitational interactions, into a new condensate for the expanded horizon volume. This produces local correlations between modes of different wavevector since the perturbation of wavevector $\vec{k}$, upon entering the horizon, is determined by the perturbations of wavevector say $\frac{1}{2}\vec{k}$ in its neighborhood. We propose this as a mechanism for the alignment of CMBR anisotropy multipoles[36] through the integrated Sachs-Wolfe (ISW) effect. Unlike CDM, the ISW effect is large in axion BEC because the Newtonian potential ψ changes entirely after entering the horizon in response to the rearrangement of the axion BEC.

References

1. M. A. Shifman, A. I. Vainshtein and V. I. Zakharov, Nucl. Phys. B **166**, 493 (1980).
2. J. E. Kim, Phys. Rev. Lett. **43**, 103 (1979)
3. A. P. Zhitnitskii, Sov. J. Nucl. **31**, 260 (1980); M. Dine, W. Fischler and M. Srednicki, Phys. Lett. B **104**, 199 (1981).
4. G. 't Hooft, Phys. Rev. Lett **37**, 8 (1976) and Phys. Rev. D **14**, 3422 (1976); R. Jackiw and C. Rebbi, Phys. Rev. Lett. **37**, 172 (1976); C. G. Callan, R. F. Dashen and D. J. Gross, Phys. Lett. B **63**, 334 (1976).
5. C. A. Baker et al., Phys. Rev. Lett. **97**, 131801 (2006).
6. M. Kobayashi and K. Maskawa, Progr. Theor. Phys. **49**, 652 (1973).
7. C. Vafa and E. Witten, Phys. Rev. Lett. **53**, 535 (1984).
8. R. D. Peccei and H. Quinn, Phys. Rev. Lett. **38**, 1440 (1977) and Phys. Rev. D **16**, 1791 (1977).
9. S. Weinberg, Phys. Rev. Lett. **40**, 223 (1978); F. Wilczek, Phys. Rev. Lett. **40**, 279 (1978).
10. Axion reviews include: J. E. Kim, Phys. Rep. **150**, 1 (1987); H.-Y. Cheng, Phys. Rep. **158**, 1 (1988); R. D. Peccei, in it CP Violation, ed. by C. Jarlskog, World Scientific Publ., 1989, pp 503-551; M. S. Turner, Phys. Rep. **197**, 67 (1990); G. G. Raffelt, Phys. Rep. **198**, 1 (1990).
11. D. Dicus, E. Kolb, V. Teplitz and R. Wagoner, Phys. Rev. D **18**, 1829 (1978) and Phys. Rev. D **22**, 839 (1980).
12. G. Raffelt and D. Dearborn, Phys. Rev. D **36**, 2211 (1987).
13. D. Dearborn, D. Schramm and G. Steigman, Phys. Rev. Lett. **56**, 26 (1986).
14. J. Ellis and K. Olive, Phys. Lett. B **193**, 525 (1987); G. Raffelt and D. Seckel, Phys. Rev. Lett. **60**, 1793 (1988); M. Turner, Phys. Rev. Lett. **60**, 1797 (1988); H.-T. Janka et al., Phys. Rev. Lett. **76**, 2621 (1996); W. Keil et al.. Phys. Rev. D **56**, 2419 (1997).
15. E.W. Kolb and M.S. Turner, *The Early Universe*, Addison Wesley, 1990.
16. E. Massó, F. Rota and G. Zsembinski, Phys. Rev. D **66**, 023004 (2002).
17. P. Sikivie, Lect. Notes Phys. **741**, 19 (2008).
18. J. Preskill, M. Wise and F. Wilczek, Phys. Lett. B **120**, 127 (1983); L. Abbott and P. Sikivie, Phys. Lett. B **120**, 133 (1983); M. Dine and W. Fischler, Phys. Lett. B **120**, 137 (1983).
19. A. Vilenkin, Phys. Rev. Lett. **48**, 59 (1982).
20. L.D. Duffy and K. van Bibber, arXiv:0904.3346.
21. P. Sikivie and J. Ipser, Phys. Lett. B **291**, 288 (1992).
22. P. Sikivie, Phys. Lett. B **432**, 139 (1998).
23. P. Sikivie, Phys. Rev. D **60**, 063501 (1999).
24. A. Natarajan and P. Sikivie, Phys. Rev. D **73**, 023510 (2006).
25. J.A. Fillmore and P. Goldreich, Ap. J. **281**, 1 (1984); E. Bertschinger, Ap. J. Suppl. **58**, 39 (1985).
26. P. Sikivie, I. Tkachev and Y. Wang, Phys. Rev. Lett. **75**, 2911 (1995); Phys. Rev. D **56**, 1863 (1997).
27. L.D. Duffy and P. Sikivie, Phys. Rev. D **78**, 063508 (2008).

28. W. Kinney and P. Sikivie, Phys. Rev. D **61**, 087305 (2000).
29. P. Sikivie, Phys. Lett. B **567**, 1 (2003).
30. P. Sikivie and Q. Yang, "Bose-Einstein Condensation of Dark Matter Axions," arXiv:0901.1106 [hep-ph].
31. D. V. Semikoz and I. I. Tkachev, Phys. Rev. Lett. **74**, 3093 (1995) and Phys. Rev. D **55**, 489 (1997). See also: S. Khlebnikov, Phys. Rev. A **66**, 063606 (2002) and references therein.
32. J. c. Hwang and H. Noh, Phys. Lett. B **680**, 1 (2009) [arXiv:0902.4738 [astro-ph.CO]].
33. L. M. Widrow and N. Kaiser, Ap. J. **416**, L71 (1993).
34. P. J. E. Peebles, Ap. J. **155**, 393 (1969).
35. C. J. Pethik and H. Smith, *Bose-Einstein Condensation in Dilute Gases*, Cambridge University Press, 2002.
36. M. Tegmark, A. de Oliveira-Costa and A. Hamilton, Phys. Rev. D **68**, 123523 (2003); A. de Oliveira-Costa, M. Tegmark, M. Zaldarriaga and A. Hamilton, Phys. Rev. D **69**, 063516 (2004); C. J. Copi, D. Huterer, D. J. Schwarz and G. D. Starkman, MNRAS **367**, 79 (2006) and references therein.

Searching for Axions and Other Exotics with Dark Matter Detectors

Maxim Pospelov

Perimeter Institute for Theoretical Physics, Waterloo, ON, N2J 2W9, Canada

I consider models of light super-weakly interacting cold dark matter, with $\mathcal{O}(10)$ keV mass, focusing on bosonic candidates such as pseudoscalars and vectors. I analyze the cosmological abundance, the γ-background created by particle decays, the impact on stellar processes due to cooling, and the direct detection capabilities in order to identify classes of models that pass all the constraints. In certain models, variants of photoelectric (or axioelectric) absorption of dark matter in direct-detection experiments can provide a sensitivity to the superweak couplings to the Standard Model which is superior to all existing indirect constraints. In all models studied, the annual modulation of the direct-detection signal is at the currently unobservable level of $O(10^{-5})$.

1. Introduction

This talk is based on my recent paper written in collaboration with Adam Ritz and Mikhail Voloshin.[1] Before I proceed, I would like to congratulate Misha Shifman with his remarkable, successful, long and productive career in theoretical physics, and wish him all the best for the future.

The evidence for the existence of non-baryonic dark matter (DM) now comes from many sources and ranges over many distance scales,[2] from the rotation curves of galaxies, the dynamics of clusters, lensing data and the characteristics of large-scale structure, to the features of the cosmic microwave background (CMB) fluctuation spectrum and the success of big bang nucleosynthesis (BBN). All of these pieces of astronomical data point to a similar cosmological density of dark matter, several times that of visible baryonic matter. However, this data only probes the gravitational interaction of dark matter and, while it presents us with one of the most compelling arguments for physics beyond the Standard Model, gaining insight into its non-gravitational interactions remains a primary experimental focus, both through underground detectors, particle colliders, and the observation of photon and neutrino fluxes from overdense regions in the galaxy and beyond.

This lack of any direct information on how dark matter may couple to the Standard Model (SM) means we are forced to rely on various theoretical expectations. In particular, the many successes of standard cosmology motivate a simple thermal mechanism for populating the universe with dark matter with a well-defined freeze-out abundance as the universe expands. This in turn requires a specific annihilation cross-section which is necessarily non-gravitational in origin. The fact that a weakly interacting particle with a weak-scale mass has an annihilation rate in the right ballpark,[2,3] combined with our expectations for new physics at the electroweak scale, has rightly led to the prevailing WIMP paradigm for cold dark matter. Nonetheless, persistent problems in understanding the small-scale gravitational clustering properties in cold dark matter simulations and on galactic scales has motivated variants of this picture where the dark matter may be somewhat lighter, with masses down to the keV range. Masses in this range imply a super-weak interaction strength between dark matter and the SM sector, indeed many orders of magnitude below weak-scale cross sections. This follows from the necessity to have early thermal decoupling of the DM sector, prior to the electroweak epoch at $T \sim 100$ GeV, in order to satisfy the conflicting requirements of not having too much energy density in dark matter, and the strong lower bounds on m_{DM} coming from the analysis of structure formation. If early decoupling can be achieved, then masses in the keV range may withstand these combined constraints and at the same time provide a rather attractive mechanism for ensuring the correct dark matter energy density.

An important feature of keV-scale dark matter is that, unlike the majority of electroweak scale WIMP models, it need not be stable against decays to light SM degrees of freedom, *e.g.* photons and neutrinos. Given the super-weak strength of its interaction such decays may be strongly inhibited, but nevertheless keV dark matter can be emitted or absorbed in astrophysical environments and in terrestrial experiments, and thus is possibly subject to additional constraints. The main question to be studied in this paper is whether the application of cosmological and astrophysical constraints leaves any realistic chance for the direct detection of such keV-scale candidates. The best studied models in this class are the sterile neutrino and gravitino, examples of fermionic superWIMPs,[4,5] where it is well-known that there are no chances for direct detection. In contrast, the present paper studies bosonic superWIMP models, which need not entail such a pessimistic conclusion. In particular, we show that the potential sensitivity of direct detection experiments to bosonic keV dark matter can indeed be competitive with astrophysical bounds as well as with the lifetime and cosmic gamma-background constraints.

The majority of underground direct-detection experiments are specifically tuned to the nuclear recoil energy regime and all γ- or β-like events are typically considered as background. Currently, the experiments with the best limits are now sensitive to picobarn-scale cross-sections for WIMP-nucleus elastic scattering, when the WIMP mass is on the order of the electroweak scale. To illustrate our main point, we present a simple estimate that compares the rate of absorption of 10 keV-mass axion-like dark matter (a-particles), coupled to electrons with strength $m_e/f_a = m_e(10^{10}\text{ GeV})^{-1}$, with the recoil signal from a 1 TeV WIMP that has an elastic scattering cross section on the nucleus of $\sigma_{el} \sim 10^{-36}\text{cm}^2$:

$$\frac{\sigma_{abs} v_{\rm DM} n_a}{\sigma_{el} v_{\rm DM} n_{\rm WIMP}} \sim \frac{m_{\rm WIMP}}{m_a} \times \frac{c}{v_{\rm DM}} \times \frac{(10^{10}\,{\rm GeV})^{-2}}{10^{-36}\,{\rm cm}^2} \sim 10^8 \times 10^3 \times 10^{-11} \sim O(1). \tag{1}$$

For this rather crude estimate we simply took $\sigma_{abs} v_{\rm DM} \sim f_a^{-2}$, and $m_a n_a \sim m_{\rm WIMP} n_{\rm WIMP} \sim \rho_{DM}$. Although the superWIMP absorption cross section is orders of magnitude below the weak cross section, it is compensated by the tremendous gain in the local number density of dark matter particles, and by the fact that the inelastic cross section σ_{abs} scales inversely with the dark matter velocity. Although rather imprecise, the estimate (1) illustrates that the absorption of 10 keV dark matter can indeed produce signals that are well within modern detection capabilities. Moreover, the scale $f_a \sim 10^{10}$ GeV is at or above the limits imposed by even the most stringent constraints on star cooling. We believe that this is an important point, and the sensitivity to new and viable dark matter scenarios can (and should) be explored by direct detection experiments. For comparison, fermionic candidates in the superWIMP class, a well-studied example of which is sterile neutrino dark matter, do not fall into the class covered by the estimate (1) for many different reasons; the primary one being that sterile neutrinos are not fully absorbed but rather converted to active neutrinos that carry away most of the rest energy.

Thus far, the issue of direct detection of keV dark matter has only been discussed by the DAMA collaboration[6] [a] in connection with an annual modulation signal initially reported by DAMA/NaI[8] and recently confirmed by

[a] As this paper was being prepared for publication, another experimental collaboration, CoGeNT,[7] reported their results on the absorption of keV-scale superWIMPs.

DAMA/Libra.[9] If indeed the annual modulation signal of DAMA were to be attributed to the absorption of keV dark matter, which would cause ionization but no significant recoil, it could provide a plausible explanation for why other front-running experiments such as CDMS and Xenon[11,12] see no signal. This was the main point of Ref. [6], and in this paper we provide a re-analysis of this possibility, reaching instead a negative conclusion: the models that were presented in [6] as explaining the modulation are in fact ruled out either by lifetime arguments, by astrophysical constraints, or directly by the large unmodulated counting rates in underground detectors.

To make the discussion sufficiently general, we focus on two generic possibilities for bosonic superWIMP dark matter: pseudoscalars and vectors.

2. Pseudoscalar Super-WIMPs

We start with the pseudoscalars a, and write the interactions as a combination of several derivative-like operators of dimension five:

$$\mathcal{L}_{\text{int}} = \frac{C_\gamma a}{f_a} F_{\mu\nu}\tilde{F}^{\mu\nu} - \frac{\partial_\mu a}{f_a}\bar{\psi}\gamma^\mu\gamma^5\psi + \cdots \tag{2}$$

where $F_{\mu\nu}$ and ψ are the electromagnetic field strength and the Dirac field of the electrons, and the ellipsis denotes possible interactions with other fermions and gauge bosons, and for simplicity we shall assume a similar strength for the a-SM couplings in those sectors. Notice that the other possible pseudoscalar coupling to the electron, $a\bar{\psi}i\gamma^5\psi$, can always be decomposed into the two operators in (2) using the equations of motion once we account for the chiral anomaly. While the dimensionful coupling f_a does regulate the overall strength of the SM-a interaction, the dimensionless coupling to photons C_γ is crucial for determining the lifetime and γ-background created by a decay. Restricting ourselves to the electron-photon sector, we expect three generic possibilities for the size of the coupling C_γ:

$$\begin{aligned} \text{A}: &\quad C_\gamma \sim \frac{\pi}{\alpha}, \\ \text{B}: &\quad C_\gamma = \frac{\alpha}{4\pi}, \\ \text{C}: &\quad C_\gamma \sim \frac{\alpha}{\pi} \times \frac{m_a^2}{m_e^2}. \end{aligned} \tag{3}$$

Case A corresponds to a pseudoscalar coupled to photons at some UV normalization scale, with couplings to electrons generated radiatively. Having normalized the electron coupling to $1/f_a$ in (2), the coupling $C_\gamma \gg 1$ in this case. Case C is the inverse of Case A. The derivative coupling to the electron axial-vector current may only lead to the $F_{\mu\nu}\tilde{F}^{\mu\nu}\partial^2 a$ operator at loop level, hence the $(m_a/m_e)^2$ suppression. Finally, Case B is intermediate, when a is initially coupled to the fermion via the $m\bar{\psi}i\gamma_5\psi$ pseudoscalar operator. Clearly any of these three choices can be realized without fine tuning given an appropriate UV completion. An additional advantage of this model is the automatic protection of the pseudoscalar mass against radiative corrections, exactly as in the conventional axion case. As we are going to see later, only option C allows for the possibility of keV-scale dark matter without imposing overly strong constraints on the size of f_a.

Given an a-boson (or s-boson) mass below the electron threshold, the decays will be almost exclusively to photons, and mediated by the 2-photon interaction in (2), with the appropriate low energy value of f_a. The decay is shown in Fig. 1 and the width is

$$\Gamma_{a\to 2\gamma} = \frac{C_\gamma^2}{4\pi f_a^2}\, m_a^3\,. \tag{4}$$

Requiring that the dark matter lifetime be at least the age of the universe implies,

$$\tau_{\rm U}\Gamma_{a\to 2\gamma} \lesssim 1 \quad\Longrightarrow\quad C_\gamma^2 \leq 2\times 10^{-6}\times\left(\frac{f_a}{10^{10}\,{\rm GeV}}\right)^2\times\left(\frac{10\,{\rm keV}}{m_a}\right)^3, \tag{5}$$

which we see is already a significant constraint for Case A in particular, and also for Case B if we stick with fiducial values for f_a and m_a. In contrast, Case C is less constrained. However, we also need to consider the γ-background created by a-decays and, as we will see in the next section, this provides a more stringent constraint.

For the derivation of the cosmological abundance and various direct and indirect limits I refer the readers to the original paper [1]. Here I only highlight the following approximate relation between the cross sections for axion absorption and the photoelectric effect:

$$\frac{\sigma_{abs}v}{\sigma_{photo}(\omega = m_a)c} \simeq \frac{3\,m_a^2}{4\pi\,\alpha\,f_a^2}\,. \tag{6}$$

Notice that ratio (6) remains finite even in the limit of $v\to 0$. For the purposes of an experimental search for pseudoscalar dark matter, it is useful to express (6) directly as a counting rate in a detector consisting of a single atomic species of atomic mass A,

$$R \simeq \frac{1.2\times 10^{19}}{A}\, g_{aee}^2\left(\frac{m_a}{\rm keV}\right)\left(\frac{\sigma_{photo}}{\rm bn}\right)\,{\rm kg}^{-1}{\rm day}^{-1}, \tag{7}$$

where we used the local dark matter density, $\rho_{DM} = 0.3$ GeV cm^{-3}, and introduced the dimensionless coupling $g_{aee} = 2m_e/f_a$ to allow for direct comparison with the existing results of DAMA[6,9] and the most recent experimental paper by the CoGeNT collaboration [7].

For completeness, we will also quote the axioelectric cross section for massless axions with energy of $\mathcal{O}$(keV), as this result is in widespread use for the solar physics of axions. Taking $\omega_a \ll m_e$, and $m_a \to 0$, we arrive at the following relation between the photo- and axio-electric cross sections:

$$\left.\frac{\sigma_{abs}}{\sigma_{photo}(\omega = \omega_a)}\right|_{m_a\to 0} \simeq \frac{\omega_a^2}{2\pi\,\alpha\, f_a^2}\,. \tag{8}$$

We observe that this result is twice as large as the formula regularly quoted in the literature (see *e.g.* [17]). The source of the discrepancy with previous calculations can be traced to the omission of the $\gamma_0\gamma_5$ contribution (see [1] for details), and this corrected formula may prove useful for solar axion searches.

Despite a reasonably sensitive counting rate (7), the constraints from the diffuse gamma ray background prove to be much stronger, and there is little prospect that modest improvements in direct pseudoscalar dark matter detection could compete with indirect constraints.[1]

3. Vector Super-WIMPs

Finally, we introduce a model of keV-scale vector dark matter. We choose the initial Lagrangian in the form identical to that studied in [10], where an extra U(1)′ gauge field is coupled to the SM via kinetic mixing with the hypercharge field strength,

$$\mathcal{L} = -\frac{1}{4}V_{\mu\nu}^2 - \frac{\kappa}{2}V_{\mu\nu}F_{\mu\nu} + \mathcal{L}_{h'} + \mathcal{L}_{\rm dim>4}\,, \tag{9}$$

with $\mathcal{L}_{h'}$ encoding the physics responsible for breaking the U(1)′, and $\mathcal{L}_{\rm dim>4}$ includes possible non-renormalizable higher-dimension interaction terms such as $H^\dagger H V_{\mu\nu}^2$, $V_{\mu\nu}^2 F_{\alpha\beta}^2$, etc. After the breaking of this secluded U(1)′, the model takes the simplest possible form,

$$\mathcal{L} = -\frac{1}{4}V_{\mu\nu}^2 + \frac{1}{2}m_V^2 V_\mu^2 + \kappa V_\nu \partial_\mu F_{\mu\nu}, + \cdots\,, \tag{10}$$

where we retained only relevant and marginal operators, and suppressed the U(1)′ Higgs sector. This is one of the simplest UV-complete extensions of the SM, and it has been addressed in connection with electroweak-scale physics on a number of occasions.[10,13] Most recently a study of cosmology in this model was performed for small sub-eV values of m_V.[14] For vanishingly small values of m_V^2, the extra sector decouples as mixing with the photon can be reabsorbed into the mass term. This leads to an additional suppression of γ-V conversion at temperatures much in excess of m_V,[14] but for keV-scale dark matter this issue is often less important, and (10) can simply be traded for

$$\mathcal{L} = -\frac{1}{4}V_{\mu\nu}^2 + \frac{1}{2}m_V^2 V_\mu^2 + e\kappa V_\nu \bar{\psi}\gamma_\mu\psi + \cdots, \tag{11}$$

where V_μ couples to the electromagnetic current. For convenience, we introduce an analogue of the electromagnetic coupling and of the fine structure constant,

$$e' = e\kappa, \qquad \alpha' = \frac{(e\kappa)^2}{4\pi}, \tag{12}$$

that necessarily appear in all rates for the emission or absorption of U(1)′ vectors by SM particles.

Obviously, the mass of the $U'(1)$ gauge boson is protected by gauge symmetry, and the question of naturalness is relegated to the corresponding U(1)′ Higgs sector. However, given the smallness of the couplings that we are going to consider, *e.g.* $\kappa \sim O(10^{-10})$, the naturalness problem is no more severe than for the SM Higgs.

The coupling of V to neutrinos is also possible via mixing with the Z-boson. This mixing, however, is further suppressed by a factor of $(m_a/M_Z)^2 \sim 10^{-14}$, and we will disregard it in the analysis of the model. Other realizations of U(1)′,[15,16] such as a gauged version of $B - L$, would allow couplings to neutrinos of the same size as the couplings to electrons.

A remarkable property of this model is the long lifetime of the vector boson in the limit $m_V \ll 2m_e$. Direct calculation of the loop-induced three-gamma decay of V gives be

$$\Gamma = \frac{17\,\alpha^3\alpha'}{2^7 3^6 5^3 \pi^3}\frac{m_V^9}{m_e^8} \approx \left(4.70\times 10^{-8}\right)\alpha^3\alpha'\,\frac{m_V^9}{m_e^8}. \tag{13}$$

An immediate application of this result is to provide a constraint on the masses and couplings such that the dark matter lifetime is at least equal to the age of the universe. We find

$$\tau_{\rm U}\Gamma_{V\to 3\gamma} \lesssim 1 \quad\Longrightarrow\quad m_V\,(\alpha')^{1/9} \lesssim 1\,\text{keV}. \tag{14}$$

Even if the lifetime constraint is satisfied, the γ background created by V decays can be detected as a diffuse cosmological γ background, or as an extra contribution to the γ background in our galaxy.

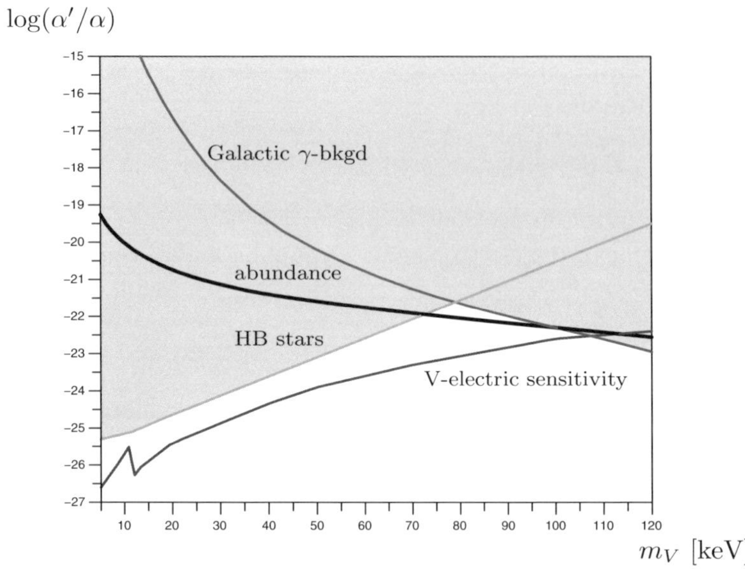

Fig. 1. We plot the direct detection sensitivity to vector DM from the V-electric cross-section on Ge, assuming a fiducial sensitivity of the detector equivalent to a 1pb cross-section for a 100 GeV WIMP. We also show the constraints from the He-burning lifetime in HB stars, and the γ-background from 3γ-decays in the Galaxy. The grey shaded region is again excluded by the indirect constraints, while the thick black line corresponds to the parameters required to reproduce the required dark matter abundance from thermal production.

Omitting again the details of the calculation of the abundance etc, I quote the absorption cross section in a dark matter detector as the function of M_V and α':

$$\frac{\sigma_{abs} v}{\sigma_{photo}(\omega = m_V)c} \simeq \frac{\alpha'}{\alpha}. \tag{15}$$

Converting this cross section into a counting rate gives

$$R \simeq \frac{4 \times 10^{23}}{A} \frac{\alpha'}{\alpha} \left(\frac{\text{keV}}{m_V}\right) \left(\frac{\sigma_{photo}}{\text{bn}}\right) \text{kg}^{-1}\text{day}^{-1}. \tag{16}$$

The summary of various constraints are given in Fig. 1. It turns out that the results of direct dark matter searches can easily compete with indirect constraints on vector super-WIMPs.

4. Suppressed modulation of the super-WIMP counting rate

In principle, the absorption of bosonic superWIMPs could also generate an annual modulation of the signal.[6] This idea, however, immediately runs aground because inelastic cross sections are typically inversely proportional to the velocity of the incoming nonrelativistic particles . In combination with the flux, this renders the counting rate independent of the velocity, and therefore un-modulated by the Earth's motion at the experimentally relevant percent level. We find that this is indeed the case for both classes of models considered in this paper, pseudoscalar and vector superWIMP dark matter, with rates given in (7) and (16). We can nonetheless estimate the degree of modulation in the absorption, noting that the dark matter velocity enters via $\exp(i\mathbf{kr})$, which in the Born approximation can be combined with $\exp(i\mathbf{pr})$ representing the wavefunction of the outgoing photo-electron. Therefore, modulation of the counting rate arises primarily due to the modulation of p. Since $k \ll p$ and the cross section is generally a smooth function of energy, we estimate that

$$\text{Modulated absorption} \sim \frac{\Delta k}{p} \sim \frac{m_{a(V)}\Delta v_{DM}}{\sqrt{m_e m_{a(V)}}c} \sim \frac{\Delta v_{DM}}{c} \times \sqrt{m_{a(V)}/m_e}. \tag{17}$$

For the characteristic energy range of the DAMA signal, $m_{a(V)} \sim 5$ keV, the modulation does not exceed 10^{-5}. One could potentially worry that $m_{a(V)}$ may turn out to be *exactly* equal to some ionization threshold where the assumption of a smooth cross section as a function of energy breaks down, and the momentum of the photoelectron is suppressed relative to its natural value. In this case the modulated part of the cross section can be enhanced, but not by the four orders of magnitude required to explain the DAMA signal. This is because the detector comprises many-electron atoms and unmodulated absorption by other shells would become important. From Eq. (17), one immediately concludes that the reported DAMA modulation signal cannot be explained by the absorption of axion-like or vector-like particles. Specifically, if the couplings are tuned in such a way that the modulated part of the axio-electric effect matches the DAMA signal, the unmodulated part will be in excess of the total number of events that DAMA observes by four orders of magnitude.

5. Concluding remarks

With vast resources now justifiably being devoted toward the direct detection of dark matter, it appears all but clear that the scientific scope of these searches should be diversified. While a characteristic elastic WIMP-nucleus recoil may remain the main benchmark scenario for these experiments, the detection possibilities for other generic classes of dark matter should certainly be exploited. In this paper, we have shown that bosonic superWIMPs represent a legitimate and feasible target.

The models analyzed in this paper can naturally produce the required relic abundance of dark matter, once the couplings α' or $(m_e/f_a)^2$ are in the superweak $\sim O(10^{-20})$ range. The feeble nature of this coupling to the SM is partially overcome by a factor of $(c/v_{DM}) \times (m_{\rm WIMP}/m_{a(V)})$ which enhances superWIMP absorption relative to WIMP scattering and renders direct detection feasible. We have analyzed models that do not require any special fine-tunings and enjoy protection for the small mass scales due to the symmetries of their interactions: gauge symmetry for the model of secluded $\mathrm{U}(1)'$, and shift symmetry for the pseudoscalar model.

The result of our analysis has revealed that for the pseudoscalar model, direct detection sensitivity may compete with the red giant, solar and SN constraints. However, we found that the model is most strongly constrained by limits on monochromatic X-ray lines in our galaxy, that rule out much of the interesting regions of the parameter space. In vector models of dark matter, the production of X-rays is strongly inhibited by gauge invariance, and for the model studied here direct searches for ionization are apparently capable of probing the most interesting range of the coupling–mass parameter space, namely that consistent with the observed dark matter energy density. Only part of this range is constrained by indirect astrophysical probes, namely the low mass region where stellar energy loss constraints become important. Finally, contrary to some existing claims in the literature, we have found that the absorption of superWIMPs does not lead to an annual modulation of the ionization signal at a level that would be of experimental interest.

References

1. M. Pospelov, A. Ritz and M. B. Voloshin, Phys. Rev. D **78**, 115012 (2008) [arXiv:0807.3279 [hep-ph]].
2. *see e.g.* G. Jungman, M. Kamionkowski and K. Griest, Phys. Rept. **267**, 195 (1996) [arXiv:hep-ph/9506380]; G. Bertone, D. Hooper and J. Silk, Phys. Rept. **405**, 279 (2005) [arXiv:hep-ph/0404175].
3. B. W. Lee and S. Weinberg, Phys. Rev. Lett. **39**, 165 (1977); M. I. Vysotsky, A. D. Dolgov and Y. B. Zeldovich, JETP Lett. **26**, 188 (1977) [Pisma Zh. Eksp. Teor. Fiz. **26**, 200 (1977)].

4. S. Dodelson and L. M. Widrow, Phys. Rev. Lett. **72**, 17 (1994) [arXiv:hep-ph/9303287]; A. D. Dolgov and S. H. Hansen, Astropart. Phys. **16**, 339 (2002) [arXiv:hep-ph/0009083].
5. J. R. Ellis, D. V. Nanopoulos and S. Sarkar, Nucl. Phys. B **259**, 175 (1985); J. R. Ellis, G. B. Gelmini, J. L. Lopez, D. V. Nanopoulos and S. Sarkar, Nucl. Phys. B **373**, 399 (1992); J. L. Feng, A. Rajaraman and F. Takayama, Phys. Rev. D **68**, 063504 (2003) [arXiv:hep-ph/0306024].
6. R. Bernabei *et al.*, Int. J. Mod. Phys. A **21**, 1445 (2006) [arXiv:astro-ph/0511262].
7. C. E. Aalseth *et al.*, arXiv:0807.0879 [astro-ph].
8. R. Bernabei *et al.* [DAMA Collaboration], Phys. Lett. B **480**, 23 (2000).
9. R. Bernabei *et al.* [DAMA Collaboration], arXiv:0804.2741 [astro-ph].
10. M. Pospelov, A. Ritz and M. B. Voloshin, Phys. Lett. B **662**, 53 (2008) [arXiv:0711.4866 [hep-ph]].
11. D. S. Akerib *et al.* [CDMS Collaboration], Phys. Rev. Lett. **96**, 011302 (2006) [arXiv:astro-ph/0509259]; Z. Ahmed *et al.* [CDMS Collaboration], Phys. Rev. Lett. **102**, 011301 (2009) [arXiv:0802.3530 [astro-ph]].
12. J. Angle *et al.* [XENON Collaboration], Phys. Rev. Lett. **100**, 021303 (2008) [arXiv:0706.0039 [astro-ph]].
13. W. F. Chang, J. N. Ng and J. M. S. Wu, Phys. Rev. D **74**, 095005 (2006) [arXiv:hep-ph/0608068]; W. F. Chang, J. N. Ng and J. M. S. Wu, Phys. Rev. D **75**, 115016 (2007) [arXiv:hep-ph/0701254].
14. J. Jaeckel, J. Redondo and A. Ringwald, Phys. Rev. Lett. **101**, 131801 (2008) [arXiv:0804.4157 [astro-ph]].
15. P. Fayet, Nucl. Phys. B **347**, 743 (1990).
16. P. Langacker, arXiv:0801.1345 [hep-ph].
17. F. T. Avignone *et al.*, Phys. Rev. D **35**, 2752 (1987); S. Dimopoulos, G. D. Starkman and B. W. Lynn, Phys. Lett. B **168**, 145 (1986).

Unparticle Physics

Howard Georgi

The Physics Laboratories
Harvard University
17 Oxford St.
Cambridge, MA 02138
georgi@physics.harvard.edu

Misha Shifman knows much more about conformal symmetry than I do, but I have been thinking hard about it, and I thought it would be fun to present some of my ideas as a little birthday present.

1. Happy Birthday Misha

It is a pleasure to help celebrate Misha Shifman's birthday. As he and his colleagues know very well, I have the greatest respect for what Misha and his group have accomplished over many, many years.

I am a little embarrassed to give this talk in front of Misha, because I have lately gotten interested is a rather technical mathematical subject in 1+1 dimensional QFT, and Misha surely knows much more about this than I do. But it has been fun, and I have had the help of two outstanding graduate students, Yevgeny Kats (who is finishing up a PhD) and Aqil Sajjad (who is just starting), so I think it may be worth telling the story as a little gift to Misha.

I used to be a particle physicist. In fact, I had a wonderful time as a particle physicist for about 40 years, ever since a teaching fellow pointed me to Feynman's "Theory of Fundamental Processes" to keep me from being bored and obnoxious in my first physics course at Harvard. I loved (and still love) the elegance and certainty of particle physics.

But for the last few years, I have been an Unparticle Physicist. I got fascinated by the idea that there might be stuff in our world that was not particles. I had always known vaguely that scale invariant field theories with nontrivial anomalous dimensions had no obvious particle interpretation, But I hadn't realized how strange this was. The mathematics is powerful and beautiful. It is only when you think about how this kind of physics might exist in our world that it becomes confusing. As soon as I thought carefully about this, I realized that there was a very precise sense in which such stuff might coexist with the ordinary particles of the standard model. This is the scenario that I call unparticle physics. Since the original formulation of unparticle physics [1,2] as an effective field theory in which the standard model couples only at high energies to a "Banks-Zaks" [3] sector that is scale-invariant with anomalous dimensions at low energies, this idea has been explored both theoretically and phenomenologically in many works and I won't have time to discuss them here (see [4] for a recent review). Here I want to talk about some theoretical issues surrounding unparticle self-interactions and illustrate them in a simple 1+1 dimensional model.[5]

This has also been a trip down memory lane. In the early 60's, my thesis adviser, Charlie Sommerfield solved exactly 1+1 electrodynamics with an explicit photon mass - that is the Schwinger model [6] with an additional mass term for the vector boson. I call this the Sommerfield model,[7–11]

$$\mathcal{L} = \overline{\psi}\,(i\,\partial\!\!\!/ - e\,A\!\!\!/)\,\psi - \frac{1}{4}F^{\mu\nu}F_{\mu\nu} + \frac{m_0^2}{2}A^{\mu}A_{\mu}\,. \tag{1}$$

My student Yevgeny Kats and I got interested in this ancient model a couple of years ago when we realized that it becomes scale-invariant at low energies, and (unlike the Schwinger model) has fractional anomalous dimensions. At low energies, below the vector boson mass, the theory reduces to the Thirring model, that is the theory of a fermion with a quartic self-interaction which is scale-invariant at all energies.[12–14] Thus this is a nice toy example of a Banks-Zaks model.

We assume that the very high energy theory includes the interaction

$$\begin{aligned}\mathcal{L}_{\rm int} &= \frac{\mu}{2}\left[\overline{\psi}(1+\gamma_5)\chi\,\phi^* + \overline{\psi}(1-\gamma_5)\chi\,\phi\right] + \text{h.c.}\\ &= \mu\,(\psi_2^*\,\chi_1\,\phi^* + \psi_1^*\,\chi_2\,\phi) + \text{h.c.}\end{aligned} \tag{2}$$

that couples the fermion ψ of the Banks-Zaks sector to a neutral complex scalar ϕ with mass $m_\phi \ll m$ that plays the role of a standard model field. The interaction is mediated by the heavy fermion χ with mass $M \gg m,\ \mu^2/m$ and the same coupling to A^μ as ψ.

The interaction preserves a global $U(1)$ symmetry with charge +1 for

ϕ^* and ψ_1^* and charge -1 for ψ_2^*. Integrating out χ we obtain[a]

$$\mathcal{L}_{\rm int} = \frac{h}{2}\left(\mathcal{O}\,\phi^{*2} + \mathcal{O}^*\phi^2\right), \qquad h \equiv \frac{2\mu^2}{M}, \tag{3}$$

where the composite operator

$$\mathcal{O} \equiv \overline{\psi}\,\tfrac{1}{2}\left(1+\gamma^5\right)\psi = \psi_2^*\psi_1 \tag{4}$$

(that has charge -2 under the global $U(1)$ symmetry) will have a fractional anomalous dimension at low energies.

Since this symmetry acts chirally on the fermionic field one may be concerned about anomalies. Indeed, the axial current $j^{5\mu} = \overline{\psi}\gamma^\mu\gamma^5\psi$ has a non-zero divergence,[15] and in the Schwinger model the symmetry transformation gives rise to an additional term in the Lagrangian proportional to

$$\mathcal{L}_\theta \propto \partial^\mu j^5_\mu = -\frac{e}{\pi}\epsilon^{\mu\nu}\partial_\mu A_\nu = -\frac{e}{2\pi}\epsilon^{\mu\nu}F_{\mu\nu}\,. \tag{5}$$

While this term is a total derivative, it cannot be eliminated in the Schwinger model due to instanton configurations in Euclidean space that do not decay fast enough at infinity and for which $\int d^2x\,\mathcal{L}_\theta$ does not vanish.[16] This leads to the existence of degenerate θ vacua[17,18] that differ in the value of the symmetry-breaking condensate $\langle\mathcal{O}\rangle \propto e^{i\theta}$. But in the Sommerfield model (unlike the Schwinger model that has gauge invariance) we can define a conserved axial current as

$$\overline{j}^{5\mu} = j^{5\mu} + \frac{e}{\pi}\epsilon^{\mu\nu}A_\nu\,. \tag{6}$$

Then, since a (non-anomalous) symmetry cannot be broken spontaneously in two dimensions,[19] there is no condensate. Note also that a non-vanishing condensate of an operator with a non-zero dimension would be in conflict with scale invariance. In the Schwinger model, the dimension of $\mathcal{O}$ at the IR fixed point is 0, but we will see that this is not the case in the Sommerfield model.

The vertices in (2) that couple the standard model field ϕ to the fermions of the Sommerfield model involve also the heavy field χ. Since χ effectively propagates over distances of order $1/M$ which are much shorter than the distances we want to consider, our effective interactions will involve two

[a] In a 4D unparticle theory, the interaction corresponding to (3) would typically be nonrenormalizable.

ϕs coupling to the leading operator in the operator product expansion (OPE) [13,20] of a product of two ψs. In particular, we will be interested in

$$\mathrm{T}\psi_\beta^*(x_2)\psi_\alpha(x_1) = c(x_2 - x_1)\,\psi_\beta^*\psi_\alpha(x_2) + \cdots \qquad (\alpha \neq \beta) \tag{7}$$

that defines the operator $\mathcal{O} = \psi_2^*\psi_1$ from (4) and its conjugate $\mathcal{O}^* = \psi_1^*\psi_2$. Using this OPE and the exact solution for the fermionic correlation functions, we can write down the correlation functions of $\mathcal{O}$ and $\mathcal{O}^*$. The coefficient function $c(x_2 - x_1)$ is relevant only to the normalization of the $\mathcal{O}$ and $\mathcal{O}^*$ operators, and will not play an important role.

The result for the 2-point function is

$$i\Delta_{\mathcal{O}}(x) \equiv \langle 0|\mathrm{T}\mathcal{O}(x)\,\mathcal{O}^*(0)|0\rangle = \frac{C(x)^4}{(2\pi)^2\,(-x^2 + i\epsilon)} \tag{8}$$

where $C(x)$ arises from the exact solution,

$$\begin{aligned} C(x) &= \exp\left[i\frac{e^2}{m^2}\left[(\Delta(x) - \Delta(0)) - (D(x) - D(0))\right]\right] \\ &= \exp\left[\frac{e^2}{2\pi m^2}\left[K_0\left(m\sqrt{-x^2 + i\epsilon}\right) + \ln\left(\xi m\sqrt{-x^2 + i\epsilon}\right)\right]\right] \end{aligned} \tag{9}$$

with

$$\xi = \frac{e^{\gamma_E}}{2} \tag{10}$$

where $\Delta(x)$ and $D(x)$ are massive and massless scalar propagators, and K_0 (which comes from $\Delta(x)$) is the modified Bessel function of the second kind. For $y \to \infty$, $K_0(y) \to \sqrt{\frac{\pi}{2y}}\,e^{-y} \to 0$. For $y \to 0$, $K_0(y) \to -\ln(\xi y)$ which cancels the log term so that in the short-distance limit, $(x_i - x_j)^2 \ll 1/m^2$, $C(x) \to 1$ and one obtains free-fermion behavior. In the large-distance limit, $(x_i - x_j)^2 \gg 1/m^2$, K_0 does not contribute, and $C(x)$ is just a power of x^2:

$$C(x) \to \left[(\xi m)^2(-x^2 + i\epsilon)\right]^{e^2/4\pi m^2} \tag{11}$$

leading to scale-invariant behavior with fractional anomalous dimensions. This is the unparticle regime. Note that m plays the role of $\Lambda_{\mathcal{U}}$ from [1].

At distances large compared to $1/m$, using (11), we see that the 2-point function is proportional to an unparticle propagator[b]

$$i\Delta_{\mathcal{O}}(x) \;\to\; i\Delta_{\mathcal{U}}(x) = \frac{1}{(2\pi)^2(\xi m)^{2a}\,(-x^2 + i\epsilon)^{1+a}} \tag{12}$$

[b] Here and below, we incorporate a dimensional factor of $1/(\xi m)^{2a}$ in the unparticle propagator so that it has the same engineering dimension as the $\mathcal{O}$ propagator.

where

$$a \equiv -\frac{e^2}{\pi m^2} = -\frac{1}{1+\pi m_0^2/e^2} \tag{13}$$

denotes the anomalous dimension of the composite operator $\mathcal{O} \equiv \psi_2^* \psi_1$ (its total dimension is $d_{\mathcal{U}} = 1 + a$). For $0 < m_0 < \infty$, a is fractional, which leads to unparticle behavior.[c] In momentum space[d]

$$i\Delta_{\mathcal{U}}(p) = \frac{iA(a)(-p^2 - i\epsilon)^a}{2(\xi m)^{2a}\sin(\pi a)} = \frac{A(a)}{2\pi(\xi m)^{2a}} \int_0^\infty dM^2 \left(M^2\right)^a \frac{i}{p^2 - M^2 + i\epsilon} \tag{14}$$

where

$$A(a) \equiv -\frac{\sin(\pi a)\,\Gamma(-a)}{2^{1+2a}\,\pi\Gamma(1+a)} = \frac{1}{2^{1+2a}\left[\Gamma(1+a)\right]^2}\,. \tag{15}$$

Since

$$\operatorname{Im}\Delta_{\mathcal{U}}(p) = -\frac{A(a)}{2(\xi m)^{2a}}\theta(p^2)\left(p^2\right)^a \tag{16}$$

the unparticle phase space is

$$\Phi_{\mathcal{U}}(p) = \frac{A(a)}{(\xi m)^{2a}}\left(p^2\right)^a \theta(p^0)\,\theta(p^2)\,. \tag{17}$$

Note that in the Schwinger model limit the phase space vanishes since $A(-1) = 0$.

Because we have the exact solution, we can write (8) for arbitrary x as

$$i\Delta_{\mathcal{O}}(x) = i\Delta_{\mathcal{U}}(x)\exp\left[-4\pi i a\Delta(x)\right] = i\Delta_{\mathcal{U}}(x)\sum_{n=0}^{\infty}\frac{(-4\pi a)^n}{n!}\left[i\Delta(x)\right]^n\,. \tag{18}$$

At distances not large compared to $1/m$, the higher terms in the sum in (18) become relevant. This will be important for studying the transition between the unparticle and particle regime.

The Sommerfield model is trivial, without any real interactions. However, the unparticle model, including the couplings to our toy standard model, is a nontrivial interacting field theory, albeit in 1+1 dimensions. I believe that we can use it to illustrate some interesting physics of unparticle theories. Let me first discuss the physical process described in figure 1a:

$$\phi + \phi \to \text{Sommerfield stuff}\,. \tag{19}$$

---

[c] For $m_0 = 0$ we obtain $a = -1$, i.e., $d_{\mathcal{U}} = 0$, which is the Schwinger model result.[21]
[d] The Fourier transform is calculated on p. 284 in Ref. [22].

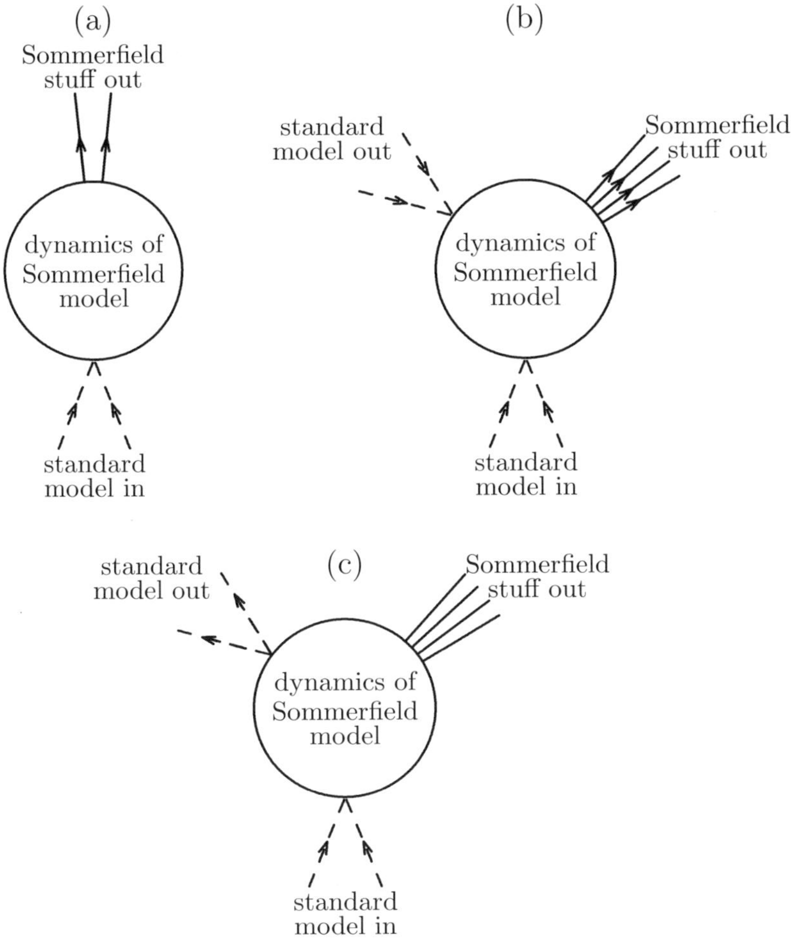

Fig. 1. (a) A disappearance process. (b) A missing charge process. (c) A missing energy process. The arrows indicate the flow of the global $U(1)$ charge.

Because ϕ^2 couples to $\mathcal{O}^*$ at low energy (from (3)), we can obtain the total cross-section for this process from the discontinuity across the physical cut in the $\mathcal{O}$ 2-point function. The cross-section for a given initial state I to scatter into any possible final state F can be obtained from the amplitude of I going back to I using the relation

$$\sum_F \int d\Phi_F \left|\mathcal{M}\left(I \to F\right)\right|^2 = -\operatorname{Disc} i\mathcal{M}\left(I \to I\right) \tag{20}$$

where Φ_F is the phase space of the final state F, and Disc refers to the discontinuity across the branch cut, $\mathcal{M}(s+i\epsilon)-\mathcal{M}(s-i\epsilon)$, where $s=E_{\rm cm}^2$. This is also known as the optical theorem, and the right-hand side of (20) is often written as $2\,{\rm Im}\,\mathcal{M}\,(I\to I)$. In our particular process, for ϕ momenta P_1 and P_2, this gives the cross-section

$$\sigma=\frac{{\rm Im}\,\mathcal{M}(P_1,P_2\to P_1,P_2)}{s}=-\frac{h^2}{s}\,{\rm Im}\,\Delta_{\mathcal{O}}(P) \tag{21}$$

where $P=P_1+P_2$ and $s=P^2$. In the unparticle limit ($\sqrt{s}\ll m$), using (16), or directly the phase space (17), we find the fractional power behavior expected with unparticle production:

$$\sigma=\frac{A(a)}{2}\,\frac{h^2}{(\xi m)^{2a}}\,\frac{1}{s^{1-a}}\,. \tag{22}$$

On the other hand, in the free-particle limit that appears at high energies $\sqrt{s}\gg m$, we have $C(x)\to 1$ in (8) and then

$$\sigma=\frac{h^2}{4}\frac{1}{s} \tag{23}$$

which is the cross-section for $\phi+\phi\to\overline{\psi}_2+\psi_1$ without the Sommerfield interaction.

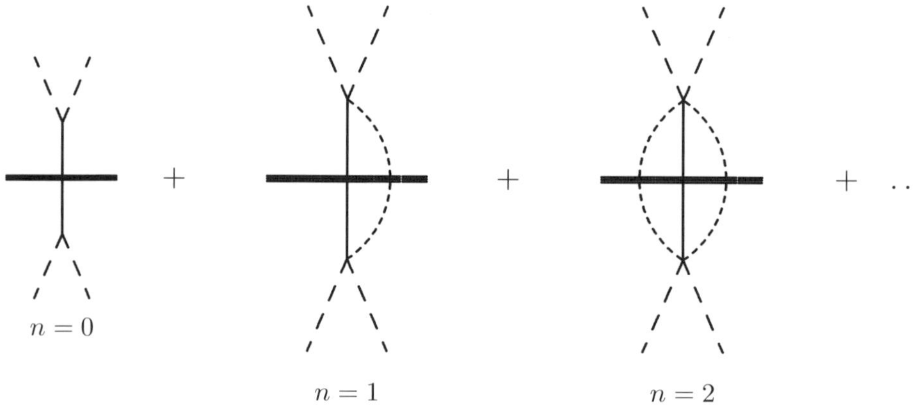

Fig. 2. The cross-section for the disappearance process, based on the expansion (25). The long-dashed lines are the standard model particles ϕ, the solid line represents the unparticle propagator while the short-dashed lines represent massive boson propagators. The thick line is a cut for computing the contribution to the cross-section.

In momentum space the 2-point function (8) [recall also (18)] becomes

$$i\Delta_{\mathcal{O}}(P) = \sum_{n=0}^{\infty} \frac{(-4\pi a)^n}{n!} \int \frac{d^2 p_{\mathcal{U}}}{(2\pi)^2}\, i\Delta_{\mathcal{U}}(p_{\mathcal{U}}) \left[\prod_{i=1}^{n} \frac{d^2 p_i}{(2\pi)^2} i\Delta(p_i)\right] (2\pi)^2 \delta^2 \left(P - p_{\mathcal{U}} - \sum_{j=1}^{n} p_j\right). \tag{24}$$

This describes a sum of two-point diagrams in which the incoming momentum P splits between the unparticle propagator and n massive scalar propagators.[e] The discontinuity gives the phase space

$$\Phi(P) = \frac{A(a)}{(\xi m)^{2a}} \sum_{n=0}^{\infty} \frac{(-4\pi a)^n}{n!} \int \frac{d^2 p_{\mathcal{U}}}{(2\pi)^2}\, \theta(p_{\mathcal{U}}^0)\, \theta(p_{\mathcal{U}}^2)\, \left(p_{\mathcal{U}}^2\right)^a \times \left[\prod_{i=1}^{n} \frac{d^2 p_i}{(2\pi)^2}\, 2\pi\delta(p_i^2 - m^2)\theta(p_i^0)\right] (2\pi)^2\delta^2\left(P - p_{\mathcal{U}} - \sum_{j=1}^{n} p_j\right). \tag{25}$$

This is the form that we used in [5] to discuss the transition from the low-energy unparticle regime and the high-energy free-fermion regime in our toy model.

The transition between the two limits can be studied using the expansion in (25) which shows that the general result can be described as the production of unparticle stuff along with an arbitrary number n of massive bosons (to the extent that this is allowed energetically), see figure 2. Along with an additional massive boson, each subsequent term is proportional to an additional power of a. One can easily obtain explicit results in the case of small a, when only the first few terms in the expansion contribute. For example, the $n = 1$ term contributes

$$\Phi^{(1)} = -a\, \theta(\sqrt{s} - m) \ln \frac{\sqrt{s}}{m} + \mathcal{O}(a^2) \tag{26}$$

which gives the total phase space at the leading non-trivial order in a as

$$\Phi = \frac{1}{2} - a \left[\ln\left(\frac{2}{e^{\gamma_E}} \frac{\xi m}{\sqrt{s}}\right) + \theta(\sqrt{s} - m) \ln \frac{\sqrt{s}}{m}\right] + \mathcal{O}(a^2)\,. \tag{27}$$

For energies $\sqrt{s} > m$, this expression reduces to

$$\Phi = \frac{1}{2} + \mathcal{O}(a^2) \tag{28}$$

[e] Similar behavior in the Schwinger model, where the unparticle stuff is absent, has been discussed in [21,23].

that is the free-fermion result (23). Thus, for $|a| \ll 1$ there is a discontinuity in $d\Phi/d\sqrt{s}$ at $\sqrt{s} = m$, where a transition occurs from pure unparticle behavior below energy m to pure free-fermion behavior above m (see figure 3).[f] Interestingly, the free-fermion behavior is obtained here as a sum of the $n = 0$ and $n = 1$ terms of figure 2.

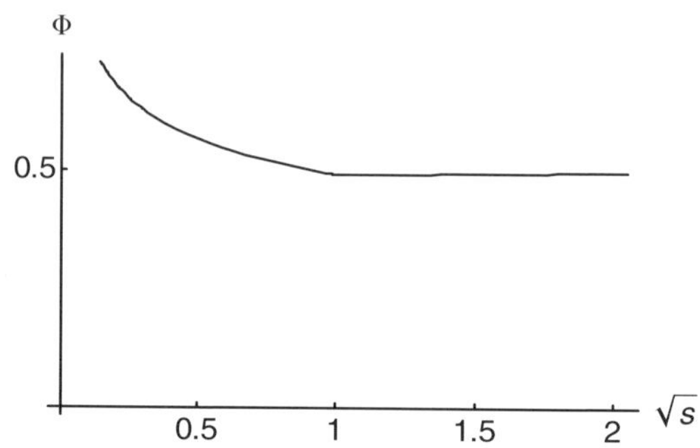

Fig. 3. Phase space Φ for the disappearance process in figure 1a as a function of the energy $\sqrt{s}$ (in units of m) for $a = -0.1$.

For larger values of $|a|$, higher powers of a will need to be taken into account in the contributions with extra bosons in order to reach the free-fermion regime at sufficiently high energies. Since each additional massive boson that we include gives a contribution with one extra power of a, we will need to include N massive bosons to cancel all a-dependent terms up to $\mathcal{O}(a^N)$. Then the free-fermion behavior will only appear at $\sqrt{s} > Nm$ (while in the range $m < \sqrt{s} < Nm$ mixed behavior will be observed, with discontinuities in $d\Phi/d\sqrt{s}$ for each multiple of m). In the limit $a \to -1$, the required value of N becomes infinitely large: the condition $|a^N| \ll 1$ implies $N \gtrsim \frac{1}{-\ln(-a)}$. Note also that for $a = -1$ the unparticle contribution disappears since $A(-1) = 0$. This limit is the Schwinger model; it has been studied in [21,23].

One moral of this analysis of the disappearance process is that we have seen precisely the sense in which the Sommerfield model is a theory of unparticle stuff and a massive vector boson. This is a rather trivial point, but I think one that is frequently misunderstood. The less trivial point is

[f] The linear approximation (27) is not valid for $\sqrt{s} \ll m$ due to large $\ln\sqrt{s}$, but we have the exact expression (17).

the structure of the phase space which is a sum of products of unparticle phase space with the phase space of whatever ordinary particles can be produced. We expect this to be valid in any unparticle theory in any number of dimensions. Now let me describe what I think is the other non-trivial important idea. Again, this is valid for any unparticle theory, but I will illustrate in the 1+1D toy model.

Higher order processes such as those shown in figure 1(b) and (c) require additional machinery since they will depend on 3- and higher n-point functions of $\mathcal{O}$. It is useful to not restrict our attention solely to the operator $\mathcal{O}$ that couples to the standard model sector. Instead we need to use more of the power of conformal invariance and consider the primary operators $\mathcal{O}_j$ of dimension d_j of the conformal field theory. Conformal invariance requires the 2-point functions to have the form[g]

$$\langle 0|\mathrm{T}\mathcal{O}_j(x)\mathcal{O}_k(0)|0\rangle \propto \frac{\delta_{jk}}{\left(-x^2+i\epsilon\right)^d} \propto \delta_{jk}\int \frac{d^Dp}{(2\pi)^D}\, e^{-ipx}\left(-p^2-i\epsilon\right)^{d-D/2}. \tag{29}$$

The δ_{jk} in (29) allows us to identify different primary operators with different kinds of unparticle stuff. For each $\mathcal{O}_j$, there is a unique phase space given by

$$\Phi_j \propto \left(p^2\right)^{d_j-D/2}\,\theta(p^0)\,\theta(p^2)\,. \tag{30}$$

The form of the 3-point function is also fixed by conformal invariance. In particular (for example), for primary scalars $\mathcal{O}_i$ with dimensions d_i:

$$\begin{aligned}&\langle 0|\mathrm{T}\mathcal{O}_1(x_1)\mathcal{O}_2(x_2)\mathcal{O}_3(x_3)|0\rangle \propto \\ &\frac{1}{\left(x_{12}^2\right)^{(d_1+d_2-d_3)/2}\left(x_{13}^2\right)^{(d_1-d_2+d_3)/2}\left(x_{23}^2\right)^{(-d_1+d_2+d_3)/2}}\end{aligned} \tag{31}$$

where $x_{ij} \equiv x_i - x_j$. This was used in [24] (in the case where all $\mathcal{O}_i$ are the same) to study processes that involve 3 **internal** standard model–unparticle vertices. Higher n-point functions are highly constrained by conformal invariance, but not completely determined.

My student Yevgeny Kats had the very nice idea of using this structure to analyze processes in which unparticle stuff is produced as an **outgoing** state. We discuss this in [4]. We show how to analyze more general inclusive processes by looking at the discontinuities across physical cuts in the

[g] If there is more than one operator with a given dimension and tensor structure, we can choose linear combinations to get the δ_{jk}. Again, we have ignored tensor structure.

n-functions of $\mathcal{O}$. And we argue that these discontinuities are related to a sum over primary conformal operators of squared "amplitudes" for the production of the corresponding unparticle stuff. These amplitudes, in turn, are determined by the coefficient functions in the conformal partial wave expansion. The resulting picture, I believe, is a step towards understanding what unparticle stuff "looks like" because it is quite analogous to way we describe the production and scattering of ordinary particles in quantum field theory, with the primary conformal operators playing the role of particles and the coefficients in the conformal partial wave expansion (and its generalization to include more fields) playing the role of amplitudes. We also illustrate this in detail in our toy 1+1D unparticle theory based on the Sommerfield model.

The conformal partial-wave expansion [25–30] (see also [31–33] and references therein in the context of AdS/CFT) is the appropriate version of the operator product expansion (OPE) for a conformal theory. For any two operators $\mathcal{O}_1(x_1)$ and $\mathcal{O}_2(x_2)$ with an arbitrary separation between them, it is possible to write

$$\mathrm{T}\mathcal{O}_1(x_1)\mathcal{O}_2(x_2)|0\rangle = \sum_k \int d^D x \; iQ_k(x|x_1,x_2)\,\mathcal{O}_k(x)|0\rangle \tag{32}$$

where T denotes time-ordering, $\mathcal{O}_k$ are the various primary operators in the theory (unlike in the OPE, their derivatives need not be included separately) and the coefficients $iQ_k(x|x_1,x_2)$ are the 3-point functions of $\mathcal{O}_1(x_1)$, $\mathcal{O}_2(x_2)$ and $\mathcal{O}_k(x)$, with the $\mathcal{O}_k$ leg amputated, namely

$$\int d^D x \; \langle 0|\mathrm{T}\mathcal{O}_k(x')\mathcal{O}_k(x)|0\rangle \, iQ_k(x|x_1,x_2) = \langle 0|\mathrm{T}\mathcal{O}_k(x')\mathcal{O}_1(x_1)\mathcal{O}_2(x_2)|0\rangle\,. \tag{33}$$

If $\mathcal{O}_1$ and $\mathcal{O}_2$ are scalars, the operators $\mathcal{O}_k$ are completely symmetric traceless tensors [28] whose amputated 3-point functions $Q_k(x|x_1,x_2)$ are fully determined, up to a constant prefactor, by the dimensions of the three operators and the tensor rank of $\mathcal{O}_k$.

Amputated 3-point functions are exactly what is needed for computing processes that produce unparticle stuff corresponding to the operator whose leg is amputated. This follows because using (32) twice we can write

$$\begin{aligned}&\langle 0|\mathrm{T}\mathcal{O}_2^*(x_2)\mathcal{O}_1^*(x_1)\mathcal{O}_1(y_1)\mathcal{O}_2(y_2)|0\rangle \\ &= \sum_k \int d^D x\, d^D y \; Q_k^*(x|x_1,x_2)\,\langle 0|\mathrm{T}\mathcal{O}_k^*(x)\mathcal{O}_k(y)|0\rangle\, Q_k(y|y_1,y_2)\,.\end{aligned} \tag{34}$$

For example, suppose we had the coupling

$$\mathcal{L}_{\rm int} \propto \phi^2\mathcal{O} \tag{35}$$

where ϕ is a standard model field and $\mathcal{O}$ is an unparticle operator. According to (34) with $\mathcal{O}_1 = \mathcal{O}_2 = \mathcal{O}$, by taking the discontinuity across the cut of the 4-point function of $\mathcal{O}$ (figure 4), which corresponds to the inclusive process

$$\phi + \phi \to \phi + \phi + \text{unparticle stuff} \tag{36}$$

we obtain the sum of cross-section for the processes

$$\phi + \phi \to \phi + \phi + \{\mathcal{O}_k \text{ stuff}\} \tag{37}$$

where $\mathcal{O}_k$ are the various primary operators in the theory (that do not necessarily couple to the standard model directly). The amplitudes $\mathcal{M}$ of these processes are the amputated 3-point functions Q_k (times factors coming from the standard model), while cutting the $\mathcal{O}_k$ propagators in (34) gives the $\mathcal{O}_k$ phase spaces in (30).

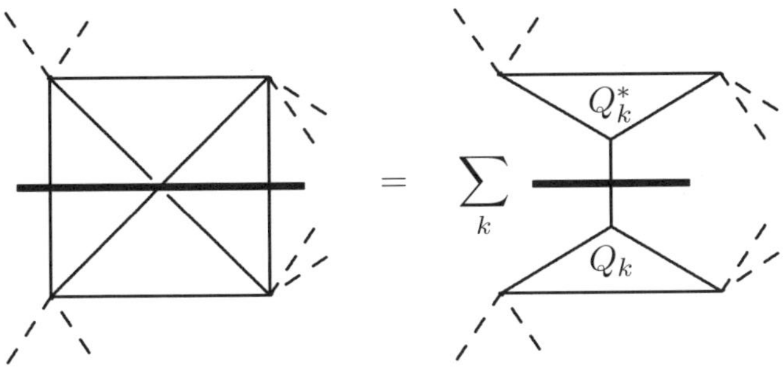

Fig. 4. On the left: the unparticle 4-point function (schematically) with a cut over the unparticle stuff for computing the cross-section of the process (36), inclusive with respect to the unparticle stuff. The dashed lines are the standard model particles ϕ. Each vertex is the interaction (35) (one should include also diagrams in which the standard model particles are attached to the 4-point function in other possible ways - see [4)]. On the right: representation (34) of the 4-point function as a sum of terms, each with two amputated 3-point functions Q_k connected by an $\mathcal{O}_k$ propagator. The cut through the $\mathcal{O}_k$ propagator allows us to interpret the terms in the sum as related to the cross-sections of the exclusive processes (37).

Away from the conformal limit, we can combine the ideas of the conformal operator product expansion with our picture of combined unparticle-particle phase space. So for each primary conformal operator that contributes in the unparticle sector, there is a corresponding sum of products of the unparticle phase space with ordinary particle phase space, as in (25).

Yevgeny and I (mostly Yevgeny) have verified these general principles in our toy unparticle model and you can read about a host of consistency checks in [4]. For this we need the general $2n$-point function of the operator $\mathcal{O}$. We calculate

$$\begin{aligned}&\langle 0|\mathrm{T}\mathcal{O}(x_1)\ldots\mathcal{O}(x_n)\,\mathcal{O}^*(y_1)\ldots\mathcal{O}^*(y_n)|0\rangle\\&=\frac{\prod_{j,k} i\Delta_{\mathcal{O}}(x_j-y_k)}{\prod_{k>j} i\Delta_{\mathcal{O}}(x_j-x_k)\,i\Delta_{\mathcal{O}}(y_j-y_k)}\,.\end{aligned} \tag{38}$$

The $2n$-point function involves the two-point function $i\Delta_{\mathcal{O}}(x)$, and also its inverse

$$i\tilde{\Delta}_{\mathcal{O}}(x)\equiv\frac{1}{i\Delta_{\mathcal{O}}(x)}\,. \tag{39}$$

We can represent the $2n$-point functions diagrammatically with $2n$ vertices

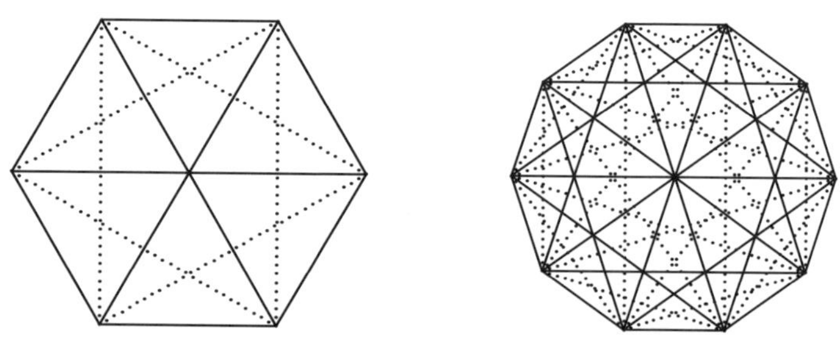

Fig. 5. The 6-point function and the 10-point function of $\mathcal{O}$.

connected by two kinds of lines: solid lines representing $i\Delta_{\mathcal{O}}(x_j-y_k)$; dotted lines representing $i\tilde{\Delta}_{\mathcal{O}}(x_j-x_k)$ or $i\tilde{\Delta}_{\mathcal{O}}(y_j-y_k)$. For example, the 2-point function is just a single solid line, and the 6- and 10-point functions are shown in figure 5. Note that the lines all involve both unparticles at low energies and the massive bosons at higher energies, so we can expand in the massive propagators to the get the full phase space.

It is important to remember that these diagrams, while useful as reminders of the structure of (38), are **not** Feynman diagrams. Eq. (38) is the sum of an infinite number of Feynman diagrams and describes the full $2n$-point function.[h]

One amusing thing to remember is that the solid lines are actually directed, carrying the conserved chiral charge from an $\mathcal{O}^*$ vertex to an $\mathcal{O}$ vertex. The dotted lines are not directed. So for example we could represent

[h] This includes also diagrams that can become disconnected in certain limits.

the 6-point function as in figure 6. Curiously, the charge carried by the solid lines depends on n. In the $2n$-point functions, each solid line carries $2/n$ units of charge. It is obvious that this conserves charge, because n lines emerge from each $\mathcal{O}^*$ vertex and n flow into each $\mathcal{O}$ vertex.

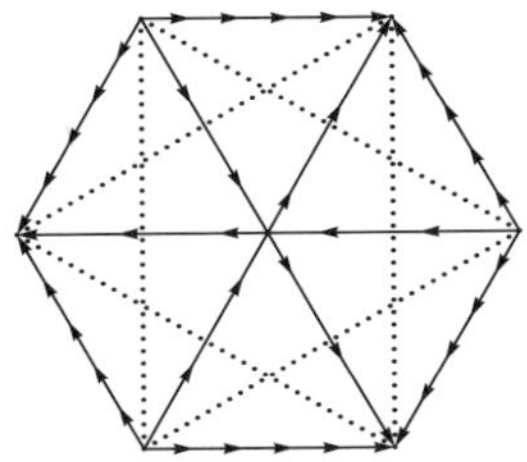

Fig. 6. The 6-point function with the direction of the charge flow indicated.

In this discussion, I have ignored the back-reaction effect of the breaking of the conformal symmetry by the coupling of the unparticle operator to the standard on the low energy conformal theory. This may be important. In the toy model, we can study this, and it does not invalidate our results. This analysis is part of a larger study with student Aqil Sajjad and the results will be written up soon I hope.

I hope that I have convinced you of the three things I advertised:

(1) That unparticle stuff in an unparticle theory comes in different types, determined by the primary conformal operators in the low energy conformal theory. At low energies, the coefficients in the conformal partial wave expansion play the role of amplitudes for production of ordinary particles.
(2) That above the unparticle scale, unparticle stuff appears along with ordinary particles in the density of final states.
(3) That that the Sommerfield model is a useful toy model of the Banks-Zaks sector. It becomes scale-invariant in the infrared, with fractional anomalous dimensions. Since all of its correlation functions can be computed exactly, we can answer explicitly many of the questions regarding the physics of the Banks-Zaks sector as seen by the (toy) standard model observer, for an arbitrary coupling strength in the conformal sector. In particular, we are able to explore the behavior of the theory away from the low-energy scale-invariant regime and to incorporate the unparticle self-interactions. Most importantly, we believe, we have used

this toy model to provide consistency checks for the two new ideas in this paper - (1) and (2).

Of course, all of this is probably obvious to Misha! Happy Birthday.

Acknowledgments

We are grateful to Aqil Sajjad and Brian Shuve for discussions. This research is supported in part by the National Science Foundation under grant PHY-0244821.

References

1. H. Georgi, "Unparticle physics," Phys. Rev. Lett. **98**, 221601 (2007) [arXiv:hep-ph/0703260].
2. H. Georgi, "Another odd thing about unparticle physics," Phys. Lett. B **650**, 275 (2007) [arXiv:0704.2457 [hep-ph]].
3. T. Banks and A. Zaks, "On the phase structure of vector-like gauge theories with massless fermions," Nucl. Phys. B **196**, 189 (1982).
4. H. Georgi and Y. Kats, "Unparticle self-interactions," arXiv:0904.1962 [hep-ph].
5. H. Georgi and Y. Kats, "Unparticle example in 2D," Phys. Rev. Lett. **101**, 131603 (2008) [arXiv:0805.3953 [hep-ph]].
6. J. S. Schwinger, "Gauge invariance and mass. II," Phys. Rev. **128**, 2425 (1962).
7. C. M. Sommerfield, "On the definition of currents and the action principle in field theories of one spatial dimension," Ann. Phys. **26**, 1 (1964).
8. L. S. Brown, "Gauge invariance and mass in a two-dimensional model," Nuovo Cim. **29**, 617 (1963).
9. W. E. Thirring and J. E. Wess, "Solution of a field theoretical model in one space- one time dimension," Ann. Phys. **27**, 331 (1964).
10. D. A. Dubin and J. Tarski, "Interactions of massless spinors in two dimensions," Ann. Phys. **43**, 263 (1967).
11. C. R. Hagen, "Current definition and mass renormalization in a model field theory," Nuovo Cim. **51A**, 1033(1967).
12. W. E. Thirring, "A soluble relativistic field theory," Ann. Phys. **3**, 91 (1958).
13. K. G. Wilson, "Operator-product expansions and anomalous dimensions in the Thirring model," Phys. Rev. D **2**, 1473 (1970).
14. H. Georgi, "Energy-momentum tensors and scale invariance in the Thirring model," Phys. Rev. D **2**, 2908 (1970).
15. H. Georgi and J. M. Rawls, "Anomalies of the axial-vector current in two dimensions," Phys. Rev. D **3**, 874 (1971).
16. C. G. Callan, R. F. Dashen and D. J. Gross, "The structure of the gauge theory vacuum," Phys. Lett. B **63**, 334 (1976).
17. S. R. Coleman, R. Jackiw and L. Susskind, "Charge shielding and quark confinement in the massive Schwinger model," Ann. Phys. **93**, 267 (1975).

18. S. R. Coleman, "More about the massive Schwinger model," Ann. Phys. **101**, 239 (1976).
19. S. R. Coleman, "There are no Goldstone bosons in two dimensions," Commun. Math. Phys. **31**, 259 (1973).
20. K. G. Wilson, "Non-Lagrangian models of current algebra," Phys. Rev. **179**, 1499 (1969).
21. A. Casher, J. B. Kogut and L. Susskind, "Vacuum polarization and the absence of free quarks," Phys. Rev. D **10**, 732 (1974).
22. I. M. Gel'fand and G. E. Shilov, *Generalized functions*, vol. 1, Academic Press, 1964.
23. A. Casher, J. B. Kogut and L. Susskind, "Vacuum polarization and the quark-parton puzzle," Phys. Rev. Lett. **31**, 792 (1973).
24. J. L. Feng, A. Rajaraman and H. Tu, "Unparticle self-interactions and their collider implications," Phys. Rev. D **77**, 075007 (2008) [arXiv:0801.1534 [hep-ph]].
25. S. Ferrara and G. Parisi, "Conformal covariant correlation functions," Nucl. Phys. B **42**, 281 (1972).
26. S. Ferrara, A. F. Grillo, G. Parisi and R. Gatto, "Covariant expansion of the conformal four-point function," Nucl. Phys. B **49**, 77 (1972).
27. V. K. Dobrev, V. B. Petkova, S. G. Petrova and I. T. Todorov, "Dynamical derivation of vacuum operator product expansion in Euclidean conformal quantum field theory," Phys. Rev. D **13**, 887 (1976).
28. G. Mack, "Convergence of operator product expansions on the vacuum in conformal invariant quantum field theory," Commun. Math. Phys. **53**, 155 (1977).
29. E. S. Fradkin and M. Ya. Palchik, "Recent developments in conformal invariant quantum field theory," Phys. Rept. **44**, 249 (1978).
30. E. S. Fradkin and M. Ya. Palchik, "New developments in D-dimensional conformal quantum field theory," Phys. Rept. **300**, 1 (1998).
31. E. D'Hoker, S. D. Mathur, A. Matusis and L. Rastelli, "The operator product expansion of $\mathcal{N} = 4$ SYM and the 4-point functions of supergravity," Nucl. Phys. B **589**, 38 (2000) [arXiv:hep-th/9911222].
32. G. Arutyunov, S. Frolov and A. C. Petkou, "Operator product expansion of the lowest weight CPOs in $\mathcal{N} = 4$ SYM_4 at strong coupling," Nucl. Phys. B **586**, 547 (2000) [arXiv:hep-th/0005182].
33. F. A. Dolan and H. Osborn, "Conformal four point functions and the operator product expansion," Nucl. Phys. B **599**, 459 (2001) [arXiv:hep-th/0011040].

The Lee-Wick Standard Model

Mark B. Wise*
Division of Physics Mathematics and Astronomy, California Institute of Technology, Pasadena, California 91125, USA
**E-mail: wise@theory.caltech.edu*

This article reviews some recent work on a version of the standard model (the Lee-Wick standard model) that contains higher derivative kinetic terms that improve the convergence of loop diagrams removing the quadratic divergence in the Higgs boson mass. Naively higher derivative theories of this type are not acceptable since the higher derivative terms either cause instabilities (from negative energies) or a loss of unitarity (from negative norm states). Lee and Wick provided an interpretation for such theories arguing that theories with higher derivative kinetic terms can be unitary and stable if the states associated with the massive propagator poles, that arise from the higher derivatives, have widths and hence decay and are not in the spectrum of the theory.

Keywords: higher derivative; Lee-Wick

1. Introduction

Its a pleasure to attend this meeting to help celebrate Misha Shifman's 60'th birthday. I have spent much of my career following up on the ideas of Misha and his collaborators and am grateful for the many things I have learned from studying their papers. I have fond memories, from my graduate student days, of waiting anxiously in the SLAC library on Friday afternoon for the new preprints to be put out and hoping there would be something from the ITEP group appearing that week. My recollection is that the paper used was roughly the same quality as toilet paper, the equations were hand written, but the physics was fantastic.

I am going to talk today about some very speculative work I have been doing over the last year or so with Benjamin Grinstein, Donal O'Connell and other collaborators. We have been studying theories with higher derivative kinetic terms. In this work the higher derivatives kinetic terms are not treated as a perturbation but re-summed into the denominator of the propogator. This improves the ultraviolet convergence of Feynman diagrams but leads to a host of potential problems. At first glance, it seems that such theories must either be unstable or not unitary. However, Lee and Wick (LW) argued that this is not necessarily the case.[1] In the next section I review the ideas of Lee and Wick using a simple toy model. Then I discuss the extension to the standard model (the LW standard model[2]) and explain why the resulting theory has no quadratic divergences in the Higgs mass and hence no hierarchy puzzle. Concluding remarks are made in the final section. There is another interpretation for theories with higher derivative kinetic terms that I will not discuss in this lecture.[3]

Another case where ultraviolet divergences cause a problem is quantum gravity. It may be that the ideas of Lee and Wick can be applied to get a sensible quantum theory of gravity.[4] The present best candidate for a quantum theory of gravity is String Theory, and most theoretical physicists believe it is the correct theory. However, there is no experimental evidence to support the conjecture that String Theory is realized in nature. I will not say anything about Lee Wick theories of gravity in this lecture taking the attitude that it is best to start off trying to understand Lee Wick versions of ordinary gauge field theories before addressing gravity.

In this talk I will try to explain some of what is known about LW-theories and the LW standard model in particular. There is some evidence a Lorentz invariant unitary S-matrix can be defined perturbatively, however, these theories have not been extensively studied and more work is needed to firmly establish that they are consistent physical theories. A very good introduction to the ideas of Lee and Wick can be found in lectures by Sidney Coleman.[5]

2. A Toy Model

Consider a model with a single real scalar field $\hat{\phi}$ and Lagrange density,

$$\mathcal{L}_{\rm hd} = \partial_\mu \hat{\phi} \partial^\mu \hat{\phi} - \frac{1}{2M^2} \left(\partial^2 \hat{\phi}\right)^2 - \frac{1}{2} m^2 \hat{\phi}^2 - \frac{1}{3!} g \hat{\phi}^3 . \tag{1}$$

This Lagrange density gives the $\hat{\phi}$ propagator,

$$D_{\hat{\phi}}(p) = \frac{i}{p^2 - p^4/M^2 - m^2} . \tag{2}$$

For $M \gg m$ the above propagator has poles at $p^2 \simeq m^2$ and also at $p^2 \simeq M^2$. The residue of the pole at $p^2 = M^2$ has the wrong sign which, naively, signals an instability or a negative norm state that causes a lack of unitarity.

Imagine, for the moment, doing classical physics with the higher derivative Lagrange density in Eq. (1) by treating the interaction term proportional to g using perturbation theory. Because of the higher derivative kinetic term specifying the usual initial conditions is not enough to get a unique solution. In addition a future boundary condition, that removes the classically unstable solutions, is imposed.

Before discussing the quantum mechanical interpretation of this theory that Lee and Wick gave it is useful to provide another formulation where an auxiliary field is introduced and the higher derivative term is absent. The field theory with two real scalars $\hat{\phi}$ and $\tilde{\phi}$ with Lagrange density

$$\mathcal{L} = \frac{1}{2}\partial_\mu \hat{\phi} \partial^\mu \hat{\phi} - \frac{1}{2} m^2 \hat{\phi}^2 - \tilde{\phi}\partial^2 \hat{\phi} + \frac{1}{2} M^2 \tilde{\phi}^2 - \frac{1}{3} g \hat{\phi}^3, \tag{3}$$

is equivalent to the field theory with the Lagrange density in Eq. (1). To see this note that the LW scalar field $\tilde{\phi}$ occurs in quadratic and linear terms in Eq. (3). Hence the functional integral over $\tilde{\phi}$ can be done exactly and this reproduces Eq. (1). Equivalently the field $\tilde{\phi}$ can be removed from Eq. (3) using its equations of motion and this also reproduces Eq. (1).

The interpretation of the fields $\hat{\phi}$ and $\tilde{\phi}$ in Eq. (3) is obscured by the fact that the kinetic terms are not diagonal. To diagonalize them we define, $\hat{\phi} = \phi - \tilde{\phi}$. In terms of the ordinary scalar ϕ and LW scalar $\tilde{\phi}$ the Lagrange density in Eq. (3) becomes,

$$\mathcal{L} = \frac{1}{2}\partial_\mu \phi \partial^\mu \phi - \frac{1}{2}\partial_\mu \tilde{\phi} \partial^\mu \tilde{\phi} + \frac{1}{2} M^2 \tilde{\phi}^2 - \frac{1}{2} m^2 (\phi - \tilde{\phi})^2 - \frac{1}{3} g (\phi - \tilde{\phi})^3. \tag{4}$$

The LW resonance that is destroyed by the field $\tilde{\phi}$ is unstable since it can decay into two ϕ particles. Its decay width is calculated by doing the standard bubble sum for the $\tilde{\phi}$ propagator. The self energy Σ is the same as for an ordinary particle. The only difference is that each LW scalar propagator comes with an additional minus sign, compared with a normal scalar propagator. Hence,

$$\begin{aligned} D_{\tilde{\phi}}(p) &= \frac{-i}{p^2 - M^2} + \frac{-i}{p^2 - M^2}\left(-i\Sigma(p^2)\right)\frac{-i}{p^2 - M^2} + \dots \\ &= \frac{-i}{p^2 - M^2 + \Sigma(p^2)}. \end{aligned} \tag{5}$$

Treating the LW scalar propagator in the narrow width approximation we see that the LW field is associated with a resonance, but its width term enters with the wrong sign, *i.e.*,

$$D_{\tilde{\phi}}(p) = \frac{-i}{p^2 - M^2 - iM\Gamma}, \tag{6}$$

with

$$\Gamma = \frac{g^2}{32\pi M}\sqrt{1-\frac{4m^2}{M^2}}. \tag{7}$$

Consider an experiment where two beams of stable particles collide in the center of mass and create an ordinary narrow resonance approximately at rest. The resonance has a probability $\Gamma \mathrm{d}t$ of decaying in the positive time interval $\mathrm{d}t$ into the stable particles that created it (we assume this is the only allowed decay channel) . The decay products are observed by detectors that can measure their momentum and energy. From those measurements we can determine the time that the resonance decayed by following the paths of the detected decay products back to the point they were produced. For a Lee-Wick resonance the flip in the sign of the width term means that the resonance appears to decay earlier than it was produced, *i.e.*, it has a probability $\Gamma \mathrm{d}t$ of decaying in the negative time interval $-\mathrm{d}t$. Hence the theory is usually called acausal. I used the phrase "appears to decay earlier than it was produced" because the future boundary condition forces the state vector for the colliding beams to contain a very small component of the decay products in the far past and this component grows with time.[6] The apparent acausal behavior is a result of the non-locality in time.

It seems that this unusual acausal behavior should lead to paradoxes. Can't you use the decay products to turn off the experiment before the collision occurred? You will get a headache if you think too hard about this issue. If the S-matrix is unitary there will not be any paradoxes in scattering experiments. You prepare initial stable particle states (in the distant past) out come final stable particle states (in the far future). The probabilities for the possible final states add up to one and, furthermore, since the S-matrix is a one-to-one map there are no paradoxes. Hence, we should pay special attention to whether the S-matrix in LW theories is unitary on the space of incoming and outgoing ordinary stable particles.

Now we go beyond the narrow resonance approximation and compare the LW scalar propagator with the propagator for a scalar field χ that corresponds to an ordinary resonance. For the ordinary resonance,

$$D_\chi(p^2) = \frac{i}{\pi}\int_{4m^2}^{\infty} \mathrm{d}s \, \frac{\rho(s)}{p^2 - s + i\epsilon} \simeq \frac{i}{p^2 - M^2 + iM\Gamma}\,. \tag{8}$$

There are no poles because the χ field destroys a resonance which is not in the spectrum of the theory. It is the stable decay products that are in the spectrum. For the LW scalar we need to get, in the narrow resonance approximation, a propagator with the opposite overall sign and pole in the opposite side of the real axis. However, the self energy is the same and so the cut piece is the same as for an ordinary resonance. Hence,

$$D_{\tilde{\phi}}(p^2) = \frac{-i}{p^2 - M_c^2} + \frac{-i}{p^2 - M_c^{*2}} + \frac{i}{\pi}\int_{4m^2}^{\infty} \mathrm{d}s\, \frac{\rho(s)}{p^2 - s + i\epsilon}, \tag{9}$$

where $M_c^2 = M^2 + iM\Gamma$. To get a unitary theory Lee and Wick advocate treating the poles at M_c^2 and M_c^{*2} as resonances that are not in the spectrum of the theory. They correspond to negative norm states and so if they were in the spectrum the theory would not be unitary. However, they are poles in the $\tilde{\phi}$ propagator so how can we have a unitary S-matrix when we exclude the LW particles from the initial and final states?

To begin with, consider tree level $\phi\phi$ scattering through a virtual $\tilde{\phi}$ resonance. The complex poles do not destroy unitarity since the incoming particles have real energies and momentum and so we cannot produce an on shell intermediate state with complex mass. Their contribution to the scattering amplitude is real and it is easy to verify that tree level exchange satisfies the optical theorem.

The real subtleties first occur at one loop in Feynman diagrams that have two LW massive particles in the intermediate state. One can get an intermediate LW two particle on shell state produced with real incoming momentum and energy . For these diagrams one must provide a definition for loop integrations of the form

$$I = \int \frac{\mathrm{d}^4p}{(2\pi)^4} \frac{-i}{(p+q)^2 - M_1^2} \frac{-i}{p^2 - M_2^2}, \tag{10}$$

where M_1 and M_2 are the complex LW masses. The p^0 integration has four poles. For time like q^0 we can go to the frame where $\mathbf{q} = 0$ and the poles are located at, $p^0 = \pm\sqrt{\mathbf{p}^2 + M_2^2}$ and $p^0 = -q^0 \pm \sqrt{\mathbf{p}^2 + M_1^2}$. At $g = 0$ the widths for the LW resonances vanish and the masses M_1 and M_2 are real. Then the contour of p^0 integration is the usual Feynman contour[a] . As g increases the contour is defined so that the poles do not cross the contour; a pole which was initially below the contour remains below for example. This leads to a well defined contour that can be Wick rotated unless poles pinch the contour. Pinching can occur because we could have $M_2^2 = M_1^{*2}$ and then for some q^0 two poles overlap. In this case we define the integral by taking the masses M_1 and M_2 to be unrelated complex mass parameters so the poles don't overlap. At the end of the calculation

[a] Since the contour can be Wick rotated the scattering amplitudes are Lorentz invariant. Some ways of defining the amplitudes are not Lorentz invariant.[7]

M_2^2 is set equal to the complex conjugate of M_1^2. One can show that this leads to a unitary Lorentz invariant scattering amplitude. This prescription is due to Cutkowski *et. al.*[8] Boulware and Gross discuss the difficulty of implementing it in a path integral formulation.[9]

It is not known if all Feynman diagrams in this theory can be treated in a similar way to get a unitary theory. I suspect there is no impediment to doing so. One reason for this is that the $O(N)$ scalar model can be made into a LW theory by adding a higher derivative kinetic term. One can show that the procedure of Cutkowski *et. al.* renders this theory unitary in the large N limit,[10] at all orders in perturbation theory.

It has been argued that $U(1)$ Lee Wick gauge theories are Lorentz invariant, unitary and can be defined nonperturbatively.[6] The argument presented by van Tonder is is a little formal and I have not worked through it.

3. LW-Gauge Bosons and the Hierarchy Puzzle

In this section we discuss LW theories with non-Abelian gauge bosons. The higher derivative Lagrange density is,

$$\mathcal{L}_{\rm hd} = -\frac{1}{2}{\rm tr}\hat{F}_{\mu\nu}\hat{F}^{\mu\nu} + \frac{1}{M_A^2}{\rm tr}\left(\hat{D}^\mu\hat{F}_{\mu\nu}\hat{D}^\lambda\hat{F}_\lambda{}^\nu\right), \tag{11}$$

where $\hat{F}_{\mu\nu} = \partial_\mu\hat{A}_\nu - \partial_\nu\hat{A}_\mu - ig[\hat{A}_\mu, \hat{A}_\nu]$, and $\hat{A}_\mu = \hat{A}_\mu^B T^B$. Now the higher derivative term contains not only the order p^4/M_A^2 contribution to the denominator of the gauge boson propagator, but also interactions. Again we remove the non-renormalizable term by introducing the auxiliary LW field $\tilde{A}_\mu$. Now the Lagrange density becomes,

$$\mathcal{L} = -\frac{1}{2}{\rm tr}\hat{F}_{\mu\nu}\hat{F}^{\mu\nu} - M_A^2{\rm tr}\tilde{A}_\mu\tilde{A}^\mu + 2{\rm tr}\hat{F}_{\mu\nu}\hat{D}^\mu\tilde{A}_\nu, \tag{12}$$

where $\hat{D}_\mu\tilde{A}_\nu = \partial_\mu\tilde{A}_\nu - ig[\hat{A}_\mu.\tilde{A}_\nu]$. To diagonalize the kinetic terms we shift the hatted gauge field,

$$\hat{A}_\nu = A_\mu + \tilde{A}_\nu. \tag{13}$$

In terms of the LW gauge field $\tilde{A}_\mu$ and the ordinary gauge field A_μ the Lagrange density now becomes,

$$\begin{aligned}\mathcal{L} = &-\frac{1}{2}\,{\rm tr}F_{\mu\nu}F^{\mu\nu} + \frac{1}{2}\,{\rm tr}\left(D_\mu\tilde{A}_\nu - D_\nu\tilde{A}_\mu\right)\left(D^\mu\tilde{A}^\mu - D^\nu\tilde{A}^\mu\right)\\ &-ig\,{\rm tr}\left(\left[\tilde{A}_\mu,\tilde{A}_\nu\right]F^{\mu\nu}\right) - \frac{3}{2}g^2\,{\rm tr}\left(\left[\tilde{A}_\mu,\tilde{A}_\nu\right]\left[\tilde{A}^\mu,\tilde{A}^\nu\right]\right)\\ &-4ig\,{\rm tr}\left(\left[\tilde{A}_\mu,\tilde{A}_\nu\right]D^\mu\tilde{A}^\nu\right) - M_A^2\,{\rm tr}\left(\tilde{A}_\mu\tilde{A}^\mu\right).\end{aligned} \tag{14}$$

In this formulation the Lagrange density is renormalizable by power counting and the LW gauge field is a vector matter field in the adjoint representation of the gauge group with very specific interactions given by Eq. (14).

Since there are massive vectors in this theory one should be concerned that the theory will become strongly coupled or violate unitarity because the scattering of the longitudinal polarizations grows with energy. After all, in the standard model with no Higgs boson but a mass term for the W-bosons the longitudinal polarization[b] scattering amplitude grows with energy as, $\mathcal{A}(LL \to LL) \sim g_2^2 E^2/M_W^2$. This bad behavior arises because the components of the longitudinal polarization vector for a vector boson with mass M, $\epsilon_L(p) = (|\mathbf{p}|, E(p)\hat{\mathbf{p}})/M$, grow with energy E. However the LW massive gauge boson has couplings constrained by an unbroken gauge invariance and so it can be shown that there is no bad growth with energy.[11] In fact, $\mathcal{A}(LL \to LL) \sim g^2 M_A^2/E^2$, $\mathcal{A}(LL \to LT) \sim g^2 M_A/E$ and $\mathcal{A}(LL \to TT) \sim g^2$. If the higher derivative Lagrange density could not be rewritten in terms of a renormalizable Lagrange density using an auxillary field we would have indeed found that the scattering amplitudes grow with energy.

We are almost ready to discuss the hierarchy problem in the LW standard model. To do that we add a scalar in the fundamental of the gauge group $\hat{\phi}$. Its interactions are governed by the Lagrange density,

$$\mathcal{L}_{\text{hd}} = \left(\hat{D}_\mu \hat{\phi}\right)^\dagger \left(\hat{D}^\mu \hat{\phi}\right) - \frac{1}{M_\phi^2} \left(\hat{D}_\mu \hat{D}^\mu \hat{\phi}\right)^\dagger \left(\hat{D}_\nu \hat{D}^\nu \hat{\phi}\right). \tag{15}$$

As in the case of the gauge fields the higher derivative term also contributes to the interactions. However, it does not destroy renormalizability because it can be eliminated by introducing the LW field $\tilde{\phi}$. After shifting $\hat{\phi} = \phi - \tilde{\phi}$ Eq. (15) becomes

$$\begin{aligned}\mathcal{L} &= (D_\mu \phi)^\dagger (D^\mu \phi) - \left(D_\mu \tilde{\phi}\right)^\dagger \left(D^\mu \tilde{\phi}\right) + M_\phi^2 \tilde{\phi}^\dagger \tilde{\phi} \\ &+ig (D^\mu \phi)^\dagger \tilde{A}_\mu^B T^B \phi + g^2 \phi^\dagger \left(\tilde{A}_\mu^B T^B \tilde{A}_\mu^C T^C\right) \phi - ig\phi^\dagger \tilde{A}_\mu^B T^B D^\mu \phi \\ &-ig \left(D^\mu \tilde{\phi}\right)^\dagger \tilde{A}_\mu^B T^B \tilde{\phi} + ig\tilde{\phi}^\dagger \tilde{A}_\mu^B T^B D^\mu \tilde{\phi} - g^2 \tilde{\phi}^\dagger \left(\tilde{A}_\mu^B T^B \tilde{A}_\mu^C T^C\right) \tilde{\phi}.\end{aligned} \tag{16}$$

First we work in the higher derivative version of the theory. The gauge is

[b] We use L to denote a Longitudinal polarization and T to denote a transverse polarization.

fixed in the usual way introducing a parameter ξ and the $\hat{A}$ propagator is,

$$\hat{D}^{AB}_{\mu\nu}(p) = -\delta^{AB}\frac{i}{p^2 - p^4/M_A^2}\left(\eta_{\mu\nu} - (1-\xi)\frac{p_\mu p_\nu}{p^2} - \xi\frac{p_\mu p_\nu}{M_A^2}\right). \tag{17}$$

We work in Landau gauge, $\xi = 0$, which the propagator falls off as $1/p^4$ at large momentum p. Now a diagram that contributes to the ϕ mass has two external ϕ legs and it is straightforward to show, using Landau gauge in the higher derivative theory, that the degree of divergence of Feynman diagrams contributing to the ϕ mass is, $d = 4 - 2L$, where L is the number of loops. At one loop, $L = 1$, there is a possible quadratic divergence but not at two loops and beyond. At one loop you can check by explicit calculation that there is no quadratic divergence. For this purpose it is easier to work in the formulation with LW fields and no higher derivative terms. In that version the propagator for the gauge bosons is,

$$D^{AB}_{\mu\nu}(p) = -\delta^{AB}\frac{i}{p^2}\left(\eta_{\mu\nu} - (1-\xi)\frac{p_\mu p_\nu}{p^2}\right), \tag{18}$$

and the propagator for the massive LW vector in the adjoint representation is,

$$\tilde{D}^{AB}_{\mu\nu}(p) = \delta^{AB}\frac{i}{p^2 - M_A^2}\left(\eta_{\mu\nu} - \frac{p_\mu p_\mu}{M_A^2}\right). \tag{19}$$

There is a remaining logarithmic divergence and so corrections to the Higgs mass in the LW standard model have the form,

$$\delta m_h^2 \sim \frac{g^2 M_{LW}^2}{16\pi^2}\ln\left(\frac{\Lambda^2}{M_{LW}^2}\right), \tag{20}$$

where Λ is some ultraviolet cutoff where new physics beyond that in the LW extension of the standard model comes in (perhaps the Planck scale). One cannot have the LW partner masses M_{LW} too much greater than a TeV and still solve the hierarchy puzzle.

In the LW standard model there are no strong constraints on the mass matrices of the LW particles from the smallness of flavor changing neutral currents.[12]

The LW standard model is not finite but it is renormalizable and the one loop gauge coupling beta functions have been computed.[13]

4. Constraints from Precision Electroweak Physics

The magnitude of the LW masses for the $SU(2) \times U(1)$ M_1 and M_2 are constrained by precision electroweak physics.[14] In the higher derivative version (neglecting the Higgs vacuum expectation value) of the LW standard model the self energies of the weak gauge bosons take the form,

$$\Pi(q^2) = q^2 - \frac{q^4}{M_{1,2}^2}. \tag{21}$$

The precision electroweak parameters, Y and W are defined by,

$$Y = \frac{M_W^2}{2}\Pi''_{BB}(0), \qquad W = \frac{M_W^2}{2}\Pi''_{33}(0). \tag{22}$$

In the LW standard model at tree level,

$$Y = -\frac{M_W^2}{M_1^2}, \quad W = \frac{M_W^2}{M_2^2}. \tag{23}$$

Precision electroweak data imply, at the 99% confidence level, that $M_{1,2} >$ 3 TeV, when we make the assumption that $M_1 = M_2$. Of course virtual LW particles impact the values of other measurable quantities including properties of the Higgs boson.[15]

5. Fermions

So far I have not mentioned how to put fermions in the LW version of the standard model. Consider a theory of a massless left handed fermion in the fundamental representation of the gauge group. The higher derivative theory in this case is,

$$\mathcal{L}_{\rm hd} = \bar{\hat{\psi}}_L i\left(\gamma^\mu \hat{D}_\mu\right)\hat{\psi}_L + \frac{1}{M_\psi^2}\bar{\hat{\psi}}_L i\left(\gamma^\mu \hat{D}_\mu\right)\left(\gamma^\nu \hat{D}_\nu\right)\left(\gamma^\eta \hat{D}_\eta\right)\hat{\psi}_L. \tag{24}$$

To eliminate the higher derivative terms we need in this case two auxiliary LW fields. One left handed ($\tilde{\psi}_L$) and the other right handed ($\tilde{\psi}_R$). With these LW-fields the fermion Lagrange density becomes,

$$\begin{aligned}\mathcal{L} = \bar{\hat{\psi}}_L i\left(\gamma^\mu \hat{D}_\mu\right)\hat{\psi}_L + M_\psi\left(\bar{\tilde{\psi}}_L\tilde{\psi}_R + \bar{\tilde{\psi}}_R\tilde{\psi}_L\right) + \bar{\tilde{\psi}}_L i\left(\gamma^\mu \hat{D}_\mu\right)\hat{\psi}_L \\ + \bar{\hat{\psi}}_L i\left(\gamma^\mu \hat{D}_\mu\right)\tilde{\psi}_L - \bar{\tilde{\psi}}_R i\left(\gamma^\mu \hat{D}_\mu\right)\tilde{\psi}_R.\end{aligned} \tag{25}$$

Using the equations of motion,

$$\tilde{\psi}_R = -i\frac{\gamma^\mu \hat{D}_\mu}{M_\psi}\hat{\psi}_L, \quad \tilde{\psi}_L = \frac{\gamma^\mu \hat{D}_\mu \gamma^\nu \hat{D}_\nu}{M_\psi^2}\hat{\psi}_L. \tag{26}$$

Eq. (25) can be seen to be equivalent to Eq. (24). Finally, the kinetic terms can be put into standard form by performing the familiar shift, $\hat{\psi}_L = \psi_L - \tilde{\psi}_L$.

Of course to be consistent the LW fermions must develop a width and decay. This occurs because of the Yukawa couplings in the theory. Some of the Yukawa's are very small, and hence some of the LW fermions are long lived (by particle physics standards). Because of this it may be possible to actually observe the acausal behavior discussed in section 2 in scattering experiments at the LHC.[16]

One popular method for giving neutrinos mass is the see-saw mechanism with heavy right handed neutrinos. It is possible to do this in the LW standard model and solve the hierarchy problem with the right handed neutrinos much heavier than the scale of LW particle masses.[17]

6. Conclusions

LW- theories are quite interesting. They have a nonlocality in time or if you prefer an acausality that implies unusual behavior in scattering experiments. The order p^4 term in the propagators means these theories have softer behavior at high momentum than ordinary quantum field theories and therefore can help with the hierarchy problem and perhaps even with quantum gravity.

If the LW standard model is the correct theory of nature you could eventually observe the unusual overall sign and perhaps the unusual sign of the width term in the LW-gluon propagator from interference between exchange of an ordinary gluon and LW-gluon in scattering amplitudes.[18]

In this talk I have considered LW theories with order p^4 terms in the propagators. It is possible to have even higher derivative terms.[19] Also the fine tuning associated with the Higgs mass can be reduced if only the particles that couple strongly to the Higgs have LW partners.[20]

In constructing LW theories some of the usual rules of quantum field theory have been changed. The amplitudes have unusual analytic properties. It is certainly possible that further study will reveal that LW-theories are inconsistent. (As David Gross emphasized to me, "the odds are not in your favor when you bet that a theory that changes the basic tenets of quantum field theory is consistent".) In my opinion the best chance for discovering a problem with LW theories is not in scattering amplitudes (which likely can be defined to all orders in perturbation theory to give a Lorentz invariant and unitary S-matrix) but in collective properties like the behavior of LW-theories at finite temperature.

The LW toy model in section 2 has been studied[21] at finite temperature, T, by Fornal *et. al.*. They found that in the energy density and pressure at $T >> M$ the order T^4 term cancelled between the LW and ordinary scalar. This gives a speed of sound in the gas that approaches unity (from below) at very high T. Currently, the authors of this paper are puzzling over the fermion case since naively there are twice as many LW fermion degrees of freedom ($\tilde{\psi}_L$ and $\tilde{\psi}_R$) as ordinary ones (ψ_L) and so the LW degrees of freedom overcompensate giving a negative energy density[c]. I hope this issue this will get resolved in the near future.

There are many more things to check and try to understand about LW-theories and the implications of the LW standard model for particle physics and for cosmology.[22] I certainly think they are interesting to study. I hope this brief presentation will encourage others to think about LW theories and perhaps one day not too far in the future more definitive statements can be made about their consistency as physical theories.

References

1. T. D. Lee and G. C. Wick, Nucl. Phys. B **9**, 209 (1969), T. D. Lee and G. C. Wick, Phys. Rev. D **2**, 1033, (1970).
2. B. Grinstein, D. O'Connell and M. B. Wise, Phys. Rev. D **77**, 025012 (2008) [arXiv:0704.1845 [hep-ph]].
3. C. M. Bender and P. D. Mannheim, arXiv:0902.1365 [hep-th], C. M. Bender and P. D. Mannheim, J. Phys. A **41**, 304018 (2008) [arXiv:0807.2607 [hep-th]], C. M. Bender and S. Boettcher, Phys. Rev. Lett. **80**, 5243 (1998) [arXiv:physics/9712001], A. M. Shalaby, Phys. Rev. D **80**, 025006 (2009) [arXiv:0812.3419 [hep-th]].
4. E. Tomboulis, Phys. Lett. B **70**, 361 (1977), E. Tomboulis, Phys. Lett. B **97**, 77 (1980), I. Antoniadis and E. Tomboulis, Phys. Rev. D **33**, 2756 (1986).
5. S. Coleman, "Acausality," In *Erice 1969, Ettore Majorana School On Subnuclear Phenomena,* A. Zichichi, Editor, Academic Pres, New York, 282 (1970).
6. A. van Tonder, arXiv:0810.1928 [hep-th].
7. N. Nakanishi, Phys. Rev. D **3**, 3235 (1971), N. Nakanishi, Phys. Rev. D **5**, 1968 (1972), N. Nakanishi, Prog. Theor. Phys. Suppl. **51** 1 (1972), K. Jansen, J. Kuti and C. Liu, Phys. Lett. B **309** 119 (1993).
8. R. E. Cutkowski, P. V. Landshoff, D. I. Olive and J. C. Polkinghorne, Nucl. Phys. B **12**, 281 (1969).
9. D. G. Boulware and D. J. Gross, Nucl. Phys. B **233**, 1 (1984).
10. B. Grinstein, D. O'Connell and M. B. Wise, Phys. Rev. D **79**, 105019 (2009) [arXiv:0805.2156 [hep-th]].
11. B. Grinstein, D. O'Connell and M. B. Wise, Phys. Rev. D **77**, 065010 (2008) [arXiv:0710.5528 [hep-ph]].
12. T. R. Dulaney and M. B. Wise, Phys. Lett. B **658**, 230 (2008) [arXiv:0708.0567 [hep-ph]].
13. B. Grinstein and D. O'Connell, Phys. Rev. D **78**, 105005 (2008) [arXiv:0801.4034 [hep-ph]].

[c] I am grateful to Jose Ramon Espinosa for pointing this issue out.

14. T. E. J. Underwood and R. Zwicky, Phys. Rev. D **79**, 035016 (2009) [arXiv:0805.3296 [hep-ph]], E. Alvarez, L. Da Rold, C. Schat and A. Szynkman, JHEP **0804**, 026 (2008) [arXiv:0802.1061 [hep-ph]].
15. F. Krauss, T. E. J. Underwood and R. Zwicky, Phys. Rev. D **77**, 015012 (2008) [arXiv:0709.4054 [hep-ph]], G. Cacciapaglia, A. Deandrea and J. Llodra-Perez, JHEP **0906**, 054 (2009) [arXiv:0901.0927 [hep-ph]], C. D. Carone and R. Primulando, "Constraints on the Lee-Wick Higgs Sector," arXiv:0908.0342 [hep-ph].
16. E. Alvarez, L. Da Rold, C. Schat and A. Szynkman, "Vertex Displacements for Acausal Particles: Testing the Lee-Wick Standard Model at the LHC," arXiv:0908.2446 [hep-ph].
17. J. R. Espinosa, B. Grinstein, D. O'Connell and M. B. Wise, Phys. Rev. D **77**, 085002 (2008) [arXiv:0705.1188 [hep-ph]].
18. T. G. Rizzo, JHEP **0706**, 070 (2007) [arXiv:0704.3458 [hep-ph]], T. G. Rizzo, JHEP **0801**, 042 (2008) [arXiv:0712.1791 [hep-ph]].
19. C. D. Carone, Phys. Lett. B **677**, 306 (2009) [arXiv:0904.2359 [hep-ph]], C. D. Carone and R. F. Lebed, JHEP **0901**, 043 (2009) [arXiv:0811.4150 [hep-ph]].
20. C. D. Carone and R. F. Lebed, Phys. Lett. B **668**, 221 (2008) [arXiv:0806.4555 [hep-ph]].
21. B. Fornal, B. Grinstein and M. B. Wise, Phys. Lett. B **674**, 330 (2009) [arXiv:0902.1585 [hep-th]].
22. S. Lee, "Lee Wick Dark Energy," arXiv:0810.1145 [astro-ph], Y. F. Cai, T. t. Qiu, R. Brandenberger and X. m. Zhang, Phys. Rev. D **80**, 023511 (2009) [arXiv:0810.4677 [hep-th]], N. Barnaby, D. J. Mulryne, N. J. Nunes and P. Robinson, JHEP **0903**, 018 (2009) [arXiv:0811.0608 [hep-th]].

SECTION 9.

NON–QCD

Flavor Mixing, Quark Masses, Neutrino Masses and Neutrino Oscillations

H. Fritzsch

University of Munich
Munich, Germany

We discuss first the flavor mixing of the quarks, using the texture zero mass matrices. Then we study a similar model for the mass matrices of the leptons. We are able to relate the mass eigenvalues of the charged leptons and of the neutrinos to the mixing angles and can predict the masses of the neutrinos. We find a normal hierarchy - the masses are 0.004 eV, 0.01 eV and 0.05 eV. The atmospheric mixing angle is given by the mass ratios of the charged leptons and the neutrinos. We find about 40°, consistent with the experiments. The mixing element, connecting the first neutrino with the electron, is predicted to be 0.05. This prediction can soon be checked by the Daya Bay experiment.

We observe three lepton-quark families in nature. The first family consists of the electron, its neutrino and the u- and d-quarks. The members of the second family are the muon, its neutrino and the c- and s-quarks. The third family consists of the tauon, its neutrino and the t- and b-quarks. It is unknown, whether there is a connection between the number of families and the number of colors in quantum chromodynamics.

In the Standard Model 28 fundamental constants have to be introduced. Their values cannot be calculated - they have to be measured in the experiments. Theses constants are:

- the constant of gravity G,
- the fine structure constant,
- the coupling constant of the weak interactions,
- the coupling constant of the strong interactions or the scale of QCD,
- the mass of the W-boson,
- the mass of the "Higgs"-boson,
- the masses of the three charged leptons,
- the masses of the three neutrinos,

- the masses of the six quarks,
- the four parameters, describing the flavour mixing of the quarks and
- the six parameters, describing the flavour mixing of the leptons.

For the masses of the quarks we assume the following values:

$$\begin{aligned}
u &: 5.3 \quad MeV, \\
d &: 7.8 \quad MeV, \\
s &: 146 \quad MeV, \\
c &: 1050 \; MeV, \\
b &: 4600 \; MeV, \\
t &: 174000 \, MeV.
\end{aligned}$$

The quark masses are scale dependent. The values given above, except the one for the t-mass, are normalized at an energy scale of 1 GeV. The experimental errors are not given, but are at least 10 %, except for the t-mass, which is known to about 1 %.

The quarks of the same charge do mix. If the u-quark interacts with a W-boson, a mixture of d, s and b appears. These mixtures are described by the CKM matrix.[1] I prefer a description, which we introduced some time ago [2]:

$$V = \begin{bmatrix} c_u & s_u & 0 \\ -s_u & c_u & 0 \\ 0 & 0 & 1 \end{bmatrix} \begin{bmatrix} e^{-i\phi} & 0 & 0 \\ 0 & c & s \\ 0 & -s & c \end{bmatrix} \begin{bmatrix} c_d & -s_d & 0 \\ s_d & c_d & 0 \\ 0 & 0 & 1 \end{bmatrix} . \tag{1}$$

In the case of three families there are three mixing angles and one phase parameter. The latter describes the CP-violation. The angles with the index u or d describe the mixing in the $u-c$ sector or the $d-s$ sector, the angle with no index describes the mixing between the (t,c)-system and the (b,s)-system.

I proposed years ago a simple texture 0 mass matrix for the quarks, given by the following matrix, valid both for the (u,c,t)-system and the (d,s,b)-system [3]:

$$M = \begin{bmatrix} 0 & A & 0 \\ A^* & C & B \\ 0 & B^* & D \end{bmatrix} . \tag{2}$$

Such mass matrices are obtained, if in the electroweak theory special symmetries are present, either discrete reflection symmetries or continuous symmetries.[3] In the Standard Model such symmetries cannot be introduced, but if the electroweak gauge group contains besides the lefthanded group also a righthanded group, such symmetries can easily be obtained. A left-right symmetric gauge group arises, for example, if one uses the group $SO(10)$ to describe the grand unification.

After diagonalization one finds the following relations for the mixing angles [4]:

$$\theta_d = \tan^{-1}\sqrt{m_d/m_s} \quad , \qquad \theta_u = \tan^{-1}\sqrt{m_u/m_c}\,. \tag{3}$$

For the angle θ_d we find from the ratio of the quark masses $13 \pm 0.4°$, the experimental value is $11.7 \pm 2.6°$. The angle θ_u is calculated to $5.0 \pm 0.7°$, the experimental value is $5.4 \pm 1.1°$. Thus these angles agree very well with the experimental data on flavor mixing, if the phase angle is assumed to be close to 90°. The experiments give for the phase angle $86 \ldots 95°$. Note that in this case the flavor mixing matrix is very simple - instead of the arbitrary phase the matrix element is just given by $-i$.

Analogously we can describe the flavour mixing in the lepton sector, which can be studied in the neutrino oscillations. The experiments, carried out in Japan, (Kamiokande detector, ref. [5]) and in Canada (SNO detector, ref. [6]), give the following results for the mass-squared differences of the three neutrino mass eigenstates:

$$\Delta m_{21}2 \simeq 8 \times 10^{-5} eV^2\,, \qquad \Delta m_{32}2 \simeq 2.5 \times 10^{-3} eV^2\,. \tag{4}$$

In neutrino oscillations only mass squared differences can be measured. No information can be obtained for the absolute magnitude of the neutrino masses. There is the possibility that the neutrino masses are nearly degenerate, e.g. masses like $m_1 = 0.94eV$, $m_2 = 0.95eV$, $m_3 = 1eV$ are possible. If one neutrino remains massless, one would have the masses $m_1 = 0eV$,$m_2 = 0.009eV$, $m_3 = 0.05eV$. In the latter case a hierarchy of the masses is present, but this hierarchy is much weaker than the mass hierarchy for the charged leptons.

Below we shall calculate the neutrino masses. Neutrino oscillations arise, since the neutrinos, produced by the weak interactions, are not mass eigenstates, but mixtures of mass eigenstates. Like for the quarks one has a 3×3 mixing matrix, which can be written as a product of three simple matrices, and a phase matrix, which is present only, if neutrinos are Majorana particles:

$$V = U \cdot P\,,$$

$$U = \begin{bmatrix} c_l & s_l & 0 \\ -s_l & c_l & 0 \\ 0 & 0 & 1 \end{bmatrix} \begin{bmatrix} e^{-i\phi} & 0 & 0 \\ 0 & c & s \\ 0 & -s & c \end{bmatrix} \begin{bmatrix} c_\nu & -s_\nu & 0 \\ s_\nu & c_\nu & 0 \\ 0 & 0 & 1 \end{bmatrix} \tag{5}$$

$$P = \begin{bmatrix} e^{i\rho} & 0 & 0 \\ 0 & e^{i\sigma} & 0 \\ 0 & 0 & 1 \end{bmatrix} .$$

Here s_ν stands for $\sin\theta_\nu$ (θ_ν : solar mixing angle), s stands for $\sin\theta$ (θ: atmospheric mixing angle), and s_l stands for $\sin\theta_l$ (θ_l : reactor mixing angle). The latter has not been measured. The Chooz experiment gives an upper limit of 13° for this angle.

The experimental results for the solar and the atmospheric mixing angles are [5,6]:

$$30^o \leq \theta_\nu \leq 39^o \,, \qquad 37^o \leq \theta \leq 53^o \,. \tag{6}$$

We assume for the lepton mass matrices the same texture 0 pattern as for the quarks.[7] Thus we obtain the same relations between the mixing angles and the mass eigenvalues:

$$\tan\theta_l = \sqrt{m_e/m_\mu} \simeq 0.07 \,, \qquad \tan\theta_\nu = \sqrt{m_1/m_2} \,. \tag{7}$$

Since the solar angle has been measured (we take 33°), we find for the mass ratio of the first two neutrinos:

$$m_1/m_2 \approx 0.42 \,. \tag{8}$$

The oscillation experiments provide us with the mass squared differences. The new relation[6] allows us to determine the neutrino masses. We find (in eV):

$$m_1 \approx 0.0036...0.0059 \,, \qquad m_2 \approx 0.0085...0.0140 \,, \qquad m_3 \approx 0.044...0.058 \,. \tag{9}$$

These neutrino masses are very small. We observe also that the masses show a rather weak hierarchy, but the mass spectrum is not inverted. The first neutrino has the smallest mass.

The atmospheric mixing angle is consistent with 45°. The parameter C in eq. (2) might be zero, as originally assumed.[3] But the high mass of the t-quark does not allow this possibility for the quarks. It might work for the leptons. In this case the atmospheric mixing angle is related to the two angles, which are given by the corresponding mass ratios:

$$\tan\theta_1 = \sqrt{m_\mu/m_\tau}\,, \qquad \tan\theta_2 = \sqrt{m_2/m_3}\,; \tag{10}$$
$$\theta_1 \approx 14^o\,, \qquad \theta_2 \approx 24^o\,.$$

The atmospheric mixing angle is given by the absolute value of the sum of the two angles, including a relative phase between the two terms. In order to get the direct sum, this phase must be 180°. In this case we have for the atmospheric mixing angle:

$$\theta = \theta_1 + \theta_2 \simeq 38^o\,. \tag{11}$$

We cannot obtain a maximal mixing (45°), but our result is consistent with the experiment.

We can predict the matrix element V_{3e} of the mixing matrix V:

$$V_{3e} = \sin\theta\sin\theta_l \approx 0.707\sqrt{m_e/m_\mu} \approx 0.05\,. \tag{12}$$

A matrix element of this magnitude could be observed in the upcoming reactor neutrino experiments.

If the neutrino masses are Majorana masses, one expects a neutrinoless double beta decay. The present limit on the Majorana mass is about 0.23 eV [8], if mixing angles are disregarded. We can estimate the effect. Only the third neutrino would play a role, however it couples to the electron with a magnitude, given by V_{3e}. Thus we find an upper limit for the effect of the order of $0.05 \times 0.05 = 0.0025$, which is a factor 100 smaller than the observed limit. The experiments on double beta decay must be improved by a factor 100 in order to see an effect. This seems extremely difficult.

References

1. M. Kobayashi and T. Maskawa, Prog. Theor. Phys. **49**, 652 (1973).
2. H. Fritzsch and Z. Xing, Phys. Lett. B **413**, 396 (1997); H. Fritzsch and Z. Xing, Phys. Rev. D **57**, 594 (1998).
3. H. Fritzsch, *Nucl. Physl.* **B155**, 189 (1979).
4. H. Fritzsch and Z. Xing, Nucl. Phys. B **556**, 49(1999).
5. J. P. Cravens *et al.* [Super-Kamiokande Collaboration], Phys. Rev. D **78**, 032002 (2008).
6. B. Aharmim *et al.*, Phys. Rev. C **72**, 055502 (2005)
7. H. Fritzsch and Z. Xing, Phys. Lett. B **634**, 514 (2006)
8. C. Arnaboldi *et al.* [CUORICINO Collaboration], Phys. Rev. C **78**, 035502 (2008).

Nature of Microscopic Black Holes and Gravity in Theories with Particle Species

Gia Dvali

CERN Theory Division, Switzerland,,
and
CCPP, Department of Physics, New York University, USA
georgi.dvali@cern.ch, gd23@nyu.edu

Relying solely on unitarity and the consistency with large-distance black hole physics, we derive model-independent properties of the microscopic black holes and of short-distance gravity in theories with N particle species. In this class of theories black holes can be as light as $M_{Planck}/\sqrt{N}$ and be produced in particle collisions above this energy. We show, that the micro black holes must come in the same variety as the species do, although their label is not associated with any conserved charge measurable at large distances. In contrast with big Schwarzschildian ones, the evaporation of the smallest black holes is maximally undemocratic and is biased in favor of particular species. With an increasing mass the democracy characteristic to the usual macro black holes is gradually regained. The lowest possible mass above which black holes become Einsteinian is $\sqrt{N}M_{Planck}$. This fact uncovers the new fundamental scale (below the quantum gravity scale) above which gravity changes *classically*, and the properties of black holes and gravity are such as if some extra dimensions open up, although no such input exists in the theory. Our observations indicate that geometry of space may be an emergent concept following from large number of species and the consistency with macro black hole physics. We apply our findings to the phenomenological properties of the micro black holes that can be observed at LHC for large N. Extrapolating our findings to small N, one may ask whether the existence of quark and lepton flavors is already an evidence for emergent extra dimensions at short distances.

1. Introduction

The main theoretical motivation for observing the physics beyond the Standard Model around the TeV scale is the Hierarchy Problem, an inexplicable smallness of the weak interaction scale relative to the Planck mass, $M_P \sim 10^{19}$ GeV. The key problem is that the former scale is determined by the Higgs mass square, which is *quadratically sensitive* to an ultraviolet (UV) cutoff of the theory. In the absence of any new stabilizing physics at intermediate scales, the natural cutoff of the theory is M_P, where the Einsteinian gravity gets strong and can no longer be neglected.

One approach to this problem [1] goes under the name of TeV scale quantum gravity, and is based on the idea that UV divergences in the Higgs mass get cut off by gravity which gets strong around the TeV energies. The original framework realizing this proposal is based on large extra dimensions, in which there are n compact extra dimensions of size R to which gravity can propagate. Then, there is the following relation between M_P and the fundamental high-dimensional Planck mass (M_*), the cutoff of the low energy theory,

$$M_P^2 \,=\, M_*^2\,(M_*R)^n\,. \tag{1}$$

This relation can be simply understood in terms of the high dimensional Gauss law, but for our purpose we shall focus on interpretation in terms of number of particle species. For such interpretation it is enough to notice that the factor $(M_*R)^n \equiv N_{KK}$ measures number of Kaluza-Klein (KK) species with mass of order or below the cutoff. Thus, the relation (1) can be rewritten as

$$M_P^2 \,=\, M_*^2\,N_{KK}\,. \tag{2}$$

Thus, in large extra dimensions the hierarchy between the M_P and the gravity cutoff M_* is set by $\sqrt{N_{KK}}$. This form of the relation is remarkable, since it allows to express the hierarchy between M_P and the gravity cutoff in a geometry-independent way, solely in terms of the number of KK species.

Relying on consistency with the large-distance black hole (BH) physics, it was understood recently [2,3] that the latter relation is much more general and holds true in the presence of particle species irrespective of their geometric meaning. Namely, in a theory with N particle species, both, the bound on their masses and on the gravitational cutoff of the theory is given by the scale

$$M_* \,\equiv\, \frac{M_P}{\sqrt{N}}\,. \tag{3}$$

Thus, any theory with large number of particle species lowers the cutoff, and can potentially address the hierarchy problem, irrespectively whether the species are massive spin-2 states organized in the tower of KK states or are simply degenerated-in-mass lower spin particles. For example, N exact copies of the Standard Model would do the same job.

The macroscopic BH physics indicates, that the masses of species are bounded by,[2]

$$M_{species} \lesssim \frac{M_P}{\sqrt{N}} \,. \tag{4}$$

Consistency of the resulting BH physics with the well-known macroscopic BH properties, such as their thermal nature, also indicates the simultaneous lowering of the gravity cutoff.[3] These non-perturbative BH arguments are in excellent agreement with the perturbative ones[3–5] that signal the breakdown of perturbation theory at the scale (3).

The simultaneous lowering of the species masses and of the gravity cutoff also makes the whole picture consistent with the low energy effective field theory, since the quadratic divergences in the scalar masses are cutoff by strong gravity at the scale M_*. Below we shall be interested in this situation.

Since gravity gets strong at M_*, the model-independent signature of this class of theories is the existence of micro black holes with the mass $M_{BH} \geq M_*$, and size $r_g \geq M_*$. In particular, such micro black holes can be produced in particle collisions at energies above M_*.

The purpose of this paper is to derive some generic properties of these micro black holes, without making any *a priory* assumptions about the nature of the underlying microscopic gravity theory. We shall rely solely on the low energy consistency requirements, such as the unitarity and consistency of the low energy theory with the well-understood large-distance black hole physics. As we shall see, these arguments uncover the existence of the second length scale, larger than the fundamental length M_*^{-1}, at which gravity starts changing the regime *classically*. Thus, before becoming quantum in UV, gravity starts departing from the Newtonian gravity classically already at larger distances. The absolute lower bound on the BH mass below which they can no longer be treated as normal Schwarzschild BHs is given by

$$M_{Schwarzschild\ BH} \geq M_P \sqrt{N} \,. \tag{5}$$

This bound is model independent, and in any theory with N species, black holes lighter than $M_P\sqrt{N}$ must be non-Einsteinian. Notice that, in some theories the departure from the Einsteinian regime can start for even heavier BHs (see below), but never below the limiting mass (5).

In the intermediate regime the properties of the BH are rather peculiar. For example, their evaporation is undemocratic in species. For the same mass BH, different species see the different horizons, so that some do not see it at all. This properties amplify with the decreasing mass. Heavier

HBs on the contrary are more and more democratic, and closer to being Schwarzschildian. By all accounts the emerging gravitational dynamics on intermediate scales is such, as if some extra dimensions open up. This fact may imply that geometry of the extra space is an emergent phenomenon, entirely forced upon us by number of four-dimensional species, and consistency of the theory with a large distance BH physics. In the other words species imply the change of short distance geometry! Extrapolating our findings to moderate values of N, one may ask whether existence of several quark-lepton species in the Standard Model can be regarded as the evidence for such emerging dimensions in UV?

2. Power of Unitarity: Small Black Holes Carry "Hair"

In general the notion of "black holeness" in Einsteinian gravity is associated with the absence of "hair". This association is based on the well-understood properties of the large classical black holes, that are known to satisfy no-hair theorems.[6] The absence of hair implies that large black holes can carry information only about the charges that can be measured at infinity, either classically or quantum mechanically.

Such are the electric or magnetic type charges associated with the massless gauge fields, corresponding classical flux of which can be measured at infinity by the Gauss law, or the charges associated with the Aharonov-Bohm type phases, which are undetectable classically but can be measured quantum-mechanically.[7]

As a result, in the absence of such detectable charges, the Hawking evaporation process of a large Schwarzschild black hole is completely democratic in species. For example, in the presence of N massless particle species, such a black hole will evaporate in each of these species with the same probability.

The nature of the micro black holes is dramatically different. As it was shown in [3], the decay of a lightest micro BH in a theory with N particle species, must be maximally undemocratic. This follows from unitarity. Let us briefly repeat the argument here. Consider a theory of N particle species Φ_j with $j = 1, 2, ...N$, that are inter-coupled only through gravity. Consider a smallest mass micro BH produced in the particle-anti-particle collision of i-th species. Such a black hole can only be produced at energy $\sim M_*$, with a characteristic mass $M_{BH} \sim M_*$. Strictly speaking, the lightest BH is still a quantum object, but it is a strongly gravitating object at the distance of order its Compton wavelength and is on the verge of becoming what one would call a quasi-classical BH. Since both the center of mass energy and an impact parameter are continuous variables, the transition from quantum to classical regime must be gradual. Since the only scale in the problem is M_*, production rate of such states is set by it and we have $\Gamma \sim M_*$. Since the BH was produced in particle-anti-particle collision it is neutral and can decay into particle-anti-particle pairs of all the species. By analogy with the classical Einsteinian BH one would expect this decay to be democratic in all the species. However, this is impossible by unitarity. By unitarity, the BH should be able to decay back into a pair of i-th species by the rate $\Gamma \sim M_*$, but then by the same unitarity, it cannot decay into the other individual species with the same partial rates.

First, if a BH could decay democratically into each of N other species with partial rates exceeding $\frac{M_*}{N}$, the lifetime of the BH would be less than its mass, which would indicate that it is not even a well defined state. In the other words, such a BH cannot exist. The latter conclusion would be impossible to reconcile with the fact that gravity is getting strong at M_* energies.

Secondly, exchange of such a virtual BH would generate an effective operator inducing the inter-species transitions

$$\Phi_i + \bar{\Phi}_i \rightarrow \Phi_j + \bar{\Phi}_j \,, \tag{6}$$

at the rate that would violate unitarity way below M_* energies.

For the consistent picture, a small BH produced in the collision of i-th species should decay predominantly back into the same species, and only with very suppressed partial rates

$$\Gamma_{BH \rightarrow other\ copy} \lesssim M_*/N \tag{7}$$

into each of the other individual species.

Notice, that for the small high-dimensional BHs in large extra dimensions this property is explicit. Consider a micro BH of mass M_* produced in particle collision on a 3 brane where Standard Model species are localized. From the point of view of the four-dimensional observer there are N_{KK} additional species in form of the KK gravitons, however the evaporation rate into them and the Standard Model species is not democratic. Decay rate into each individual KK graviton is

$$\Gamma_{BH \rightarrow\ each\ KK} \sim \frac{M_*}{N} \,, \tag{8}$$

which gives the total evaporation rate $\sim M_*$.

As a second example consider species that are localized on a distant brane. Consider a small BH produced in particle collision on "our" brane.

If the high-dimensional gravitational radius is smaller than the inter-brane distance, the BH cannot evaporate into the species that are localized on the distant brane.

In both examples, the non-democracy has a clear geometric meaning, that can be understood as a consequence of locality in the extra space. However, our discussion shows, that the geometric interpretation may be viewed as completely secondary, and non-democracy of the small BHs holds true in any theory with many species.

Coming back to the four-dimensional setup with N species, we can repeat the same unitarity argument for a BH produced in particle-anti-particle collision of each of the N species, and the situation is completely reciprocal. Thus, we are lead to the conclusion that the smallest BHs are highly non-democratic, and have a memory of belonging to a particular species. Thus, small BHs come in N distinct flavors. In the other words, the microscopic BHs carry hair. This hair is not associated with any conserved charge measurable at large distances either classically of quantum-mechanically, and is microscopic in nature. But its presence does not violate standard no hair theorems, as they are not applicable. Notice that micro BH are not Einsteinian, for example their mass to gravitational radius dependence is inevitably very different, from the Schwarzschildian ones.

We shall now show that the above fact has the profound consequences for understanding the nature of micro BHs. We shall first derive the model-independent properties, and then apply to a concrete case when species are identical copies of each other and are related by symmetry.

As we shall see, although no extra dimensions are postulated in our analysis, the unitarity and other consistency arguments tell us that gravitational physics in the presence of species is such, as if the space of species is a true dimension. For example, the above mentioned memory of the micro BHs can be interpreted as a notion of locality in the imaginary space of species, even though, species were introduced as particle species in four dimensions. This interesting fact may indicate that the locality and geometry may be the emergent phenomena, for accommodating consistency with BH physics in the presence of the multiple particle species.

3. Inevitability of the Larger Crossover Length Scale

Non-democracy of the smallest BHs, inevitably leads to the existence of a second length scale, which we shall denote by R. The key point is that R is larger than the fundamental Planck length M_*^{-1}, and marks the crossover distance, beyond which the black holes become the normal Einsteinian black holes. That is, the BHs larger than R "loose" all the hair associated with the species flavor, and their evaporation become democratic in all the species.

Indeed, we can prove inevitability of this second scale, by performing the following thought experiment. Consider a micro BH produced in collision process of particles and anti-particles of i-th species. The smallest such black holes will start getting created for center of mass energies of order M_* and the impact parameter $\sim M_*^{-1}$. At this point BH are still quantum and as argued above, by unitarity must have a complete memory of the species of origin. Such BHs will quickly decay into their own species, and with $1/N$- suppressed probability to most of the other individual species.

Now, rising the center of mass energy in this thought experiment, we produce heavier and bigger black holes, which gain more classicality, and correspondingly, more inter-species democracy in their evaporation process. This is inevitable for the consistent matching with the well known physics of large Einsteinian BHs.

Indeed, for example had we increased the center of mass energy up to a solar mass, while keeping the impact parameter small (at least smaller than 3 km, the solar Schwarzschild radius), we would start producing macroscopic BHs of the solar mass. Obviously, the evaporation process of such a BH must be completely democratic. The evaporation rates in all the thermally-available species (with masses below the Hawking temperature) are equal.

Thus, we see that by gradually increasing the mass of a micro BH, we make transition from the maximal "favoritism" to the full democracy. The critical size of the BH horizon, at which this transition is complete, we shall denote by R. By default, R is related to the mass of the BH via the usual Schwarzschild relation $R = 2M_{BH}\,G_{Newton}$. Obviously, in the same time R marks the length scale beyond which, for distances $r \gg R$ gravity is a normal Einsteinian gravity with the usual $1/r$ Newtonian potential.

Thus, consistency of the black hole physics provides an important information about the nature of the short distance gravity in theories with N-species. Namely, we learn that gravity starts departing from the Einsteinian (and Newtonian) gravity already at distances that are still larger than the fundamental length M_*^{-1}. The important fact is, that at intermediate distances

$$M_*^{-1} \ll r \ll R\,, \tag{9}$$

gravity is still classical, but the micro BHs of size (gravitational radius) r_g in the interval $M_*^{-1} \ll r_g \ll R$ are *not* Einsteinian BH, and are characterized by an r_g-dependent level of non-democracy with respect to the particle species. As said above, the black holes of sizes $r_g \sim M_*^{-1}$ and $r_g \gtrsim R$ are extreme cases, representing maximally undemocratic and maximally democratic objects.

Again, the distance R in theories with extra dimensions has a very simple physical meaning. It is the size of the compact extra dimensions. But, what we are discovering is that existence of such crossover scale is a generic property of theories with many species. This is remarkable, since at no pint in our discussion we made any assumption or reference about the underlying extra dimensional physics. Yet we are finding an outcome which normally would be interpreted as a signal of an underlying geometry.

Notice that in principle the distance scale R can parametrically be close to M_*^{-1}. In extra dimensional example, this is the case when number of extra dimensions is large, since $R \to M_*^{-1}$ as $n \to \infty$. This possible proximity is however an illusion, since the separation of the relevant energy scales is controlled by N. Indeed, to probe the length scales between M_*^{-1} and R, the BH mass must interpolate within the interval $M_*^{-1} < M_{BH} < M_P^2 R$. Even for the lowest possible value $R \sim M_*^{-1}$ this corresponds to the N-fold change in the energy scale. In the other words, the absolute lower bound on the Einsteinian BH mass is given by (5). The shortest possible mass interval in which BH are non-Einsteinian, therefore is

$$\frac{M_P}{\sqrt{N}} \leq M_{Non-Enisteinian\ BH} \leq M_P\sqrt{N}\,. \tag{10}$$

At this point of the discussion we know the properties of the BHs, such as their masses and decay rates into various species, only at the endpoints of the non-Einsteinian interval. We know that for $r_g \sim M_*^{-1}$ and $r_g \sim R$, the masses of the black holes are $M_{BH} \sim M_*^{-1}$ and $M_{BH} \sim RM_P^2$ respectively. However, we do not know what is the dependence of M_{BH} and of the BH evaporation rate on r_g within the interval.

We now wish to investigate these questions. First, notice that, since in the interval of interest the properties of the BHs are not democratic with respect to all the species, the notions such as BH horizon require certain care. In the other words, horizon with respect to some species, may not be seen as a horizon for the others. This simply follows from the fact that intermediate size BH are not evaporating universally to all the thermally-available species. By consistency, the species that are not produced in the evaporation of a given BH, cannot see the horizon of this BH. Such species can go through the given BH untrapped.

A simple explicit example is provided by species that live on different branes separated by a large extra dimension. Obviously, horizon of a BH created by species on a given brane, will not be seen as a horizon by species on the neighboring brane, if the high-dimensional Schwarzschild radius of a BH is smaller than the inter-brane separation. Even if four-dimensional coordinates of particles on the second brane are within the four-dimensional projection of the BH gravitational radius, they cannot be captured by the BH. Correspondingly the BH will not be able to evaporate into these species. What we are discovering, however, is that irrespectively whether we wish for it, some sort of extra dimensional behavior emerges whenever we talk about many species.

What the above fact tells us is that at the intermediate distances the gravitational force between two particles depends on which species they come from. The two particles (or the localized sources) of the same species with the center of mass energy $\sim M_*$ should gravitate strongly (with $\sim M_*$ gravitational potential), when separated by the distances $\sim M_*^{-1}$. This is not necessarily true when the two particles belong to the different species. Again, this is obvious from the extra dimensional example, since the two point-like sources of mass M_* localized on the two largely-separated 3-branes, will gravitate weakly even if their four-dimensional coordinates coincide within the radius M_*^{-1}.

Therefore, let us first analyze gravity of sources and probes belonging to the same i-th flavor of species. Consider a Newtonian gravitational interaction between the two point-like sources of masses M_i and m_i separated by a distance r, where index i indicates that sources are made out of i-th species. Since we are making no assumption about the existence of any extra dimensions, r will always refer to the distance in our three-dimensional space. When r changes from R to M_*^{-1}, this Newtonian interaction interpolates between the usual Newtonian force and the strong gravity. This interpolation can be parameterized by the following gravitational potential (below everywhere we shall ignore numbers of order one)

$$V(r) = \frac{M_i m_i}{M_*^2} \frac{1}{r\,\nu(M_* r)_{ii}}\,, \tag{11}$$

where $\nu(M_* r)_{ii}$ is some smooth function, such that

$$\nu \sim 1 \quad \text{for} \quad r \sim M_*^{-1}, \quad \text{and} \quad \nu \simeq \frac{M_P^2}{M_*^2} \quad \text{for} \quad r > R\,. \tag{12}$$

The indexes ii indicate that gravitational force is measured between the two sources belonging to the i-th species. Since at intermediate distances gravity is weakly coupled, gravitational radii of the sources can be determined in

Newtonian approximation, from the condition that gravitational potential becomes order one.

Thus, the effective gravitational radius (r_g^{ii}) of the source of mass M_{BHi} seen by the particles of the same i-th species can be estimated from the equation

$$1 = \frac{M_{BHi}}{M_*^2} \frac{1}{r_g^{ii}\, \nu(M_* r_g^{ii})_{ii}} \,. \tag{13}$$

We shall refer to this radius as the self-species horizon.

Crudely speaking, r_g^{ii} marks a distance from the source at which the Newtonian escape velocity for a massive particle of the same species becomes order one, and thus r_g^{ii} can be thought as the BH horizon for the same species. However, as was argued above, r_g^{ii} is not necessarily a horizon for all the other species. Only the species that can be produced in the evaporation process of a given BH, can see its horizon. For any BH belonging to i-th site, the number of such species is a function of the black hole mass, or equivalently of its self-species gravitational radius. $\mathcal{N}(M_* r_g^{ii})_i$. Again, the index i labels the site of origin in the space of species. The boundary properties of this function are similar to (12)

$$\mathcal{N}(1)_i \sim 1, \quad \mathcal{N}(M_* R)_i = \frac{M_P^2}{M_*^2} \,. \tag{14}$$

In general, if we are interested in gravity between the two sources of masses M_i and M_j belonging to the two different i-th and j-th species, we should consider the function $\nu(M_* r)_{ij}$. For $i \neq j$ the only model independent boundary condition is $\nu(MR)_{ij} = M_P^2/M_*^2 = N$, which follows from the definition of R. The necessary condition for the particles of j-th species to be produced in the evaporation process of the BH of mass M_{BHi} made out of i-th species, is that j particles see the horizon of the i-th BH. The corresponding horizon we shall call r_g^{ij}. Generalizing notion of r_g^{ii}, we can try to estimate this quantity again in Newtonian approximation as the radius where Newtonian gravity of the i-th BH exerted on the j-th probe source becomes strong. That is,

$$\frac{M_{BHi}}{M_*^2} \frac{1}{r_g^{ij}\, \nu(M_* r_g^{ij})_{ij}} = 1 \,. \tag{15}$$

Then r_g^{ij} marks the gravitational radius (horizon) of i-th BH with respect to the particles from the j-th species. Then, the function $\mathcal{N}_i$ counts the solutions of (15).

The species-dependence of functions ν_{ij} have a well defined meaning in large extra dimensional example. There the species localized at different extra coordinates will exert different force on each other. Correspondingly, they will see different four-dimensionally-projected horizons of a given high-dimensional BH. This is defined by the metric in the extra dimensional space, and by sites of localization of species. Small BH cannot "reach out" to the species localized on the distant branes. Again, what we are finding is that some notion of metric emerges in the "space of species", irrespectively of any input.

This phenomenon is remarkable because of the following reason. Of course, since species are the orthogonal states, formally one can always think of them as being eigenstates of either coordinate or momentum operators in some mathematical space. That is, the label j can formally correspond to some extra coordinate y_j or a momentum $p_j \equiv -i\frac{\partial}{\partial_{y_j}}$. This is a triviality. The non-trivial part however is that this space inevitably modifies laws of gravity, and thus is not just a mathematical formality.

Coming back to the definition of r_g^{ij}, the following subtlety arises. Is could happen that for a given species r_g^{ij} is defined in a probabilistic sense. This will happen if i and j are not in the same (coordinate or momentum) representations in the space of species. For example, say Φ_i is an eigenvalue of y_i, whereas Φ_j is an eigenvalue of p_j. Of course, one can try to go in the same representation by taking a new superposition of species

$$\sum_j c_j \Phi_j, \tag{16}$$

but the new states will no longer be mass eigenstates in general. So to work with mass eigenstates, besides r_g^{ij}-s we have to introduce the probabilities of capture P_{ij} within the radius r_g^{ij}. By unitarity this is related to the emission probability. The black hole evaporation rate then will be given by

$$\frac{dM_{BHi}}{dt} = -\sum_j (r_g^{ij})^{-2} P_{ij} \, . \tag{17}$$

The summation is over thermally available states only (or else one has to explicitly include Boltzmann-suppression factors). This expression looks singular for $r_g^{ij} = 0$, but it is not, since in this case $P_{ij} = 0$. In the other words, the particles that cannot see the horizon, cannot be produced. The non-zero quantities $(r_g^{ij})^{-2} P_{ij}$ cannot exceed a given fixed value, and for a static black hole presumably must be equal by thermodynamics arguments. Notice that the above expression correctly reproduces the evaporation of the high-dimensional BH in KK theories. Applied to evaporation of a BH of gravitational radius $r_g \ll R$ in n extra dimensions of size R, we get for each KK graviton

$$(r_g^{ij})^{-2}\, P_{ij} \,=\, r_g^{-2} \left(\frac{r_g}{R}\right)^n . \tag{18}$$

Summation up to the thermally available species gives an extra factor $\left(\frac{r_g}{R}\right)^{-n}$, which gives the correct evaporation rate

$$\frac{dM_{BHi}}{dt} \,=\, -\, r_g^{-2} . \tag{19}$$

We shall consider now the class of theories in which the form of these functions is highly restrictive. This is the situation when N species are exact copies of each other, related by an exact permutation symmetry.

4. Micro Black Holes in N-copies of Species

We shall now consider a special case in which the species are identical copies of each other and are related by some exact permutation symmetry. Namely we require that physics is identical as seen from each copy. In this case, by the symmetry the quantities such as r_g^{ii}, ν_{ii} and $\mathcal{N}_i$ are independent of i. Moreover for $N \gg 1$, if we require some analog of flatness in the space of species, then functions ν and $\mathcal{N}$ can be approximated by

$$\nu \,=\, \mathcal{N} \,=\, \left(\frac{r}{R}\right)^n \frac{M_P^2}{M_*^2} , \tag{20}$$

where n is an arbitrary number (not necessarily an integer). The boundary condition $\mathcal{N}(1) = 1$ fixes $R \,=\, (M_P^2/M_*^2)^{\frac{1}{n}} M_*^{-1}$, and thus,

$$\nu \,=\, \mathcal{N} \,=\, (rM_*)^n . \tag{21}$$

In this case gravity is such as if the species are equally spaced in n extra dimensions.

Let us compute the lifetime of a neutral BH of mass M_{BH} produced in the collision of particle and antiparticle of a given species. Gravitational radius of such a BH can be determined from (13). Taking into the account (21), we get

$$r_g \,=\, M_*^{-1} \left(\frac{M_{BH}}{M_*}\right)^{\frac{1}{1+n}} . \tag{22}$$

Not surprisingly, the relation between r_g and M_{BH} are such as if the space has n flat extra dimensions (with n not necessarily an integer), although no extra dimensions have ever been postulated. The change of mass due to Hawking evaporation for any BH of Hawking temperature T_H, is given by

$$\frac{dM_{BH}}{dt} \,=\, -\, T_H^4(M_{BH}) r_g^2\, \mathcal{N}(M_{BH}) , \tag{23}$$

where both the Hawking temperature and the number of available species are expressed as functions of M_{BH}. Since $T_H = r_g^{-1}$, using (21), we get

$$\frac{dM_{BH}}{dt} = M_*^2 \left(\frac{M_*}{M_{BH}}\right)^{\frac{2-n}{1+n}}, \tag{24}$$

which for the BH lifetime gives

$$\tau_{BH} \simeq M_*^{-1} \left(\frac{M_{BH}}{M_*}\right)^{\frac{3}{n+1}} \frac{n+1}{3}. \tag{25}$$

For $n = 0$ this reproduces the lifetime of an usual Schwarzschild BH, as it should. Notice, that this lifetime is by a factor $\left(\frac{M_{BH}}{M_*}\right)^{\frac{n}{n+1}}$ shorter than the lifetime of a BH in n flat extra dimensions with a single 3-brane, and is approximately the lifetime of a BH in n flat extra dimensions in which there are N 4-dimensional species localized at uniformly distributed sites (3-branes). Again, none of this was input in our discussion. Yet, we see that requiring exact symmetry between the species, effectively leads to the behavior as if the species are localized on N uniformly distributed branes in n flat large extra dimensions.[a]

It is also clear that throughout the evaporation process of a heavy BH ($M_{BH} \gg M_*$) produced by particles of i-th copy, only a fraction

$$\frac{E_{BH \to i-\text{th copy}}}{E_{BH \to \text{all copies}}} \sim \left(\frac{M_*}{M_{BH}}\right)^{\frac{n}{n+1}} \tag{26}$$

of the total energy will be released in i-th species. The rest will be distributed over $\mathcal{N} = \left(\frac{M_{BH}}{M_*}\right)^{\frac{n}{n+1}}$ extra copies.

5. 10^{32} Standard Model Copies

We are now ready to apply our findings to the situation of $N = 10^{32}$ Standard Model copies. With this choice the fundamental Planck mass becomes $M_* \equiv M_P/\sqrt{N} \sim$TeV, as needed for the solution of the hierarchy problem.

Consider now an experiment with particle collision performed in our copy of the Standard Model (which we can call simply the Standard Model). By default, around TeV energies and TeV^{-1} impact radius, the micro BHs

[a] It was noted by Michele Redi, that had we gone beyond the requirement of identical physics as seen from each site, and assumed a stronger symmetry under the full permutation group, this would correspond to a maximal departure from the flatness in the species space.

start being created. According to our findings, the smallest mass BHs will decay back to the Standard Model states, and with an extremely suppressed probability into the other copies. It will be virtually impossible to distinguish these lightest BH-s from some new unstable particles, which promptly decay back into the standard model states, plus some new invisible states. However, once the center of mass energy climbs above the fundamental scale, the BH properties will quickly unfold. Both the lifetime, and the fraction of the missing energy will rise according to (25) and (26) respectively. In the same time, the amplitudes must soften, with the typical momenta of the scattering products being cutoff by the inverse BH size r_g^{-1} given by (22). This is a characteristic BH behavior. Notice, that the expected softening of the scattering products is a general qualitative property for the processes with quasi-classical black hole formation regardless of number of species (see [8], and references therein).

By lifting the requirement of exact symmetry, one effectively introduces "warping" in the space of species. This may affect form of the functions ν and $\mathcal{N}$, but for small deformations the qualitative picture should remain as discussed above.

What about the constraints on parameter n? As we saw, just postulating N exact copies of species, gives rise to the gravitational behavior characteristic of extra dimensions. So it is tempting to identify n with (generically fractional) number of extra dimensions. However, irrespectively whether we wish to do so, the important physics is that N species do imply new classical gravitational physics at the length scale R that is related to n the way radius of the extra space would be. This is enough for putting constraints of n. The immediate phenomenological constraint $n > 2$, comes from the absence of new gravitation forces at the distances down to fraction of a millimeter [9]. However, unlike the standard extra dimensional case, we have no argument against non-integer n. So in principle n could be any fractional (or even irrational) number exceeding 2 or so.

6. Discussions and Outlook

We have seen that the low energy consistency requirements, such as unitarity and consistency of theory with large-distance BH physics, are powerful tools in understanding the general properties of microscopic gravity and BHs in the presence of N particle species.

Our analysis uncovers the existence of the second scale, below the gravitational cutoff of the theory, above which gravity starts departing from the Einsteinian (Newtonian) gravity *classically*, even before reaching the quantum regime. Correspondingly, the model-independent lower bound on the mass of Einsteinian BHs is set by (5).

In the window between the two scales, BH gradually depart from the known classical properties, such as democracy in the evaporation process and the absence of hair. Smallest BHs therefore are maximally undemocratic and come in the same variety as the species do.

The overall picture is such as if some new dimensions open up, although nothing like this was input in the theory. This fact may indicate that geometry is a secondary concept, which emerges in order to accommodate consistency of microscopic physics with the low energy unitarity and macro BH physics, in the presence of species. The natural question then is whether our findings can be extrapolated for moderately small values of N, such as the number of quark lepton flavors in the standard model. If yes, then the existence of flavors may be a direct evidence for emerging new dimensions in UV. Of course, if the number of species is limited by the Standard Model alone, corresponding length scale will be small, but this is conceptually unimportant.

On the other hand if the solution of the hierarchy problem is related to the number of species, LHC should observe BH production in particle collisions, irrespectively of the precise origin of the species. The versions of the theory in which species are organized in a symmetric way are especially predictive. The two opposite extremes are, flat large extra dimensions with $N_{KK} = 10^{32}$ KK states, and $N = 10^{32}$ copies of the standard model. The properties of the BHs, such as their evaporation rate and the fraction of missing energy, are different in the two cases.

Acknowledgments

We thank Oriol Pujolas, Michele Redi, Sergey Sibiryakov and Sergey Solodukhin for discussions and comments. This work is supported in part by David and Lucile Packard Foundation Fellowship for Science and Engineering, and by NSF grant PHY-0245068.

References

1. N. Arkani-Hamed, S. Dimopoulos and G. Dvali, Phys. Lett. B **429**, 263 (1998) [hep-ph/9803315]; Phys. Rev. D **59**, 086004 (1999) [hep-ph/9807344].
I. Anatoniadis, N. Arkani-Hamed, S. Dimopoulos and G. Dvali, Phys. Lett. B **436**, 257 (1998) [hep-th/9804398].
2. G. Dvali, "Black Holes and Large N Species Solution to the Hierarchy Problem," arXiv:0706.2050 [hep-th].

For generalization to deSitter space, see G. Dvali and D. Lust, "Power of Black Hole Physics: Seeing through the Vacuum Landscape," arXiv:0801.1287 [hep-th].
3. G. Dvali and M. Redi, Phys. Rev. D **77**, 045027 (2008) [arXiv:0710.4344 [hep-th]].
4. G. Dvali and G. Gabadadze, Phys. Rev. D **63**, 065007 (2001) [hep-th/0008054].
5. G. Veneziano, JHEP **0206**, 051 (2002) [arXiv:hep-th/0110129].
6. W. Israel, Phys. Rev. **164**, 1776 (1967); Commun. Math. Phys. **8**, 245 (1968); B. Carter, Phys. Rev. Lett. **26**, 331 (1971); J. Hartle, Phys. Rev. D **3**, 2938 (1971); J. Bekenstein, Phys. Rev. D **5**, 1239 (1972); Phys. Rev. D **5**, 2403 (1972); Phys. Rev. Lett. **28**, 452 (1972);
C. Teitelboim, Phys. Rev. D **5**, 294 (1972).
7. BH quantum hair under discrete gauge symmetries was first discussed in L.M. Krauss and F. Wilczek, Phys. Rev. Lett. **62**, 1221 (1989); J. Preskill and L.M. Krauss, Nucl. Phys. B **341**, 50 (1990); S. Coleman, J. Preskill and F. Wilczek, Phys. Rev. Lett. **67**, 1975 (1991).
Axionic quantum hair of BHs was discovered in
M. J. Bowick, S .B. Giddings, J.A. Harvey, G. T. Horowitz, A. Strominger, Phys. Rev. Lett. **61**, 2823 (1988); T.J. Allen, M.J. Bowick, A. Lahiri, Phys. Lett. B **237**, 47 (1990).
Massive spin-2 quantum hair of BHs was found in
G. Dvali, Phys. Rev. D **74**, 044013 (2006) [hep-th/0605295]; "Black holes with flavors of quantum hair?," arXiv:hep-th/0607144.
8. G. Veneziano, JHEP **0411**, 001 (2004) [hep-th/0410166].
9. D. J. Kapner, T. S. Cook, E. G. Adelberger, J. H. Gundlach, B. R. Heckel, C. D. Hoyle and H. E. Swanson, Phys. Rev. Lett. **98**, 021101 (2007) [hep-ph/0611184].

Symmetries in the Gross-Neveu Phase Diagram

Gerald V. Dunne

Physics Department, University of Connecticut, Storrs, CT, 06269, USA

The existence of crystalline condensates in the temperature and chemical potential phase diagram of the Gross-Neveu models can be traced to intricate symmetries of the associated inhomogeneous gap equation. The gap equation based on the Ginzburg-Landau expansion is precisely the mKdV or AKNS hierarchy of integrable nonlinear equations for the Gross-Neveu model with discrete or continuous chiral symmetry, respectively. The former model also has a dense-dilute symmetry that is due to the energy-reflection duality of the underlying quasi-exactly soluble spectral operators.

Keywords: gap equation; phase diagram; crystalline condensates; integrability

1. Introduction

The phase diagram of interacting fermion systems at finite density and temperature is a general problem with applications in a wide range of physical contexts. The Gross-Neveu models [1–4] are $(1+1)$-dimensional interacting fermionic quantum field theories that possess some important features in common with QCD; namely asymptotic freedom, dynamical mass generation, and chiral symmetry breaking. The (T, μ) phase diagram of these models have only recently been established,[5,6] and the interesting new feature is that they possess regions in which the system prefers a crystalline condensate to a uniform one. The resulting phase diagram differs significantly from that based on an assumption of a uniform condensate.[7] The [massless] GN models can be solved analytically in the large N_f limit; in this talk I explain why this is possible, and trace it to remarkable symmetry and integrability properties of the associated inhomogeneous gap equation. The simplest Gross-Neveu model, the GN_2 model, has a scalar four-fermion interaction term:

$$\mathcal{L}_{\text{GN}} = \bar{\psi}\, i\, \partial\!\!\!/ \psi + \frac{g^2}{2}\left[\left(\bar{\psi}\psi\right)^2\right] \quad , \tag{1}$$

and has a *discrete* chiral symmetry: $\psi \to \gamma^5 \psi$. The chiral GN, or "NJL_2", model, has both scalar and pseudoscalar four-fermion interaction terms:

$$\mathcal{L}_{\rm NJL} = \bar{\psi}\, i\, \partial\!\!\!/ \psi + \frac{g^2}{2}\left[(\bar{\psi}\psi)^2 + (\bar{\psi} i\gamma^5 \psi)^2\right] \quad , \tag{2}$$

and has a *continuous* chiral symmetry: $\psi \to e^{i\gamma^5 \alpha}\psi$. We suppress flavor indices, but investigate these models in the large N_f limit where the semiclassical approximation applies, and chiral symmetry breaking can be studied.

2. Solving the Inhomogeneous Gap Equation

By a Hubbard-Stratonovich transformation, the four-fermion interaction terms can be expressed in terms of scalar and pseudo-scalar bosonic condensate fields, Σ and Π (respectively), which are conveniently expressed in terms of a *complex* condensate field: $\Delta = \Sigma - i\Pi$. For GN_2 we only have Σ, and so the condensate field Δ is real. The general NJL_2 system can be described equivalently by the effective Lagrangian:

$$\mathcal{L} = \bar{\psi}\left[i\, \partial\!\!\!/ - \frac{1}{2}(1-\gamma^5)\Delta - \frac{1}{2}(1+\gamma^5)\Delta^*\right]\psi - \frac{1}{2g^2}|\Delta|^2, \tag{3}$$

which is now quadratic in the fermion fields. The corresponding single particle fermionic Hamiltonian is

$$\begin{aligned} H &= -i\gamma^5 \frac{d}{dx} + \gamma^0 \left(\frac{1}{2}(1-\gamma^5)\Delta - \frac{1}{2}(1+\gamma^5)\Delta^*\right) \\ &= \begin{pmatrix} -i\frac{d}{dx} & \Delta(x) \\ \Delta^*(x) & i\frac{d}{dx} \end{pmatrix} \quad , \end{aligned} \tag{4}$$

with the choice of the Dirac matrices: $\gamma_0 = \sigma_1$, $\gamma_1 = -i\sigma_2$ and $\gamma^5 = \sigma_3$. This Hamiltonian is also known as the Bogoliubov-de Gennes (BdG) Hamiltonian in the superconductivity literature.[8] Integrating out the fermion fields in (3) we obtain an effective action (per fermion flavor) for the condensate field:

$$S_{\rm eff}[\Delta] = -\frac{1}{2g^2 N_f}\int d^2x |\Delta|^2 - i \ln\det\left[i\, \partial\!\!\!/ - \frac{1}{2}(1-\gamma^5)\Delta - \frac{1}{2}(1+\gamma^5)\Delta^*\right] \tag{5}$$

The gap equation for the condensate field is obtained by looking for the stationary points of $S_{\rm eff}[\Delta]$:

$$\begin{aligned} 0 &= \frac{\delta S_{\rm eff}}{\delta \Delta^*(x)} \\ &= -\frac{\Delta(x)}{2g^2 N_f} - i\frac{\delta}{\delta\Delta(x)^*}\ln\det\left[i\, \partial\!\!\!/ - \frac{1}{2}(1-\gamma^5)\Delta(x) - \frac{1}{2}(1+\gamma^5)\Delta(x)^*\right] \end{aligned} \tag{6}$$

For a *static* condensate, the gap equation (6) can also be written as

$$\Delta(x) = -iN_f g^2 \, \mathrm{tr}_{D,E} \left[\gamma^0 \left(\mathbf{1} + \gamma^5\right) R(x;E)\right] \quad , \tag{7}$$

where $R(x;E)$ is the diagonal Gorkov Green's function, or resolvent:

$$R(x;E) \equiv \langle x | \frac{1}{H-E} | x \rangle \quad . \tag{8}$$

It is straightforward to solve the gap equation when the condensate field Δ is uniform,[7] but it is more technically challenging to solve it for an inhomogeneous condensate field $\Delta(x)$. The general solution has the form of a "twisted kink crystal",[9] described below.

At nonzero temperature and chemical potential, the gap equation (6) is equivalent to minimizing the thermodynamic grand potential

$$\Psi = -\frac{1}{\beta}\int_{-\infty}^{\infty} dE\, \rho(E)\, \ln\left(1+e^{-\beta(E-\mu)}\right) + \frac{1}{2N_f g^2}\frac{1}{L}\int_0^L dx |\Delta(x)|^2 \quad , \tag{9}$$

with respect to $\Delta^*(x)$, where $\beta = \frac{1}{T}$ is the inverse temperature, μ is the chemical potential, and $\rho(E)$ is the single-particle spectral function for the BdG hamiltonian (4), for quantized fermions in the presence of the background condensate field $\Delta(x)$. $\rho(E)$ is expressed in terms of $R(x;E)$ as

$$\rho(E) = \frac{1}{\pi}\, \mathrm{Im}\, \mathrm{Tr}_{D,x}\left[R(x;E+i\epsilon)\right] \quad , \tag{10}$$

where the trace is a Dirac trace as well as a spatial trace. It is clear from (9) and (10) that the dependence of the grand potential Ψ on the condensate $\Delta(x)$ is encoded in the spectral function $\rho(E)$, via the resolvent $R(x;E)$.

The key to solving the gap equation is to note that the form of the 2×2 matrix $R(x;E)$ is highly constrained by elementary differential equations considerations.[8,9] These considerations imply the nontrivial constraints

$$R^\dagger = R \quad , \quad \mathrm{tr}_D(\sigma_3\, R(x;E)) = 0 \quad , \quad \det R = -\frac{1}{4} \quad . \tag{11}$$

And $R(x;E)$ must satisfy a simple first order matrix differential equation

$$\frac{\partial}{\partial x} R(x;E)\sigma_3 = i[\begin{pmatrix} E & -\Delta(x) \\ \Delta^*(x) & -E \end{pmatrix}, R\sigma_3] \quad , \tag{12}$$

known as the Eilenberger equation.[8–10] For example, if $\Delta = M$ is a constant, we have the familiar solution

$$R(x;E) = \frac{1}{2\sqrt{M^2-E^2}}\begin{pmatrix} E & M \\ M & E \end{pmatrix} \quad . \tag{13}$$

The inhomogeneous gap equation (6) can be solved by the following simple resolvent ansatz (motivated by the gap equation [9)]:

$$R(x;E) = \mathcal{N}(E) \begin{pmatrix} a(E) + |\Delta|^2 & b(E)\,\Delta - i\Delta' \\ b(E)\,\Delta^* + i\Delta^{*\prime} & a(E) + |\Delta|^2 \end{pmatrix} \quad , \tag{14}$$

where $\mathcal{N}(E)$, $a(E)$ and $b(E)$ are functions of the energy E, determined by the solution. The ansatz (14) automatically solves the diagonal part of the Eilenberger equation (12), while the off-diagonal part requires that the condensate field Δ satisfy the complex nonlinear Schrödinger equation (NLSE):

$$\Delta'' - 2|\Delta|^2\,\Delta + i\,(b(E) - 2E)\,\Delta' - 2\,(a(E) - E\,b(E))\,\Delta = 0\,. \tag{15}$$

The advantage of this ansatz approach is that the NLSE (15) has a general bounded solution, expressed in closed-form in terms of elliptic functions.[9] This solution has the form of a twisted kink crystal, whose *phase* winds though a given angle each period, and whose *amplitude* is also periodically modulated, as shown in Figure 1. The explicit form is

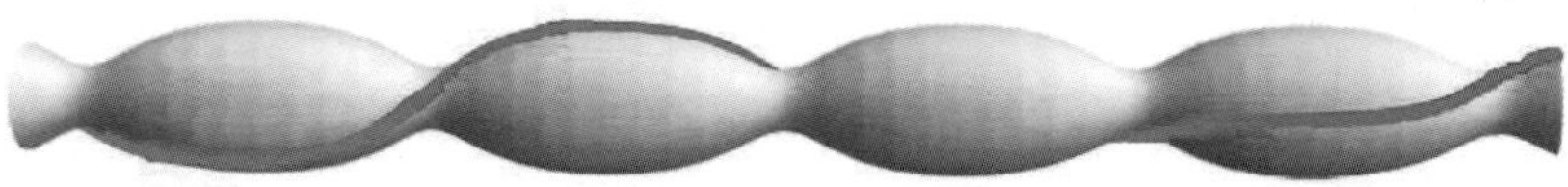

Fig. 1. The twisted kink crystal condensate of (16), shown as the red curve. The blue skeleton surface is shown just to illustrate the periodic amplitude modulation and phase winding.

$$\Delta(x) = -\lambda\, e^{2iqx}\, A\, \frac{\sigma(\lambda A\, x + i\mathbf{K}' - i\theta/2)}{\sigma(\lambda A\, x + i\mathbf{K}')\sigma(i\theta/2)} \\ \times\; \exp\left[i\lambda A\, x\,(-i\,\zeta(i\theta/2) + i\,\mathrm{ns}(i\theta/2)) + i\,\theta\eta_3/2\right] \tag{16}$$

where the functions σ and ζ are the Weierstrass sigma and zeta functions,[11] chosen to have real and imaginary half-periods: $\omega_1 = \mathbf{K}(\nu)$, and $\omega_3 = i\,\mathbf{K}' \equiv i\mathbf{K}(1-\nu)$. Both periods are therefore controlled by a single [real] elliptic parameter $0 \le \nu \le 1$. Note that $\eta_3 = \zeta(i\mathbf{K}')$ is pure imaginary. The angular parameter θ takes values in the range $\theta \in [0, 4\mathbf{K}'(\nu)]$. The [real] constant A is a function of θ and the elliptic parameter ν: $A = A(\theta,\nu) = -2i\,\mathrm{sc}(i\theta/4;\nu)\,\mathrm{nd}(i\theta/4;\nu)$.

The twisted kink crystal solution (16) depends on four parameters: the overall scale λ, the elliptic parameter ν that determines the period, the angular parameter θ that determines through what angle the kink winds over each period, and the parameter q that determines the overall offset of the fermion spectrum. This reflects the important fact that the associated Dirac spectrum has precisely two gaps. [Recall that the generic situation in a one dimensional periodic system is an *infinite* number of gaps; special condensates $\Delta(x)$ lead to a "finite-gap" system, and these are the periodic analog of the "reflectionless" non-periodic systems found in the inverse scattering approach.[2–4] The solution (16) can alternatively be characterized by the four band edges of this two-gap fermion spectrum: E_i, $i = 1, 2, 3, 4$. Furthermore, the associated energy functions $\mathcal{N}(E)$, $a(E)$ and $b(E)$ are simple functions of E, with coefficients determined by E_i. For details see [6,9].

An important special case corresponds to the GN_2 model, where the condensate is real [as there is no pseudoscalar condensate], and so the BdG hamiltonian has an additional symmetry, $\{H, \sigma_2\} = 0$, which implies that the single-particle fermion spectrum is necessarily symmetric about 0. This occurs when $q = 0$ and $\theta = 2\mathbf{K}'$, leaving just two parameters, λ and ν. This is consistent, since with this additional symmetry the two-gap system

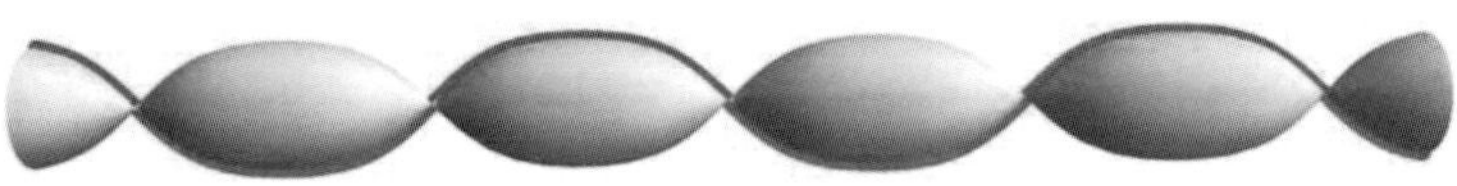

Fig. 2. The real kink crystal condensate of (17), shown as the red curve. The blue skeleton surface is shown just to illustrate the periodic amplitude modulation and phase winding. For this real kink crystal, the amplitude vanishes each period, and the kink rotates through π [i.e., changes sign] each period.

is characterized by just two band edges. Indeed, the general solution of the (real) NLSE in this case is

$$\Delta(x) = \lambda\,\tilde{\nu}\,\frac{\mathrm{sn}(\lambda\,x;\tilde{\nu})\,\mathrm{cn}(\lambda\,x;\tilde{\nu})}{\mathrm{dn}(\lambda\,x;\tilde{\nu})} \quad , \quad \tilde{\nu} \equiv \frac{4\sqrt{\nu}}{(1+\sqrt{\nu})^2} \tag{17}$$

which is shown in Figure 2. This real kink crystal describes the crystalline phase of the GN_2 model.[5,6]

Another important special case is the *spiral* crystal, when the amplitude of the condensate is constant, while the period winds with constant pitch:

$$\Delta = \lambda\,e^{2iqx} \tag{18}$$

as shown in Figure 3. This spiral crystal describes the crystalline phase of the NJL_2 model.[6]

Fig. 3. The spiral condensate of (18), shown as the red curve. The blue skeleton surface is shown just to illustrate the periodic phase winding. In contrast to the twisted kink crystal in Figure 1, for the spiral, the amplitude is constant.

3. Thermodynamic Phase Diagrams

Given the exact solution $\Delta(x)$ in (16) to the gap equation (6), it is also possible to find the exact single-particle eigenstates of the BdG hamiltonian (4). The corresponding spectral function $\rho(E)$ is known analytically, so we can analyze the thermodynamical properties of this model precisely. Since the general solution to the inhomogenous gap equation depends on a finite number of parameters, the problem of minimizing the grand potential now reduces to minimization with respect to these parameters. This is done for GN_2 in [5], and for NJL_2 in [6].

In the case of the GN_2 model (1), there are just two parameters, and the minimization is leads to the phase diagram shown in Figure 4. All tran-

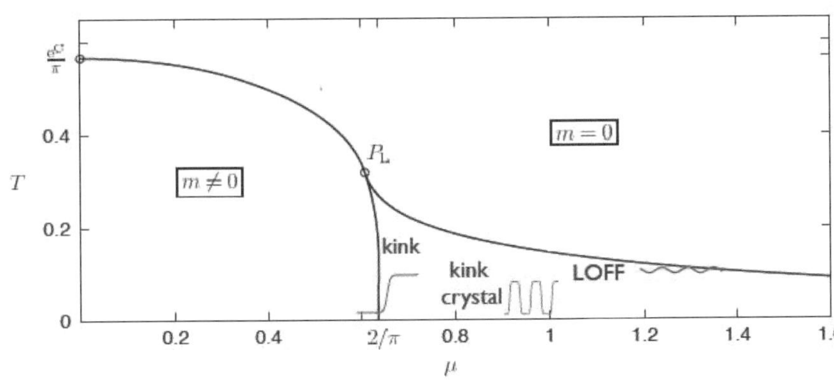

Fig. 4. The phase diagram of the GN_2 model, showing the crystalline phase at low T and large μ. The transitions are discussed in the text.

sitions are second order, and there is a tricritical point at $T_{tc} = 0.318$, $\mu_{tc} = 0.608$. It is interesting to consider the transition from the massless

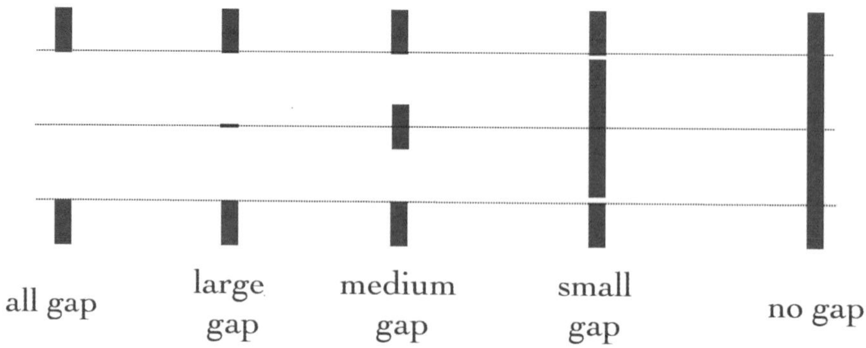

Fig. 5. The evolution [from right to left] of the two-gap fermion spectrum as the system crosses from the massless phase through the crystal phase and into the homogeneous massive phase.

region to the massive region in terms of the form of the condensate and also in terms of the form of the associated single-particle spectrum, as depicted in Figure 5. The transition from the massless phase to the massive phase is driven by the one dimensional Peierls instability,[5] which favors the opening of a gap in the spectrum. Since the spectrum must the symmetric about 0, the simplest way to open a gap is to have just one gap centered at 0, and this corresponds to the low μ situation where one crosses directly from the massless to the massive phase. On the other hand, one could also open a gap away from $E = 0$, but the symmetry then requires that this gap have a partner gap, reflected about $E = 0$. In this way one has a two-gap system, whose condensate must be of the real kink crystal form in (17). This corresponds to crossing the boundary from the massless phase into the crystal phase, going from right to left in Figure 5. Near the phase boundary, the amplitude of the condensate is weak, and the solution behaves like a LOFF condensate $\Delta(x) \approx \lambda \sin(qx)$, characterized by the elliptic parameter $\nu \approx 0$. Here the two gaps are small. As one moves deeper into the crystal phase region, the gaps increase, which means that the shape of the crystal changes: the period and amplitude change with T and μ. Finally, as one approaches the other boundary, to cross from the crystalline region into the homogeneous massive region, the elliptic parameter approaches 1 and the crystal looks like a very dilute collection of isolated baryons, the kink-antikink DHN baryons.[2] The baryon mass is $2/\pi$ in units of the renormalized fermion mass, explaining the $T = 0$ intercept in Figure 4. Here, the gaps have become very large, and the central band shrinks and disappears, as on the left of Figure 5, leading to the single-gap spectrum of a massive fermion.

The GN_2 phase diagram has another special symmetry. The phase transition line between the massless and the crystal phase is characterized by elliptic parameter $\nu = 0$, while the phase transition line between the crystal phase and the massive phase is characterized by elliptic parameter $\nu = 1$. There is a perturbative/nonperturbative duality between these two limits obtained by the transformation $\nu \to (1-\nu)$, which relates the single-particle spectra of these two physically different regions.[12] Small ν corresponds to a weak sinusoidal LOFF form of the condensate, while $\nu \approx 1$ corresponds to a low density crystal of well-separated DHN baryons.

In the case of the NJL_2 model (2), there are four parameters, and the minimization[6] leads to the phase diagram shown in Figure 6. The transition from the massless to the crystal region is a straight line at $T = e^{\gamma}/\pi = 0.5669$, and the crystalline condensate has the spiral form (18), with $\lambda = \lambda(T)$, and $q = \mu$.

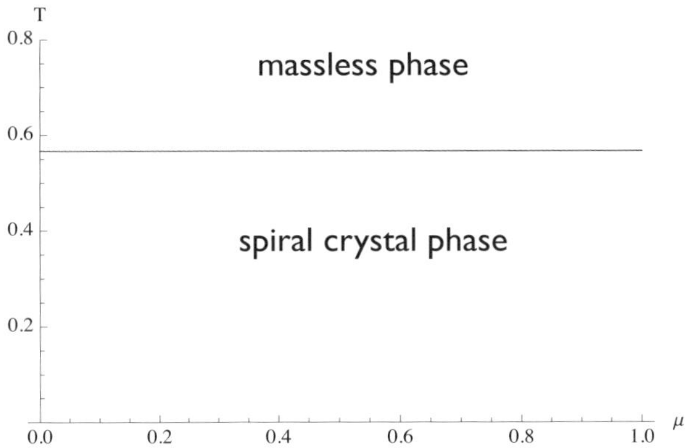

Fig. 6. The phase diagram of the NJL_2 model, showing the spiral crystalline phase.

4. Ginzburg-Landau Expansion and Integrable Hierarchies

It is highly instructive to compare these exact solutions of the inhomogeneous gap equation for the Gross-Neveu models with the general approach based on the Ginzburg-Landau (GL) expansion.[9,13] In the GL expansion, the thermodynamic grand potential Ψ is expanded in powers of the condensate $\Delta(x)$ and its derivatives. The precise form of this expansion, and the dependence of the coefficients on T and μ, is determined by the symmetries of the microscopic interacting fermion system. The GL expansion is derived from an asymptotic expansion of the resolvent: $R(x;E) = \frac{1}{2}\sum_{n=0}^{\infty} \hat{r}_n(x)/E^n$. Inserting the corresponding expansion of the spectral function $\rho(E)$ into the grand potential (9), and performing the energy integrals, we obtain the GL expansion:

$$\Psi = \sum_{n=0}^{\infty} \alpha_n(T,\mu)\, \hat{G}_n \quad , \qquad \hat{G}_n \equiv \frac{1}{2}\int dx\, \mathrm{tr}_D\left(\hat{r}_n(x)\right) \quad . \tag{19}$$

The T and μ dependent coefficients $\alpha_n(T,\mu)$ are known functions.[5,9] The explicit form of the $\hat{G}_n$ is generated by inserting the resolvent expansion into the Eilenberger equation (12) and matching powers of E. Using the constraints (11), we write

$$\hat{r}_n(x) = \begin{pmatrix} \hat{g}_n(x) & \hat{f}_{n-1}(x) \\ \hat{f}^*_{n-1}(x) & \hat{g}_n(x) \end{pmatrix} \quad , \tag{20}$$

and obtain simple recursion formulas for the $\hat{f}_n$ and $\hat{g}_n$:

$$\hat{f}_n = -\frac{i}{2}\hat{f}'_{n-1} + \Delta\, \hat{g}_n \quad , \quad \hat{f}_{-1} = 0 \quad , \tag{21}$$

$$\hat{g}'_n = i\left(\Delta^* \hat{f}_{n-1} - \Delta\, \hat{f}^*_{n-1}\right) \quad , \quad \hat{g}_0 \quad . \tag{22}$$

The initial values follow from the $\Delta = M$ (constant) case above. It is easy to generate these terms recursively. The first few off-diagonal terms are:

$$\begin{aligned} \hat{f}_0 &= \Delta \quad , \\ \hat{f}_1 &= -\frac{i}{2}\Delta' \quad , \\ \hat{f}_2 &= -\frac{1}{4}\left(\Delta'' - 2|\Delta|^2\Delta\right) \quad , \\ \hat{f}_3 &= \frac{i}{8}\left(\Delta''' - 6|\Delta|^2\Delta'\right) \quad , \\ \hat{f}_4 &= \frac{1}{16}\left(\Delta^{(\mathrm{iv})} - 8|\Delta|^2\Delta'' - 2\Delta^2\Delta^{*\prime\prime} - 4|\Delta'|^2\Delta - 6\Delta^*\Delta'^2 + 6|\Delta|^4\Delta\right) \quad . \end{aligned} \tag{23}$$

The first few diagonal terms are:

$$
\begin{aligned}
\hat{g}_1 &= 0 \quad , \\
\hat{g}_2 &= \frac{1}{2}|\Delta|^2 \quad , \\
\hat{g}_3 &= \frac{i}{4}\left(\Delta\Delta'^* - \Delta'\Delta^*\right) \quad , \\
\hat{g}_4 &= \frac{1}{8}\left(3|\Delta|^4 + 3|\Delta'|^2 - \left(|\Delta|^2\right)''\right) \quad , \\
\hat{g}_5 &= \frac{i}{16}\left(\Delta'''\Delta^* - \Delta\Delta^{*'''} + \Delta'\Delta^{*''} - \Delta''\Delta^{*'} + 6|\Delta|^2(\Delta^{*'}\Delta - \Delta'\Delta^*)\right) \quad .
\end{aligned}
\tag{24}
$$

Note that in principle there are integration constants appearing in the $\hat{g}_n$, but these all vanish by virtue of the normalization condition in (11).

These recursively generated quantities $\hat{f}_n$ and $\hat{g}_n$ define the AKNS hierarchy of nonlinear integrable equations.[14,15] In the special case where Δ is real, only the even-indexed terms survive, and the AKNS hierarchy reduces to the mKdV hierarchy. These hierarchies have many interesting properties. For example, the functional variation of $\hat{G}_n = \int dx \hat{g}_n(x)$ is related to $\hat{f}_{n-2}$:

$$
\frac{\delta \hat{G}_n}{\delta \Delta^*(x)} = \frac{(n-1)}{2}\,\hat{f}_{n-2}(x) \quad . \tag{25}
$$

The physical reason for this remarkable relation is the following. The variation of the grand potential (9) can clearly be expressed as

$$
\frac{\delta \Psi}{\delta \Delta^*(x)} = \sum_{n=0}^{\infty} \alpha_n(T,\mu)\,\frac{\delta \hat{G}_n}{\delta \Delta^*(x)} \quad . \tag{26}
$$

On the other hand, the grand potential (9) can alternatively be written, including the thermodynamical Fermi factor in the energy trace, as

$$
\Psi = \int \frac{dE}{2\pi}\frac{1}{e^{\beta(E-\mu)}+1}\,\mathrm{Tr}_{D,x}\,\ln\left[\gamma^0\left(E-H\right)\right] \quad . \tag{27}
$$

Using the BdG Hamiltonian H in (4), we find the variation of the grand potential with respect to Δ^* can be written as

$$
\begin{aligned}
\frac{\delta \Psi}{\delta \Delta^*(x)} &= \int \frac{dE}{2\pi}\frac{1}{e^{\beta(E-\mu)}+1}\,\mathrm{tr}_D\left[\gamma^0(\mathbf{1}+\gamma^5)\,R(x;E)\right] \\
&= \sum_{n=2}^{\infty} \alpha_n(T,\mu)\,\frac{(n-1)}{2}\hat{f}_{n-2}(x) \quad .
\end{aligned}
\tag{28}
$$

Comparing (26) and (28) we find the non-trivial relation (25).

GL expansion of NJL_2 gap equation	$\leftrightarrow$	AKNS integrable hierarchy,
GL expansion of GN_2 gap equation	$\leftrightarrow$	mKdV integrable hierarchy.

This implies that the gap equation of the NJL_2 model is precisely an equation of the AKNS integrable hierarchy, in the sense that the GL expansion to a particular order leads to a gap equation $\sum_n (n-1)\alpha_n \hat{f}_{n-2} = 0$, which is the AKNS equation at that hierarchy order. For example, expanding up to $n = 4$, we obtain from (23) the NLSE (15). The integrability properties of the AKNS hierarchy imply that a solution to AKNS at order N satisfies AKNS at order M, *for all* $M > N$. This remarkable fact explains why it was possible to solve exactly the gap equation using the solution to the NLSE equation. This solution solves the hierarchy to all orders, and when the expansion coefficients are resummed, they give the exact solution contained in the anstaz (14). In the GN_2 model where the condensate is real, we only have the even terms in the hierarchy, and this is known as the mKdV hierarchy.

It is interesting to note that the thermodynamic minimization of the grand potential of the GN_2 model selects the real kink crystal condensate in (17), which has the form of the solution to the real NLSE, the lowest nontrivial member of the mKdV hierarchy. Similarly, in the NJL_2 model, the thermodynamic minimization of the grand potential selects the spiral crystal condensate in (18), which has the form of the solution to the lowest nontrivial member of the AKNS hierarchy.

5. Conclusion

The phase diagrams of the Gross-Neveu models, with either discrete or continuous chiral symmetry, have crystalline phases, and the corresponding crystalline condensates can be found analytically as solutions to the inhomogeneous gap equation. These condensates have the form of the simplest finite-gap solutions of the associated integrable hierarchies, mKdV and AKNS, respectively. The Ginzburg-Landau expansions of these models generate precisely these integrable hierarchies through the gap equation. Furthermore, in the GN_2 model, the phase diagram has a dense-dilute duality symmetry that can be identified with the energy-reflection duality symmetry studied in [12].

Acknowledgements: This talk is based on work done with G. Basar, F. Correa, M. Plyushchay, M. Thies and M. Shifman, each of whom I thank. It is a particular pleasure to thank Misha Shifman on this happy occasion. This work is supported in part by DOE grant DE-FG02-92ER40716.

References

1. D. J. Gross and A. Neveu, "Dynamical Symmetry Breaking In Asymptotically Free Field Theories," Phys. Rev. D **10**, 3235 (1974).
2. R. F. Dashen, B. Hasslacher and A. Neveu, "Semiclassical Bound States In An Asymptotically Free Theory," Phys. Rev. D **12**, 2443 (1975).
3. S. S. Shei, "Semiclassical Bound States In A Model With Chiral Symmetry," Phys. Rev. D **14**, 535 (1976).
4. J. Feinberg, "All about the static fermion bags in the Gross-Neveu model," Annals Phys. **309**, 166 (2004) [arXiv:hep-th/0305240].
5. M. Thies, "From relativistic quantum fields to condensed matter and back again: Updating the Gross-Neveu phase diagram," J. Phys. A **39**, 12707 (2006) [arXiv:hep-th/0601049].
6. G. Basar, G. V. Dunne and M. Thies, "Inhomogeneous Condensates in the Thermodynamics of the Chiral NJL_2 model," Phys. Rev. D **79**, 105012 (2009), [arXiv:hep-th/0903.1868].
7. U. Wolff, "The Phase Diagram Of The Infinite N Gross-Neveu Model At Finite Temperature And Chemical Potential," Phys. Lett. B **157**, 303 (1985).
8. I. Kosztin, S. Kos, M. Stone and A. J. Leggett, "Free energy of an inhomogeneous superconductor: A wave-function approach", Phys. Rev. B **58**, 9365 (1998); S. Kos and M. Stone, "Gradient expansion for the free energy of a clean superconductor", Phys. Rev. B **59**, 9545 (1999).
9. G. Basar and G. V. Dunne, "Self-consistent crystalline condensate in chiral Gross-Neveu and Bogoliubov-de Gennes systems," Phys. Rev. Lett. **100**, 200404 (2008), [arXiv:hep-th/0803.1501]; G. Basar and G. V. Dunne, "A Twisted Kink Crystal in the Chiral Gross-Neveu model," Phys. Rev. D **78**, 065022 (2008), [arXiv:/hep-th0806.2659].
10. G. Eilenberger, "Transformation of Gorkov's Equation for Type II Superconductors into Transport Like Equations", Z. Phys. **214**, 195 (1968).
11. D. F. Lawden, *Elliptic Functions and Applications*, (Springer-Verlag, New York, 1980).
12. G. V. Dunne and M. Shifman, "Duality and Self-Duality (Energy Reflection Symmetry) of Quasi-Exactly Solvable Periodic Potentials," Annals Phys. **299**, 143 (2002). [arXiv:hep-th/0204224].
13. F. Correa, G. V. Dunne and M. S. Plyushchay, "The Bogoliubov/de Gennes system, the AKNS hierarchy, and nonlinear quantum mechanical supersymmetry," arXiv:0904.2768 [hep-th], Ann. Phys., in press.
14. M. J. Ablowitz, D. J. Kaup, A. C. Newell and H. Segur, "The Inverse scattering transform fourier analysis for nonlinear problems," Stud. Appl. Math. **53**, 249 (1974).
15. F. Gesztesy and H. Holden, *Soliton Equations and their Algebro-Geometric Solutions*, (Cambridge University Press, 2003).

Strongly Coupled Condensate of High Density Matter

Gregory Gabadadze

Center for Cosmology and Particle Physics, Department of Physics, New York University, New York, NY 10003, USA

Arguments are summarized, that neutral matter made of helium, carbon, etc., should form a quantum liquid at the above-atomic but below-nuclear densities for which the charged spin-0 nuclei can condense. The resulting substance has distinctive features, such as a mass gap in the bosonic sector and a gap-less spectrum of quasifermions, which determine its thermodynamic properties. I discuss an effective field theory description of this substance, and as an example, consider its application to calculation of a static potential between heavy charged impurities. The potential exhibits a long-range oscillatory behavior in which both the fermionic and bosonic low-energy degree of freedom contribute. Observational consequences of the condensate for cooling of helium-core white dwarf stars are briefly discussed.

Foreword

Like many in the audience, I first met Misha on the pages of journal publications, before meeting him in person. While working on an undergraduate thesis at Moscow University, I came across Misha's review paper "Anomalies and Low-Energy Theorems of Quantum Chromodynamics" [1]. Impressions of that work were very distinct – a clear exposition of subtle field theory aspects of the quantum anomalies, culminating in creative applications to low-energy hadron phenomenology. The work stood out by its originality, depth, inspiration and balance of the formalism and applications – the remarkable signatures of Misha's enormous contribution to theoretical physics at the forefront of both field theory and particle phenomenology.

I met Misha in person in Minneapolis in 1998. The discussion with him was very inspiring. Soon, in Aspen, we started to work on a project. A bit later I ceased the opportunity to get exposed to two years of the unique FTPI experience. We continued to work on and off on various projects since then. I value those works very highly, and feel privileged, as I'm sure many of you do too, for having such a collaborator.

Happy 60th Birthday Misha!

Description of charged condensate

Consider a neutral system of a large number of nuclei each having charge Z, and neutralizing electrons. If average inter-particle separations in this system are much smaller than the atomic scale, $\sim 10^{-8}\ cm$, while being much larger than the nuclear scale, $\sim 10^{-13}\ cm$, neither the atomic nor nuclear effects will play any significant role. Moreover, the nuclei can also be treated as point-like particles.

In what follows we focus on spin-0 nuclei with $Z \leq 8$ (helium, carbon, oxygen), and consider the electron number-density in the interval $J_0 \simeq (0.1 - 5\ MeV)^3$. Then the electron Fermi energy will exceed the electron-electron and electron-nucleus Coulomb interaction energy. Moreover, at temperatures below $\sim 10^7\ K$, which are of interest here, the system of electron represents a degenerate Fermi gas.

Since the nuclei (we also call them ions below) are heavier, temperature at which they'll start to exhibit quantum properties will be lower. Let us define the "critical" temperature T_c, at which the de Broglie wavelengths of the ions begin to overlap

$$T_c \simeq \frac{4\pi^2}{3m_H d^2}, \qquad d \equiv \left(\frac{3Z}{4\pi J_0}\right)^{1/3}, \tag{1}$$

where, m_H denotes the mass of the ion (the subscript $''H''$ stands for heavy), and d denotes the average separation between the ions[a].

Somewhat below T_c quantum-mechanical uncertainties in the ion positions become greater than an average inter-ion separation. Hence the latter concept looses its meaning as a microscopic characteristic of the system; the ions enter a quantum-mechanical regime of indistinguishability. Then, the many-body wavefunction of the spin-0 ions should be symmetrized, and this would unavoidably lead to probabilistic "attraction" of the bosons to condense, i.e., to occupy one and the same quantum state. We refer the system of condensed nuclei and electrons as charged condensate.

In the condensate the scalars occupy a quantum state with zero momentum. Moreover, small fluctuations of the bosonic sector happen to have a mass gap, $m_\gamma = (Ze^2 J_0/m_H)^{1/2}$, which exceeds T_c by more than an order of magnitude. Therefore, once bosons are in the charged condensate, their

[a] The de Broglie wavelength above is defined as $\lambda_{dB} = 2\pi/|\mathbf{k}|$, where $\mathbf{k}^2/2m_H = 3k_BT/2$. We define T_c as the temperature at which $\lambda_{dB} \simeq d$. Note that this differs by a numerical factor of $\sqrt{2\pi/3}$ from the standard definition of the *thermal* de Broglie wavelength, $\Lambda \equiv \sqrt{2\pi/mk_BT}$, that appears in the partition function of an ideal gas of number-density n in the dimensionless combination $\Lambda^3 n$.

phonons cannot be thermally excited. However, the gap-less fermionic degrees of freedom are thermally excited, and carry the most of the entropy of the entire system.[2–5]

For further discussions it is useful to rewrite the expression for T_c in terms of the mass density $\rho \equiv m_H J_0$ measured in g/cm^3:

$$T_c = \rho^{2/3} \left(\frac{3.5 \cdot 10^2}{Z^{5/3}} \right) K \,, \tag{2}$$

where the baryon number of an ion was assumed to equal twice the number of protons, $A = 2Z$ (true for helium, carbon, oxygen...). Thus, for $\rho = 10^6 \ g/cm^3$ and helium-4 nuclei we get $T_c \simeq 10^6 \ K$, while for the carbon nuclei with the same mass density $T_c \simeq 2 \cdot 10^5 \ K$.

Temperature at which the condensation phase transition takes place, $T_{condens}$, need not coincide with T_c. Moreover, we would expect $T_{condens} \lesssim T_c$. Calculation of $T_{condens}$ from the fundamental principles of this theory is hard. However, we can obtain an interval in which $T_{condens}$ should fit. For this we introduce the following parametrization:

$$T_{condens} = \zeta \, T_c \,, \tag{3}$$

where ζ is an unknown dimensionless parameter that should depend on density more mildly than T_c does. Numerically, however, this parameter should vary in the interval $0.1 \ll \zeta \lesssim 1$: The point $\zeta = 0.1$ would corresponds to the temperature of the Bose-Einstein (BE) condensation of a free gas for which, $T^{BE}_{condens} \simeq 1.3/m_H d^2$, is known from the fundamental principles. The condensation temperature in our system should be higher than $T^{BE}_{condens}$ since the repulsion makes easier for the condensation to take place.[6] In our case, repulsive interactions between the bosons are *strong* – the Coulomb energy is at least an order of magnitude greater that any other energy scale in the system. Hence, we should expect $\zeta \gg 0.1$. On the other hand, given the definition of T_c, the parameter ζ cannot be greater than unity. In what follows we will retain ζ in our expressions, but use $\zeta \simeq 1$ when it comes to numerical estimates.

The condensation will take place after gradual cooling, only if $T_{condens}$ is greater than the temperature at which the substance could crystallize. A classical plasma crystallizes when the Coulomb energy becomes about ~ 180 times greater than the average thermal energy per particle.[7–9] This gives the following crystallization temperature[b]

$$T_{\rm cryst} \simeq \rho^{1/3} \left(0.8 \cdot 10^3 Z^{5/3}\right) \, K\,. \tag{4}$$

Note that the density dependence of T_c is different from that of $T_{\rm cryst}$ – for higher densities T_c grows faster, making condensation more and more favorable! One can define the "equality" density for which $T_{condens} = T_{\rm cryst}$:

$$\rho_{\rm eq} = \left(\frac{2.3}{\zeta}\right)^3 Z^{10}\, g/cm^3\,. \tag{5}$$

For helium, $Z = 2$, and $\rho_{\rm eq} \simeq 10^4\ g/cm^3$, while for carbon, $Z = 6$, and $\rho_{\rm eq} \simeq 10^9\ g/cm^3$ (as mentioned above, we use $\zeta \simeq 1$). These results are very sensitive to the value of ζ; for instance, $\rho_{\rm eq}$ could be an order of magnitude higher if $\zeta \simeq 0.5$. Irrespective of this uncertainty, however, the obtained densities are in the right ballpark of average densities present in helium-core white dwarfs $\sim 10^6\ g/cm$, (for carbon dwarfs, they're closer to those expected in high density regions only.[5])

Is the charged condensate a ground state of the system at hand? For the higher values of the density interval considered, the crystal would not exist due to strong zero-point oscillations. At lower densities, the crystalline state has lower free energy (at least near zero temperature) due to more favorable Coulomb binding. Hence, the condensate can only be a metastable state. The question arises whether after condensation at $\sim T_{condens}$ the system could transition at lower temperatures $\sim T_{\rm cryst}$ to the crystal state, as soon as the latter becomes available.

In the condensate, the boson positions are entirely uncertain while their momenta equal to zero. In order for such a system to crystallize later on, each of the bosons should acquire energy of the zero-point oscillations of crystal ions. As long as this energy, $\sim (Ze^2 J_0/m_H)^{1/2}$, is much greater than $T_{\rm cryst}$, no thermal fluctuations can excite the condensed bosons to transition to the crystalline state. The latter condition is well-satisfied for all the densities considered in this work. There could, however, exist a spontaneous transition of a region of size R_c to the crystallized state via tunneling.

---

[b] The presented formula for the crystallization temperature is entirely *classical*. The temperature scale that determines the classical versus quantum nature of the crystallization transition is the Debye temperature $\theta_D \simeq 4 \cdot 10^3 \rho^{1/2}\ K$. Often, θ_D may significantly exceed $T_{\rm cryst}$ [10]. In such cases, quantum zero-point oscillations should be taken into account. This seems to delay the formation of quantum crystal, lowering $T_{\rm cryst}$ from its classical value at most by about $\sim 10\%$ [11]. Since this is a small change, we will ignore it in our estimates.

The value of R_c, and the rate of this transition, will be determined, among other things, by tension of the interface between the condensate and crystal state, which is hard to evaluate. However, for estimates the following qualitative arguments should suffice: the height of the barrier for each particle is $(Ze^2 J_0/m_H)^{1/2} = m_\gamma$, while the number of bosons in the R_c region $\sim R_c^3 J_0/Z$. Hence, the transition rate should scale as $\exp(-m_\gamma J_0 R_c^4/Z)$. Since we expect that $R_c > 1/m_\gamma$, the rate is strongly suppressed for the parameters at hand.

Effective field theory description

We use a low-energy effective field theory description to study the charged condensate. Even though realistic temperatures in the system may be well above zero, we focus on the zero-temperature limit. The relevance of this limit is justified *a posteriori* and goes as follows: the spin-0 nuclei undergo the condensation to the zero-momentum state; their phonons cannot be excited since their gap, m_γ, is greater than T_c. On the other hand, gapless near-the-Fermi-surface quasielectrons will be excited. Therefore, all the thermal fluctuations will end up being stored in the fermionic quasiparticles. For the latter, however, the finite temperature effects aren't significant since their Fermi energy is so much higher, $T/J_0^{1/3} \ll 10^{-2}$. We note that the finite temperature effects, in a general setup with condensed bosons, were calculated in Refs. [12,13].

We begin at scales that are well below the heavy mass scale m_H, but somewhat above the scale set by $\max[\mu_f, m_e]$, where μ_f and m_e are the electron chemical potential and mass respectively. Hence the electrons are described by their Dirac Lagrangian, while for the description of the nuclei we will use a charged scalar *order parameter* $\Phi(x)$. As it was shown in [4], in a non-relativistic approximation for the nuclei, the effective Lagrangian proposed by Greiter, Wilczek and Witten (GWW) [14] in a context of superconductivity, is also applicable here, given that an appropriate reinterpretation of its variables and parameters is made.

The construction of the GWW Lagrangian is based on the following fundamental principles: it is consistent with the translational, rotational, Galilean and the global $U(1)$ symmetries, preserves the algebraic relation between the charged current density and momentum density, gives the Schrödinger equation for the order parameter in the lowest order, and is gauge invariant.[14] Combined with the electron dynamics the GWW Lagrangian reads (we omit for simplicity the Maxwell term):

$$\mathcal{L}_{eff} = \mathcal{P}\left(\frac{i}{2}(\Phi^* D_0\Phi - (D_0\Phi)^*\Phi) - \frac{|D_j\Phi|^2}{2m_H}\right) + \bar{\psi}(i\gamma^\mu D^f_\mu - m_f)\psi, \quad (6)$$

where we use the standard notations for covariant derivatives with the appropriate charge assignments: $D_0 \equiv (\partial_0 - iZeA_0)$, $D_j \equiv (\partial_j - iZeA_j)$, $D^f_\mu = \partial_\mu + ieA_\mu$, while $\mathcal{P}(x)$ stands for a general polynomial function of its argument. The coefficients of this polynomial, $\mathcal{P}(x) = \sum_{n=0}^{\infty} C_n x^n$, are dimensionful parameters that are inversely proportional to powers of a short-distance cutoff of the effective field theory.[c]

Once the basic Lagrangian is fixed, we introduce the electron chemical potential term $\mu_f\psi^+\psi$, where $\mu_f = \epsilon_F = [(3\pi^2 J_0)^{2/3} + m_f^2]^{1/2}$. This is the only term that at the tree level sets a frame in which the electron total momentum is zero.

There exists a homogeneous solution of the equations of motion that follow from the effective Lagrangian (6) [3]:

$$Z|\Phi|^2 = J_0\,, \quad A_\mu = 0, \quad \mathcal{P}'(0) = 1\,. \quad (7)$$

(We use the unitary gauge $\Phi = |\Phi|$). The condition $\mathcal{P}'(0) = 1$ is satisfied by any polynomial functions $\mathcal{P}(x)$ for which the first coefficient is normalized to unity

$$\mathcal{P}(x) = x + C_2 x^2 + \ldots\,. \quad (8)$$

The above solution describes a neutral system of negatively charged electrons of charge density $-eJ_0$, and positively charged scalar condensate of charge density $Ze\Phi^+\Phi = eJ_0$ [4,5].

Calculation of the spectrum of small perturbations is straightforward. The Lagrangian density for the fluctuations in the quadratic approximation

---

[c] In general one should also add to the Lagrangian terms $\mu_{NR}\Phi^*\Phi$, $\lambda(\Phi^*\Phi)^2/m_H^2$, $\lambda_1(\Phi^*\Phi)\bar{\psi}\psi/(m_H J_0^{1/3})$, and other higher dimensional operators that are consistent with all the symmetries and conditions that lead to (6) (the Yukawa term is not). Here μ_{NR} denotes a non-relativistic chemical potential for the scalars. These terms are not important for the low-temperature spectrum of small perturbations we're interested in, as long as $\lambda, \lambda_1 \lesssim 1$ and $J_0 \ll m_H^3$. However, near the phase transition point it is the sign of μ_{NR} that would distinguish between the broken and symmetric phases, so these terms should be included for the discussion of the symmetry restoration. We also note that the scalar part of (6) is somewhat similar to the Ginzburg-Landau (GL) Lagrangian for superconductivity. However, there are significant differences between them, one such difference being that the coherence length in the GL theory is many orders of magnitude greater than the average interelectron separation, while in the present case, the "size of the scalar" Φ is smaller that the average interparticle distance.

reads [2]

$$\mathcal{L}_2 = -\frac{1}{4}F_{\mu\nu}^2 + \frac{1}{2}m_0^2 A_0^2 - \frac{1}{2}m_\gamma^2 A_j^2 + \frac{1}{2}\, A_0 \frac{(2m_H m_\gamma)^2}{-\Delta} A_0 \,, \tag{9}$$

where Δ denotes the Laplacian, and the last term emerged due to mixing of A_0 with the fluctuation of the $|\Phi|$, which we integrated out. As before,

$$m_\gamma^2 \equiv \frac{Ze^2 J_0}{m_H}\,, \tag{10}$$

and $m_0^2 \equiv C_2 e^2 J_0^2$. At this stage we retained the fermionic fluctuations only in the Thomas-Fermi approximation [3]; an important refinement of this approximation, discussed in [4], will be included below.

That there are no pathologies in (9), such as ghost and/or tachyons, can be seen by calculating the Hamiltonian density:

$$\mathcal{H} = \frac{\pi_j^2}{2} + \frac{F_{ij}^2}{4} + \frac{1}{2}(\partial_j \pi_j)\left(m_0^2 + \frac{4M^4}{-\Delta}\right)^{-1}(\partial_j \pi_j) + \frac{1}{2}m_\gamma^2 A_j^2 \,. \tag{11}$$

Here, $M^2 \equiv m_H m_\gamma$ and $\pi_j \equiv -F_{0j}$. The Hamiltonian is positive semi-definite. Moreover, the spectrum has a mass gap determined by m_γ (10). There are two transverse polarizations of a massive photon, as well as the longitudinal mode, the phonon, with the same mass m_γ [2].

The massive bosonic collective excitations give rise to exponentially suppressed contributions to the value of specific heat of the charged condensate since typically $m_\gamma \gg T_c$. The suppression scales as $\exp(-m_\gamma/T)$, where $T \lesssim T_c$. This is in contrast with the crystal, where the dominant contribution to the specific heat comes from a gap-less phonon, and scales with temperature as T^3.

As to the electrons, their behavior is similar in both crystal and condensate cases. At temperatures of interest they form a degenerate Fermi gas with gap-less excitations near the Fermi surface. Their contribution to the specific heat scales linearly with temperature. In the case of crystallized substance this is sub-dominant to the specific heat due to the crystal phonon. For the charged condensate, however, the (quasi)electron fluctuations are the dominant contributors to the specific heat.

To study the effects of collective bosonic and fermionic modes, as an interesting example, we look at a potential between two impurity nuclei (say hydrogen, or helium-3) of charge Q_1 and Q_2. The calculation of the propagator that involves the light collective modes (for relativistic fermions) gives the following result [4]:

$$V_{stat} = \alpha_{\rm em} Q_1 Q_2 \left(\frac{e^{-Mr}}{r} \cos(Mr) + \frac{4\alpha_{\rm em}}{\pi} \frac{k_F^5 \sin(2k_F r)}{M^8 r^4} \right) . \qquad (12)$$

The first, exponentially suppressed term modulated by a periodic function, is due to cancellation between the screened Coulomb potential and that of a phonon.[4] The fact of such a cancellation, and that it could give rise to the oscillatory behavior of the exponentially screened potential was pointed out before in Ref. [15] in the context of superconductivity.[d]

Most important, however, is the second term in (12) that has a long-range.[4] It dominates over the exponentially suppressed term in (12) for scales of physical interest, and exhibits the power-like behavior modulated by a periodic function.

The potential (12) is a generalization of the Friedel potential to the case when in addition to the fermionic excitations there are also collective modes due to the charged condensate. The long-range oscillating term in (12) is also a result of a subtraction between the conventional Friedel term and the long-range oscillating term due to a phonon. As a result, its magnitude is suppressed compared to what it would have been in a theory without the condensed charged bosons[4] (see, [16] for the discussion of the conventional Friedel potential, and Ref. [13] for its recent detailed study in the presence of the charged condensate at finite temperature.)[e]

The potential (12) is not sign-definite. In particular, it can give rise to attraction between like charges; this attraction is due to collective excitations of both fermionic and bosonic degrees of freedom. This represents a generalization of the Kohn-Luttinger[18] effect to the case where on top of the fermionic excitations the collective modes of the charged condensate are also contributing.[f]

---

[d] I'd like to thank Ki-Myeong Lee who recently brought the paper [15] to my attention.

[e] Note that for spin-dependent interactions the same effects of the charged condensate would give a generalization of the Ruderman-Kittel-Kasuya-Yosida (RKKY) potential.[17]

[f] In the charged condensate Cooper pairs of electrons can also be formed, however, the corresponding transition temperature, and the magnitude of the gap, are suppressed by a factor $\exp(-1/e_{eff}^2)$, where e_{eff}^2 is proportional to the value of the inter-electron potential that contains both screened Coulomb and phonon exchange. The fact that this potential has attractive domain, but is very small, is suggested by the static potential found in [4] (see also eq. (12) above); the latter is suppressed by a power of a large scale M. Furthermore, taking into account the frequency dependence of the propagator in the Eliashberg equation does not seem to change qualitatively the conclusion on a strong suppression of the Green's function and pairing temperature.
Hence, even though the bosonic sector (condensed nuclei) is superconducting at reason-

Applications to White Dwarfs

The above described system of electrons and nuclei constitutes cores of white dwarf stars. Up to a factor of a few, these are roughly Earth's size solar-mass objects; their mass density may range over $\sim (10^6 - 10^{10})\ g/cm^3$, most of them being near the lower edge of this interval. Since the dwarf stars exhausted thermonuclear fuel in their cores already, they evolve by cooling [19]; the ones that we consider in this work cool from $\sim 10^7\ K$ down to lower temperatures.

As a typical dwarf star cools down, the Coulomb interaction energy in a classical plasma of charged nuclei will significantly exceed their classical thermal energy, and the nuclei, in order to minimize energy, would organize themselves into a crystal lattice.[20] In most of these cases quantum effects of the nuclei should be negligible; for instance, the Debye temperature should be less than the temperature at which crystallization takes place, and the de Broglie wavelengths of the nuclei should be much smaller than the average internuclear separations. This indeed is the case in majority of white dwarf stars, the cores of which are composed of carbon and/or oxygen nuclei and span the interval of mass densities around $\sim (10^6 - 10^8)\ g/cm^3$.

However, there exists a class of dwarf stars in which the nuclei enter the quantum regime before the classical crystallization process sets in [10,11]. Among these, furthermore, there is a relatively small subclass of the dwarf stars for which the temperature T_c, is higher than the would-be crystallization temperature T_{cryst} [5]. In such dwarf cores the charged condensation should be expected to take place.

White dwarfs composed of helium constitute a smaller sub-class of dwarf stars (see, [21,22] are references therein); they exhibit best conditions for the charged condensation. Most of helium dwarfs are believed to be formed in binary systems, where the removal of the envelope off the dwarf progenitor red giant by its binary companion happened before helium ignition, producing a remnant that evolves to a white dwarf with a helium core. Helium dwarf masses range from $\sim 0.5\ M_\odot$ down to as low as $(0.18-0.19)\ M_\odot$, while their envelopes are mainly composed of hydrogen.

Using the approach of [23], and following [5] we will consider an over-simplified model of a reference helium star of mass $M = 0.5\ M_\odot$ with the

ably high temperatures $\lesssim 10^6\ K$, interactions with gap-less fermions could dissipate the superconducting currents. Only at extremely low temperatures, exponentially close to the absolute zero, the electrons could also form a gap leading to superconductivity of the whole system. In the present work we consider temperatures at which electrons are not condensed into Cooper pairs, and ignore the finite temperature effects.

atmospheric mass fractions of the hydrogen, and heavy elements (metallicity) respectively equal to

$$X \simeq 0.99, \qquad Z_m \simeq (0.0002 - 0.002)\ . \tag{13}$$

The lower value of the metallicity $Z_m \simeq 0.0002$ is appropriate for the recently discovered 24 He WDs in NGC 6397 [22], but for completeness, we consider a wider range for this parameter.

It is straightforward to find the following expression for the cooling time of a star in the classical regime[23]

$$t_{He} = \frac{k_B}{CAm_u}\left[\frac{3}{5}(T_f^{-\frac{5}{2}} - T_0^{-\frac{5}{2}}) + Z\frac{\pi^2}{3}\frac{k_B}{E_F}(T_f^{-\frac{3}{2}} - T_0^{-\frac{3}{2}})\right], \tag{14}$$

where T_f and T_0 denote the final and initial core temperatures. The first term in the bracket on the right hand side corresponds to cooling due to classical gas of the ions and the second term corresponds to the contribution coming from the Fermi sea. The latter is sub-dominant in the range of final temperatures we are interested in (the factor Z in front of this term is due to Z electrons per ion). Since $T_f \ll T_0$, the age of a dwarf star typically doesn't depend on the initial temperature. Neglecting the fermion contribution, we find time that is needed to cool down to critical temperature $T_f = T_c$

$$t_{He} = \frac{3}{5}\frac{k_B T_c M}{Am_u L(T_c)} \simeq (0.76 - 7.6)\ \text{Gyr}\,. \tag{15}$$

Where an order of magnitude interval in (15) is due to the interval in the envelope metallicity composition given in (13). We also find the corresponding luminosities

$$L(T_c) \simeq (10^8\ erg/s)\frac{M}{M_\odot}\left(\frac{T_c}{\mathrm{K}}\right)^{7/2} \simeq 1.5\cdot(10^{-4} - 10^{-5})L_\odot\,, \tag{16}$$

which are in the range of observable luminosities ($L_\odot \simeq 3.84 \cdot 10^{33}\ erg/s$).

After the condensation, specific heat of the system dramatically drops as the collective excitations of the condensed nuclei become massive and "get extinct". A contribution from the Fermi sea, which is strongly suppressed by the value of Fermi energy, becomes the dominant one. The phase transition itself would take some time to complete, and the drop-off in specific heat will not be instantaneous.

In the zeroth approximation, we can regard the transition to be very fast, and retain only the fermion contribution to specific heat below T_c. Then, the expression for the age of the star for $T_f < T_c$, reads as follows

$$t'_{He} = \frac{k_B}{CAm_u}\left[\frac{3}{5}(T_c^{-\frac{5}{2}} - T_0^{-\frac{5}{2}}) + Z\frac{\pi^2}{3}\frac{k_B}{E_F}(T_f^{-\frac{3}{2}} - T_0^{-\frac{3}{2}})\right]. \tag{17}$$

Notice the difference of (17) from (14) – in the former $T_f < T_c$ and it is T_f that enters as final temperature in the fermionic part, while T_c should be taken as the final temperature in the bosonic part.

From the ratio of ages, $\eta = t_{He}/t'_{He}$, for two identical helium dwarf stars, with and without the interior condensation, we deduce that the charged condensation substantially increases the rate of cooling– the age could be twenty times less than it would have been without the condensation phase.[5]

The condensation of the core would induce significant deviations from the classical curve for helium white dwarfs. What is independent of the uncertainties involved in these discussions, is the fact that the luminosity function (LF) will experience a significant drop-off after the charged condensation phase transition is complete. This is due to the "extinction" of the bosonic quasiparticles below the phase transition point. In fact, the LF will drop by a factor of ~ 200. This may be relevant for an explanation of the observed termination of a sequence of the 24 He WD's found in Ref. [22]. See Ref. [5] for more details.

Finally, the magnetic properties of the charged condensate, which are similar to those of type II superconductor, and in particular admit the presence of Abrikosov's vortices, were studied in Ref. [24]. As was shown there, only very strong magnetic fields, $\gtrsim 10^7$ *Gauss*, will be able to penetrate the dwarf cores in the vortices, while weaker fields will be entirely expelled from it.

Acknowledgments

The above-reported results constitute a part of the work done in collaboration with Rachel A. Rosen and David Pirtskhalava in Refs. [2–5], [24]. I'd like to thank Nima Arkani-Hamed, Paul Chaikin, Daniel Eisenstein, Leonid Glazman, Andrei Gruzinov, Stefan Hofmann, Andrew MacFadyen, Juan Maldacena, Aditi Mitra, Slava Mukhanov, Hector Rubinstein, Malvin Ruderman, Daniel Stein and Arkady Vainshtein for useful discussions and correspondence on these topics. The work was supported by the NSF grant PHY-0758032.

References

1. M.A. Shifman, "Anomalies and Low-Energy Theorems of Quantum Chromodynamics," Phys. Rept. **209**, 341 (1991) [Sov. Phys. Usp. **32**, 289 (1989)].
2. G. Gabadadze and R. A. Rosen, Phys Lett. B **658**, 266 (2008); *ibid.* **666**, 277 (2008).

3. G. Gabadadze and R. A. Rosen, JCAP **0810**, 030 (2008).
4. G. Gabadadze and R. A. Rosen, JCAP **0902**, 016 (2009).
5. G. Gabadadze and D. Pirtskhalava, JCAP **0905**, 017 (2009).
6. K. Huang, Phys. Rev. Lett. **83**, 3770 (1999).
7. D. Q. Lamb and H. M. Van Horn, Ap. J. **200**, 306 (1975).
8. S. Ichimaru, Rev. Mod. Phys. **54**:1017 (1982)
9. G. S. Stringfellow, H. E. DeWitt and W. I. Slattery, Phys. Lett. A **41**, 1105 (1990).
10. G. Chabrier, N. W. Ashcroft and H. E. DeWitt, Nature, **360**, 48 (1992).
11. G. Chabrier, Ap. J. **414**, 695 (1993).
12. A. D. Dolgov, A. Lepidi and G. Piccinelli, JCAP **0902**, 027 (2009).
13. A. D. Dolgov, A. Lepidi and G. Piccinelli, "Screening effects in plasma with charged Bose condensate," arXiv:0905.4422 [hep-ph]
14. M. Greiter, F. Wilczek and E. Witten, Mod. Phys. Lett. B **3**, 903 (1989).
15. K. Lee and O. Tchernyshyov, Mod. Phys.Lett. A **13**, 987 (1998) [cond-mat/9707202].
16. A. L. Fetter, J. D. Walecka, *"Quantum Theory of Many-Particle Systems"*, McGraw-Hill, 1971.
17. M. A. Ruderman and C. Kittel, Phys. Rev. **96**, 99 (1954); T. Kasuya, Prog. Theor. Phys. **16**, 45 (1956); K. Yosida, Phys. Rev. **106**, 893 (1957).
18. W. Kohn, J. H. Luttinger, Phys. Rev. Lett. **15**, 524 (1965).
19. L. Mestel, MNRAS, **112**, 583 (1952).
20. L. Mestel and M. A. Ruderman, MNRAS, **136**, 27 (1967).
21. J. Liebert, P. Bergeron, D. Eisenstein, H.C. Harris, S.J. Kleinman, A. Nitta and J. Krzesinski, Ap. J. **606**, L147, (2004), [astro-ph/040429].
22. R. R. Strickler, A. M. Cool, J. Anderson, H. N. Cohn, P. M. Lugger and A. M. Serenelli, "Helium-Core White Dwarfs in the Globular Cluster NGC 6397" arXiv:0904.3496 [astro-ph.GA].
23. S.L. Shapiro and S. A. Teukolsky, *Black Holes, White Dwarfs, and Neutron Stars*, John Wiley & Sons, (1983).
24. G. Gabadadze and R. A. Rosen, JHEP **0907**, 093 (2009) [arXiv:0905.2444 [hep-th]].

Exotic Superfluidity in Cold Atoms

Elena Gubankova*

Tufts University, Boston, MA
E-mail: elena1@mit.edu

We derived the low energy effective action for the collective modes in asymmetric fermionic systems with attractive interaction. We obtained the phase diagram in terms of the chemical potentials. It features a stable gapless superfluidity with one Fermi surface on the BEC side of the resonance. Also we predict a sharp increase in outer core of a vortex, i.e. vortex size, upon entering into the gapless phase. This may serve as a signature of a gapless phase.

Keywords: BCS-BEC crossover; asymmetric ultracold Fermi systems; gapless superfluidity.

1. Introduction

Experiments in cold atomic gasses renewed the interest in fermionic pairing.[1,2] In particular, much effort was made to understand superfluidity in imbalanced fermionic gasses. Imbalanced systems have unequal number of particles of two fermionic species that pair. The system consists of fermions of two different species, which correspond to two hyperfine states of a fermionic atom such as ^{6}Li. These fermions have opposite spin and interaction between them can be tuned by using a Feshbach resonance.[3]

For zero imbalance, the system properties are qualitatively well understood using mean field.[4] In weak coupling the system lives in a weakly coupled BCS state and crosses over to a strongly coupled BEC state through the resonance region. While the extreme BCS and BEC regimes are in good control in mean field theory, quantitative understanding of the phases close to resonance comes mainly from Monte-Carlo calculations.[5] At resonance, fluctuations change mean field only quantitatively. Situation is completely different at nonzero polarization, where fluctuations change mean field results around the resonance qualitatively; and many features of the phase

* This work is done in collaboration with Massimo Mannarelli and Rishi Sharma.

diagram are not caught by the mean field. For imbalanced systems, a qualitatively complete picture of the phase diagram has not been established yet. At small asymmetry, and weak coupling, there is standard BCS superfluidity, which breaks down at Clogston limit. At large asymmetry, proposed possibilities are phase-separation,[6] gapless (or breached) superfluidity,[7–10] deformed Fermi sea pairing,[11] and non-homogeneous or LOFF pairing.[12]

The phase diagram of the system at T=0 as a function of scattering length and the chemical potential difference has been explored. In the mean field approximation,[13] it was found that on the BCS side of the resonance there are no stable gapless superfluid phases with net polarization; and at the resonance the first order phase transition from superfluid to normal phase takes place as asymmetry is increased, without any intervening gapless superfluids. In experiment with trapped atoms this leads that the system phase separates, because the gapped phase can not feature a net polarization, unpolarized superfluid is in the central region of the trap and a polarized normal fluid is at the exterior.[13,14] There was an effort done to go beyond the mean field approximation by using Monte-Carlo simulations,[15] an expansion in $\epsilon = d - 4$ space dimensions,[16] $1/N$ loop expansion around the BCS-BEC solution,[17,18] a superfluid local density approximation.[19,20]

The authors of [16] propose a phase diagram in the plane of scaled polarization $\eta \sim \delta\mu$ and diluteness parameter $\kappa = -1/na^3$ which features a splitting point near the resonance at nonzero mismatch where the homogeneous superfluid, a LOFF like inhomogeneous phase, and the gapless superfluid phase coexist. They also find stable gapless fermionic modes with one and two Fermi surfaces on the BCS side of the resonance. This picture is supported by Monte-Carlo simulations.[15]

In order to understand how fluctuations in the condensate around the mean field value affect the phase diagram of cold atomic gasses with asymmetry, we study the effective theory describing these fluctuations.

2. Effective low energy theory

We consider a non-relativistic system consisting of two species of fermions ψ_1 and ψ_2 of equal mass m but different chemical potentials μ_1 and μ_2. We assume that the Feshbach interaction between fermions of different species, given via the scattering length, can be modeled by a point like four Fermi interaction with coupling $g = 1/(k_F a_s)$, $n = k_F^3/(2\pi^2)$. BCS is at $g \to -\infty$, $a_s < 0$; BEC is at $g \to \infty$, $a_s > 0$. The effect of the attractive interaction between fermions is to produce a difermion condensate. Condensate spontaneously breaks $U(1)$ associated with conservation of total fermion number (fermions are not charged); hence there is one Goldstone mode and the system is superfluid. By gauging $U(1)$, condensate spontaneously breaks local $U(1)$; hence the gauge boson becomes massive, i.e. there is Meissner mass and the system is a superconductor. We established a correspondence between coefficients of effective theory for Goldstone mode and screening masses of the gauge field. Mean field Lagrangian is quadratic in fermion fields,

$$\mathcal{L} = \Psi^\dagger \begin{pmatrix} i\partial_t - \xi(\mathbf{p}) + \delta\mu\sigma^3 & -\Delta(x)\varepsilon \\ \Delta^*(x)\varepsilon & i\partial_t + \xi(\mathbf{p}) - \delta\mu\sigma^3 \end{pmatrix} \Psi - \frac{|\Delta(x)|^2}{\lambda}, \tag{1}$$

Ψ stands for the four component Nambu-Gorkov spinor. We consider a homogeneous condensate, independent of x, i.e. $\Delta(x) = \Delta$, as a mean-field solution. The quasiparticle dispersion law is $\epsilon_\pm = \pm\delta\mu + \sqrt{\xi(\mathbf{p})^2 + \Delta^2}$. At some momenta it features zero energies, describing spherical Fermi surfaces in momentum space.

We include fluctuations of the condensate around the mean field solution, by introducing the field $\eta(x)$, $\Delta(x) = \Delta + \eta(x)$. Partition function contains fermion field Ψ and fluctuation of the condensate fields,

$$Z = \int \mathcal{D}\eta^* \mathcal{D}\eta \mathcal{D}\Psi^\dagger \mathcal{D}\Psi e^{-\mathcal{S}[\Psi^\dagger, \Psi, \eta, \eta^*]}\,. \tag{2}$$

To get an effective action for the fluctuations, we integrate out the fermion fields; we get the fermion determinator,

$$\begin{aligned} S[\eta, \eta^*] = \int d^4x \left\{ \frac{1}{g} |\Delta + \eta(x)|^2 \right\} \\ - \left\{ \frac{1}{2} \operatorname{Tr} \log \begin{pmatrix} -\partial_{x^4} - \xi(\mathbf{p}) + \delta\mu & -(\Delta + \eta(x)) \\ -(\Delta + \eta^*(x)) & -\partial_{x^4} + \xi(\mathbf{p}) + \delta\mu \end{pmatrix} \right\}. \end{aligned} \tag{3}$$

We expand the fermion determinator in fluctuations and their derivatives, $\mathcal{S}[\eta, \eta^*] = \mathcal{S}^{(0)} + \mathcal{S}^{(1)} + \mathcal{S}^{(2)} + \ldots$. As a result we get an effective theory of collective modes (Goldstone and Higgs) away from unitarity. To the leading order, $S^{(0)}$ is the free energy of the system without fluctuations. Due to the gap equation the next term $S^{(1)}$ vanishes (stationary point of the action). The lowest nontrivial term in the expansion is $S^{(2)}$,

$$\mathcal{S}^{(2)} = -\frac{1}{2}\frac{T}{V} \sum_k (\lambda(-k)\theta(-k)) \begin{pmatrix} M_{11}(k) & M_{12}(k) \\ M_{21}(k) & M_{22}(k) \end{pmatrix} \begin{pmatrix} \lambda(k) \\ \theta(k) \end{pmatrix}. \tag{4}$$

with $\eta = (\lambda + i\theta)/\sqrt{2}$. It is UV finite; UV divergent contributions cancel exactly by employing the gap equation. (Note, the gap equation is UV divergent and it is regulated by the cutoff Λ, usually the Fermi momentum, which is the highest scale in the problem). We separate magnitude and the phase of the fluctuations, $\eta = (\lambda + i\theta)/\sqrt{2}$. The physical meaning of the real and complex components of the fluctuation field is easy to understand in small fluctuation and long wavelength limit,

$$\mathcal{S}^{(2)} \rightarrow \begin{pmatrix} -C + Dk_0^2 - \frac{E}{3}k^2 & ik_0 F \\ -ik_0 F & Ak_0^2 - \frac{B}{3}k^2 \end{pmatrix} . \tag{5}$$

Let us first consider the phase of condensate, but not its magnitude, fluctuates, $\Delta \rightarrow \Delta e^{i\phi(x)}$. The field ϕ represents the Goldstone mode associated with the spontaneous breaking of the total fermion number, $n_1 + n_2$. There is no explicit breaking, hence the mass of Goldstone boson is zero. In the low energy Lagrangian density,

$$\mathcal{L}_\phi = \Delta^2 \left[A(\partial_t \phi(x))^2 - \frac{B}{3}(\partial_i \phi(x))^2 \right] , \tag{6}$$

A is the kinetic energy, B is the spatial variation of the Goldstone mode. Negative A or B means instability towards the growth of phase fluctuations. If A and B are positive, the system is stable and the speed of sound is $\sqrt{B/3A}$. In our analyses we find $A > 0$ always, it can be $B < 0$.

The case when magnitude fluctuates corresponds to the Higgs mode, $\Delta \rightarrow \Delta + \lambda(x)/\sqrt{2}$, with the Lagrangian of a massive bosonic field

$$\mathcal{L}_\lambda = -\frac{1}{2} C\lambda(x)^2 + \frac{1}{2} D(\partial_t \lambda(x))^2 - \frac{E}{6}(\partial_i \lambda(x))^2 \tag{7}$$

with all positive coefficients, C, D, $E > 0$. $m_H^2 = C/D$ is the mass squared, squared of the gap in the excitation spectrum. $C\lambda^2$ is change in free energy $S^{(0)}$ due to change in magnitude of condensate; C is proportional to the curvature of the potential, $C > 0$ is local minimum (stable/meta-stable) and $C < 0$ is local maximum (unstable). When $D < 0$ or $E < 0$, mean field solution is unstable with respect to time or space-dependent fluctuations of the magnitude; $D > 0$ always.

$B < 0$, $E < 0$ mean growth of spatially non-uniform fluctuations, unstable to phase and magnitude modulation of the condensate. However, these instabilities point to different ground state of the condensate. $B < 0$: condensate develops phase modulation which carries a current, balanced by a counter-propagating current carried by gapless fermions. $E < 0$: formation of a spatial modulation in the magnitude of the condensate, which does not carry a current.

F mixes Higgs and Goldstone components. F vanishes in weak coupling.

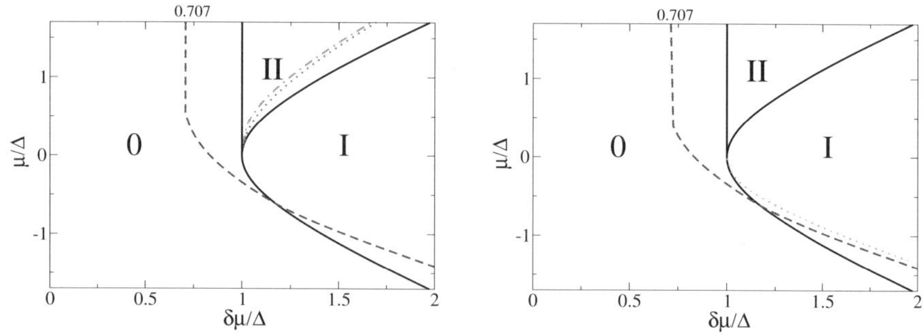

Fig. 1. Left panel. The $(\delta\mu, \mu)$ plane is divided into three regions corresponding to different number of gapless surfaces in momentum space, marked 0, I, II (solid lines). Unstable regions according to criteria 1 and 3: speed of sound for Goldstone and Higgs modes should be real (local criteria), free energy of the superfluid phase is lower than the free energy of the normal state (global criteria; $0.707 = 1/\sqrt{2}$ is Chandrasekhar-Clogston limit of breaking BCS superfluidity at weak coupling). Right panel. Unstable regions according to criteria 2 and 3: superfluid solution should be a minimum of the free energy (local criteria), pressure of the superfluid state is higher than the pressure of normal state (global criteria). Requirement 2 excludes states above doted line. Criteria 2 is the strongest among local criteria, and it leaves gapless states only with one Fermi surface. Local minimum of free energy ($C > 0$) is equivalent that the number susceptibility matrix is positive definite.

A, B, C, D, E involve Matsubara sums over bosonic frequencies $\omega_n = (2n+1)\pi T$ and 3-d integration, which is done numerically. Analytic expressions can be obtained in weak coupling, using BCS hierarchy of scales, $\delta\mu, \Delta << \mu$ (integration in thin shell around μ). Concluding, in the weak coupling gapless phases, $\delta\mu > \Delta$, are unstable (for both Goldstone and Higgs mode $B < 0$, $E < 0$, $C < 0$). In the weak coupling, the first order phase transition superfluid to normal state takes place, at $\delta\mu = \Delta/\sqrt{2}$, before the gapless state develops, $\delta\mu = \Delta$; i.e. for the window $\Delta/\sqrt{2} < \delta\mu < \Delta$ Higgs and Goldstone modes are fluctuations around the meta-stable solution. To find stable gapless phases one should explore the strong coupling regime.

3. Analysis of stability at $T = 0$

Using the quasiparticle dispersion law, we identify regions in $(\delta\mu, \mu)$ where the system has gapless excitations, i.e. momenta where energy is zero. Gapless fermionic excitations live at one and two spherical surfaces in momen-

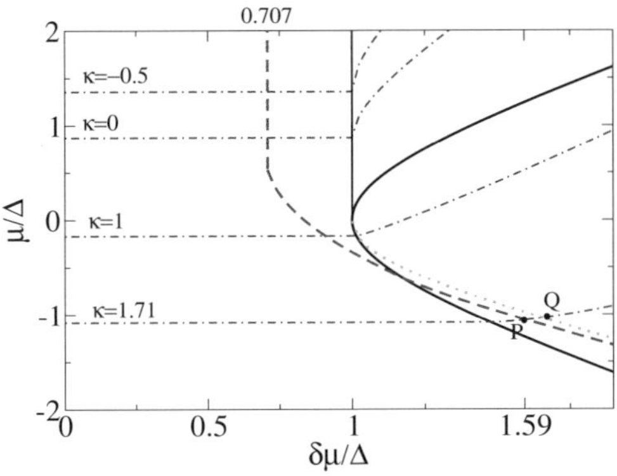

Fig. 2. The four dot-dashed lines show how μ varies as function of $\delta\mu$ for four different values of the dimensionless variable $\kappa = \pi/2\sqrt{2m\Delta}a_s$. Values $\kappa << -1$, $a_s < 0$, correspond to being deep in the BCS regime, while $\kappa >> 1$, $a_s > 0$, deep in the BEC regime. Region between curves 2 and 3, local and global minimum of free energy, correspond to meta-stable gapless state. Moving along $\kappa = 1.71$ (BEC), point P correspond to the largest $\delta\mu$ where superfluidity is globally stable, $\delta\mu/\Delta \approx 1.59$; after point Q, $\delta\mu/\Delta \approx 1.66$, it is locally unstable. Curves 2 and 3 run closer as $\delta\mu$ increases, and converge asymptotically for $\delta\mu >> \Delta$. Higgs mass is zero along curve 2. Hence Higgs mass is small along curve 3 and gets smaller as $\delta\mu$ increases.

tum space. We map regions with gapless excitations and regions where coefficients A, B etc. are negative, indicating instabilities to the $(\delta\mu, \mu)$ space.

Stability criteria are

1. Real speed of sound for Goldstone and Higgs modes $(B, E > 0)$,
2. SF is a local minimum $(C > 0)$,
3. SF is a global minimum $(\Omega_s - \Omega_n < 0)$,

which are equivalent to

1. Real speed of sound is equivalent to positive screening masses,
2. Local minimum is equivalent to positive number susceptibility.

At the Fig. 1, the $(\delta\mu, \mu)$ plane is divided into three regions corresponding to different number of gapless surfaces in momentum space; also unstable regions according to criteria 1 and 3 are shown.

Requirement 1 renders all states above the dotted and dashed lines unstable, which leaves gapless states with one and two Fermi surfaces as possibility. Requirement 3 renders all states above solid line unstable.

At the Fig. 1, unstable regions according to criteria 2 and 3 are shown.

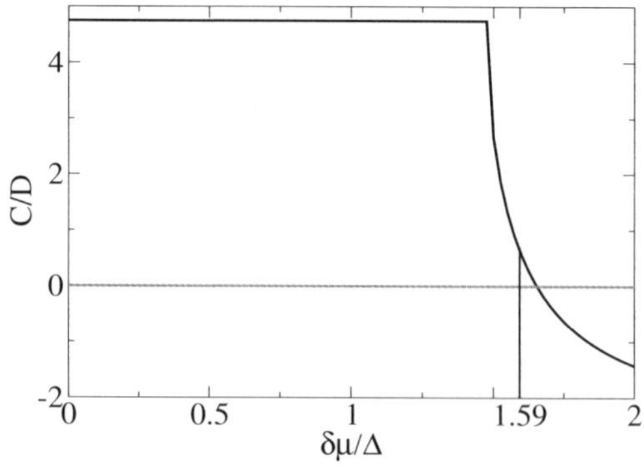

Fig. 3. Higgs mass, $m_H^2 = C/D$, along $\kappa = 1.71$ ($g = 1/k_F a_s \approx 1.31$) curve as $\delta\mu$ increases. For small $\delta\mu$, the Higgs mass is constant in the gapped superfluid phase. Indeed, in the gapped phase free energy of superfluid is independent of $\delta\mu$, superfluidity is favored over normal and it is locally stable ($C > 0$), hence m_H is independent of $\delta\mu$. In gapless phase, Higgs mass decreases. Superfluidity wins over the normal, hence the smallest m_H in this regime (C describes oscillations about the global minimum). Between P and Q, there is meta-stable region; below Q, it is locally unstable ($C < 0$, m_H is imaginary). From $\delta\mu = 0$ to P, m_H drops by a factor of 7.5.

However, we still learn from criteria 1 about the tendency of the system to go to another state. Real speed of sound for the Goldstone mode, which is equivalent that Meissner mass is real (or current instability), is more stringent than the real speed of sound of Higgs mode. Both criteria point towards a non-homogeneous phase. We learn, that LOFF like state is possible only in weak coupling at the BCS side.

The global minimum of free energy is the strongest requirement. There is a sliver of parameter space corresponding to meta-stable superfluid state (it is a local minimum, according to 2, but not a global one of free energy, according to 3). Gapless superfluid state with one spherical Fermi surface is at BEC side ($\mu < 0$).

4. Parameters of the Higgs Lagrangian

To be more concrete, we solve the gap equation for various scattering lengths and see where we land in the parameter space. At the Fig. 2, the four dot-dashed lines show how μ varies as function of $\delta\mu$ for four different values of the dimensionless variable $\kappa = \pi/2\sqrt{2m\Delta}a_s$. Region between

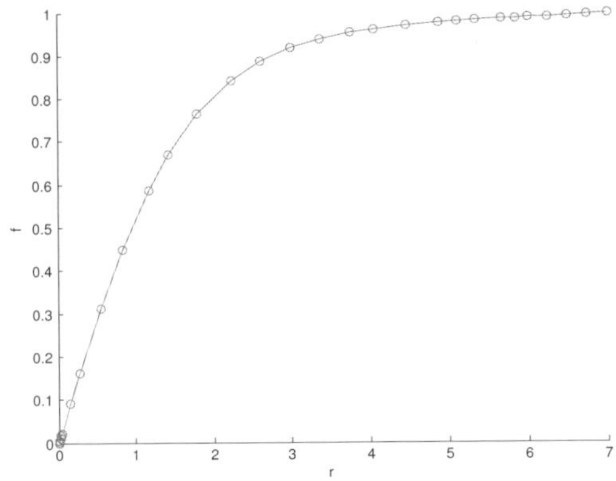

Fig. 4. Numerical solution of the ODE, Eq. 8, for the condensate in a vortex configuration. Vortex center is at $r = 0$. $\Delta(r)$ increases from 0 to Δ, as we go away from the inner core of the vortex, over a length scale $r_0 = \sqrt{E/3C}$. Units are in Δ at both axis.

curves 2 and 3, local and global minimum of free energy, correspond to meta-stable gapless state. Moving along $\kappa = 1.71$ (BEC), point P correspond to the largest $\delta\mu$ where superfluidity is globally stable, $\delta\mu/\Delta \approx 1.59$; after point Q, $\delta\mu/\Delta \approx 1.66$, it is locally unstable. Higgs mass is zero along curve 2. Hence Higgs mass is small along curve 3 and gets smaller as $\delta\mu$ increases. Thus, there is a light Higgs in the gapless BEC superfluid state.

At the Fig. 3, we depict Higgs mass, $m_H^2 = C/D$, along $\kappa = 1.71$ curve as $\delta\mu$ increases. For small $\delta\mu$, the Higgs mass is constant in the gapped superfluid phase. In gapless phase, Higgs mass decreases. Between P and Q, there is meta-stable region; below Q, it is locally unstable ($C < 0$, m_H is imaginary). From $\delta\mu = 0$ to P, m_H drops by a factor of 7.5.

At point P:

a. At P, there is the smallest Higgs mass in the regime where superfluidity is favored over normal state.

b. Note about reliability of mean field. At point P, $g = 1.31$ and as κ increases, g increases even further. In BCS $g < 0$ (large), BEC $g > 0$ (large), since $g = 1/k_F a_s$ large g corresponds to small a_s, which means that the mean field is reliable.

c. Effective theory near P: gapless fermions living on one spherical surface in momentum space, massless fluctuations in the phase of the condensate and massive but light fluctuations in magnitude of condensate. It is interesting to probe the spectrum.

d. Quantum corrections may alter small Higgs mass (beyond mean field).

It is interesting to explore part c., but even before a detailed study there is another striking consequence of our results.

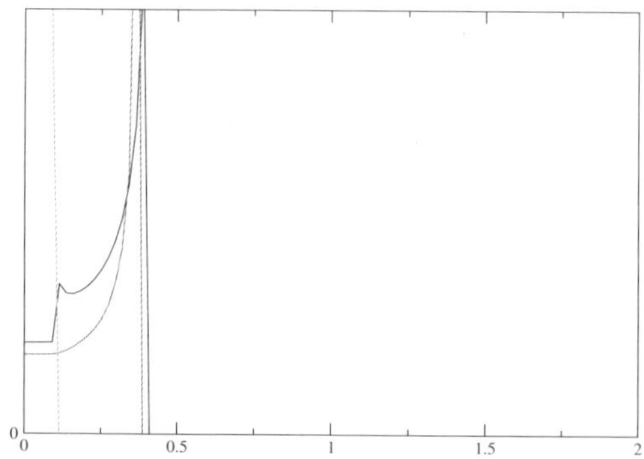

Fig. 5. Square of vortex radius r_0^2 as a function of a distance R from the center of a trap ($R \sim \delta\mu$). Two slightly different curves correspond to two ways of calculation. Bottom, smooth curve, first 3-d momentum integrals are taken numerically followed by taking a limit of small momentum. Top, curve with a casp, the order of integration and small momentum limiting procedure is interchanged, which is apparently wrong.

5. Vortex in a trap

The correlation length r_0, or the typical length scale at which the magnitude of condensate Δ varies in field configurations that arise when the system is excited, is inversely proportional to the mass of the Higgs mode of the system, $r_0\ 1/m_H$. For example, r_0 governs the size of the outer core of a vortex configuration in a superfluid phase. This can be seen from the classical field equations for $\Delta(r) = f(r)e^{i\phi(\varphi)}$, $f(r) = \Delta + \rho(r)$, for static configuration, where Δ is the ground state value and $\rho(r)$ is fluctuation of the condensate, (r, ϕ) are cylindrical polar coordinates, and the vortex is at $r = 0$,

$$\Delta(r) - r_0^2 \nabla^2 \Delta(r) = const\,,$$
$$f(r \to 0) = 0\,, \quad f(r \to \infty) = \Delta\,. \tag{8}$$

ODE is solved numerically, Fig. 4, with boundary conditions: zero condensate at the center of the core, and it tends to a constant ground value at infinite r. For a vortex configuration, ϕ winds around by a multiple of 2π as we traverse a loop around the vortex. $\Delta(r)$ increases from 0 to Δ, as we go away from the inner core of the vortex, over a length scale $r_0 = \sqrt{E/3C}$. The fact that C (Higgs mass) is numerically small close to the point P in parameter space, will manifest itself in an increased size for the outer core of the vortex.

To construct an actual vortex solution, it will be important to include the fourth order term in the effective action. However, the coefficient of this term is dimensionless and does not introduce a new scale, hence our basic arguments remain valid.

To see the effect quantitatively, we plot square of vortex radius r_0^2 as a function of a distance R from the center of a trap (harmonic trap in BEC; distance from the center of the trap is proportional to mismatch in chemical potentials, $R \sim \delta\mu$). At the Fig. 5, in the gapped region the outer radius of the vortex is constant, and in the gapless region it increases monotonically on increasing R (or $\delta\mu$). r_0^2 formally diverges at some point in the exterior of the trap, which corresponds to the transition from superfluid to normal state. In reality this divergency is absent, since there are no vortices in the normal state, i.e. our formula does not apply then. Two slightly different curves correspond to two ways of calculation. At the bottom smooth curve, 3-d momentum integrals are taken numerically first followed by taking a limit of small momentum. At the top curve with a casp, the order of integration and small momentum limiting procedure is interchanged, which is apparently wrong. Here the long wavelength limit is not well defined, hence the order of operations matters.

We estimate r_0 from the fact that it is the length scale at which the condensation energy is comparable to the kinetic energy of a superfluid element close to the vortex, $E_k = E_{cond}$. Taking into account the velocity of superfluid matter near a vortex, which diverges towards the center of a vortex $v = \frac{1}{2mr} e_\theta$, we estimate the kinetic energy, $E_k = n\frac{mv^2}{2}$. Condensation energy is proportional to a condensation energy density, $E_{cond} = n\varepsilon_{cond}$. We get for the vortex radius $r_0 \sim 1/\sqrt{\varepsilon_{cond}}$. Therefore, in the gapped phase, r_0 is constant, since condensation energy density does not change; in the gapless phase condensation energy density decreases with increasing $\delta\mu$, hence r_0 increases steeply in the gapless phase; r_0 diverges at the transition from superfluid to normal state, because this formula does not work in the normal phase (there are no vortices). In conclusion, r_0 continuously increases in the gapless phase. Vortex size grows steeply by entering into the gapless phase.

We suggest to tune parameters of the trap, the number of particles of the two species, N_1 and N_2, and the scattering length, a_s, so that there is a sufficiently wide region in position space where the atomic system is in the gapless BEC phase. We predict, that effect of sharp increase in vortex size can be seen. Probably, it requires very flat traps to realize this phenomena in a wide enough region to be observed.

6. Conclusion

We obtained an effective theory for collective modes, Goldstone and Higgs, for the system of interacting two fermionic species. There is an interesting low energy content of effective theory, which includes gapless fermions, massless Goldstone mode, and very light Higgs mode. From our phase diagram, gapless state is stable at the BEC side of the resonance, with gapless fermions residing on one spherical Fermi surface in momentum space. Instability towards non-homogeneous LOFF state occurs only at weak coupling, i.e. LOFF is favored away from the resonance.

We would like to study the low energy effective theory further. Apart from that, we suggest a possible experimental evidence/signature of the gapless phases. We predict a sharp increase in outer core of a vortex, i.e. vortex size, upon entering into the gapless phase. To observe this dramatic effect will require tuning of the parameters of the trap to a gapless BEC phase.

It will be interesting to study the core structure of the vortex, including inner and outer core, as well as interaction between two vortices. This requires solving Bogoluibov-De Ginnas equation and/or constructing the Landau-Ginsburg functional to the fourth order. However, we believe that our prediction with respect to vortex size will not change qualitatively. According to our estimate, the mean field treatment is reliable in the region of interest. Probably, the main improvement will come from incorporating momentum/energy dependent Fermi interaction.

References

1. M. W. Zwierlein, A. Schirotzek, C. H. Schunck and W. Ketterle, Science **311**, 492 (2006) [arXiv:cond-mat/0511197]; M. W. Zwierlein and W. Ketterle, "Comment on "Pairing and Phase Separation in a Polarized Fermi Gas" by G. B. Partridge, W. Li, R. I. Kamar, Y. Liao, R. G. Hulet, Science 311, 503 (2006)," arXiv:cond-mat/0603489; Y.-Il Shin, C. H. Schunck, A. Schirotzek and W. Ketterle, "Phase diagram of a two-component Fermi gas with resonant interactions," arXiv:0709.3027.
2. G. B. Partridge, W. Li, R. I. Kamar, Y.-a. Liao and R. G. Hulet, Science **311**, 503 (2006) [arXiv:cond-mat/0511752].

3. E. Tiesinga, B. J. Verhaar and H. T. C. Stoof, Phys. Rev. A **47**, 4114 (1993); S. Inouye *et al.*, Nature **392**, 151 (1998); P. Courteille *et al.*, Phys. Rev. Lett. **81**, 69 (2004); J. L. Roberts *et al.*, Phys. Rev. Lett. **81**, 5109 (1998); E. Timmermans, P. Tommasini, M. Hussein and A. Kerman, Phys. Rep. **315**, 199 (1999).
4. D. M. Eagles, Phys. Rev. **186**, 456 (1969); A. J. Leggett in *Modern Trends in the theory of condensed matter*, edited by A. Pekalski and J. Przystawa (Springer-Verlag, Berlin, 1980).
5. J. Carlson and S. Reddy, Phys. Rev. Lett. **95**, 060401 (2005) [arXiv:cond-mat/0503256].
6. P. F. Bedaque, H. Caldas and G. Rupak, Phys. Rev. Lett. **91**, 247002 (2003) [arXiv:cond-mat/0306694].
7. G. Sarma, Phys. Chem. Solids **24**, 1029 (1963).
8. W. V. Liu and F. Wilczek, Phys. Rev. Lett. **90** (2003) 047002 [arXiv:cond-mat/0208052].
9. E. Gubankova, W. V. Liu and F. Wilczek, Phys. Rev. Lett. **91** (2003) 032001 [arXiv:hep-ph/0304016].
10. M. M. Forbes, E. Gubankova, W. V. Liu and F. Wilczek, Phys. Rev. Lett. **94**, 017001 (2005) [arXiv:hep-ph/0405059].
11. H. Muther and A. Sedrakian, Phys. Rev. Lett. **88**, 252503 (2002) [arXiv:cond-mat/0202409].
12. P. Fulde and R. A. Ferrell, Phys. Rev. **135**, A550 (1964); A. I. Larkin and Y. N. Ovchinnikov, Sov. Phys. JETP **20**, 762 (1965).
13. D. E. Sheehy and L. Radzihovsky, Phys. Rev. Lett. **96**, 060401 (2006) [arXiv:cond-mat/0508430]; Annals of Physics, **322**, 1790, 2007 [arXiv:cond-mat/0607803v2].
14. P. Pieri and G. C. Strinati, Phys. Rev. Lett. **96**, 150404 (2006), [arXiv:cond-mat/0512354]; For a recent review on the properties of trapped fermi gases, and more references, see S. Giorgini, L. P. Pitaevskii and S. Stringari, arXiv:0706.3360v2.
15. J. Carlson and S. Reddy, Phys. Rev. Lett. **100**, 150403 (2008) [arXiv:cond-mat/0711.0414].
16. D. T. Son and M. A. Stephanov, Phys. Rev. A **74**, 013614 (2006) [arXiv:cond-mat/0507586].
17. P. Nikolic and S. Sachdev, Phys. Rev. A **75**, 033608 (2007).
18. M. Y. Veillette, D. E. Sheehy and L. Radzihovsky, Phys. Rev. A **75**, 043614 (2007).
19. A. Bulgac and M. M. Forbes, Phys. Rev. A **75**, 031605 (2007) [arXiv:cond-mat/0606043].
20. A. Bulgac, Phys. Rev. A **76**, 040502 (2007) [arXiv:cond-mat/0703526].

Double Well Potential: Perturbation Theory, Tunneling, WKB (Beyond Instantons)

Alexander V. Turbiner*

Instituto de Ciencias Nucleares, UNAM, A.P. 70-543, Mexico D.F., 04510 Mexico

A simple approximate solution for the quantum-mechanical quartic oscillator $V = m^2x^2 + gx^4$ in the double-well regime $m^2 < 0$ at arbitrary $g \geq 0$ is presented. It is based on a combining of perturbation theory near true minima of the potential, semi-classical approximation at large distances and a description of tunneling under the barrier. It provides 9-10 significant digits in energies and gives for wavefunctions the relative deviation in real x-space less than $\lesssim 10^{-3}$.

Needless to say that since the creation of quantum mechanics the one-dimensional quartic anharmonic oscillator

$$\mathcal{H} = -\frac{d^2}{dx^2} + m^2x^2 + gx^4 , \tag{1}$$

always attracted a lot of attention being among the most celebrated problem of quantum mechanics. The interest to these problems ranges from various branches of physics, from quantum field theory to chemistry and biology. It is especially true for the case when $m^2 < 0$ and the potential has two minima. This problem is known in literature as the *double-well potential.* It was studied in hundreds papers, appeared practically in all books on quantum mechanics. A special emphasis was made to a domain $m^2 \to \infty$ - a domain where the barrier penetration is described by instantons (see e.g. [1–3]) that gives rise to the instanton physics.

The first detailed study [4] of (1) carried out by Bender and Wu at 1969-1973 revealed in this seemingly simple Hamiltonian the extremely rich analytic structure which looks intrinsic for any non-trivial eigenvalue problem of quantum mechanics and even though for quantum field theory. In fact, one of the most important unwritten conclusions was that in no way this problem can be solved exactly. The goal of the present talk is to give an

*E-mail: turbiner@nucleares.unam.mx

approximate solution valid, actually, for any $g > 0$ and m^2. The solution is given in a form of the fairly simple expression for the ground state and the first excited state wavefunctions, which for any real x and for $g \geq 0$, real m^2 differs from the exact wavefunction for not more than a small number δ,

$$\left|\frac{\Psi_{approximate} - \Psi_{exact}}{\Psi_{approximate}}\right| \leq \delta \ .$$

In our case the $\delta \approx 10^{-6}$. Evidently, it implies that *any* quantity related with the first two eigenstates like expectation values can be calculated with accuracy δ^2.

There are three basic analytic approaches to study the spectra in quantum mechanics: (i) perturbation theory, (ii) WKB method and (iii) instanton calculus.[a] Each approach has its domain of applicability and usually these domains do not overlap. We attempt to combine (incorporate, unify) all three approaches into one by making interpolation. The most convenient object to incorporate (i) and (ii) is the *logarithmic derivative of the wavefunction.* While the suitable object to incorporate the property (iii) is the wavefunction. A final form of the approximation depends on a few free parameters. Roughly speaking, their behavior as a function of g, m^2 is rather smooth and simple. They can be fixed variationally, although it is not very important: a small variation of the parameters does not lead to dramatic loss of accuracy.

As a first step to approach the problem let us remind the Symanzik rescaling for eigenvalues and eigenfunctions

$$E(m^2, g) = g^{1/3} E\left(\frac{m^2}{g^{2/3}}\, , 2\right) \ , \ \Psi(x; m^2, g) = \Psi\left(x g^{1/6}; \frac{m^2}{g^{2/3}}\, , 2\right) \ .$$

It manifests that the original problem (1) is in fact one-parametric. The Hamiltonian (1) can be rewritten in the form

$$\mathcal{H} \ = \ -\frac{d^2}{dx^2} \ + \ a x^2 + 2 x^4 \ , \quad x \in (-\infty, +\infty) \ , \tag{2}$$

where $a \equiv \frac{m^2}{g^{2/3}}$. This is the form of the Hamiltonian we are going to study with the real parameter a varying from negative to positive values. The

---

[a] It must be noted that, in fact, any one-dimensional Schrödinger equation can be solved numerically with any desirable accuracy. However, it is not true for multidimensional case.

Schrödinger equation for (2) reads

$$-\frac{d^2\Psi}{dx^2} + ax^2\Psi + 2x^4\Psi = E\,\Psi\,, \qquad \int_{-\infty}^{+\infty} |\Psi|^2 dx < \infty\,. \tag{3}$$

Eigenfunctions of (3) are sharply changing functions in $x \in \mathbf{R}$ being characterized by a power-like behavior at $|x| \to 0$ and an exponentially-decaying one at $|x| \to \infty$. Following the oscillation (Sturm) theorem the nth eigenfunction has n simple (real) zeros. It seems natural to introduce the representation for eigenfunctions as follows

$$\Psi(x) = p_n(x)e^{-\varphi(x)}\,, \tag{4}$$

where the phase $\varphi(x)$ is a slow-changing smooth function and $p_n(x)$ is a polynomial of nth degree with real coefficients which has n real roots.[5] Recently, it was obtained a remarkable result [6]: Any eigenfunction $\Psi(x)$ of (2) for any real a is entire function and it has infinitely-many simple complex zeros all situated on imaginary axis symmetrically! It implies that the phase $\varphi(x)$ has infinitely-many logarithmic branch points in complex x-plane and has no singularities at real x.

After substitution of (4) into (3) we get the following equation

$$y' - y^2 - \frac{p_n'' - 2yp_n'}{p_n} = E - ax^2 - 2x^4\,, \quad y = \varphi' = (\log\Psi(x))'\,,\; y(0) = 0\,. \tag{5}$$

In order to define the problem (5) we impose two conditions that y has no simple poles at real x (i) and it grows at $|x| \to \infty$ not faster than polynomial (ii). The condition (i) implies that the coefficients of the polynomial $p_n(x)$ are those that the residues in the simple poles in the third term in l.h.s. of (5) vanish. The condition (ii) assures squire-integrability of the wavefunction (4). It is evident that the polynomial $p_n(x)$ has parity $p = (-)^n$, hence, can be written as $p_n(x) = x^p P_{[\frac{n}{2}]}(x^2)$, and $y(x)$ is odd, $y(x) = y(-x)$ (for discussion, see [7]).

From the analysis of (5) it is easy to find asymptotic behavior of the phase, [b]

$$\varphi = \frac{2^{1/2}}{3}x^2|x| + \frac{a}{2^{3/2}}|x| + (n+1)\log|x| + \frac{8E + a^2}{2^{9/2}}\frac{1}{|x|} + \left(\frac{a}{2} - 4A\right)\frac{1}{x^2} + \dots\,, \tag{6}$$

[b] The phase is defined up to additive constant which is fixed following a normalization of the wavefunction. We will omit it.

at $|x| \to \infty$, where A is sum of squared of nodes, while

$$\varphi = \frac{E}{2}x^2 + \frac{E^2 - a}{12}x^4 + \frac{2E(E^2 - a) - 6}{90}x^6 + \ldots \quad \text{at } |x| \to 0 \ , \qquad (7)$$

(see [8]). It is important to note that the first two terms in (6) are defined by the equation (5) with omitted y' term, which is actually the Hamilton-Jacobi equation. Therefore, these two terms coincide with first two terms of the asymptotics of the classical action at $|x| \to \infty$. The third term in (6) is also reproduced in the expansion of the classical action but with a wrong coefficient. The correct coefficient can be obtained if the first correction to the classical action is taken into account (quadratic fluctuations). The first three terms in (6) grow when $|x|$ tends to ∞. In fact, they characterize a singularity at $|x| = \infty$. These terms do not depend on energy E and are found explicitly. For the single well potential $a \geq 0$ the expansion (7) is nothing but the (divergent) perturbation theory series near the minimum of the potential. It is an expansion around the true vacuum.

Now let us construct a simplest function for phase (4) which interpolates small and large distance expansions, reproducing *exactly* the first three (growing) terms in (6). It has the form,

$$\varphi_{int} = \frac{A + (D^2 + 3a)x^2 + 4x^4}{6(D^2 + 2x^2)^{1/2}} - \frac{n+1}{2} \log\left(D^2 + 2x^2\right) . \qquad (8)$$

The corresponding wavefunction (4) for $n \equiv (2k + p)$ excited state (for $k = 0, 1, \ldots$ and parity $p = 0, 1$) is equal to

$$\psi_0^{(k,p)} = \frac{x^p P_k(x^2)}{(D^2 + 2x^2)^{k + \frac{p+1}{2}}} \exp\left\{ - \frac{A + (D^2 + 3a)x^2 + 4x^4}{6(D^2 + 2x^2)^{1/2}} \right\} , \qquad (9)$$

(cf. [8] at $k = 0$), where A, D are free parameters and P_k is a polynomial with real coefficients having positive roots only. There are two ways to find the polynomial P_k: either imposing the orthogonality conditions to the functions with smaller quantum numbers $k = 0, 1, \ldots (k-1)$, or requiring the absence of simple poles in the third term in l.h.s. of (5). Surprisingly, in concrete calculations these two conditions lead to polynomials whose coefficients coincide with high accuracy. Each function $\psi_0^{(k,p)}$ is characterized by two free parameters A, D, which can be fixed if the function (9) is taken as variational trial function. If $k = 0$ in (9) both ground ($p = 0$) and the first excited ($p = 1$) states occur, respectively. Concrete calculations of the variational energies of these states with (9) taken as a trial function lead to unprecedented accuracy [8] (see below). Furthermore, if in the exponential in (9) in the term $4x^4$ (this term governs the asymptotic behavior of the wave function at large distances where the wave function is exponentially small) the factor 4 is replaced by a parameter, Wx^4, minimization of such a trial function leads to a value of W which is equal to 4 with accuracy 10^{-4}!

In the case $a < 0$ the vacuum at $x = 0$ becomes the false vacuum and two classically degenerate vacua at $x_\pm = \pm a^{1/2}/2$ appear. The expansion (7) becomes the expansion around false minimum (maximum), it is not relevant physically. The expansion around one or another true vacuum should be considered instead, as a relevant one. This expansion can be easily derived and we skip it.

In this case a new physical phenomenon of the quantum mechanical tunneling (barrier penetration) occurs. There is a probability to meet the particle under the barrier, near $x \approx 0$. It decays exponentially when $a \to -\infty$. This phenomenon is absent in (9). A prescription how to describe tunneling it is given in the celebrated Landau-Lifschitz book.[9] It employs a linear superpositions of the wavefunctions centered at different minima which has positive parity for the ground state and similar one of negative parity for the first excited state. From the viewpoint of construction of the interpolation it is similar to making an interpolation between the expansion at one minimum and another one, and at $|x| \to \infty$. Finally, such an interpolating function is a linear superposition of two off-centered functions (9), which can be written as

$$\psi_0^{(k,+)} = \frac{q_k^+(x^2)}{(D_+^2+2x^2)^{k+\frac{1}{2}}} \exp\left\{-\frac{A_+ + (D_+^2+3a)x^2+4x^4}{6(D_+^2+2x^2)^{1/2}}\right\} \cosh\frac{\alpha_+ x}{(D_+^2+2x^2)^{1/2}}, \tag{10}$$

for the states of positive parity and

$$\psi_0^{(k,-)} = \frac{q_k^-(x^2)}{(D_-^2+2x^2)^{k+1}} \exp\left\{-\frac{A_- + (D_-^2+3a)x^2+4x^4}{6(D_-^2+2x^2)^{1/2}}\right\} \sinh\frac{\alpha_- x}{(D_-^2+2x^2)^{1/2}}, \tag{11}$$

for the states of negative parity. Parameter $\alpha_+(\alpha_-)$ 'measures' a displacement of the peaks of wave function from the origin, $x = 0$. If $\alpha_+ = 0$ the function (10) becomes (9). Here $(q_k^\pm)$ is a polynomial of degree k with real coefficients having positive roots only. At fixed k any function depends on three free parameters α, A, D.

In order to proceed further I need to remind two important and poorly known results: (i) a special form of perturbation theory in QM sometimes called 'Logarithmic Perturbation Theory' or 'Non-linearization Method', and (ii) a connection between variational calculation and perturbation theory.

(i) A special form of perturbation theory is a certain iterative procedure developed for solving the Riccati equation (5) instead of the Schrödinger equation. For finding the wave function it is a multiplicative perturbation theory unlike a standard additive Rayleigh-Schrödinger perturbation theory. Such a multiplicative perturbation theory was developed for the first time by Price[10] and then it was numerously rediscovered (for early history and discussion see [7] and references therein). In presentation we follow closely to [11]. For simplicity we will consider the eigenstates for which the nodes are absent (ground state) or nodal positions are known.

As a first step to develop the perturbation theory we make a choice of some square-integrable function Ψ_0 and calculate its logarithmic derivative

$$y_0 = (\log \Psi_0)' = \frac{\Psi_0'}{\Psi_0} \ . \tag{12}$$

It is clear that Ψ_0 is the exact eigenfunction of the Schrödinger operator with a potential

$$V_0 = \frac{\Psi_0''}{\Psi_0} = y_0^2 - y_0' \ , \tag{13}$$

where without a loss of generality we put their eigenvalue equals to zero, $E_0 = 0$. It is nothing but a choice of the reference point for eigenvalues. Now we can construct a perturbation theory for Riccati equation taking Ψ_0 and y_0, V_0 as zero approximation, which characterizes the unperturbed problem. One can write the original potential $V = m^2x^2 + gx^4$ as a sum,

$$V = V_0 + (V - V_0) \ \equiv \ V_0 + V_1 \ , \tag{14}$$

thus, taking a deviation of the original potential from the potential of the zero approximation as a perturbation. We always can insert a formal parameter λ in front of V_1 and develop a perturbation theory in powers of λ,

$$E = \sum_{k=0}^{\infty} \lambda^k E_k \ , \qquad y = \sum_{k=0}^{\infty} \lambda^k y_k \ , \tag{15}$$

putting $\lambda = 1$ afterwards. Perhaps, it is worth emphasizing that in spite of the fact that we study iteratively the equation (5), in general, this perturbation series has nothing to do with a standard WKB expansion. By substituting (15) into (5) we arrive at the equations which defines iteratively the corrections

$$y_k' - 2y_0 y_k = E_k - Q_k \ , \tag{16}$$

where

$$Q_1 = V_1 \ ,$$
$$Q_k = -\sum_{i=1}^{k-1} y_i \cdot y_{k-i} \ , \quad k = 2, 3, \dots \ .$$

It is interesting that the operator in the l.h.s. of (16) does not depend on k, while Q_k in the r.h.s. can be interpreted as a perturbation on the level k. The solution of (16) can be found explicitly and is given by

$$E_k = \frac{\int_{-\infty}^{\infty} Q_k \Psi_0^2 \, dx}{\int_{-\infty}^{\infty} \Psi_0^2 \, dx} \ , \tag{17}$$

$$y_k = \Psi_0^{-2} \int_{-\infty}^{x} (E_k - Q_k) \Psi_0^2 \, dx' \ . \tag{18}$$

It is easy to demonstrate that if the first correction y_1 is bounded,

$$|y_1| \leq \text{Const} \ , \tag{19}$$

it provides a sufficient condition for this perturbation theory (15) to be convergent.[7] Note that this condition is very rough and very likely can be strengthened.

(ii) The first two terms in the expansion of energy (15) in the above-described perturbation theory admit an interpretation in the framework of the variational calculus.[5] Let us assume that our variational trial function $\Psi_0(x)$ is normalized to 1. We can calculate the potential V_0 where $\Psi_0(x)$ is the ground state eigenfunction and even put $E_0 = 0$ (see a discussion above). Formally, we construct the Hamiltonian $H_0 = p^2 + V_0$ for which $H_0 \Psi_0(x) = 0$. The variational energy is equal to

$$E_{var} = \int \psi_0 H \psi_0 = \underbrace{\int \psi_0 H_0 \, \psi_0}_{=E_0} + \underbrace{\int \psi_0 \underbrace{(H - H_0)}_{V - V_0} \psi_0}_{=E_1}$$
$$= E_0 + E_1(V_1 = V - V_0) \geq E_{exact} \ . \tag{20}$$

Of course, $\Psi_0(x)$ could depend on free parameters. In this case both V_0 and V_1 depend on parameters as well. Minimization of E_{var} with respect to the parameters can be performed and the variational principle guarantees that E_{var} gives upper bound to the ground state energy. This simple interpretation (20) reveals a fundamental difference between perturbation theory and variational calculus. Variational estimates can be obtained independently on the fact that the perturbation theory associated with trial function $\Psi_0(x)$ is convergent or divergent. However, it seems natural to remove this difference by requiring a convergence of the perturbation series. In this case by calculating the next terms $E_2, E_3, \ldots$ in (15) one can estimate the accuracy of variational calculation from one side and improve it iteratively from another side. An immediate criteria how to choose $\Psi_0(x)$ in order to get a convergent perturbation theory is to have the perturbation potential V_1 to be subordinate with respect to the non-vanishing potential of zero approximation V_0,

$$\left|\frac{V_1}{V_0}\right| < 1 \quad , \qquad \text{for } |x| > R \quad . \tag{21}$$

From this point of view any function (9), (10) or (11) taken as an entry leads to a convergent perturbation theory. An open question is how to estimate the radius of convergency.

The requirement (21) has a non-trivial physical implication: in order to guarantee a convergence of perturbation theory a domain where the wavefunction is exponentially small (classically-prohibited domain) should be reproduced as precise as possible. The same time a description of a domain where the wavefunction is of the order 1 is not important. It contradicts to a straightforward physics intuition and underlying idea of variational calculus which, in particular, requires a precise description of the domain where the wavefunction is of the order 1. Needless to say that namely the latter domain gives a dominant contribution to the integrals which define the energy in the variational calculations. Similar conclusion was presented in [12].

1. Results

1.1. *Ground state*

$$\boldsymbol{a = 1}$$

This is the case of a single-well potential (anharmonic oscillator). The variational parameters in (10) are

$$D \ = \ 4.33441 \,, \quad A \ = \ -9.23456 \,, \quad \alpha \ = \ 2.74573 \,.$$

Ground state energy is

$$E_{var} \ (\equiv \ E_0 + E_1) \ = \ 1.607541302594 \,,$$

while the first correction to it is

$$\Delta E_{var}\ (\equiv\ E_2)\ =\ -1.2552\times 10^{-10}\ .$$

Eventually,

$$\tilde{E}_{var} = E_{var} + \Delta E_{var}\ =\ 1.607541302469\ , \tag{22}$$

where all 13 digits are correct, since the next correction $E_3 \sim 10^{-14}$. The rate of convergency seems extremely high, $\sim 10^{-4}$! The dependence the energy on α is very weak: the results are almost unchanged if $\alpha = 0$ (and (10) becomes (9)). It is the case for $a \geq 0$.

$$\boldsymbol{a = -1}$$

This is already the case of a double-well potential. The variational parameters in (10) are

$$D\ =\ 4.059888\ ,\quad A\ =\ -12.4816\ ,\quad \alpha\ =\ 3.07041\ .$$

In comparison with the case $a = 1$ the parameters D, α are slightly changed unlike the parameter A which decreased in $\sim 40\%$. Ground state energy is

$$E_{var}\ (\equiv\ E_0 + E_1)\ =\ 1.029560832093\ ,$$

while the first correction to it is

$$\Delta E_{var}\ (\equiv\ E_2)\ =\ -1.0382\times 10^{-9}\ .$$

Eventually,

$$\tilde{E}_{var} = E_{var} + \Delta E_{var}\ =\ 1.029560831054\ , \tag{23}$$

where all 13 digits are correct, since the next correction $E_3 \sim 10^{-13}$. Similar to the case $a = 1$ the rate of convergency seems extremely high, $\sim 10^{-4}$!

On Fig.1 one can see the behavior of logarithmic derivative y_0 vs $x \geq 0$. It is a very smooth function. The first correction $|y_1|$ (see Fig.2) has very interesting behavior: it is of the order of 10^{-4} at $x \lesssim 1$ (in the domain which give a dominant contribution to the energy integral (20)), then it starts to grow and reaches the maximum ~ 0.006 at $x \sim 3.9$ in the domain which gives negligibly small contribution to the energy integral, $y_0(x = 3.9) \sim 15.2$ and $\Psi_0(x\ =\ 3.9) \sim 10^{-9}$. In this case $\delta \sim max|y_1| \sim 0.006$ (see p.1). Notably, $y_1 \propto 1/x^2$ at $x \gg 1$. Similar behavior is demonstrated by the higher corrections $|y_n|, n = 2, 3$: they are very small at $x \lesssim 1$, its maximum is reached for $x > 1$ but its position is systematically reduced with n.

$$\boldsymbol{a = a_{crit}}$$

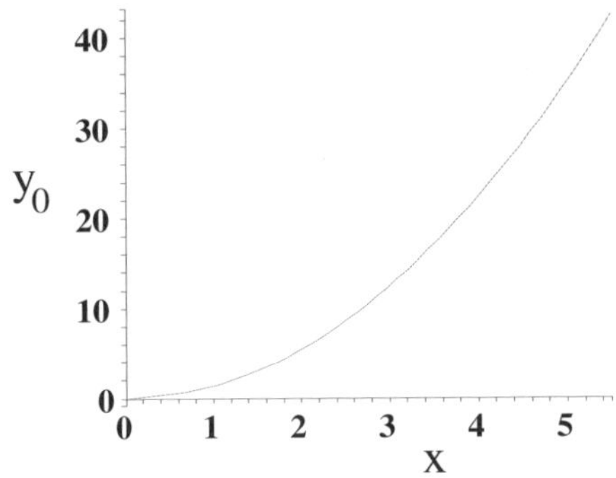

Fig. 1. Logarithmic derivative $y_0 = \varphi'_{int}$ (see (8)) as function of x for double-well potential (2) with $a = -1$

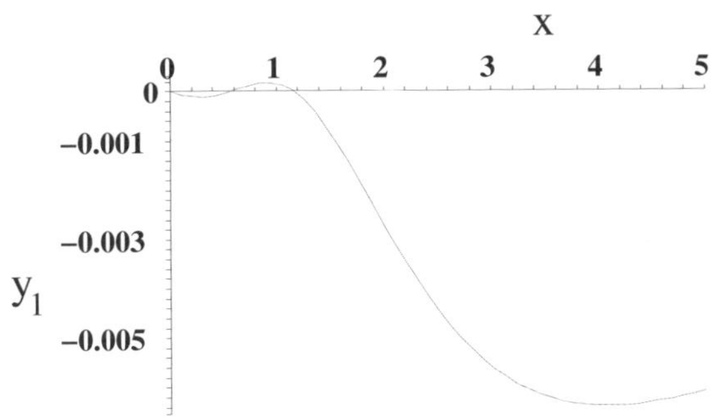

Fig. 2. The first correction y_1 for $a = -1$ (cf. Fig.1)

Let us consider the Hamiltonian (1). The ground state function is symmetric w.r.t. $x \to -x$. Hence, it has to have an extremum at $x = 0$. For any fixed g there exists a value $m^2_{crit} < 0$ such that for $m^2 > m^2_{crit}$ this extremum is a maximum, otherwise a minimum. It is easy to find out that the critical point m^2_{crit} corresponds to the vanishing ground state energy, $E = 0$.[c] Using the function (10) it was calculated the critical value $m^2_{crit} = -2.2195970861$ for $g = 1$. For the Hamiltonian (2)

$$E(a_{crit} = -3.523390749) \ = \ 0 \ .$$

It is quite interesting from physical point of view that for a family of double-well potentials with fixed g there exists a domain $0 > m^2 > (m^2)_{crit}$ where

[c] In classical case, it corresponds to the particle stopping on the top of the barrier.

the ground-state eigenfunction has the maximum at the origin, which corresponds to the position of the unstable equilibrium similar to what takes place for the single-well case. It implies that the particle in such a potential with the ground state energy above the barrier, $E > 0$, somehow does not feel the existence of two minima. In this domain WKB consideration at $x \sim 0$ is not valid.

$a = -20$

This is the case of a double-well potential. The variational parameters in (10) are

$$D = 6.765663\,, \quad A = -286.6456\,, \quad \alpha = 49.6136\,.$$

Ground state energy is

$$E_{var}\ (\equiv\ E_0 + E_1) = -43.7793127\,,$$

while the first correction to it is

$$\Delta E_{var}\ (\equiv\ E_2) = -3.81 \times 10^{-6}\,.$$

Eventually,

$$\tilde{E}_{var} = E_{var} + \Delta E_{var} = -43.7793165\,, \tag{24}$$

where all 9 digits are correct, since the next correction $E_3 \sim 10^{-8}$. Similar to the cases $a = \pm 1$ the rate of convergency seems still high, $\sim 10^{-2}$!

1.2. *First excited state*

$a = -20$

This is the case of a double-well potential. The variational parameters in (11) are

$$D = 5.584376\,, \quad A = -246.64375\,, \quad \alpha = 38.82768\,.$$

The parameters seems quite close to those for $a = 20$ for the ground state. Ground state energy is

$$E_{var}\ (\equiv\ E_0 + E_1) = -43.77931637\,,$$

while the first correction to it is

$$\Delta E_{var}\ (\equiv\ E_2) = -9.3618 \times 10^{-8}.$$

Eventually,

$$\tilde{E}_{var} = E_{var} + \Delta E_{var} = -43.77931646\,, \tag{25}$$

where all 10 digits are correct, since the next correction $E_3 \sim 10^{-10}$. Similar to the cases $a = 20$ for the ground state the rate of convergency seems still high, $\sim 10^{-2}$!

1.3. *Energy Gap*

By definition the energy gap is

$$\Delta E \;=\; E_{first\ excited\ state} \;-\; E_{ground\ state}\ .$$

For the double-well potential (2) at $a \to -\infty$ it can be calculated and in one-instanton approximation it reads[9,13,14]

$$\Delta E \;=\; \frac{2^{11/4}}{\sqrt{\pi}}|a|^{5/4}\exp\left[-\frac{\sqrt{2}|a|^{3/2}}{6}\right]\left(1-\frac{71}{12}\frac{1}{\sqrt{2}|a|^{3/2}}-\frac{6299}{288}\frac{1}{2|a|^{3}}\right.$$
$$\left.-\frac{2691107}{10368}\frac{1}{2\sqrt{2}|a|^{9/2}}-\frac{2125346615}{497664}\frac{1}{4|a|^{6}}\cdots\right)\ . \qquad (26)$$

The first term in the expansion is a common knowledge. All other coefficients in the expansion are due to J Zinn-Justin, 1981-2005 (see [13] and references therein), they were obtained using the so called *exact Bohr-Sommerfeld quantization condition*. The coefficient 71/12 was independently calculated (and confirmed) in two-loop instanton calculation by E. Shuryak (see [14] and references therein). It seems highly desirable to perform three-loop calculation to check the next coefficient in the expansion. It is an asymptotic expansion.[13]

It is interesting to compare the energy gap ΔE at $a = -20$ calculated in the (convergent) perturbation theory with (10) and (11) as zero approximations and with use of the (asymptotic) expansion (26). Subsequent expressions show how ΔE evolves from pure variational results to ones with the first corrections E_2 taken into account and then to ones with the second corrections E_3 involved

$$\Delta E_{var} \;=\; 1.03282\times 10^{-7}\ ,$$

$$\Delta E_{var}^{(1)} \;=\; 1.06529\times 10^{-7}\ ,$$

$$\Delta E_{var}^{(2)} \;=\; 1.06525\times 10^{-7}\ . \qquad (27)$$

In the last expression $\Delta E_{var}^{(2)}$ all six significant digits are correct. Now how the energy gap ΔE looks like as function of a number of corrections to one-instanton result included (see (26))[d]:

$$one - instanton = 1.12154 \times 10^{-7} \qquad (5.3\%\ deviation)$$

$$one - instanton + 1st\ correction = 1.06908 \times 10^{-7}\ (0.36\%\ deviation)$$

$$one - instanton + 1st\ and\ 2nd\ corrections = 1.06754 \times 10^{-7}\ (0.22\%\ deviation)$$

$$one - instanton + four\ corrections = 1.06738 \times 10^{-7}\ (0.20\%\ deviation) \tag{28}$$

The numbers in brackets are relative deviations from (27). The fact that a deviation stays almost the same after adding 2nd, 3rd and 4th corrections likely indicates the maximal accuracy based on a use of asymptotic expansion is reached and in any moment the result can blow up. A comparison of two values (27) and (28) shows that they do not agree in the 4th digit.

In a conclusion I have to say that a similar consideration based on interpolation of phase between small and large distances was done for sextic oscillator and the Zeeman effect on hydrogen. In both cases the exceptionally high accuracies were obtained.

Acknowledgments

Author thanks E. Shuryak, A.I. Vainshtein, J.C.L. Vieyra for valuable discussions. The work is supported in part by grants: CONACyT *58942-F*, DGAPA *IN115709-3*.

References

1. A. M. Polyakov, Nucl. Phys. B **122**, 429 (1977).
2. S. R. Coleman, Subnucl. Ser. **15**, 805 (1979).
3. M. A. Shifman, “Instantons in Gauge Theories” (World Scientific, Singapore, 1994).
4. C. M. Bender, T. T. Wu, Phys. Rev. **184**, 1231 (1969); Phys. Rev. D **7**, 1620 (1973).
5. A. V. Turbiner, Zh. Eksp. Teor. Fiz. **79**, 1719-1734 (1980), Sov. Phys. – JETP **52**, 868-876 (1980) (English Translation).
6. A. Eremenko, A. Gabrielov, B. Shapiro, Ann. Inst. Fourier, Grenoble **58**, 603-624 (2008).
7. A. V. Turbiner, Usp. Fiz. Nauk **144**, 35-78 (1984), Sov. Phys. – Uspekhi **27**, 668-694 (1984) (English Translation).

[d] Three and more instanton contributions are negligibly small for $a = -20$.

8. A.V. Turbiner, Letters in Mathematical Physics **74**, 169-180 (2005).
9. L. D. Landau and E. M. Lifshitz, Quantum Mechanics, Pergamon Press (Oxford - New York - Toronto - Sydney - Paris - Frankfurt), 1977.
10. P. J. Price, Proc. Phys. Soc. London **67**, 383 (1954).
11. A.V. Turbiner, Pis'ma Zh. Eksp. Teor. Fiz. **30**, 379-383 (1979). JETP Lett. **30**, 352-355 (1979) (English Translation).
12. R. P. Feynman, "Difficulties In Applying The Variational Principle To Quantum Field Theories," in Proceedings of Int. Workshop on Variational Calculus in Quantum Field Theory, Wangerooge, West Germany, Sept. 1-4, 1987 (World Scientific, Singapore, 1987) pp. 28-40.
13. J. Zinn-Justin and U. D. Jentschura, Annals Phys. **313**, 269-325 (2004); quant-ph/0501137 (updated, January 2005)
14. C. E. Wöhler and E. Shuryak, Phys. Lett. B **333**, 467–470 (1994).

SECTION 10.

GLIMPSES OF THE CONFERENCE

in total

Photos courtesy of: Marie Larson, Arkady Vainshtein and Mikhail Voloshin.

LIST OF THE PARTICIPANTS

Nima Arkani-Hamed, Institute for Advanced Study
Roberto Auzzi, Swansea University
Ian Balitsky, Old Dominion University, Jefferson Lab
Viatcheslav Belyaev, Institute for Theoretical and Experimental Physics
Zvi Bern, University of California, Los Angeles
Tirthabir Biswas, Penn State University
Stefano Bolognesi, University of Minnesota, FTPI
Pavel Bolokhov, University of Pittsburgh
Thomas Cohen, University of Maryland
Xiaoyi Cui, University of Minnesota
Sergiy Dubynskiy, University of Minnesota
Gerald Dunne, University of Connecticut
Joshua Erlich, College of William and Mary
Harald Fritzsch, Ludwig-Maxmilians University, Munich
Gregory Gabadadze, New York University
Marcos Alejandro, Garcia Garcia Instituto de Ciencias Nucleares, UNAM
Stephen Gasiorowicz, University of Minnesota, FTPI
Howard Georgi, Harvard University
Ivan Gordeli, University of Minnesota
Elena Gubankova, Tufts University
Emir Gumrukcuoglu, University of Minnesota
Amihay Hanany, Imperial College, London
Burak Himmetoglu, University of Minnesota
Ken Intriligator, University of California, San Diego
Joe Kapusta, University of Minnesota
Alexander Khodjamirian, University of Siegen
Igor Klebanov, Princeton University
Kenichi Konishi, University of Pisa
Gregory Korchemsky, Institut de Physique Théorique
Peter Koroteev, University of Minnesota
Alex Kovner, University of Connecticut

Lev Koyrakh, St. Jude Medical
Frieder Lenz, Inst. f. Theoretische Physik III, Univ. Erlangen-Nürnberg
Juan Carlos Lopez Vieyra, Instituto de Ciencias Nucleares, UNAM
Alexander Monin, University of Minnesota
Keith Olive, University of Minnesota, FTPI
Serkay Olmez, University of Minnesota
Sener Ozonder, University of Minnesota
Marco Peloso, University of Minnesota
David Politzer, California Institute of Technology
Maxim Pospelov, University of Victoria
Jose A. R. Cembranos, University of Minnesota, FTPI
Anatoly Radyushkin, Old Dominion University, Jefferson Lab
Pierre Ramond, University of Florida
Adam Ritz, University of Victoria
Serge Rudaz, University of Minnesota
Gregory Ruffa, University of Minnesota
Chris Savage, University of Minnesota, FTPI
John Schwarz, California Instutute of Technology
Nathan Seiberg, Institute for Advanced Study
Mikhail Shifman, University of Minnesota, FTPI
Edward Shuryak, Stony Brook University
Pierre Sikivie, University of Florida
Dam Son, University of Washington
Mikhail Stephanov, University of Illinois, Chicago
Raman Sundrum, The Johns Hopkins University
Tonnis ter Veldhuis, Macalester College
David Tong, University of Cambridge
Arkady Tseytlin, Imperial College, London
Alexander Turbiner, Instituto de Ciencias Nucleares, UNAM
Mithat Unsal Stanford University, SLAC
Alex Ushveridze, Capella University
Arkady Vainshtein, University of Minnesota, FTPI
Misha Voloshin, University of Minnesota, FTPI
Mark Wise, California Institute of Technology
Ariel Zhitnitsky, University of British Columbia